Element	Symbol	Atomic number	Atomic weight	Element	Symbol	Atomic number	Atomic weight
Mercury	Hg	80	200.59	Samarium	Sm	62	150.4
Molybdenum	Mo	42	95.94	Scandium	Sc	21	44.9559
Neodymium	Nd	60	144.24	Selenium	Se	34	78.96
Neon	Ne	10	20.179	Silicon	Si	14	28.086
Neptunium	Np	93	237.0482	Silver	Ag	47	107.868
Nickel	Ni	28	58.71	Sodium	Na	11	22.9898
Niobium	Nb	41	92.9064	Strontium	Sr	38	87.62
Nitrogen	N	7	14.0067	Sulfur	S	16	32.06
Nobelium	No	102	(254)	Tantalum	Ta	73	180.9479
Osmium	Os	76	190.2	Technetium	Tc	43	98.9062
Oxygen	O	8	15.9994	Tellurium	Te	52	127.60
Palladium	Pd	46	106.4	Terbium	Tb	65	158.9254
Phosphorus	P	15	30.9738	Thallium	Tl	81	204.37
Platinum	Pt	78	195.09	Thorium	Th	90	232.0381
Plutonium	Pu	94	(242)	Thulium	Tm	69	168.9342
Polonium	Po	84	(210)	Tin	Sn	50	118.69
Potassium	K	19	39.102	Titanium	Ti	22	47.90
Praseodymium	Pr	59	140.9077	Tungsten	W	74	183.85
Promethium	Pm	61	(147)	Uranium	U	92	238.029
Protactinium	Pa	91	231.0359	Vanadium	V	23	50.9414
Radium	Ra	88	226.0254	Xenon	Xe	54	131.30
Radon	Rn	86	(222)	Ytterbium	Yb	70	173.04
Rhenium	Re	75	186.2	Yttrium	Y	39	88.9059
Rhodium	Rh	45	102.9055	Zinc	Zn	30	65.37
Rubidium	Rb	37	85.4678	Zirconium	Zr	40	91.22
Ruthenium	Ru	44	101.07				

79714   QD
31
Weller   W5

Chemistry: elementary principles

Date Due

JA 23 '73	MAR 12 '79	NOV 18 '85	FEB 08 1992
FE 9 '73	MAR 21 '79	JAN 30 '86	
FE 23 '73	JUN 19 '79	AUG 15 '86	JAN 30 '96
JE 12 '73	AUG 8 '79	OCT 07 '86	MAR 21 '97
JE 12 '73	OCT 18 '83	OCT 10 '86	AUG 6 '98
FE 17 '74	OCT 11 '83	MAR 05 '87	OCT 10 2002
	NOV 15 '83	MAR 16 '88	OCT 28 2002
FEB 21 '75	FEB 23 '84	MAY 26 1988	
JUL 29 '75	MAR 6 '84	OCT 11 '89	
OCT 11 '75	MAR 07 '84	DEC 5 '89	
MAY 26 '76	OCT 4 '85	MAY 8 1991	

CHEMISTRY: ELEMENTARY PRINCIPLES

PAUL F. WELLER and JEROME H. SUPPLE
State University of New York, College at Fredonia

▲▼▼ **ADDISON-WESLEY PUBLISHING COMPANY**
Reading, Massachusetts · Menlo Park, California · London · Don Mills, Ontario

This book is in the
ADDISON-WESLEY SERIES IN CHEMISTRY

Francis T. Bonner
Consulting Editor

PREFACE

This text is an outgrowth of lectures given in two introductory chemistry courses at the State University of New York at Fredonia, New York. One was a course for nonscience majors, most of whom had not had high school chemistry, and the other a one-semester course for nursing students, all of whom had had chemistry in high school. It was our feeling that nonscience majors, as well as nurses, should be exposed to organic and biochemistry along with inorganic chemistry. We have therefore tried to introduce the necessary principles of chemistry in a manner that will make this book usable by beginning students of chemistry, as well as by those who are not prepared for—or are not interested in—becoming majors in chemistry or the natural sciences.

In developing the text we have been very conscious of two common, and related, problems. The first involves the level of difficulty of the concepts covered. All too often, in an attempt to make the material more understandable to students, authors of chemistry texts either omit entirely the areas of chemistry that are conceptually more difficult, or oversimplify these areas to such an extent that their presentation is unrealistic or misleading. We have tried to avoid this pitfall, by holding to the principle that all college students have the intellectual capacity needed to understand —at least in a qualitative sense—the fundamental concepts of chemistry. In our experience, one of the main differences between science and nonscience students is in problem-solving—or quantitative—abilities. We have consequently kept the mathematical requirements and approaches at relatively low levels, but we have stressed problem-solving and quantitative data-handling in several sections, since involvement with science *is* involvement with quantitative concepts.

The second problem that we considered in developing the text was the common

tendency to emphasize the *facts* of chemistry nearly to the exclusion of the *process* of chemistry. We felt that the emphasis should be placed more on the *way* a chemist solves a problem than on the solution itself. The facts of chemistry are more than just collections of data, compendia of theories, physical data, and chemical reactions. The facts of chemistry are there to be used, to be applied to solutions of new problems, to be expanded into a greater understanding of our world. Unless we understand how chemistry can go about solving new problems, how the concepts of chemistry can lead to a greater knowledge of the physical universe, we do not know what chemistry is. We have therefore tried to present chemistry in a way that emphasizes the procedures of chemistry, the interactions of ideas, observations, hypotheses, experimental results, and so on; in other words, in a way that emphasizes how a chemist actually does chemistry.

Perhaps this last point touches on the most fundamental point concerning a text such as this—or possibly any chemistry textbook. Why is it important to study chemistry? Particularly, is it important for a historian, an artist, or a linguist to know something about chemistry? Why should we try to get a nonscientist to understand chemistry?

We can consider these questions by reflecting on some of the recent accomplishments of chemistry. One of the things that distinguishes our times from pre-World War II years is the extensive use of plastics and synthetic fibers. A great advantage of plastic materials is their durability. Most are impervious to reaction with air, sunlight, water, and microorganisms—the enemies of the paper, wood, and natural fibers for which plastics substitute. Generally durability is desirable, and there are many instances in which synthetic materials do a job that is far superior to that of natural substances. This same durability, however, creates new problems. How do you dispose of the items? Must we have a countryside littered with plastic containers? Is there a solution that will allow us to have the advantages of synthetic polymers without the drawbacks?

The importance of chemistry is also very apparent in the field of medicine. The cooperative efforts of the chemist, the pharmacologist, and many others have led to nearly incredible progress in the treatment of disease. Yet this, too, is not without bad side effects. The deformed babies resulting from thalidomide use, the possibility of blood clots resulting from the use of birth control pills, the controversy over food additives such as calcium cyclamate and glutamic acid, the widespread, indiscriminate use of drugs such as tranquilizers, alcohol, barbiturates, amphetamines, LSD, and so forth, all make us realize that the remarkable progress in the medical field has also been accompanied by undesirable side effects.

One of the most emotional issues today is environmental pollution. Everyone agrees that a solution must be found, but beyond that there is little agreement. Currently laws are being passed to ban the use of DDT. DDT is very resistant to biological degradation, and because of this it becomes concentrated in animals such as birds and fish in unacceptably high levels. The answer seems simple: Ban DDT. But we must have some control over insects. We can't in good conscience, for example, permit the death rate due to malaria in Asia and India to return to pre-DDT levels. Can we be sure that a substitute for DDT will be ecologically more acceptable?

The pollution of the lakes and waterways of the world is a problem of which we —living on the shores of Lake Erie—are personally very aware. Some of the pollution can be controlled by very strict regulations on the contents of the effluents of factories and homes. It is likely that we can find an economical substitute for the phosphates in household and industrial detergents, for example. A harder problem, and a critical one in Lake Erie, concerns the runoff from farmlands. The chemical industry produces fertilizers and insecticides that have greatly increased crop yields, and new fertilizer factories are constantly being built in areas of the world in which starvation is a serious problem. However, fertilizers are effective on algae as well as on crop growth, and when they are washed out of the farmlands into the lakes by rainfall, they stimulate the growth of algae in the lakes. When the algae die and decompose, they consume the oxygen in the lake and kill the fish. Thus when this algae growth is unchecked, it leads to the biological death of the lakes.

Solutions to these and many other problems are urgently needed, and it seems clear that chemistry will play a crucial role in formulating them. However, the magnitude of these problems is so great that solving them is vital to every living person. Because of this, these problems go far beyond the scope of chemistry, or even the scope of all of science and technology.

The problems, then, become political problems, and the answers will require a fiscal and philosophical commitment on the part of the leaders of this country and the world. We find ourselves in a situation in which the future of mankind will be decided largely by the judgment of nonscientists in matters that require a knowledge of science. This is the way it should be. Politicians—if they are to serve the best interests of their constituencies—should be able to listen knowledgeably to the advice of scientists. These communications that are so necessary between scientists and politicians, between scientists and laymen, between politicians and their constituents, and indeed between voters will be possible only if everyone develops the best possible understanding of chemistry (and science) and of what a chemist actually does.

In writing this text, we tried to establish this understanding of "doing chemistry" by emphasizing, throughout the early chapters, the relationships between observations, experiments, hypotheses, and the ideas of individual chemists. In these initial chapters (largely repetitious for students with high school chemistry backgrounds), we intentionally made the pace slow and the conceptual material less taxing or presented in simplified form. We hope that this will enable the newcomer to accustom himself to the symbolism, nomenclature, and approach of chemistry, and will provide needed review for students with poor backgrounds. Atomic theory, for example, is developed with an emphasis on the importance of the interpretation of experiments. The emphasis on experimental observation is continued until, in Chapter 6, we discuss the crucial topic of chemical bonding on a conceptually current but nonmathematical level. We use the same technique later in discussing rates of reactions. The level of difficulty increases as the student progresses.

▶ Triangles are used throughout the text to set off material which is of special importance or interest. These asides are meant to clarify and explain concepts, or else to bring to the student's attention the practical implications of the subject matter. ◀

We believe that the approach used in this text should be applicable to many beginning college chemistry courses. In particular the two-year colleges should find the text usable in courses for students with no—or poor—backgrounds in chemistry. The material as presented should bring the student to an understanding of chemistry beyond that generally obtained in high school and will serve as a preparation for more rigorous presentations of introductory chemistry. For a one-semester or two-quarter course for liberal-arts students, the instructor should be able to omit many of the more difficult sections placed at the ends of chapters, or even omit several complete chapters without destroying the continuity.

For example, the chapters on Rates of Reactions, Metals and Nonmetals, Stereochemistry and Reaction Mechanisms, Digestion and Metabolism, Vitamins, Hormones, and Drugs, and Nuclear and Radiochemistry might require too much time for such a short course. In the one-semester nonscience majors' course at Fredonia, we omit or only briefly mention these chapters, as well as much of the material covered in Sections 5–8, 5–9, 6–5, 6–6, 9–6, 10–6, 11–5, 12–6, 13–7, 13–11, 13–12, and 16–12. In the nonscience majors' full-year course, the instructor may wish to add some more descriptive inorganic chemistry or otherwise embellish the course with outside readings and other materials. For students in paramedical programs, who generally have taken high school chemistry, the first chapters could be omitted or rapidly reviewed and the emphasis rightly placed on the latter chapters. We hope that the book will be flexible enough to meet the requirements of these several groups of beginning chemistry students.

At various points in the text we have briefly discussed some problems that are currently important or certain applications relating to chemistry, such as air pollution, the differences between basic and applied chemistry, the replication of living systems, the chemical structure of hormones and drugs, and so on. It is our belief, however, that students of whatever persuasion get a better feel for what chemistry is and for what chemists do by studying the *principles* of chemistry. This philosophy underlies the entire presentation.

We would like to thank the many people who encouraged and helped us in writing this text. Sincere thanks go to the staff of Addison-Wesley for their constant aid, and to Dr. Francis Bonner, of the State University of New York at Stony Brook, N.Y., for his many constructive comments on the manuscript. Mrs. Frances Granata typed the entire manuscript, and her cheerful help and cooperation are greatly appreciated. We would also like to thank our teachers at Cornell University (PFW) and at the University of New Hampshire and Boston College (JHS). Particular thanks are due Professor Michell J. Sienko (from PFW), who is a teacher beyond compare, and a continuing source of inspiration and stimulation.

And, of course, without the year-in, year-out encouragement of our families, this book could never have been completed. Special thanks therefore to Gail and Cathy.

Fredonia, New York   P.F.W.
November 1970   J.H.S.

CONTENTS

1. OBSERVATIONS AND MEASUREMENTS

1-1 THE STUDY OF CHEMISTRY

What are the differences between table salt and sugar? What are their compositions? How can they be prepared from other substances? Are they pure? Why do they dissolve in water? What is water? Why do large chunks of ice or wood float on water while a small diamond sinks? Why is a diamond colorless and a ruby brilliant red? These are the kinds of questions that confront chemists, the kinds of problems involved in the study of chemistry. Some of the problems are very difficult and complex, while others are—or at least appear to be—much less involved. But they all have one common ingredient: the challenge of the unknown and the desire for greater understanding. This pursuit of increased knowledge is one of the major motivations for all chemists. Very often, the problems selected for study do not appear to have an immediate application or technological use. Projects are undertaken for their intrinsic interest or their fundamental importance. Such work is generally called fundamental or basic research.

▶ As a crude analogy for fundamental or basic studies (research), we can consider a child interested in building things with blocks. During the course of his play (studies or research), he might discover that it is beneficial to have large blocks located near the bottom of a structure with small and irregularly shaped blocks at the top, or

1

that tall buildings need firm foundations, or that curved objects are hard to construct from blocks with flat sides, and so on. Knowledge, interest, and ideas gained from working with blocks might encourage more sophisticated ventures, say with erector sets or more complicated building materials. From these further experiences the child can learn more fundamental principles about construction. Note that his activities have no direct application; the results are of no immediate use to the child, or anyone else. He has chosen to play with blocks and similar materials because they are intrinsically interesting to him. It is fun to build things and to learn how to improve them. ◀

Problems of a fundamental nature that are investigated by research chemists—such as the preparation of a new material, or understanding how a substance is formed, or discovering why a certain material exhibits a characteristic property—are as varied as the chemists themselves and are unlimited in number. The particular problems that are studied do, of course, vary with time, as knowledge grows and situations change. For example, 200 years ago the idea of chemical elements was not firmly established; oxygen had not been discovered. One hundred years ago the periodic properties of the elements were being recognized for the first time; concepts of atomic structure were still undeveloped. Today, chemists are struggling with problems concerning the submicroscopic structure of vitamins and proteins, the composition of the surface of the moon, the mechanisms of energy transfer between molecules, the effects of high energy radiation on the formation of chemical compounds and on their reactions, the unusual chemical properties of the surfaces of various substances, the purity of materials and how the properties of these materials are affected by minute amounts of contaminants, and many more. The variety and number of problems under study today is huge, and yet there appears to be an ever-increasing number of *new* problems that need investigation. It almost seems, sometimes, that the more we learn about our problems, the more we find that we do not understand.

Along with the study of fundamental chemical problems, there is an equally important facet of chemistry: the application of fundamental chemical concepts, knowledge and discoveries to specific practical and technological problems.

To understand the impact that applied chemistry has on all our lives we need only consider the amazing advances in recent years in the development and production of high-speed computers, the introduction of color television, the control of many crippling diseases, and the incredible explorations in outer space. For society in general, these technological advances are of primary importance, but we must always remember that the foundations of application lie in our store of fundamental knowledge. Advances in the understanding of underlying principles often lead to increased applications; technological advances often make possible more elaborate and sophisticated experimental methods and techniques which lead to increased fundamental knowledge.

▶ To illustrate applied studies (research), we can continue the analogy of a child and building blocks used to describe fundamental research. The information that the

child learned while playing with (experimenting with) different types of construction toys might remain dormant for some time, possibly until the young man begins doing odd jobs at home. Then some of the fundamental knowledge he obtained as a child might be applied to real and specific problems. For example, how can a good, solid workbench be made? The ideas concerning heavy, strong, well-braced construction learned many years prior to the particular problem can then be applied. Note that the original interest in building blocks, etc. (and the resulting basic knowledge) was not stimulated by the applied problem of constructing a workbench.

In comparing the studies and problems involved in basic chemistry as opposed to those in applied chemistry, as we have presented them, you might conclude that there is no clear-cut distinction between them. Indeed there is no sharp dividing line between fundamental and applied science. This is especially true, and apparent, when we try to label a particular research project as basic or applied; opinions often differ sharply. On the other hand, when the two broad categories of basic research and applied research are used in a very general way, they are commonly understood as we have described them above. As an example of the differences between fundamental and applied research, we can consider the case of the gas neon (a chemical element with the symbol Ne), along with the other members of a family of chemical elements called the noble gases. As you no doubt know, neon gas emits a red glow when it is placed in a light-bulb-type arrangement (neon tubes or lights). This intriguing effect was investigated by an experimental technique called spectroscopy, i.e., very careful determination of the type of light emitted by the Ne, and the resulting information compared to and understood in terms of available ideas of the atom. Similar information was obtained for the other noble gases. These spectroscopic studies would generally be considered basic or fundamental research, since they were primarily conducted for scientific information and had no immediate application.

In the early 1960's, however, the discovery of gas lasers made this fundamental information about the spectroscopic properties of the noble gases quite useful. Knowledge of these properties was applied to the requirements of a gas laser to produce new gas lasers and gas lasers with different properties. The lasers developed were then used for various purposes, such as retinal surgery, minute spot welding, unusual light sources for use in microscopes and in Raman spectroscopy. These development and application studies would generally be considered applied research.

But the use of the lasers in microscopy and spectroscopy improves these experimental techniques and permits measurements that yield new and different fundamental information about not only the noble gases but many other substances as well. Then further basic studies and fundamental knowledge have been made possible by the applied studies conducted on the noble gases. ◀

No matter what variety of chemical problem is being solved—whether it is small or large, experimental or theoretical, fundamental or applied—the requirements for success are: hard work, enthusiasm, imagination, patience, ideas, and money. The practicing scientist, of course, furnishes the necessary individual qualities. Since the funds required for the equipment, supplies, and other expenses involved in solving

sophisticated chemical problems can become very large, they are generally supplied by the government, industries, or universities. This is especially true for applied chemical problems and for the advancement of technology, which consumes millions, or even billions, of dollars each year. Projects designed for the purpose of eliminating fatal diseases and physical ills, of constructing artificial body organs, of overcoming water and air pollution, of feeding, clothing, and housing all peoples of the world, of manned exploration of outer space are of primary importance to us all. But projects such as these take an overwhelming amount of money. Consequently, priorities among the various desirable projects must always be established. Decisions must be made concerning which projects can be supported. And among those projects selected, we must decide how the available funds should be distributed. These decisions are difficult and complex; and they are made every day. They are made by company executives, boards of directors, stockholders, groups of businessmen; by congressmen, senators, the President and his advisors—politicians in general; and by the votes of all our citizens.

It is therefore necessary for the politicians, the businessmen, the stockholders, the voters to have some basic understanding of the ways of science, of its needs, of its possibilities, of its costs, of its implications, and of its limitations. Intelligent, well-considered decisions cannot be made without such a foundation.

We believe that the most direct approach to this basic understanding is through the study of chemistry as it actually exists, as chemists themselves see it and use it. A study of chemistry in this way will reveal the importance of experiments and the observations of experimental results, of the many ideas and hypotheses of individual chemists that are required to solve seemingly simple problems, of the interrelationship among experiments, observations, and hypotheses in solving problems, of the necessary quantitative aspects of chemistry and scientific problems in general. An understanding of, and feeling for, these aspects of chemistry will provide a firm foundation for an understanding of science.

1-2 METHODS OF SCIENCE

Chemists study materials; their compositions, their properties, their interactions. There are, of course, many materials and many chemists. Hence there are many modes of study. But there does seem to be one unifying aspect in these investigations: in one way or another, they involve ideas, hypotheses or models, experiments and observations. Discovery is often a complex mixture of these ingredients. Sometimes it appears as though certain observations are most important in starting an investigation, while it also is often the case that a particular experiment can lead to new and unexpected knowledge. And at other times it is a proposal from an individual scientist that is the impetus for discovery. Irrespective of the starting point in a scientific study, there is always an underlying theme of verification. The process of verification generally combines experiments, observations, and hypotheses and has no particular or distinct order or method. But there is one thing about the process, which might be called revision-repetition, that seems to be almost self-generating and is quite im-

portant, since it accommodates our errors and misunderstandings. This revision-repetition aspect is nothing more than a commonsense approach to problem-solving. An idea is proposed to explain some observed behavior. In some way, this hypothesis is questioned or tested, and the results of these tests are then compared to those expected or predicted from the original hypothesis. If the test results do not agree with those predicted, then the original hypothesis must undergo revision until it is consistent with experimental observation.

Our everyday life is an unconscious but continuous involvement with similar problem-solving conditions. As an example, we can consider the problem of growing radishes in a small garden. Our first proposed idea or hypothesis might be that the radishes will grow better and can be tended more easily if we planted them in a few short rows running across the garden. But after the plants are an inch or two tall, a simple observation indicates that many of the plants need to be pulled out so that some may grow to maturity. Using this observation, we might then question the original hypothesis. Would it be better to simply scatter the seeds over an area of the garden rather than carefully separating the radishes into rows? This procedure should at least cut down on the number of plants that need to be pulled out.

A second planting, using the scattered-seed procedure, at the other end of the garden, will provide an experimental answer to this question. Happily you might find that the second planting of scattered seeds eliminated some thinning and produced a superior crop of radishes. You might even feel so good about your newfound discovery that you tell a friend about it and suggest that his next radish planting be of the scattered-seed variety. Three weeks later, much to your chagrin and dismay, the friend reports that your suggested method of radish planting was poor advice indeed. His first planting, by rows, produced 30–40 good radishes, while his second planting, by the scattered method, produced only 3 edible radishes.

Now, with these experimental results, the hypothesis that it is better to plant radishes by scattering the seeds as opposed to placing them in rows needs to be reconsidered. Evidently there are other more important things that affect the growth of radishes than just the pattern in which they are planted. Upon considering and discussing the radish plantings of your friend along with your own, another and rather different idea might occur to you. Maybe good radish production depends more on *when* than *how* the radishes are planted.

This new hypothesis can be tested experimentally the following year, but it can also be considered by consulting other gardeners as well as magazines, books, and other references pertaining to gardening. This process will lead to constant revision of the hypotheses to include the fertility of the soil, the temperature of the days and nights after planting, the amount of sunshine and moisture available to the plants, and so on.

The revised hypotheses enable gardeners to improve their radish crops as well as understand the important variables involved in raising radishes. And this acquired knowledge that pertains specifically to radishes can then be applied to other plants, and to gardening procedures in general. With each new application, the hypotheses will again need to be tested, revised, retested, and refined until they fit each different new situation.

▶ Note how the original hypothesis concerning planting of radishes in rows has been completely abandoned. This often happens. Our first ideas are not always correct, but through testing, discussion, thinking, and revising, we arrive at a more correct and more productive understanding of our problems.

From all of the variables that are listed as being important in growing radishes, a rule can develop: When radishes are planted early in the spring, after the last frost, the yield tends to be high. When similar kinds of summarizing statements or general descriptions of observed behaviors evolve in science, they are often called *laws*; for example, the law of gravity. When an explanation for some observed behavior is proposed in the form of a hypothesis or model which is then supported by considerable experimental evidence, the hypothesis is often dignified by the title *theory*. ◀

Solutions to problems in science take routes similar to those described above for growing radishes. There are many wrong or only partly true hypotheses, many ideas and questions that are productive, and some that are misleading, and many observations, tests, experiments, and measurements that need to be made to provide a better understanding of our known problems and to uncover an ever-increasing number of problems and areas of investigation. Some pertinent observations and measurements necessary to the study of chemistry will be considered in the following pages.

1–3 PHYSICAL PROPERTIES: OBSERVATIONS AND MEASUREMENTS

Substances. We shall first consider some observations of physical properties of substances. You undoubtedly have some feeling for what a *substance* is, but a more specific definition is that it is a material which has characteristic properties and a particular chemical compostition. Some examples of substances are: water (H_2O), table salt (sodium chloride, NaCl), sugar, oxygen gas, dry ice (solid carbon dioxide, CO_2), ethyl alcohol, and so on (Fig. 1–1).

When the *physical properties* of a substance are observed or altered, there is no change of one substance into another. For example, when ice melts to form liquid water, the same substance is present before and after melting; only its physical form is changed. Magnesium (symbol Mg) is a shiny, metallic-appearing, strong but lightweight substance when it is present as sheets or bars, but when it is in powdered form, it is grayish and no longer shiny, and yet it is still magnesium. Sulfur (symbol S) is a substance that can exist in several physical forms, such as fine yellow powder, a yellow granular powder, yellow chunks or yellow sticks; all of these forms show the characteristic composition and properties of sulfur. Dry ice—a white, cold solid—is a form of the substance carbon dioxide, and transforms to gaseous carbon dioxide under normal room conditions. Figures 1–2a and 1–2b further illustrate physical properties.

These examples of physical properties illustrate the possible differences in the properties of a substance when it exists in solid or liquid or gaseous form. A *solid* is a material that has characteristic size and shape. An ice cube, for example, has its own particular size and shape, as does a solid iron nail. When ice melts, however,

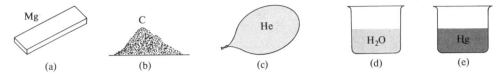

Fig. 1–1 Some examples of substances: (a) a shiny metallic strip of magnesium (Mg), (b) graphite a form of carbon (C), (c) helium (He), a very light gas, (d) water (H_2O), a colorless liquid, (e) mercury (Hg), a shiny, metallic liquid.

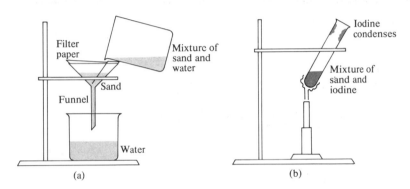

Fig. 1–2 (a) Separation of the two materials, sand and water, using differences in their physical properties. Sand is solid and collects on the filter paper, while liquid water passes through the filter paper into the beaker below. (b) Separation of the two materials, sand and iodine, using differences in their physical properties. Solid iodine is converted to iodine gas by the heat from the Bunsen burner. The gaseous iodine condenses in the cooler (top) section of the test tube. The sand is not volatile and remains at the bottom of the tube.

liquid water is formed and no longer does the water have a characteristic shape. A *liquid* is a material that takes the shape, but not necessarily the volume, of its container. The temperature at which a solid converts to a liquid is called the *melting point* of the substance. When a change from liquid to solid is observed, this conversion point is called the *freezing or crystallization point*.

When liquid water or solid dry ice is allowed to stand open under normal room conditions, liquid water changes to gaseous water or water vapor and dry ice converts to carbon dioxide gas. A *gas* is a material that occupies both the shape and volume of its container. The temperature at which a gas converts to a liquid or a gas converts to a solid is called the *condensation point* of the substance.

These three states—solid, liquid, and gas—are often called the *three states of matter* (Fig. 1–3).

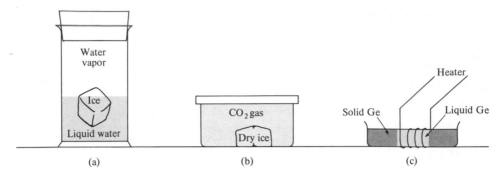

Fig. 1–3 (a) The three states of matter. Water (H_2O), existing as a solid (ice), a liquid, and a gas. Note that the gas takes the shape and volume of its container, the liquid the shape but not the volume, and the solid takes neither, but has its own shape and volume. (b) Carbon dioxide (CO_2), existing as a solid (dry ice)) and as a gas. (c) Germanium (Ge), existing as a solid and as a liquid. Zone refining process of Ge is used to produce very pure samples for use in transistors.

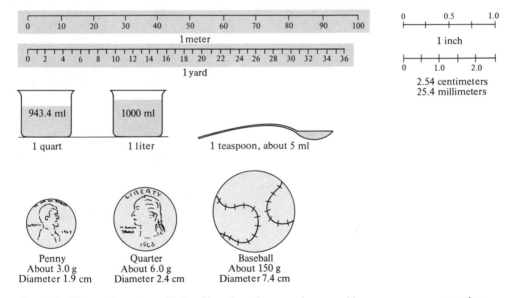

Fig. 1–4 The metric system. Units of length, volume, and mass, with some common examples.

Matter. Substances such as sugar, iron, water, and oxygen exhibit the properties of *matter*, which is anything that occupies space and has mass. These two properties of matter are readily demonstrated by considering an iron (Fe) bar. The bar obviously occupies space, but this can be shown experimentally in a more quantitative fashion by measuring the dimensions of the bar with a ruler or meter stick.

Table 1–1. Common prefixes of the metric system

Prefix	Decimal notation	Exponential notation
kilo-	1000	1×10^3
deci-	0.1	1×10^{-1}
centi-	0.01	1×10^{-2}
milli-	0.001	1×10^{-3}
micro-	0.000001	1×10^{-6}

Table 1–2. Common units of the metric system for length, volume, and mass

	Abbreviation	Conversion factors	
Length			
meter	m		
centimeter	cm	0.010 m	1.0×10^{-2} m
millimeter	mm	0.0010 m	1.0×10^{-3} m
kilometer	km	1000 m	$1.0 \times 10^{+3}$ m
Volume			
liter	ℓ		
milliliter	ml	$0.0010\ \ell$	$1.0 \times 10^{-3}\ \ell$
cubic centimeter	cm^3 or cc	$0.0010\ \ell$	$1.0 \times 10^{-3}\ \ell$
Mass			
gram	g		
milligram	mg	0.0010 g	1.0×10^{-3} g
kilogram	kg	1000 g	$1.0 \times 10^{+3}$ g

In order to measure the bar, we need a system of units. In making measurements of length or of volume or of mass, we shall use the metric system. This is the most common system of units used in science and throughout most of the world. In the United States, however, there is still no general understanding of, or feeling for, the metric system. To aid in the development of a feeling for metric units, we give several comparisons in Fig. 1–4, prefixes in Table 1–1, and the common units of length, volume, and mass in Table 1–2. Appendix B also lists many metric system units and conversion factors for changing from one unit to another.

In making measurements of length, the units most often used are meter (m), centimeter (cm) and millimeter (mm). As shown in Tables 1–1 and 1–2, these units are related by factors of ten; that is, 1.0 m = 100 cm = 1000 mm, or we can say that there are 100 cm in 1.0 m, 1000 mm in 1.0 m, and 10 mm in 1.0 cm. A very convenient method of expressing such numbers and handling their interconversions is through

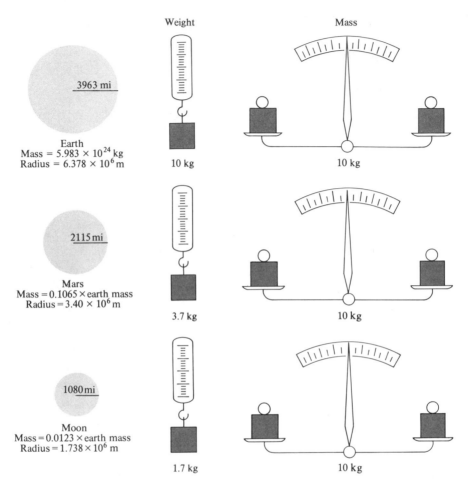

Fig. 1–5 Comparisons of mass and weight. We can determine the weight of a sample body by using a spring scale and the mass by using an equal-arm balance, as shown. The weight of a body depends on its mass and its distance from a second body. At the North Pole on earth, the sample body shown has a mass and a weight of 10 kg (at higher elevations on the earth, the weight would be less but the mass would remain 10 kg). On Mars the same sample would have a weight of about 3.7 kg, and its mass would be 10 kg. On the moon, the sample weight would be down to about 1.7 kg, but its mass would still be 10 kg.

exponential notation: $100 = 1.0 \times 10^2$, $0.010 = 1.0 \times 10^{-2}$, and so on, as illustrated in Tables 1–1 and 1–2. This notation will be used throughout the text and should be mastered as quickly as possible. An extended treatment of exponentials is given in Appendix A.

If the three dimensions—length, width, and thickness—of the Fe bar are measured, the volume of the bar can be calculated. For example, if the bar measured $5.0 \text{ cm} \times 2.0 \text{ cm} \times 1.0 \text{ cm}$, the volume would be 10 cm^3. The most common units of

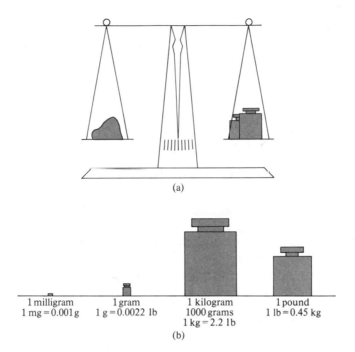

(a)

1 milligram	1 gram	1 kilogram	1 pound
1 mg = 0.001 g	1 g = 0.0022 lb	1000 grams	1 lb = 0.45 kg
		1 kg = 2.2 lb	

(b)

Fig. 1-6 (a) An equal-arm chemical balance with an unknown mass on the left pan and a known mass (weights) on the right pan. (b) Relative sizes of different masses.

volume in the metric system are liter (ℓ) and milliliter (ml) or cubic centimeter (cm^3 or cc). These units are also related by factors of ten; that is, $1.0 \, \ell = 1000$ ml $= 1000$ cm^3.

The second property of matter, that of mass, can be demonstrated by lifting the bar of Fe. The bar has a certain weight and also a certain mass. *Mass* is an inherent property of matter and is a measure of the actual quantity of matter in a body. *Weight*, on the other hand, is a relative property and is a measure of the attractive force between two bodies. Your own body weight depends, for example, on your distance from the centre of the earth. Figure 1-5 illustrates the differences between mass and weight, and indicates that the inherent property, mass, can be measured by the relative property, weight. For our purpose the differences between mass and weight will generally be neglected (most of our experiences and experiments are confined to the earth); we shall use the terms interchangeably.

The mass of a substance can be determined experimentally by *weighing* the substance on a balance, as diagrammed in Fig. 1-6. The units of mass in the metric system are gram (g), milligram (mg), and kilogram (kg). As with length and volume units, these units for mass are also related by factors of ten; that is, $1.0 \, \text{g} = 1000$ mg $= 0.0010$ kg.

1–4 CHEMICAL PROPERTIES: OBSERVATIONS AND EXPERIMENTS

Chemical reactions. All the observations considered up to this point have been of physical properties. When the physical properties of a substance are changed, it is not converted into a different substance. On the other hand, when the *chemical properties* of a substance are changed, the properties of the original substance are altered; one substance is transformed into another.

▶ The difference between physical and chemical properties can be illustrated by using a burning house as an analogy. Rapid entry to the house sometimes requires chopping down one of the wooden doors. The physical properties of the wood are changed when the door is chopped apart. The wood exists in a different physical form, but it has not been changed into another substance. It is still wood, but now it exists in a few small, irregularly shaped pieces. The physical properties of the wood have been altered. On the other hand, another door of the house that burns down is changed in a completely different way. It no longer exists as wood; it has been changed into other substances. As the door burns, gases are formed and the wood is transformed into a completely different substance, which we sometimes call charcoal. The chemical properties of the wood have been altered. ◀

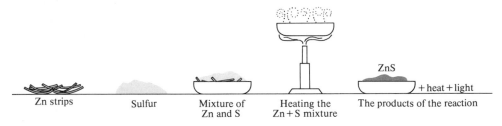

| Zn strips | Sulfur | Mixture of Zn and S | Heating the Zn + S mixture | The products of the reaction |

Fig. 1–7 Heating of the elements Zn and S produces a chemical change; a new substance, the compounds ZnS, is formed. Chemical energy is changed into heat and light energy.

To illustrate chemical properties and some typical observations that can be made during chemical change, we shall consider what happens when the substances zinc and sulfur are mixed together and heated, as illustrated in Fig. 1–7. Zinc and sulfur are represented by the chemical symbols Zn and S, respectively, and are called elements. An *element* is a substance that cannot be decomposed by ordinary chemical reaction. That is, under any ordinary circumstances an element cannot be divided or changed into other elements or simpler substances. When we mix these two elements, we can easily identify their individual properties. Small strips of Zn are shiny, metallic appearing, and can be bent and twisted into various shapes. The Zn strips can be easily separated from the bright yellow, granular S powder with which it is mixed.

These observable properties are greatly changed, however, after the mixture has been heated. When the proper conditions of heating are attained, the element mixture very rapidly begins to sputter and fume; even more heat is generated by the chemical

changes occurring; very bright, white light is emitted from the mixture. After it has cooled, inspection of the material present shows only a grayish, powdery solid. None of the material has the properties of the original elements in the mixture. The metallic-like element Zn has combined with the yellow powder of the element S to form a new substance, the gray, powdery solid, which apparently is composed of Zn and S in some way but no longer is simply a mixture of the two individual elements.

That Zn and S are actually combined in some fashion in the gray solid formed by the chemical change can be shown by several methods of chemical investigation. One relatively simple experimental procedure would be to place the gray solid product in a container and then evacuate and heat the container. Under these experimental conditions, the gray powder would be observed to decompose into a shiny, metallic-appearing substance with other properties, such as melting point and density (see Section 1–6) identical to the original element Zn, and into a bright yellow powder with properties identical to the original element S.

This chemical change can be written in a more concise form,

$$Zn + S \rightarrow ZnS.$$

This shorthand form of describing a chemical reaction is called a *chemical equation*. The substances to the left of the arrow are called *reactants* and those to the right of the arrow are called *products*. In the language of the chemist, we say that zinc reacts with sulfur to produce zinc sulfide. (Some common rules of nomenclature are given in Appendix C.) We could also include heat and light as products of the reaction, along with the ZnS. Generally, however, only the chemical substances involved are included in the equation.

Chemical compound. The reactants in our example are the elements Zn and S. The ZnS (zinc sulfide) product obtained after reaction is called a compound and contains both the elements zinc and sulfur.

A *compound* is a substance containing more than one element, which can be decomposed into two or more substances. Although the compound ZnS contains both zinc and sulfur, its properties are completely different from those of either zinc or sulfur. The elements Zn and S have been combined in some fashion to give a new substance with its own characteristic physical and chemical properties. When the compound ZnS decomposes, the two substances Zn and S are formed.

▶ We have written the compound zinc sulfide as ZnS in chemical symbols. It is not obvious that the ratio of Zn to S should be 1 : 1, as given by the ZnS formula, nor is the determination of this ratio always easily done experimentally. As your study of the next few chapters continues, it will become more evident why different chemical compounds have different combining ratios. We should always remember, however, that these ratios are not obvious and that they are experimentally determined. ◀

Other examples of common compounds are: water (H_2O), sodium chloride (NaCl, table salt), carbon dioxide (CO_2) and calcium carbonate ($CaCO_3$, limestone). All these compounds are composed of more than one element, as we can see from their

chemical formulas, and they can all be decomposed into at least two other substances. Water, for example, can be prepared by reacting hydrogen and oxygen gases
under the proper experimental conditions. The chemical compound H_2O is composed of the elements hydrogen and oxygen. If H_2O is electrolyzed,* it can be decomposed into its two constituent elements, reforming hydrogen and oxygen gas.

Sodium chloride, the compound NaCl, can also be prepared by reacting elements.
Exposure of solid Na metal to chlorine gas forms NaCl, a solid compound of the two
substances sodium and chlorine. The compound CO_2, carbon dioxide, contains the
elements carbon and oxygen. It can be decomposed into the two substances (both
gases) carbon monoxide, CO, and oxygen by vigorous heating. Carbon monoxide
is also a chemical compound, since it contains the two elements carbon and oxygen
and can be decomposed into these two substances. Limestone, $CaCO_3$, represents a
slightly more complex chemical compound, since it is composed of three elements,
calcium (Ca), carbon (C), and oxygen (O), and upon decomposition forms at least
two other compounds, calcium oxide (CaO) and carbon dioxide (CO_2):

$$CaCO_3 \rightarrow CaO + CO_2.$$

The two products CaO (lime) and CO_2 are also compounds, since they too can be decomposed into two other substances, calcium and oxygen in the case of CaO.

Note that in all cases the chemical formula shows the elements that are present in
the compound and gives the ratio in which the elements are present in the compound.
In ZnS this ratio is $1 : 1$; in H_2O the ratio is $2 : 1$; in $CaCO_3$ the ratio is $1 : 1 : 3$; and
so on. The understanding and experimental determination of these combining ratios
is one of the prime concerns of chemists. This topic will be considered throughout
this entire text.

▶ Formation of chemical compounds is somewhat analogous to the formation of
athletic teams. There are many different types of teams—baseball, football, basketball, soccer, hockey, etc.—just as there are different chemical compounds. The teams
are composed of different types and numbers of fundamental units, or elements, such
as pitchers, catchers, shortstops, etc., or centers, forwards, and guards, or ends,
quarterbacks, fullbacks, etc. There is no team unless each of its constituent units is
present. Each team has a purpose, form, and identity all its own. All the teams can
be broken up (decomposed) into at least two other types of units. For example, the
nine players on a baseball team might, under the proper experimental conditions such
as the off-season between November and March, form a basketball team of five members, and become two golfers and two insurance salesmen. The particular decomposition products or units depend on the kind of team under consideration and the
type of fundamental unit, or element, composing the team, or compound. ◀

---

*   We shall describe this experimental technique in Chapter 11. If this is an unfamiliar term and
procedure, you need not consider it further now.

1–5 ENERGY

Forms and interconversions. In the chemical reaction between Zn and S above, heat was given off. Heat is a form of energy. *Energy* can be defined as the capacity to do work, that is, the capacity to move matter against an opposing force. To see that heat does fulfill this definition of energy, we can consider the case of heating water to produce steam, as in a steam engine. In this case the expanding steam can be used to move objects and do useful work.

You are, no doubt, aware of other forms of energy; there are many. You can drive a large stake into the ground by striking it with a sledgehammer. In this case the hammer supplies the required energy. It possesses energy because it is a substance in motion. This type of energy is called *kinetic energy*. An automobile, for example, has no kinetic energy when it is parked, but it has a large kinetic energy when it is traveling at 60 miles an hour. From these two examples of kinetic energy you might be able to predict that the kinetic energy of a body is higher if the mass of the body is made larger and that the kinetic energy also increases as the speed of movement of the body increases. That is, a large truck traveling at 60 mph has more kinetic energy than a motorcycle traveling at 60 mph, and a truck going 60 mph has more kinetic energy than the same truck at 30 mph.

▶ The mathematical relationship relating the kinetic energy and the mass and speed of a body is $KE = \frac{1}{2} mv^2$, where m is the mass of the body and v is the velocity at which the body is moving. ◀

Another very important form of energy is *potential energy*. In this case, a substance possesses energy because of its position. For example, a large book can crack a peanut shell if the book is dropped on the shell from three or four feet above the shell. On the other hand, the large book might not crack the peanut if it fell from only two inches above the shell, or if a very small, paperback book was used, the shell might not crack even if the book fell from 10 feet. The potential energy of a body, then, seems to depend both on the mass of the body and on its distance above its final position. An increase in this distance or the mass of the body will increase the potential energy of the body.

▶ The exact relationship between potential energy, mass, and distance is: $PE = mgh$, where m is the mass of the body, g is the universal constant of gravitation, and h is the height of the object above its final position. ◀

One of our most important forms of energy is *electric energy*. All electric motors and electric appliances run by using electric energy.

As mentioned above, *chemical energy* is also available for use. A lead storage battery commonly used in automobiles is a typical example of a source of chemical energy, as are dry-cell batteries for use in flashlights or for firing flashbulbs in cameras.

It is evident from this last example that forms of energy are constantly being interconverted. Chemical energy from a dry-cell battery, for example, is converted

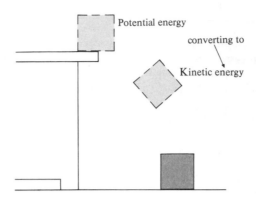

Fig. 1–8 The potential energy of an Fe block resting above the floor is converted to kinetic energy as the block is falling from the table to the floor.

to electrical energy, which is then converted to heat and light energy in a flashlight or in a flashbulb. The chemical energy from a battery in an automobile is converted into electrical energy, which is used to start the motor. The potential energy of an Fe block resting on the edge of a table above the floor is continuously converted to kinetic energy when the block falls from the table (see Fig. 1–8). When the block strikes the floor, its kinetic energy is converted into heat energy, into sound energy, and into kinetic energy for the composite parts composing the floor. If the block is pushed across the floor, its kinetic energy is converted to heat energy (because of friction, as the block rubs against the floor).

These few examples of some of the forms of energy and the ways in which they can be interconverted should indicate that energy is a very broad, very important, but a rather difficult and elusive concept. The more times you encounter energy in its various forms, see conversions between one form and another, and see its effects and uses, the better will be your feeling for this complex and important term.

1–6 OBSERVATIONS, MEASUREMENTS, AND CALCULATIONS

Heat and temperature.   In the above chemical reaction between Zn and S, heat energy was produced by the reaction. A chemical reaction that produces heat is called an *exothermic* reaction; a reaction that absorbs heat is an *endothermic* reaction.

The intensity of heat, hotness, or the direction in which heat flows, from hot to cold, is indicated by the *temperature*. There are three common scales of temperature: Fahrenheit, Celsius (also called Centigrade), and Kelvin (also called Absolute). The Celsius and Kelvin scales are most generally used in science and are similar in that the size of the degree is the same on both scales. There are 100 divisions or degrees between the freezing and boiling points of water. The difference between the two scales is in the assignment of temperature values to standard reference points (see Fig. 1–9). On the Celsius scale the freezing and boiling points of water are given the values of 0 °C and 100 °C, respectively. These same two points on the Kelvin scale

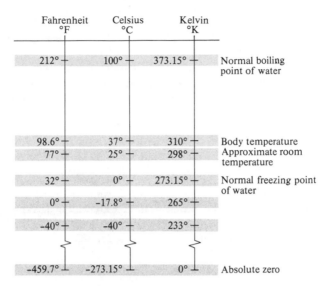

Fig. 1–9 Three temperature scales. Common temperatures on the Fahrenheit, Celsius (centigrade), and Kelvin (absolute) scales. The temperatures within each horizontal block are identical.

Table 1–3. Some representative temperatures on the Celsius and Kelvin scales

Observation of:	t, °C	T, °K
Sun's surface	6000	6273
Oxyacetylene flame	3500	3773
Melting tungsten, W	3380	3653
Boiling gold, Au	2660	2933
Melting gold	1063	1336
Melting aluminum, Al	660	933
Melting tin, Sn	232	505
Boiling water, H_2O	100	373
Room temperature	25	298
Melting ice	0	273
Subliming dry ice, CO_2	−78	195
Boiling liquid nitrogen, N_2	−196	77
Boiling liquid helium, He	−269	4
Absolute zero	−273	0

are 273.15 °K and 373.15 °K (see Table 1–3). Note that negative temperatures are common on the Celsius scale, while on the Kelvin scale 0 °K is generally taken to be the lowest possible temperature attainable (more about this later, in Chapter 8).

In order to interconvert Celsius and Kelvin temperatures, we must consider the different values assigned to the standard points of reference, the freezing and boiling points of water. The Kelvin temperatures are higher than the Celsius temperatures by 273° (this approximation is good enough for our purposes). Then, to convert °C to °K, we simply account for this different definition of reference points by adding 273° to the temperature given in °C. In order to convert °K to °C, we subtract 273° from the Kelvin temperature.

Example 1-1. Room temperature on the Celsius scale is around 25 °C. What is this temperature in °K?

The freezing point of water on the Celsius scale is 0 °C, while on the Kelvin scale it is 273 °C. The Celsius temperature given in the problem is 25° above the freezing point, that is, 0° + 25° = 25 °C. On the Kelvin scale, this temperature is also 25° above the freezing point of water. The temperature in °K is, therefore, 273° + 25° = 298 °K.

Example 1-2. The normal melting point of sodium chloride (common table salt that has the chemical formula NaCl) is 801 °C. If a batch of NaCl is heated to a temperature of 1080 °K, is the NaCl in the liquid state or the solid state?

To determine whether the sodium chloride is solid or liquid, we have to find out if the NaCl melting point has been exceeded. Has the NaCl sample been heated to 801 °C or above? We can, therefore, convert 1080 °K to °C and compare this temperature with 801 °C. To convert °K to °C, we subtract 273° from 1080°, which gives 807°. Since this 807 °C temperature exceeds the 801 °C melting point of NaCl, the NaCl sample will be melted and in the liquid state.

To interconvert Celsius and Fahrenheit temperatures, we must consider the difference in the size of the degree as well as different reference point values. This is illustrated in Example 1-3. We can make the following summarizing statements.

1. To convert °C to °F: $\frac{9}{5}$(the temperature in °C) + 32° = the temperature in °F.

2. To convert °F to °C: (the temperature in °F − 32°)$\frac{5}{9}$ = the temperature in °C.

Example 1-3. A frying pan coated with nonstick Teflon should not be heated above 250 °C. What is this temperature in °F?

The two standard reference points are 0 °C and 32 °F for the freezing point of water and 100 °C and 212 °F for the boiling point of water. To convert the Celsius to the Fahrenheit scale, we first determine the number of degrees above the Celsius freezing point: 250° − 0° = 250 °C. Then we must consider the size of the degree. Since there are 100 divisions (degrees) between the 0 °C and 100 °C reference points on the Celsius scale and 180 divisions between these same reference points on the Fahrenheit scale (212° − 32° = 180 °F), each Fahrenheit degree is smaller than each Celsius degree by

$$\frac{100}{180} \quad \text{or} \quad \frac{5\,°C}{9\,°F}.$$

Therefore a temperature that is 250°C above the freezing point of water is

$$250 \; ^{\circ}\mathcal{C} \left(\frac{9 \; ^{\circ}F}{5 \; ^{\circ}\mathcal{C}} \right) \quad \text{or} \quad 450 \; ^{\circ}F$$

above the freezing point of water. This temperature on the Fahrenheit scale is 32 °F, so that the final °F temperature is 450 °F + 32 °F = 482 °F. Thus one should not exceed the temperature of 250 °C or 482 °F when one is using Teflon-coated utensils.

Density. Why does a large tree branch float on water while a small brass ring sinks immediately? Why can a fisherman use small lead (symbol Pb) shot as sinkers and a large cork as a float for his fishing line and hook? It is obvious that the total weight of the objects is not the determining factor; the tree branch has a vastly larger total weight than the ring. In some way, water completely supports substances such as wood and cork, while other substances such as Pb and brass sink instead of float. Since the total weight of objects is not critical, we might propose another explanation (hypothesis) for these observations: The weight of a specific volume of a substance determines whether it will float or sink in a liquid such as water.

We can readily test this hypothesis experimentally by taking small blocks of each substance, say 1.0 cm × 1.0 cm × 1.0 cm, and weighing these blocks to determine their mass. We can then compare the masses of each block of 1.0 cm³ (1.0 cm × 1.0 cm × 1.0 cm = 1.0 cm³). When we do this, we find that 1.0 cm³ blocks of those substances that float in water have weights less than 1.0 gram, while those that sink in water have weights greater than 1.0 gram. Interestingly, when we measure the weight of 1.0 cm³ of water, we find that it is 1.0 gram. That is, each 1.0 cm³ of water weighs 1.0 g, or there is 1.0 g of water in each 1.0 cm³ of water. This is called the density of water: 1.0 g per 1.0 cm³ or 1.0 g/cm³ or 1.0 g/cc.

The *density* of a substance is defined as the mass of the substance that is contained in a unit volume. Written in the form of an equation,

$$\text{Density} = \frac{\text{Mass}}{\text{Volume}}.$$

The units of density are the units of mass (grams) divided by the units of volume (cc), grams over cc or grams per cc or g/cc.* Then, to determine the density of a substance, we need to know the mass of a given amount of the substance and the volume that it occupies. As described above, these two properties are easily determined experimentally. The mass of an object can be obtained by weighing it on a balance. The volume of the object can be obtained, for example, by measuring its length, width, and thickness, and then multiplying the three dimensions together. The density can then be calculated by dividing the mass by the volume.

---

*   If you are unfamiliar with this notation, more examples are given in Appendix A.

Example 1–4. An aluminum (symbol Al) bar measures 5.0 cm × 4.0 cm × 1.0 cm and weighs 54 g. What is the density of Al?

We need to find the density D of the Al bar. Density is

$$D = \frac{\text{Mass}}{\text{Volume}}.$$

We must therefore find the mass and the volume of the bar. The mass is given as 54 g. We can calculate the volume by multiplying the three dimensions together.

(5.0 cm)(4.0 cm)(1.0 cm) = 20 cm³ or 20 cc. [*Note:* (cm)(cm)(cm) = cm³ = cc.] Now we can calculate the density.*

$$D = \frac{54 \text{ g}}{20 \text{ cc}} = 2.7 \text{ g/cc.}$$

Note the units, mass divided by volume (g/cc).

Example 1–5. Suppose that we have an iron (Fe) bar and that 15.7 g of Fe are needed for an experiment. Fe has a density of 7.86 g/cc. What is the volume of the bar needed to give the desired weight of Fe?

In this problem we need to find: the volume of the Fe bar. We know: the mass of the bar, 15.7 g, and the density of Fe,

$$D = \frac{\text{Mass}}{\text{Volume}} = 7.86 \text{ g/cc.}$$

Since we know the density D and the mass, we can calculate the volume from the equation for density:

$$D = \frac{\text{Mass}}{\text{Volume}} = \frac{m(\text{g})}{V(\text{cc})},$$

or, by rearranging the equation,

$$V(\text{cc}) = \frac{m(\text{g})}{D(\text{g/cc})} = \frac{15.7 \text{ g}}{7.86 \text{ g/cc}} = 2.00 \text{ cc.}$$

Units. Note that, even if we did not remember the formula for calculating density, we could still solve Example 1–5. The *units* indicate the proper solution. A volume is desired; the solution, then, must yield a volume unit, cc for example. Given are the mass in g and the density in g/cc. The question is, how do we manipulate g and g/cc to obtain cc? These units can be treated just like numbers in mathematical cal-

---

*   Remember that the process of division converts the denominator to unity, i.e., to 1.

culations.* Therefore the volume unit, cc, can be obtained if g is divided by g/cc:

$$\frac{g}{g/cc} = cc.$$

This calculation is similar to one in which the number 5 is divided by the fraction $\frac{5}{27}$. That is,

$$\frac{5}{5/27} = 27.$$

Here the 5's cancel just as the g's do in the calculation with g and g/cc.

One final word about units. They provide a check on the mathematical operations involved in the problem and should *always* be included in problem-solving. The calculations in Example 1–5 can be used to illustrate this idea. If we decided, incorrectly, that the proper solution was to divide the density by the mass,

$$\frac{7.86 \text{ g/cc}}{15.7 \text{ g}} = 0.500/cc \qquad \text{or} \qquad 0.500 \text{ per cc,}$$

the units would immediately indicate that this is not the necessary method of calculation. The volume unit cc is needed, not /cc (per cc or reciprocal cc). If we decided to multiply the mass and density,

$$(15.7 \text{ g})(7.86 \text{ g/cc}) = 123 \text{ g}^2/cc,$$

the units would again indicate the error; g^2/cc is not a volume.

This example is, of course, rather simple, but it illustrates an important idea. Scientists measure things, lengths, volumes, masses, temperatures, etc., and not just numbers. *Units are therefore extremely important and useful in problem-solving.* As problems become more involved, the necessity of carrying out the calculations with the units as well as with the numbers will become more apparent.

Significant figures. When we are conducting experiments and making observations, we are constantly measuring things in one way or another. Samples need to be weighed; lengths need to be measured; volumes need to be determined; temperatures need to be read. Some of the measurements that are made are rather crude, while others are made to a very high degree of precision. The data that we collect, report, and use in computations must reflect the precision of the measurements. For example, we can measure the length of a copper (symbol Cu) tube to the nearest centimeter, 32 cm long, or to the nearest millimeter, 32.4 cm long, as shown in Fig. 1–10a. In the first case the tube could be between 31 and 33 cm long, but we estimate that it is closer to 32 cm. The answer, 32 cm, has only two significant figures because there is some

---

* A description of the use of units as fractions is given in Appendix A.

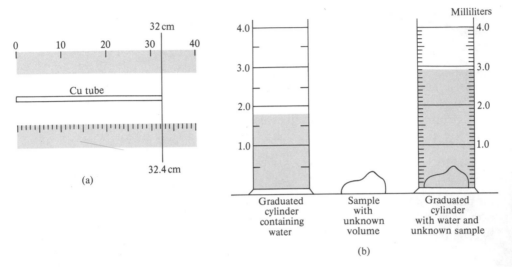

Fig. 1–10 (a) Measurement of a Cu tube, showing how the number of significant figures reflects the precision of an experimental determination. (b) Often when we are determining the density of an unknown sample, we obtain the volume of the sample by water displacement, as shown. Note that there are fewer graduated markings on the first cylinder. We can determine the volume of water in the second cylinder more precisely, i.e., to more significant figures.

doubt that 32 is a better figure than 31 or 33. In the second case we measure the tube with greater precision. Now we can say that it is between 32.3 and 32.5 cm long and that it is probably 32.4 cm in length. There are three significant figures in the answer because all three figures give information that is at least estimated to be true. *Significant figures* are the reasonably reliable digits in a number.

Another example is given in Fig. 1–10b. To determine the density of the irregularly shaped object in Fig. 1–10b, we need to measure its mass and volume. We can obtain its mass by weighing the object on a balance (such as the one in Fig. 1–6). We can obtain its volume by immersing the object in a liquid such as water, and recording the increase in the liquid level. (The object must not dissolve in the liquid.)

Note that in Fig. 1–10b the two cylinders have a different number of graduated markings. That means that we can determine the liquid level to different degrees of precision. In the first cylinder the volume of water is 1.8 ml; in the second cylinder the water volume can be measured more precisely, and is 2.88 ml. With the first cylinder only two significant figures can be obtained; with the second cylinder three significant figures are obtained.

The change in liquid level—that is, the volume of the object—can be calculated by subtracting these two numbers, 2.88 ml = 1.8 ml = 1.08 ml or 1.1 ml, to the proper number of significant figures. In addition or subtraction, the answer must be rounded off* to contain the same number of significant figures after the decimal point as in the least accurately known figure used.

These two experimental measurements indicate another thing. The mass of the object in Fig. 1-10b need be determined to only two significant figures. This is true because in multiplication or division answers must contain the same number of significant figures as in the least accurately known figure used in the calculation. If the object was weighed and found to be 2.2 g, then the density would be

$$D = \frac{2.2 \text{ g}}{1.1 \text{ ml}} = 2.0 \text{ g/ml} \quad \text{or} \quad 2.0 \text{ g/cc.}$$

The answer contains two significant figures. Note that the answer would be unchanged if the mass were determined more precisely, say 2.234 g. Then

$$D = \frac{2.234 \text{ g}}{1.1 \text{ ml}} = 2.0 \text{ g/ml.}$$

The answer reflects the precision of *all* the experimental measurements involved.

There is often confusion about zeros as significant figures. In general, we can say that zeros are not significant unless they are preceded in the number by a figure that is not zero. For example, 0.00224 has only three significant figures, while 0.22400 has five, since the zeros in the first number are not significant. Often when we are dealing with large numbers, we add zeros to indicate the decimal place. We might say that 50,000 people attended a football game. This is an ambiguous number, since we do not know this figure to the nearest person. The ambiguity can be removed by using exponential notation. We can write 5.0×10^4 or 5.00×10^4, which indicates accuracy to the nearest 1000 or 100, that is, to two and three significant figures, respectively. Further examples of significant figures are given in Appendix A.

QUESTIONS

1. Define or discuss the following: Basic research, applied chemistry, technology, scientific method, hypothesis, theory, variables, physical properties, chemical properties, substance, matter, solid, liquid, gas, melting point, matter, mass, weight, volume, milliliter, centimeter, kilogram, exponential numbers, element, compound, chemical equation, reactants and products, kinetic energy, potential energy, heat energy, Celsius, Kelvin, density, significant figures, metric system.

2. What observations could be made to prove that: (a) oxygen is a gas; (b) water is a liquid; (c) ice is a solid; (d) steam is a gas and is composed of water vapor?

---

* In rounding off numbers, if the number to be dropped is: (1) less than 5, it is simply dropped; (2) greater than 5, the preceding number is increased by 1; (3) equal to 5, the preceding number is increased by 1, if it is odd, and remains unchanged, if it is even.

3. Under normal conditions, in which state of matter are the following substances?

   a) H_2O
   b) Ice
   c) Na
   d) O_2
   e) Air
   f) Hydrogen
   g) Al
   h) Nitrogen
   i) CO_2
   j) MgO
   k) Hg
   l) Fe

4. Write the name and symbol for the first 20 elements, as given in the Periodic Table on the inside back cover of the text.

5. What are the major differences between a gas, a liquid, and a solid?

6. (a) Describe some observations of physical properties of a mixture of Al metal and S powder. (b) What observations of physical and chemical properties could be made if Al and S combined to form a compound, Al_2S_3?

7. Write, in your own words without using common definitions, your understanding of the terms (a) space, (b) energy, (c) heat, (d) temperature, (e) work.

8. Discuss the differences that might be found between the mass and the weight of an astronaut: (a) on earth, (b) on the moon.

9. (a) What experiments can be performed and what observations made to prove that H_2O is a compound and not an element? (b) Do these experiments prove that H_2 and O_2 are elements?

10. Classify the following substances as elements or compounds.

    a) Na
    b) NaCl
    c) MgO
    d) O
    e) O_2
    f) Fe
    g) Fe_2O_3
    h) CuS
    i) C
    j) CaF_2
    k) N_2
    l) He
    m) H_2S
    n) Cl_2
    o) K

11. Mercury (Hg) is a liquid at 25°C. Its freezing point is 234 °K. (a) Does Hg at room temperature need to be warmed or cooled to be frozen? (b) What is its freezing point in °C? (c) When Hg is at -50 °C, is it liquid or solid? (d) When Hg is at 255 °K, is it melted or frozen?

12. For the following equations, indicate which of the elements are reactants and which of the compounds are products:

    a) $Cd + S \rightarrow CdS$
    b) $Ca + PbO \rightarrow CaO + Pb$
    c) $CuO + H_2 \rightarrow Cu + H_2O$
    d) $2H_2 + O_2 \rightarrow 2H_2O$

13. Calculate the density of a Mg bar that measures 1.0 cm × 30.0 mm × 0.040 m and that weighs 20.9 g.

14. What is the weight of an iron (Fe) bar 0.050 cm × 1.0 m × 2.0 cm? The density of Fe is 7.86 g/cc.

15. What is the volume, in cc, of a cylindrical rod of calcium fluoride (CaF_2) that weighs 6.36×10^3 mg and has a density of 3.18×10^3 g/ℓ?

16. Calculate the density of common table salt (NaCl) to the proper number of significant figures from the following data:

    Volume of NaCl sample: 0.00354ℓ; weight of NaCl sample: 7.659 g

17. Calculate the density of a liquid from the following experimental data:

Weight of container	25.345 g
Weight of container plus liquid	37.549 g
Volume of container	25 cc
Volume of liquid	7.93 cc

18. Outline an experimental procedure to determine the density of a pile of powdered graphite, which is a form of carbon.

19. Make the following conversions:

   a) 12.32 g to mg   b) 0.72 mg to g   c) 0.96 cm to mm
   d) 122.9 mm to cm   e) 35.4 cm to m   f) 0.67 m to cm
   g) 121 mm to m   h) 23 m to mm   i) 1.72 cc to ℓ
   j) 0.743 ℓ to ml   k) 5.34 ml to cc   l) 35.7 ml to cm^3

20. Convert the following temperature to the Kelvin scale, °K:

   a) 100 °C   b) 0 °C   c) 212 °F
   d) 32 °F   e) 273 °C   f) 1167 °C
   g) −90 °C   h) −270 °C   i) 3370 °C
   j) −21 °C

21. The following substances are listed with their approximate boiling points in °C. Convert these temperatures to °K and to °F.

   a) Hg, 357 °C   b) H_2O, 100 °C   c) CO_2, −78 °C
   d) N_2, −196 °C   e) H_2, −253 °C   f) He, −269 °C

22. (a) A book weighs 1.5 lb. How many grams is this? (b) Calculate the weight of a 150-lb man in grams and kilograms.

23. (a) A ruler is 1.0 ft long. How long is it in cm? (b) How many cm are there in one inch? (c) How many mm are there in one inch and in one foot? (d) What is the volume in cm^3 of a cube 1.0 mm on an edge? (e) How many cm^3 are there in 1.0 m^3?

24. How many significant figures are there in the following numbers?

   a) 2.54   b) 6.02×10^{23} c) 0.00064
   d) 900.0   e) 900   f) 9.0×10^2
   g) 0.10071   h) 0.030500   i) 10.03050
   j) 1.501×10^{-3} k) 2.90×10^4 l) 29,000

25. Write the following numbers in exponential form.

   a) 22.4   b) 238.03   c) 1000
   d) 0.0232   e) 345,000,000   f) 0.1008
   g) 507,000,000,000,000   h) 0.0000000000104   i) 100,000
   j) 10.002   k) 1.0   l) 12

26. How many significant figures are there in the answers to the following problems?

   a) 15.4 + 7.968 + 8.97   b) 1.352 − 0.2463   c) 6.43 × 4.2
   d) $(6.02 \times 10^{23})(4.2 \times 10^{-2})$ e) 44 ÷ 22.4   f) 36.00 ÷ 6.02×10^{23}

79714

2. ATOMS AND
   ATOM COMBINATIONS

2–1 CHEMICAL REACTIONS: SOME EXPERIMENTS

Chemical changes. In Chapter 1 we considered some chemical changes illustrating the differences between elements and compounds. When a mixture of metallic strips of the element Zn and yellow powder of the element S are ignited, a very rapid chemical change occurs, producing considerable heat, light, and a new substance: the compound ZnS. Evidence for this chemical transformation comes from experimental examination of the product, which is a grayish powder and which shows none of the properties, such as melting point or density, of the elements in the original mixture.

Similar kinds of chemical changes are observed when other elements are mixed and subjected to the appropriate reaction conditions, such as an elevated temperature or a completely closed container. The results of experiments of this type, producing many different compounds, can lead to some general, summarizing statements concerning chemical changes and compound formation.

The law of conservation of mass. The development of one of these general statements can be illustrated with a common example: the use of a flashbulb in taking a photograph. A flashbulb is a completely closed container holding two elements, fine magnesium (symbol Mg) wire and oxygen gas. Ignition of these two elements in the presence of one another produces a rapid chemical change, with the formation of a

white, powdery solid and the production of large amounts of heat and light. But another interesting observation can be made if we weigh the flashbulb, shoot a picture and then reweigh the bulb. The mass before the reaction is found to be equal to the mass after the reaction; that is, the bulb weighs the same before and after taking the picture, even though large amounts of heat and light were liberated by the chemical change.

Beginning with the work of Lavoisier around 1770, experiments were conducted with many element combinations and their resulting compounds. One of the important measurements in each of these experiments was the determination of the total weight of the reactants and the total weight of the products that were involved in the chemical reactions. For example, Lavoisier conducted a series of ingenious experiments with mercury (Hg), a shiny, metallic-appearing liquid under normal room conditions, and oxygen, a colorless, odorless gas. He combined a weighed amount of Hg with a definite quantity of oxygen and found that the weight of the chemical compound produced, HgO,* was identical to the initial weight of the Hg and oxygen reactants. Lavoisier then heated a weighed sample of HgO, which decomposes into its constituent elements at about 500 °C. He collected the oxygen gas and the free liquid Hg that were produced by the decomposition. Their combined weight was equal to the weight of the initial HgO sample.

Experiments similar to those of Lavoisier were conducted by many scientists working with various element pairs and their resulting compounds. Each time, within experimental error, mass was conserved, that is, the masses before and after reaction were equal. The correlation of all these observations finally led to the *law of conservation of mass*, which states that, within experimental error, mass can neither be created nor destroyed in ordinary chemical reactions.

▶ As an analogy to the law of conservation of mass, we might consider mixing six red clothespins with six white clothespins to form six red and white clothespin-couples. If the six red pins and the six white pins are weighed individually before they are combined and then the six red and white pin-couples are weighed, the two weights are equal. No mass is lost or gained when the pins are combined. Similarly, if the red and white pin-couples are separated into six red pins and six white pins and these two groups reweighed, the total weight of the separate pins will equal their original total weight as well as the weight of the six red and white pin-couples. ◀

The law of constant composition. Another interesting observation can be obtained from the consideration of chemical reactions such as the one between cadmium (Cd) and sulfur, producing cadmium sulfide (CdS):

$$Cd + S \rightarrow CdS.$$

What do you think might happen if we combined differing weight ratios of Cd and S?

---

*   Mercuric oxide is a red-orange powder with physical and chemical properties that are completely different from those of Hg and oxygen.

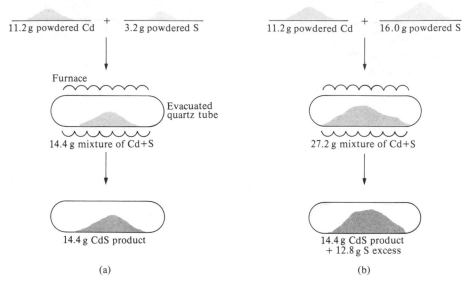

Fig. 2–1 The law of constant composition.

An easy way to do this experimentally is to combine different weights of S with the same weight of Cd, as shown in Fig. 2–1. Table 2–1 records some possible experimental data for the chemical reaction between varying weight ratios of Cd and S.

Table 2–1

Weight of Cd taken, g	Weight of S taken, g	Weight of CdS produced, g	Reactant excess, g
11.2	1.6	7.2	5.6 g Cd
11.2	3.2	14.4	none
11.2	16	14.4	12.8 g S
11.2	160	14.4	156.8 g S

Using these data, we see that after a certain weight of sulfur (3.2 g) is taken, or exceeded, and reacted with a given weight of cadmium (11.2 g), only a fixed weight of cadmium sulfide product is formed. Adding more than 3.2 g of S does not increase the amount of CdS formed; the excess S is simply left over, unreacted and unchanged. But we must add at least 3.2 g of S or 14.4 g of CdS will not be produced and Cd will be in *excess*. These data indicate that Cd and S react in a definite, fixed, weight ratio. That is, if a specific weight of Cd is used, there is only one weight of S that will combine exactly with it.

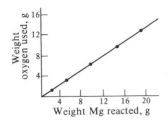

Fig. 2–2 In the reaction between Mg and oxygen that produces magnesium oxide (MgO), the same combining weight ratio of oxygen to magnesium is always found. This is shown here by the straight line of constant slope which is obtained by plotting the experimentally determined combining weights of the reactants against one another.

The reaction between Mg and oxygen gas can also be used to illustrate this principle of constant combining ratios. We can study and carry out the chemical combination between Mg and oxygen experimentally in a closed container similar to the flashbulb mentioned previously. In this set of experiments, a given weight of oxygen can be placed in the container with a known weight of Mg and the reaction be allowed to take place. After reaction the vessel can be inspected for any excess Mg left unreacted. If none is found, the experiment can be repeated, using the same weight of oxygen but increasing the weight of Mg slightly. This experimental process can be continued until the exact weight of Mg required to just react with the weight of oxygen used is determined. Then the initial weight of oxygen in the container can be increased and another set of experiments run to determine the new weight of Mg required for exact reaction.

Figure 2–2 is a plot of the results of a set of these experiments. Note that as more oxygen is used, more Mg is required for the reaction. But the weight ratio of grams of oxygen to grams of Mg remains constant,* irrespective of the weight of oxygen used. For example, from Fig. 2–2 we can determine that an initial sample of 8.00 g oxygen requires 12.1 g Mg for exact reaction. The weight ratio, therefore, is 8.00 g oxygen/ 12.1 g Mg = 0.66. This same ratio of 0.66 is obtained for any other pairs of oxygen-Mg reacting weights plotted in Fig. 2–2. Just as in the Cd-S case above, there is only one weight of Mg that will combine exactly with a given weight of oxygen. The required weight ratio of oxygen to Mg is 0.66, while the required weight ratio of S to Cd is 3.2/11.2 = 0.29.

Constant combining-weight ratios can also be studied by decomposing a weighed quantity of a chemical compound, such as HgO, and determining the weights of the elements produced. As mentioned above, HgO decomposes to its constituent elements, Hg and oxygen. To conduct this group of experiments, we need to decompose a weighed amount of HgO in an apparatus that enables us to collect the oxygen gas and liquid Hg produced. The liberated oxygen and Hg must then be weighed.

Data corresponding to a group of these experiments are shown in Fig. 2–3 for various starting weights of HgO. Parts (a) and (b) give the weights of Hg and oxygen obtained from various HgO sample weights, and part (c) shows the relative weights of

---

*   This constant ratio is indicated by the constant slope of the straight line in the graph in Fig. 2–2.

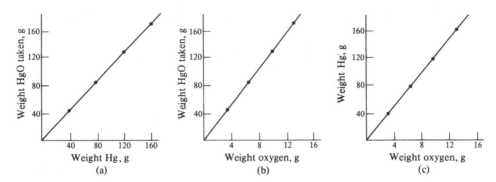

Fig. 2-3 Possible experimental data for the decomposition of mercuric oxide (HgO), to give Hg and oxygen. (a) A constant weight ratio of Hg to HgO is obtained in the reaction. (b) A constant weight ratio of oxygen to HgO is obtained in the reaction. (c) The products, Hg and oxygen, are formed in a constant weight ratio. These data indicate that HgO is composed of a constant Hg-to-oxygen ratio, irrespective of the absolute weight of HgO (or Hg plus oxygen) present.

Hg and oxygen produced in the experiments. From the graphs we can see that an 86.6 g HgO sample yields 80.2 g Hg and 6.4 g oxygen, giving a Hg/HgO weight ratio of $80.2/86.6 = 0.928$, an oxygen/HgO weight ratio of $6.4/86.6 = 0.074$, and an oxygen/Hg weight ratio of $6.4/80.2 = 0.080$. These same combining-weight ratios are obtained for any HgO weight given in Fig. 2-3, and indicate that the compound HgO is composed of a specific reacting weight ratio of Hg and oxygen.

The results of many experiments similar to those given above eventually led to the correlating statement called the *law of constant composition*, which states that a given weight of one element combines with only one definite weight of another element when they form a compound.

▶ The analogy of red and white clothespins can also be applied to the law of constant composition. An experimental procedure might be as follows. Weigh 10 red pins and then 10 white pins, and combine the obtained 200 g of red pins with the 300 g of white pins (note that each white pin is heavier than each red pin). The total weight of red and white pin-couples is 500 g. In a second experiment combine 200 g of red pins with 360 g of white pins. Again 500 g of red and white pin-couples would be formed (why?) with 60 g of white pins (equal to how many pins?) remaining uncombined. A third experiment combining 200 g of red pins with 600 g of white pins would yield 500 g of red and white pin-couples with 300 g of white pins left over. A complete series of experiments with different weights of red and white pins would lead to a conclusion: There is a definite combining-weight ratio of 2/3 for red pins/white pins when they form red and white pin-couples. If 200 g of red pins are used, 300 g of white pins are needed; if 66 g of red pins are used, 99 g of white pins are needed ($200/300 = 2/3$ and $66/99 = 2/3$). Formation of the compound (red and white pin-couples) requires that a given weight of one element (200 g of red pins) combine with only one weight of the other element (300 g of white pins). ◀

Example 2–1. We know from experimental data that 6.5 g Zn metal combines exactly with 3.2 g S powder. How many grams of S will combine with 21.3 g Zn?

What do we want to calculate? The number of grams of S. What information do we have? Zinc and sulfur combine in a definite ratio by weight, 6.5 g Zn with 3.2 g S (the law of constant composition is followed); 21.3 g Zn are available.

If we could find, from this information, the number of grams of S that combine with just 1.0 g of Zn, then just 21.3 times that many grams of S would combine with 21.3 g of Zn. We can do that in the following way:

$$6.5 \text{ g Zn react with } 3.2 \text{ g S,}$$

$$\frac{6.5}{6.5} = 1.0 \text{ g Zn reacts with } \frac{3.2}{6.5} \text{ g S,}$$

$$21.3\,(1.0) = 21.3 \text{ g Zn react with } 21.3\left(\frac{3.2}{6.5}\right) = 10.5 \text{ g S.}$$

According to the law of constant composition, 21.3 g Zn combines exactly with 10.5 g S. The weight of ZnS formed must be $21.3 + 10.5 = 31.8$ g, according to the law of conservation of mass.

Example 2–2. How much carbon dioxide (CO_2) can be prepared from 15 g of carbon (C) and 15 g of oxygen, given that CO_2 is 27% C and 73% O by weight?

In working with weight-percent problems, it helps to assume a 100 g sample. Then we can say that 27 g C combine with 73 g O to produce 100 g CO_2; that is, the laws of constant composition and conservation of mass are obeyed. Knowing the combining-weight ratio between C and O, we can proceed as in Example 2–1:

$$73 \text{ g O react with } 27 \text{ g C,}$$

$$\frac{73}{73} = 1.0 \text{ g O reacts with } \frac{27}{73} \text{ g C,}$$

$$15\,(1.0) = 15 \text{ g O react with } 15\left(\frac{27}{73}\right) \text{ g C} = 5.6 \text{ g C,}$$

$$15 \text{ g O react with } 5.6 \text{ g C to produce } 15 + 5.6 = 20.6 \text{ g } CO_2.$$

Note that: (1) The two combining ratios considered between C and O,

$$\frac{73}{27} = 2.7 \quad \text{and} \quad \frac{15}{5.6} = 2.7,$$

are the same. (2) Some C, $15 - 5.6 = 9.4$ g, remains uncombined. We say that C is *in excess*. We could have selected the following procedure:

$$27 \text{ g C react with } 73 \text{ g O,}$$

$$\frac{27}{27} = 1.0 \text{ g C reacts with } \frac{73}{27} \text{ g O,}$$

$$15\,(1.0) = 15 \text{ g C react with } 15\left(\frac{73}{27}\right) \text{ g O,}$$

$$15 \text{ g C react with } 41 \text{ g O.}$$

But we have *only* 15 g O available. This immediately indicates that C is in excess. We must then go through the first procedure selected to find the amount of C that will be used.

▶ These problems are relatively simple and can be solved by standard mathematical manipulation techniques. Rather than applying a set formula to obtain an answer to our problems, we prefer to think the problem through. At first, this might seem like extra effort. Stick with it! As problems become more involved, the advantages of thinking about them, rather than applying a pet method to them, will become increasingly apparent. ◀

2–2 THE ATOMIC THEORY

Speculations about the composition of matter began 300–400 years B.C. when two ideas were first proposed: (1) Matter is continuous, and (2) matter is discontinuous, that is, composed of small, separate units. The submicroscopic nature of matter was considered and debated at length until 1803, when John Dalton proposed that substances were composed of tiny, discrete particles called *atoms*. Dalton's hypothesis succeeded in explaining the laws of conservation of mass and constant composition, and has been supported by so many experimental results that it is now called *the atomic theory*.

In stating his hypothesis, Dalton made several basic assumptions. These assumptions, or postulates, can be summarized as follows:

1. All matter is ultimately composed of tiny, indivisible particles called *atoms* that cannot be created, destroyed, or interconverted.
2. Atoms of any particular element are identical in all properties and are different from atoms of other elements in these properties.
3. Chemical change is a union, separation, or rearrangement of atoms.
4. If the experimental conditions of chemical reaction are changed, the combining ratio of one element with another element may also change.

2–3 APPLICATIONS OF THE ATOMIC THEORY

One of the major requirements of an acceptable theory or hypothesis is that it explain known observations. Dalton's atomic theory, therefore, must provide an explanation for the laws of conservation of mass and constant composition. We shall consider some examples to show that these two laws are explained by Dalton's theory, that is, they arise naturally from Dalton's idea of chemical reactions and how they occur.

The law of constant composition. Let us consider the reaction between carbon and oxygen atoms at temperatures above 1000 °C which produces the compound carbon monoxide (CO). We can apply Dalton's assumptions to this chemical change. First we assume that the combining elements, C and O, are composed of atoms and that these C and O atoms cannot be divided or converted one into another (postulate 1 of Dalton's theory). Second, we assume that the chemical reaction between C and O is a union of C and O atoms (postulate 3). Then we can say that 1 atom of C unites with 1 atom of O to form 1 unit of CO. (We have made a further assumption here that C and O combine in a 1 : 1 atom ratio. This is true and can be shown experimentally.

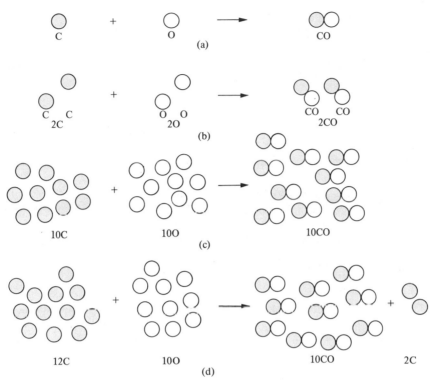

Fig. 2–4 Chemical reaction between the elements carbon (C), and oxygen (O), to produce the compound carbon monoxide (CO). Atoms of C and O combine to form molecules of CO. Note in part (d) that two C atoms are in excess and remain uncombined.

In Dalton's day, however, the choice of combining atom ratios was very difficult and, in most cases, was simply a guess.) We can write the equation,

$$1 \text{ atom C} + 1 \text{ atom O} \rightarrow 1 \text{ unit CO} \qquad \text{or} \qquad C + O \rightarrow CO.$$

The unit of CO is called a molecule of CO. A *molecule* is a neutral aggregate of two or more atoms that forms the smallest particle of a substance showing the properties of the substance.

Figure 2–4 shows this reaction schematically. Also pictured is the reaction between two C atoms and two O atoms, producing two molecules of CO, and the reaction between 10 C atoms and 10 O atoms, giving 10 CO molecules. Note what happens if different numbers of C and O atoms are used. If 12 C atoms are mixed with 10 O atoms, then only 10 molecules of CO can be formed. There will be 2 C atoms left uncombined. This is necessary since the atoms cannot be subdivided, and combine in only one ratio under the same reaction conditions. We can write this last result in equation form:

$$12C + 10O \rightarrow 10CO + 2C.$$

We can say, then, that a definite number of C atoms combine with a definite number of O atoms. Using Dalton's postulate 2, that each C atom has its own particular weight, we see that a definite weight of C combines with a definite weight of O in forming the compound carbon monoxide. Note that any excess of one substance will remain unreacted and unchanged. Thus, the law of constant composition is explained naturally by Dalton's atomic theory.

The law of conservation of mass. In the process of explaining the law of constant composition using Dalton's theory, we have also explained the law of conservation of mass. Consider again the reaction between 12 atoms of C and 10 atoms of O; only 10 molecules of CO are formed with 2 atoms of C unreacted (see Fig. 2–4). There are 12 atoms of C and 10 atoms of O present before the reaction takes place. After the reaction has occurred, there are 10 atoms of C combined with 10 atoms of O in the 10 molecules of CO, plus 2 atoms of C uncombined. That is, the number of atoms present before reaction is equal to the number of atoms present after reaction. This follows directly from Dalton's postulate 1, that atoms cannot be destroyed. Since the same number of atoms of each element is present before and after reaction, and since each atom has a characteristic mass (postulate 2), then the mass before the reaction is equal to the mass after the reaction. This is the *law of conservation of mass*.

Supporting experiments. Some possible experimental data supporting these two laws are given in Table 2–2 for the reaction between Ga and As, producing gallium arsenide,

$$Ga + As \rightarrow GaAs.$$

Table 2–2

Reactants		Product	Excess
g Ga	g As	g GaAs	
6.97	7.49	14.46	None
6.97	14.98	14.46	7.49 g As
6.97	37.45	14.46	29.96 g As
6.97	74.9	14.46	67.41 g As

Note that, just as in the case of cadmium reacting with sulfur, which we considered earlier in this chapter, there is a fixed amount of As that combines with a given weight of Ga. All excess weight is recovered; there is no gain or loss in mass during the reaction. The weight of GaAs produced is controlled by the amount of Ga present; only 14.46 g of GaAs is formed, no matter how much As is present in excess of the 7.49 g As needed. Now, assuming that Dalton's theory is valid and that 1 atom of Ga combines with 1 atom of As, then there must be the same number of atoms of

Ga in 6.97 g Ga as there are atoms of As in 7.49 g As. This is an important observation. Can you determine its use? More about this in Section 2–5.

▶ A 1 : 1 ratio between Ga and As atoms has been assumed. That this is the actual ratio in the compound gallium arsenide is not at all obvious. The chemical formula GaAs must be determined by experimental means. The proper experimental methods, as well as theoretical justification for the formula, will be considered later. ◀

2–4 A PREDICTION FROM THE ATOMIC THEORY

The law of multiple proportions. Remember that not only does an acceptable hypothesis or theory have to explain known data, it should also predict something unknown at the time. Dalton's atomic theory explained to a great extent many of the known laws and observations of his time. But it did more. Dalton assumed (postulate 4) that atoms of some elements do not always combine in the same ratio. That is, under different reaction conditions (combining substances at different temperatures or under different pressures or with different amounts present, etc.), it might be possible to combine one atom of element X with different numbers of atoms of element Z. Then, under different reaction conditions, the elements X and Z could combine to form compounds such as XZ, XZ_2, X_3Z_4, etc.

To investigate and illustrate this prediction, let us consider a well-known example. If we burn a sample of carbon in oxygen gas at temperatures below 1000 °C and collect the gaseous product, we find, from chemical analysis, that the gas is predominantly carbon dioxide, CO_2. On the other hand, if the reaction is carried out at temperatures above 1000 °C, chemical analysis shows that the gas produced is principally carbon monoxide, CO.

These experimental results are pictured in Fig. 2–5, using the atomistic point of view proposed by Dalton. In the first case, one atom of C combines with two atoms of O; in the second case, one atom of C combines with one atom of O. In the first case, the ratio of O atoms to C atoms is 2 : 1. In the second case, the ratio of O atoms to C atoms is 1 : 1. These two cases are related by a simple whole-number ratio:

$$\frac{2/1}{1/1} = \frac{2}{1}.$$

Note that, since each C and each O atom have characteristic masses or weights, the weight ratio of O to C in the two compounds is related by the same 2:1 simple whole-number ratio.

We can illustrate this simple whole-number ratio between combining weights with the following experimental data. In one set of experiments, conducted below 1000 °C, it is found that 2.70 g O combine with 1.00 g C. However, under different reaction conditions, at temperatures above 1000 °C, 1.35 g O combine with 1.00 g C. In the first case the oxygen-to-carbon combining-weight ratio is 2.70 g O/1.00 g C

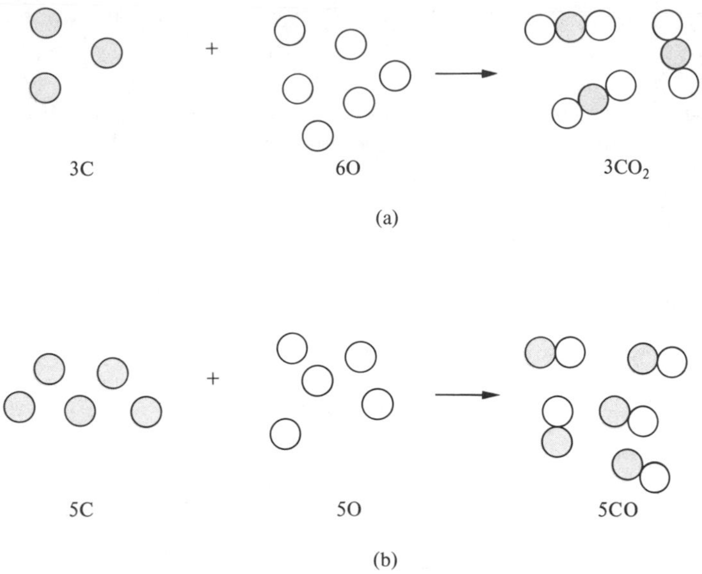

Fig. 2–5 Chemical reactions between the elements carbon (C) and oxygen (O) can produce two different compounds. (a) Below 1000 °C, molecules of carbon dioxide (CO_2) are formed. (b) Above 1000 °C, molecules of carbon monoxide (CO) are formed. These two reactions illustrate Dalton's prediction of the law of multiple proportions.

(2.70 g O per unit weight of C). In the second case the ratio is 1.35 g O/1.00 g C. These two weight ratios are related by the small whole-number ratio,

$$\frac{2.70/1.00}{1.35/1.00} = \frac{2}{1}$$

This simple combining-weight multiple between C and O in forming two compounds, CO_2 and CO, illustrates the type of experimental evidence that was needed, and found, to support Dalton's prediction.

Many experiments of a similar nature were conducted with various pairs of elements. The results of these experiments supported Dalton's assumption of varying atomic ratios in compounds of the same elements prepared under different experimental conditions and led to the *law of multiple proportions*: If two elements, X and Z, form a series of compounds, the ratio of the weights of elements X and Z in one compound will be a simple multiple of the ratio of the weights of elements X and Z in another compound.

▶ The previous analogy for the laws of conservation of mass and constant composition using red and white clothespins leads to the same explanation of these two laws as does Dalton's atomic theory. The reason that constant combining-weight ratios

are found when red and white pins are combined to form red and white pin-couples is that one red pin combines with one white pin in forming the couple. Each red pin weighs the same amount, 20 g; each white pin weighs the same amount, 30 g, but has a different weight than a red pin. Then a red and white pin-couple must weigh $20 + 30 = 50$ g, and the combining red-to-white-pin weight ratio must be 20/30 or 2/3. If there is an excess of red pins or an excess of white pins when they are combined, the excess will simply remain uncombined. No mass is lost or gained, since the total number of pins remains constant.

The law of multiple proportions has a similar explanation, except that the number of red and white pins that can combine with one another can change. If one red and one white pin combine to form a red and white pin-couple, the pin ratio is 1 red pin/1 white pin. Since each pin has a definite weight, the combining-weight ratio is also fixed, 20 g red pin/30 g white pin. If, on the other hand, two red pins combine with one white pin, the pin ratio is 2 red pins/1 white pin and the weight ratio is 40 g red pins/30 g white pins. These separate ratios form a simple ratio with one another:

$$\frac{1 \text{ red pin}/1 \text{ white pin}}{2 \text{ red pins}/1 \text{ white pin}} = \frac{1}{2} \quad \text{or} \quad \frac{20 \text{ g red pins}/30 \text{ g white pins}}{40 \text{ g red pins}/30 \text{ g white pins}} = \frac{20}{40} = \frac{1}{2}. \quad \blacktriangleleft$$

Example 2–3. In preparing a compound of copper (Cu) and oxygen, one finds that 12.7 g Cu combine exactly with 3.20 g O. Under different experimental preparation conditions, one finds that 50.8 g Cu combine with 6.40 g O. Show that these data illustrate the law of multiple proportions.

The copper-to-oxygen combining-weight ratio in the first compound is 12.7 g Cu/3.20 g O; the combining ratio in the second compound is 50.8 g Cu/6.40 g O. The law of multiple proportions states that one ratio must be a simple multiple of the other, that is, there is a simple ratio between the two individual ratios. Then the first compound ratio divided by the second compound ratio should yield a simple multiple:

$$\frac{12.7 \text{ g Cu}/3.20 \text{ g O}}{50.8 \text{ g Cu}/6.40 \text{ g O}} = \frac{12.7}{50.8} \times \frac{6.40}{3.20} = \frac{1}{4} \times \frac{2}{1} = \frac{2}{4} = \frac{1}{2}.$$

This illustrates the law of multiple proportions, since the two weight ratios are related by the simple multiple $\frac{1}{2}$. What information would be needed to determine the chemical formulas of the two compounds of copper and oxygen?

Example 2–4. It is known that iron (Fe) and sulfur (S) react to form two compounds. In forming FeS, 5.59 g Fe react with 3.20 g S. What is the atom ratio (chemical formula) of the second compound, in which 11.18 g Fe combine with 9.60 g S?

Using the law of multiple proportions, we know that the weight ratios of Fe to S in the two experiments must have a simple whole-number ratio,

$$\frac{5.59/3.20}{11.18/9.60} = \frac{5.59}{11.18} \times \frac{9.60}{3.20} = \frac{1}{2} \times \frac{3}{1} = \frac{3}{2}.$$

The atom ratio must be the same as the weight ratio. Since we know that the atom ratio in FeS is

one atom of Fe to one atom of S and that the law of multiple proportions gives a ratio of 3/2, we can calculate the atom ratio in the second component:

$$\frac{\text{Compound 1: atoms Fe/atoms S}}{\text{Compound 2: atoms Fe/atoms S}} = \frac{1/1}{\text{Fe/S}} = \frac{3}{2}$$

or

$$\text{Compound 2: Fe/S} = 2/3 \times 1/1 = 2/3.$$

The ratio of Fe : S atoms in compound 2 must be 2Fe/3S to fulfill the equation and the law of multiple proportions. The chemical formula for the second compound is therefore Fe_2S_3.

2–5 ATOMIC WEIGHTS

You probably already have the feeling that it would help if we understood more about atoms. And, since almost all the experiments and observations up to this point have involved weights of substances, it would be quite useful to know how much atoms weigh (what their masses are). Ordinarily this might be an easy process. Just place an atom on one pan of a balance and a known counterbalancing mass on the opposite pan. There is only one hitch. Atoms are very small! They cannot be seen, picked up, or easily separated into piles of just 10 or 100 (see Table 2–3).

Table 2–3. Some representative sizes

Observation	Measurement
Distance to nearest star	4×10^{18} cm
Distance to sun	1.5×10^{13} cm
Diameter of sun	1.4×10^{11} cm
Diameter of earth	1.3×10^{9} cm
Height of Mt. Everest	8.9×10^{5} cm
Height of average man	1.7×10^{2} cm
Diameter of penny	1.9 cm
Diameter of pinhead	1.5×10^{-1} cm
Diameter of a hair	2×10^{-2} cm
Diameter of smallest virus	1×10^{-6} cm
Diameter of hydrogen atom	1×10^{-8} cm
Diameter of hydrogen atom nucleus	1.2×10^{-13} cm

To give you a feeling for this incredible smallness, we shall jump slightly ahead of the story and indicate the known masses of the lightest and one of the heaviest atoms: hydrogen (H) and uranium (U), respectively.

One atom H $= 1.67 \times 10^{-24}$ g; one atom U $= 3.95 \times 10^{-22}$ g. Perhaps you would gain a better appreciation for these numbers if we wrote 3.95×10^{-22} g in standard form:

$$0.000000000000000000000395 \text{ g.}$$
$$\uparrow$$

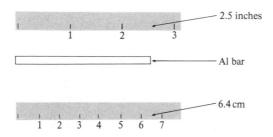

Fig. 2–6 Other relative standards: The length of an Al bar is measured relative to a yardstick (36 inches) and a meter stick (100 cm).

Typical chemical balances can weigh to the decimal point indicated by the arrow, that is, to a tenth of a milligram (0.0001 g). Note how very much smaller the U atom is. We cannot, therefore, select just 1 or 10 or 100,000 atoms and weigh them to determine the masses of the individual atoms of all of the elements.

What can be done, then? We previously, in Section 2–3, hinted at a possible solution. A relative scale of atom masses, commonly called *atomic weights*, can be established. That is, from experimental results we can determine the relative weights of two elements. Then if we assign (arbitrarily) a certain weight to an atom of one element, to be used as a standard of reference, and compare all the determined weights with the same reference, we can construct a relative scale of the atomic weights of the elements.

Similarly, relative reference standards for lengths are given in Fig. 2–6. The aluminum (Al) bar has a certain length. If we define a standard of reference, the length of the bar can be compared with the standard. When we use the yard as a defined standard of reference, the Al bar measures 2.5 inches in length. When we use another defined reference standard, the meter, the Al bar is 6.4 cm long. The length of the bar does not, of course, change. The different numbers that are obtained are the result of two different reference standards that have been chosen to measure length.

Atomic weight standard. For many years, the standard reference element for atomic weights was oxygen. Oxygen was a logical choice for a reference element because it combines readily with most other elements. Hence relative weights were easily determined.

There was one difficulty with the oxygen standard, however, which illustrates the fallacy of one of the assumptions in the atomic theory of Dalton. Naturally occurring oxygen is composed of a mixture of oxygen atoms that have three different weights. Hence Dalton's assumption that all atoms of a particular element are identical is not true.* In developing a scale of atomic weights, physicists assigned a reference weight to only one of the three oxygen atoms, while chemists assigned a reference atomic

---

*   Atoms of the same element that have different masses are called *isotopes*. We shall consider them further in Chapter 3.

weight to the mixture of three atoms present in a naturally occurring sample of oxygen. This led to two scales of atomic weights with slightly different values for all the elements, and to some confusion.

In January 1961, a new standard reference element was selected to eliminate these dual scales. Now there is only one scale of atomic weights, with the standard of reference being a particular atom of carbon, carbon-12 or ^{12}C, with an assigned atomic weight of 12.0000 atomic mass units, amu. With this atomic weight standard, one *amu* is defined as $\frac{1}{12}$ the weight of one ^{12}C atom. When all types of C atoms are considered, along with their relative percent abundance (see Question 24 at the end of this chapter), the average atomic weight for C is 12.0112 amu. Atomic weights for all the elements are listed on the inside front cover and given in the Periodic Table on the inside back cover.

Experimental determination. To acquire a better feeling for these relative atomic weights, we can consider one possible method for determining them. If we were to combine carbon and oxygen under certain reaction conditions and collect the resulting product, a chemical analysis would show the product to be 42.9% C and 57.1% O. Then there is

$$\frac{57.1}{42.9} \quad \text{or} \quad 1.33$$

times as much weight contributed by the O as by the C in the product. If we knew from other experimental data that one atom of C combined with one atom of O to form this reaction product, CO, then each O atom would be 1.33 times as heavy as each C atom.

Now, to determine the numerical atomic weights, we make use of the standard of reference. With $^{12}C = 12.0000$ amu, the average atomic weight of C is 12.011 amu. Since an atom of O is 1.33 times as heavy as an atom of C, the atomic weight of O is 1.33(12.011 amu) = 16.0 amu.

2–6 FORMULA WEIGHTS

Now that the atomic weights of elements (the weights of single atoms) are available, the weights of compounds can be calculated. Some typical examples, such as those shown schematically in Fig. 2–7, will illustrate this principle.

In the reaction between zinc and sulfur, zinc sulfide is formed:

$$Zn + S \rightarrow ZnS.$$

From a table of atomic weights, we find that one atom of Zn weighs 65.37 amu and one atom of sulfur weighs 32.06 amu. Since one *formula unit* of the compound ZnS consists of one atom of Zn and one atom of S, the weight of ZnS is

$$65.37 \text{ amu} + 32.06 \text{ amu} = 97.43 \text{ amu}.$$

(See Table 2–4.) This weight is called the *formula weight* or the *molecular weight* of

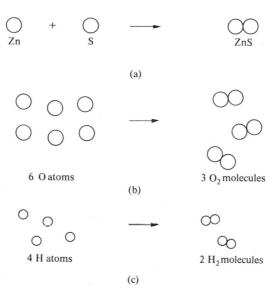

(a)

6 O atoms

3 O_2 molecules

(b)

4 H atoms

2 H_2 molecules

(c)

Fig. 2-7 (a) The formation of zinc sulfide (ZnS) from atoms of the elements. The weight of one formula unit of ZnS is the sum of the weights of one atom of Zn and one atom of S. (b) and (c). The formation of molecules of elements. *Diatomic molecules* (molecules containing 2 atoms) of O_2 and H_2 have molecular weights of 2 × 16 = 32 amu and 2 × 1.0 = 2.0 amu, respectively.

Table 2-4. Relations between atomic and formula weights

Reactants		Product
Pb + S		PbS
Element Element		Compound
Atom Atom		Formula unit
Atomic weight Atomic weight		Formula weight =
207.2 amu 32.1 amu		207.2 + 32.1 = 239.3 amu

Reactants		Product
N + N		N_2
Element Element		Element
Atom Atom		Molecule (or formula unit)
Atomic weight Atomic weight		Molecular (or formula) weight =
14.0 amu 14.0 amu		14.0 + 14.0 = 28.0 amu

the compound; that is, it is the weight of one formula unit or one molecule of the compound. (Some chemists prefer to use formula weight, since some compounds, such as NaCl or KF, etc., do not form true molecules, as we have defined them.) Note that the weight before the reaction, 65.37 amu Zn + 32.06 amu S, is equal to the weight

after the reaction, 97.43 amu ZnS. This is an illustration of the laws of conservation of mass and constant composition.

In some cases atoms of the same element react with one another to produce molecules of the element. The molecular (or formula) weight of these molecules is the sum of the atomic weights of the atoms composing the molecule. Oxygen is a very common example of an element existing as a diatomic molecule,

$$O \; + \; O \; \rightarrow \; O_2$$

O	+	O	→	O_2
16 amu	+	16 amu		16 amu
				16 amu
		32 amu		32 amu

This equation can also be written

$$2O \rightarrow O_2$$

since two atoms of oxygen combine to form one diatomic molecule of oxygen, O_2. The molecular (or formula) weight of O_2 is 32 amu.

Chlorine is another example of an element that generally exists as a molecule:

$$Cl + Cl \rightarrow Cl_2 \quad \text{or} \quad 2Cl \rightarrow Cl_2.$$

The atomic weight of Cl is 35.5 amu. The molecular (or formula) weight of the diatomic Cl_2 molecule is therefore 71.0 amu.

2–7 BALANCING SIMPLE CHEMICAL EQUATIONS

The examples of chemical reactions between Zn and S, between O atoms and between Cl atoms given above illustrate the use of simple chemical equations. In all the equations, the number of atoms of each element present before and after the reaction is the same. In the case of the formation of ZnS, there is one Zn atom and one S atom present before reaction. After combination, there is still one Zn atom and one S atom, but now the Zn and S are united and form a new substance. When this equality of atoms of each element before and after the chemical reaction is fulfilled, we say that the equation is *balanced* (see Fig. 2–8).

This balancing of equations is a direct consequence of the laws of conservation of mass and constant composition, and is readily explained by Dalton's theory. (Chemical reaction is a uniting of atoms, which cannot be divided, created, or destroyed.)

Very often the substances added together, called *reactants*, do not combine in a 1 : 1 ratio to form new substances, called *products*. The chemical equation must reflect this fact. We can illustrate this using the reaction between indium metal (In) and sulfur as an example. Under proper conditions of temperature and pressure indium sulfide, In_2S_3, is formed. To describe this reaction we would write

$$In + S \; \rightarrow \; In_2S_3.$$
$$\text{Reactants} \quad \text{Product}$$

Now to balance the equation (to ensure conservation of mass) we have to equalize the number of atoms present before and after the reaction. The product, In_2S_3, contains

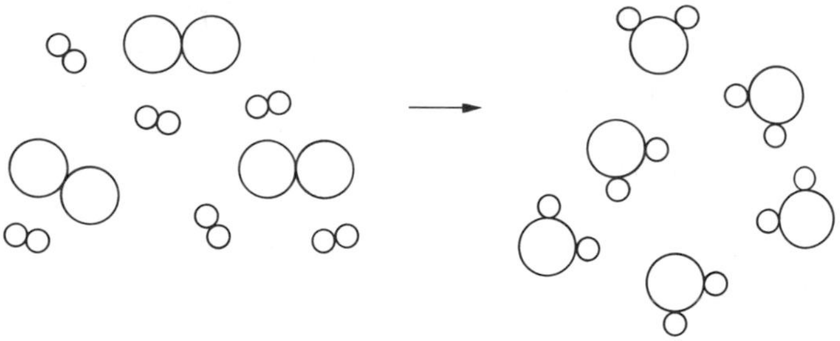

$$6H_2 + 3O_2 \longrightarrow 6H_2O$$

Fig. 2-8 Molecules of the elements hydrogen (H_2), and oxygen (O_2), react to form molecules of the compound water, H_2O. Note the rearrangements of the H and O atoms in forming the H_2O molecules; twice as many H_2 as O_2 molecules are needed. The formula H_2O indicates that two H atoms are combined with one O atom. The coefficients in the equation indicate the number of H_2 and O_2 molecules necessary to produce a certain number of H_2O molecules.

two atoms of In and three atoms of S. We must, therefore, have as reactants two In atoms and three S atoms. To accomplish this, we insert the proper coefficients in front of the appropriate symbols in the equation,

$$2In + 3S \rightarrow In_2S_3.$$

The equation is then balanced, and would read: Two atoms of In plus three atoms of S react to give one formula unit of In_2S_3. The formula weight of In_2S_3 would be $2 \times$ atomic weight of In $+ 3 \times$ atomic weight of S, or:

$$2(114.82) + 3(32.06) = 229.64 + 96.18 - 325.82 \text{ amu.}$$

It is important to understand the meanings of chemical formulas and chemical equations. The formulas tell how many atoms of each element compose the formula unit. A balanced equation indicates the number of formula units that combine with one another and that are produced in a chemical reaction.

▶ The formation of a basketball team presents a crude analogy to chemical formulas, equations, and equation balancing. A basketball team is composed of a center, Ct, two forwards, Fo, and two guards, Gu. Then its formula can be written as Gu_2Fo_2Ct. The formation of a basketball team, in equation form, is:

$$Ct + 2Fo + 2Gu \rightarrow Gu_2Fo_2Ct.$$

One center plus two forwards plus two guards yield one basketball team. Note that the equation relates the *number* of guards, the *number* of forwards and the *number* of

centers that combine to form the team. The combining weights, that is, the weight of guards, the weight of forwards, and the weight of the center, that are selected to produce the team depend on the weights of the individual players and the number of players involved. Similarly, with chemical equations, combining weights depend on the type of reactants and the number of units of each reactant being combined. Product weights are governed similarly, by the types of products and the number of units of each product formed. ◀

2–8 THE MOLE

In order to carry out chemical reactions in an exact manner, one must select definite numbers of atoms of elements or molecules of compounds for combination with one another. For example, in the formation of a compound such as ZnS, one atom of Zn reacts with one atom of sulfur; or 100 atoms of Zn react with 100 atoms of S; and so on. We must choose the same number of Zn atoms as we do S atoms; the ratio of Zn/S atoms must be 1 : 1.

All chemical reactions require us to select these certain, definite reacting ratios of atoms or molecules. Another common example is the reaction of carbon with oxygen gas to produce carbon monoxide,

$$2C + O_2 \rightarrow 2CO.$$

The balanced equation shows that:

2 atoms C react with 1 molecule O_2 to produce 2 molecules CO

or

20 atoms C react with 10 molecules O_2 to produce 20 molecules CO

or

2000 atoms C react with 1000 molecules O_2 to produce 2000 molecules CO

or

2×10^{10} atoms C react with 1×10^{10} molecules O_2 to produce 2×10^{10} molecules CO.

The ratios are always—and necessarily—constant, irrespective of the total number of atoms or molecules involved.

If we accept the fact that definite numbers of atoms and molecules must be selected to carry out chemical reactions, what is the easiest method of selection? One of the most precise and most easily made experimental measurements in a chemical laboratory is the weighing of a substance. It would be very convenient, therefore, to be able to select known numbers of atoms or molecules by weighing a certain number of grams of the substance. But we have already seen that atoms are incredibly small (see Fig. 2–9), and cannot be weighed individually. That means that we need some large, standard measure of atoms, so that each time we weigh a certain number of grams of a substance, we select a specific, known number of atoms. We could then weigh a sample of a substance and tell exactly how many atoms the sample contained.

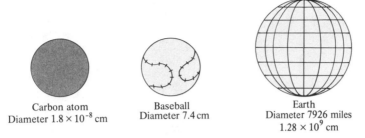

Fig. 2–9 A carbon atom enlarged 10^8 times. On the same scale, a base-ball would have a diameter of 7.4×10^8 cm, or about half that of the earth.

This situation is somewhat like trying to weigh a baseball on a bathroom scale. The scale simply is not sensitive enough. However, if we place a large number of baseballs on the scale at the same time, we can determine the total weight of the entire sample. For example, if we place 12 baseballs (we can count them) on the scale, we can determine their weight; it would be about 1800 g (or about 4 lb). Commonly we would say that 1 dozen baseballs weigh about 1800 g. If 10 dozen baseballs are re-quired for some special purpose, we no longer need to count out 120 balls. We simply dump baseballs onto a scale until a weight 10 times the weight of one dozen balls is obtained, about 18,000 g (or about 40 lb).

The same thing is done for atoms. A single atom of, for example, C cannot be weighed. If we place enough C on a balance pan, we can determine the total weight of the sample. Let us use a sample of C that weighs exactly 12.0112 g. We can deter-mine how many C atoms are contained in the 12.0112 g by using experimental tech-niques such as x-ray diffraction.* (This is not as easy as counting baseballs, but it can be done very accurately. A short discussion of x-rays is given in Section 5–2.) We find that 6.023×10^{23} atoms of C are necessary to give 12.0112 g C. This very large number is called *Avogadro's number*. Just as we often call 12 things one dozen, we call 6.023×10^{23} things one *mole*. One dozen baseballs is 12 baseballs; one mole of baseballs is 6.023×10^{23} baseballs (see Fig. 2–10). One mole of C atoms is 6.023×10^{23} atoms; 6.023×10^{23} C atoms weigh 12.0112 g. Therefore one mole of C atoms weighs 12.0112 g. Now we can select a specific, known number of atoms simply by weighing the required substances. In the case of C, exactly 12.0×10^{23} atoms of C can be selected by weighing 24.0 g C; that is, two moles of C atoms weigh 24.0 g.

Remember that a relative scale of atomic weights has been established with ^{12}C as the standard reference point. The same relative scale holds for moles of atoms (6.02×10^{23} atoms) as for single atoms. The atomic weight of ^{12}C is 12.000 amu; the

---

*   Other techniques, such as radioactivity, electrolysis, etc., are also used.

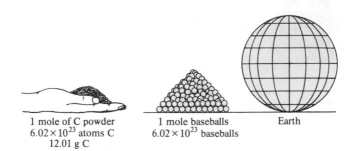

1 mole of C powder   1 mole baseballs   Earth
6.02×10^{23} atoms C 6.02×10^{23} baseballs
12.01 g C

Fig. 2–10 A mole is a large number. A mole of baseballs would occupy a volume about half that of the earth. But it takes an entire mole of C atoms (6.02×10^{23} C atoms) to give a handful of C powder.

weight of one mole of ^{12}C atoms is 12.000 g. The atomic weight of sulfur is 32.01 amu; the weight of one mole of S atoms is 32.01 g. The atomic weight of iron is 55.85 amu; the weight of one mole of Fe atoms is 55.85 g, etc. The weight of one mole of atoms, or the atomic weight expressed in grams, is often called the *gram-atomic weight*.

Note what happens when atoms of elements are combined to form compounds. In the case of

$$Ga + As \rightarrow GaAs,$$

1 atom Ga reacts with 1 atom As to produce 1 formula unit GaAs

or

1000 atoms Ga react with 1000 atoms As to produce 1000 formula units GaAs

or

6.02×10^{23} atoms Ga react with 6.02×10^{23} atoms As to produce 6.02×10^{23} formula units GaAs

or

1 mole Ga atoms reacts with 1 mole As atoms to produce 1 mole GaAs formula units.

We see that 1 mole GaAs is composed of 1 mole of Ga atoms and 1 mole of As atoms. The weight in grams of one mole of GaAs must therefore be equal to the formula weight in amu. The weight of one formula unit of GaAs is 144.64 amu; the weight of one mole of GaAs formula units is 144.64 g. The weight of one mole of formula units, or the formula weight expressed in grams, is often called the *gram-formula weight*.

A mole is:

1. Avogadro's number—6.02×10^{23}—units or things.

2. The atomic or formula weight expressed in grams.

▶ The formula weight expressed in grams is more exactly called a *gram mole*. In normal usage the gram mole is called a mole and any other definition of a mole, such as the formula weight expressed in pounds, is given a special name, a *pound mole*. ◀

2–9 MOLES AND THEIR USE

We can use moles to select known numbers of atoms or molecules simply by weighing the required amounts on a balance. A few simple examples will illustrate the use, and necessity, of working with moles of atoms and molecules.

Example 2–5. The elements nickel (Ni) and sulfur (S) combine according to the following equation:

$$Ni + S \rightarrow NiS.$$

How many moles of S atoms combine with 2.0 moles of Ni atoms? How many moles of NiS are formed?

The equation tells us that one atom of Ni combines with one atom of S to produce 1 formula unit of NiS. Then it will also be true that:
6.02×10^{23} atoms Ni combine with 6.02×10^{23} atoms S to give 6.02×10^{23} formula units NiS.

Or: 1 mole Ni atoms combines with 1 mole S atoms to give 1 mole NiS formula units.

And: 2.0 moles Ni atoms combine with 2.0 moles S atoms to give 2.0 moles NiS formula units. (Note that all the ratios are $1 : 1 \rightarrow 1$.)

We can easily select 2.0 moles of Ni atoms by weighing out 117.4 g Ni, twice the atomic weight, and 2.0 moles of S atoms by weighing out 64.2 g S, twice the atomic weight.

Example 2–6. How many grams of magnesium (Mg) are needed to react with 6.4 g of oxygen gas (O_2), given that two atoms of Mg combine with one molecule of O_2?

The approach to the problem can be summarized as follows:

We can calculate the number of moles of O_2 available, since we know the number of grams contained in one mole of O_2. From the $2 : 1$ combining ratio we know that we need twice that many moles of Mg atoms. Then we can calculate the number of grams of Mg, since we know how many grams one mole of Mg atoms weighs.

The number of moles of O_2 available:

One mole of O_2 molecules is 32.0 g O_2. (Two O atoms compose each O_2 molecule. The atomic weight of O is 16.0 amu.) The 6.4 g O_2 is therefore a fraction of one mole of O_2:

$$\frac{6.4 \text{ g } O_2}{32.0 \text{ g } O_2/\text{mole } O_2} = 0.20 \text{ mole } O_2.$$

(Note how the units indicate that the arithmetical operation was the proper one—the *unit mole* O_2 is obtained.)

The number of moles of Mg atoms necessary:

1 molecule O_2 combines with 2 atoms Mg,

1.0 mole O_2 molecules combines with 2.0 moles Mg atoms,

0.20 mole O_2 molecules combines with (0.20)2.0 moles Mg atoms,

0.20 mole O_2 molecules combines with 0.40 mole Mg atoms.

The number of grams of Mg:

One mole of Mg atoms is the atomic weight expressed in grams, 24.3 g. We do not need an entire mole of Mg, only 0.40 mole. Therefore we take only 0.40 as many grams as there are in one mole, or

$$(0.40 \ \text{mole Mg})(24.3 \ \text{g Mg/mole Mg}) = 9.7 \ \text{g Mg.}$$

(Note how the units again give the proper calculation method.) Then 9.7 g Mg react exactly with 6.4 g O_2.

Example 2–7. Determine the number of grams of In_2S_3 that can be formed from 9.60 g S and 23.50 g In, given that In and S combine according to the following equation:

$$2In + 3S \rightarrow In_2S_3.$$

The equation shows that the combining ratio between In and S is 2 atoms to 3 atoms or 2 moles In to 3 moles S. Since atoms have characteristic weights, there is also a certain In : S combining-weight ratio. This weight ratio, however, is determined by the reacting-mole ratio. We cannot, therefore, simply add the number of grams of In and S present in the original sample to find the number of grams of In_2S_3 formed. First it must be determined whether all the In and S will be used.

We know the following things:

1. There are 9.60 g S and 23.50 g In available.

2. One mole of In atoms weighs 114.82 g and one mole of S atoms weighs 32.01 g.

3. Two moles of In combine with 3 moles of S.

The weight of In_2S_3 can be found from the number of grams of In and S that combine, which is determined by the number of moles of In and S atoms present. There are 32.01 g S in one mole S atoms. Since only 9.60 g S are available, we have only a fraction of a mole,

$$\frac{9.60 \ \text{g S}}{32.0 \ \text{g S/mole S}} = 0.300 \ \text{mole S.}$$

There are 114.82 g In in one mole of In atoms. Since only 23.50 g In are available, we have only a fraction of a mole,

$$\frac{23.50 \ \text{g In}}{114.8 \ \text{g In/mole In}} = 0.204 \ \text{mole In.}$$

2 moles In combine with 3 moles S.

0.200 mole In combines with 0.300 mole S. Since there is 0.204 mole of In available and only 0.200 mole can be used, 0.004 mole In will be uncombined, or will be in excess.

The weight of In_2S_3 will then be the sum of the weights of 0.200 mole In and 0.300 mole S atoms or

$$(0.200 \ \text{mole In}) \ (114.82 \ \text{g In/mole In}) = 22.96 \ \text{g In plus 9.60 g S} = 32.6 \ \text{g } In_2S_3.$$

Note that $23.50 - 22.96 = 0.54$ g In is left unreacted and unchanged. It remains as In metal, in excess.

Example 2–8. What is the weight of one atom of aluminum (Al)?

The atomic weight of Al is 26.98 amu. The weight of one mole of Al atoms is the atomic weight expressed in grams: 26.98 g.

$$1 \text{ mole Al atoms} = 26.98 \text{ g} = 6.023 \times 10^{23} \text{ atoms}.$$

The number of grams of Al in one atom of Al (g Al/atom Al) is:

$$\frac{26.98 \text{ g Al/mole Al}}{6.023 \times 10^{23} \text{ atoms Al/mole Al}} = 4.479 \times 10^{-23} \text{ g Al/atom Al}.$$

One Al atom weighs 4.479×10^{-23} g.

The interconversions of moles and grams using atomic or formula weights are important; they should be mastered. If you know any two of the three quantities— grams, moles, and formula weight—you can calculate the third quantity. You should become familiar with these conversions. In all cases you should include the units in the calculations; this eliminates memorizing standard formulas or methods.

QUESTIONS

1. Define or discuss the following: The law of conservation of mass, the law of constant composition, chemical compound, the law of multiple proportions, Dalton's postulates, atom, molecule, atomic weight, amu, ^{12}C, formula unit, formula weight, chemical equation, reactants and products in a chemical reaction, balanced equation, Avogadro's number, mole, 6.023×10^{23} atoms ^{12}C, one mole CO_2 molecules.

2. Differentiate between the terms: atom, molecule, mole.

3. What is the relation between: grams, formula weight, mole?

4. a) What do we mean when we say that atomic weights are relative weights?
   b) What kinds of units are associated with atomic weights?

5. Calculate the formula weight of the following compounds:

   a) C_2H_4, ethylene   b) CO_2, carbon dioxide
   c) NaOH, sodium hydroxide   d) NH_4Cl, ammonium chloride
   e) CH_2O, formaldehyde   f) C_6H_6, benzene
   g) $Ba(NO_3)_2$, barium nitrate   h) $K_3Fe(CN)_6$, potassium ferricyanide
   i) $LaCl_3 \cdot 6H_2O$, lanthanum chloride hexahydrate   j) $CdIn_2S_4$, cadmium indium sulfide

6. Consider the list of compounds in question 5.

   a) How many atoms of each element are present in one formula unit of each compound?
   b) How many moles of atoms of each element are present in one mole of each compound?

7. In a reaction between one atom of cadmium metal (Cd) and two molecules of hydrogen fluoride gas (HF), one molecule of hydrogen gas (H_2) and one formula unit of cadmium fluoride (CdF_2) are produced.

   a) Write the chemical equation describing the reaction.
   b) How many moles of Cd atoms react with 4.0 moles of HF molecules?
   c) How many moles of H_2 gas can be produced from 3.0 moles of Cd and 4.0 moles of HF?

8. Show that the law of conservation of mass is obeyed in the following example: 11.2 g of cadmium (Cd) react exactly with 2.54×10^4 mg of iodine (I_2), to form 3.66×10^{-2} kg of cadmium iodide (CdI_2).

9. Based on the law of conservation of mass, how many grams of phosphorus (P) are needed to combine with 11.5 g of indium (In) in forming 14.6 g of indium phosphide (InP)?

10. Devise an experiment or set of experiments that will demonstrate the law of conservation of mass.

11. Gallium (Ga) and arsenic (As) combine to form a compound GaAs, gallium arsenide. Experiment has shown that 6.97 g Ga react with 7.49 g As.

    a) Write the chemical equation for the reaction.
    b) How many grams of GaAs will be formed if 6.97 g Ga and 7.4 g As are combined?
    c) What law is illustrated in part (b)?
    d) How many grams of As will combine exactly with 17.4 g Ga?
    e) How many grams of GaAs will be formed from 13.9 g Ga and 15.3 g As?
    f) What law is illustrated by parts (d) and (e)?

12. The compound aluminum oxide (Al_2O_3) is 53% Al and 47% O by weight.

    a) How many grams of Al will react with 9.4 g of oxygen?
    b) How many grams of Al_2O_3 can be formed from 16.5 g Al and 14.1 g oxygen?
    c) What law is illustrated by part (b)?
    d) Write the equation for the reaction between Al metal and O_2 gas.

13. a) Calculate the percentage of Cd and the percentage of I by weight in CdI_2 in question 8.
    b) Calculate the percentage of In and the percentage of P by weight in InP in question 9.

14. Balance the following equations.

    a) $C + O_2 \rightarrow CO$ b) $C + O_2 \rightarrow CO_2$ c) $Ca + Cl_2 \rightarrow CaCl_2$
    d) $Na + Cl_2 \rightarrow NaCl$ e) $Al + Cl_2 \rightarrow AlCl_3$ f) $Fe + S \rightarrow Fe_2S_3$
    g) $In + As \rightarrow InAs$ h) $Ge + O_2 \rightarrow GeO_2$ i) $W + O_2 \rightarrow WO_3$
    j) $S + O_2 \rightarrow SO_3$ k) $S + O_2 \rightarrow SO_2$

15. Consider the following chemical reaction:

$$Fe_2O_3 + 3CO \rightarrow 2Fe + 3CO_2.$$

    a) What do the subscripts in the chemical formulas mean?
    b) What do the coefficients placed before the chemical formulas mean?
    c) What relations are given by the chemical equation?

16. The metallic element nickel (Ni) reacts with fluorine gas (F_2) according to the following equation:

$$Ni + F_2 \rightarrow NiF_2.$$

    a) How many grams of F_2 react with 5.9 g Ni?
    b) What law(s) made possible the calculation in part (a)?
    c) Explain the result in part (a), using Dalton's atomic theory.

17. In the reaction between Cd and S to form cadmium sulfide (CdS), 11.2 g Cd combine exactly with 3.2 g S.

    a) How many grams of CdS are formed?
    b) What law(s) made possible the calculation in part (a)?
    c) Explain the result in part (a), using Dalton's atomic theory.

18. In one experiment 23.8 g of tin metal (Sn) reacted with 3.2 g O_2. In a second experiment, 11.9 g Sn reacted with 3.2 g O_2.

    a) Show how these data illustrate the law of multiple proportions.
    b) How can the results be explained on the basis of atomic theory?

19. Tell how many grams are contained in the following:

    a) 2.0 moles Na atoms   b) 12.0×10^{23} atoms of Na
    c) 3.01×10^{22} molecules O_2 d) 6.02×10^{24} molecules CO_2
    e) 0.25 mole HCl

20. How many moles are there in each of the following?

    a) 12.0×10^{22} atoms Hg   b) 4.0 g Ca
    c) 0.44 g CO_2 d) 24.0×10^{23} molecules H_2O
    e) 6.02×10^{20} formula units NaCl

21. An element of unknown atomic weight combines with oxygen (O_2) to form a compound containing one atom of the element and one atom of O. Given that 8.0 g O_2 reacts with 20.0 g of the element, what is the atomic weight of the unknown element?

22. Given the following reactions:

    1. $Li + F_2 \rightarrow LiF$ 2. $Mg + F_2 \rightarrow MgF_2$ 3. $Y + F_2 \rightarrow YF_3$
    4. $Si + F_2 \rightarrow SiF_4$ 5. $U + F_2 \rightarrow UF_6$

    a) Balance the equations.
    b) In each reaction, how many moles of the metal reactant (the first symbol given in the equations) combine with 3.80 g F_2?
    c) In each reaction, how many grams of the metal reactant combine with 19.0 g F_2?
    d) In each reaction, how many moles of product (the compound formed in the reaction) can be produced from 38.0 g F_2?
    e) In each reaction, how many grams of product can be produced from 9.5 g F_2?

23. In the reaction between tungsten (W) and sulfur (S), the compound WS_2 forms. How many grams of S are needed to react with 18.4 g W?

24. The reaction of tantalum metal (Ta) with O_2 produces a compound with a ratio of 2 Ta atoms to 5 oxygen atoms. How many grams of O_2 react with 0.40 mole Ta?

25. Given that 37.2 g niobium (Nb) react with 16.0 g O_2, what is the atom ratio (simplest chemical formula) of the reaction product?

26. Nickel (Ni) and O_2 react according to the following equation:

$$2Ni + O_2 \rightarrow 2NiO.$$

    The reaction is carried out using 12.5 g Ni and 3.20 g O_2.

    a) Which element is in excess? How many grams are uncombined?
    b) How many moles of NiO are formed?
    c) How many grams of NiO are formed?

27. If 20.0 g C and 20.0 g O_2 were mixed, how many grams of CO_2 could be produced?

28. The metal aluminum (Al) reacts with phosphorus (P) to produce aluminum phosphide, AlP.

    a) How many grams of Al are needed to produce 5.80 g AlP?
    b) How many grams of P are needed to produce 5.80 g AlP?
    c) In parts (a) and (b), how many moles of Al atoms and P atoms are needed?

d) In parts (a) and (b), how many atoms of Al and P are needed?

e) How many moles of AlP are produced in parts (a) and (b)?

f) How many formula units of AlP are produced in parts (a) and (b)?

29. An atom of an unknown element combines with O_2 to form a compound which contains one atom of the element and one atom of O. Given that 8.1 g of the compound is formed from 3.2 g O_2, what is the atomic weight of the unknown element?

30. Any ordinary sample of carbon contains two kinds of C atoms. One type, ^{12}C, comprises 98.89% of the atoms and has an atomic weight of 12.0000 amu. The second type is only 1.11% abundant and has an atomic weight of 13.0034 amu. What is the average atomic weight of these C atoms?

31. The atomic weight of tungsten (W) is 183.85 amu. What is the weight in grams of one atom of W?

32. The atomic weight of iodine (I) is 126.9 amu. What is the weight in grams of one I_2 molecule?

33. How many atoms of Cu are there in 12.7 g Cu?

34. Which of the following contains the largest number of atoms?

a) 5.89 g Co

b) 0.008 mole C_2H_6

c) 6.0×10^{22} molecules CO_2

d) 1.8×10^{22} atoms K

35. Given the compound C_6H_6, benzene:

a) How many moles of C_6H_6 in 7.8 g C_6H_6?

b) How many moles of C atoms in 0.25 mole C_6H_6?

c) How many atoms of C in 0.50 mole C_6H_6?

d) How many molecules of C_6H_6 in 10.0 moles C_6H_6?

e) How many molecules of C_6H_6 in 39 g C_6H_6?

f) How many grams of C_6H_6 in 1.5 moles of C_6H_6?

g) How many grams of C_6H_6 in 1.20×10^{24} molecules C_6H_6?

h) How many grams of C in 0.50 mole C_6H_6?

i) How many grams of H in 1.2×10^{22} molecules C_6H_6?

j) How many atoms of H in 3.9 g C_6H_6?

k) How many grams of H in 3.9 g C_6H_6?

3. THE STRUCTURE OF THE ATOM

3-1 BUILDING OUR KNOWLEDGE

In the first two chapters we emphasized the importance of experimentation in science. We described experiments that were performed over a period of many years by many different investigators. These experiments involved weighing reactants, weighing products, analyzing products, in short, making many observations of experimental results. From these observations came the laws of conservation of mass and constant composition, as well as many questions. What causes these laws? Why are they true? Are they always true?

Some answers to these questions were provided by Dalton. He proposed a model (hypothesis) for chemical reactions, assuming that small, individual particles—called atoms—could enter into chemical combinations. His ideas were verified through extensive experimentation, and Dalton's model became known as the atomic theory. As is true with all useful theories, Dalton was able to make a prediction using his atomic theory. His assumption that atoms combine in more than one proportion was strongly supported by experimental evidence, and led to the law of multiple proportions.

As is also true with any hypothesis or theory, Dalton's atomic theory ran into trouble. Many questions arose and were stimulated by the theory that could not be

satisfactorily answered. Some of the more obvious questions that can be posed are:

1. Are atoms composed of yet smaller particles?
2. Why do atoms react chemically, i.e., why do they "stick together"?
3. Why do atoms combine in different atomic ratios?

These questions were not answered by Dalton's atomic theory. Dalton simply assumed that atoms *were* the smallest particles and that they *did* combine chemically. If we wish to know the "whys" behind these assumptions, the model must be revised and improved.

The development of a new model for the atom required a better understanding of the composition of the atom. This understanding was obtained only after many new ideas and experiments, directed at a more detailed description of the atom, were considered or conducted. Several very important experiments concerned electricity and the electrical nature of matter.

3–2 DISCHARGE TUBE EXPERIMENTS

The Crookes tube. During the late 1800's scientists were conducting experiments with electricity and evacuated glass tubes called *discharge tubes*. In 1879 Sir William Crookes carried out experiments using a simple Crookes discharge tube, as shown in Fig. 3–1. Two metal disks, called electrodes, are sealed into a glass tube that can be evacuated, i.e., most of the air can be removed. The electrodes are attached to a high voltage supply (this can be thought of as a large battery, similar to those found in automobiles but much more powerful). When enough air is removed from the tube, the region between the two metal electrodes begins to glow. It looks as though a ray is being directed from one electrode to the other. The glow in the Crookes tube is similar to the spark discharge that can be produced by connecting wires to the positive and negative terminals of an automobile battery and touching the wires together. Hence the name *discharge tube*.

The direction of propagation of the ray (or glow discharge) in the tube—that is, whether the ray originated at the negatively charged electrode called the *cathode* or at the positively charged electrode called the *anode*—was determined by placing a screen at a diagonal angle between the electrodes, as shown in Fig. 3–2. The screen was coated with a material that glowed when struck by the ray (ZnS, zinc sulfide, was often used, since it emits bright yellow-green light when struck by a source of high energy, such as this ray). A narrow beam was produced by placing a metal shield with a small slit in front of the cathode. This narrow beam was readily seen on the ZnS screen, but only on the side facing the cathode, indicating that the ray was coming from the cathode and not from the anode. The ray was therefore called a *cathode ray*.

The electron. Several scientists experimenting with discharge tubes similar to the Crookes tube found that a cathode ray beam could be deflected in a magnetic field and in an electric field.

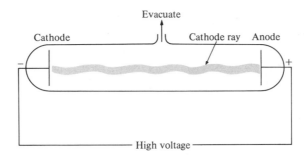

Fig. 3–1 A Crookes tube. A glow discharge, called a cathode ray, was produced in an evacuated tube using a powerful battery (high voltage).

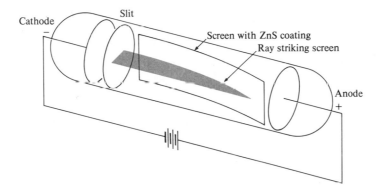

Fig. 3–2 A Crookes tube with an angled luminescent (ZnS) screen to show the direction of propagation, from cathode to anode, of the cathode ray.

We can demonstrate these observations by placing the north pole of a magnet near the side of the discharge tube pictured in Fig. 3–2. The cathode ray will bend down on the ZnS screen. The ray will be deflected up by the south pole of the magnet. If electrically charged plates are used, the cathode ray will be attracted by the positive plate and repelled by the negative plate.

These types of experiments with cathode rays, and the effects of magnetic and electric fields, were carried out in the laboratory of J. J. Thomson. Along with observing the direction of travel of the cathode ray (from negative cathode to positive anode), Thomson also observed the direction of deflection of the cathode ray in magnetic and electric fields. He found that the beam was deflected downward by the north pole of a magnet and upward by the south pole.

In an electric field the cathode ray was attracted by the positively charged plate, i.e., bent toward it, and repelled by the negatively charged plate (see Figs. 3–3a and 3–4b and c). Thomson observed that: (1) Cathode rays appear to travel from the

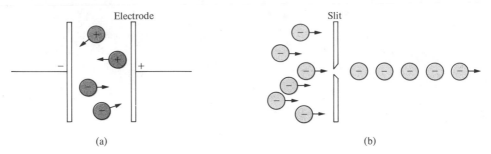

(a)   (b)

Fig. 3-3 (a) Like charges repel one another; opposite charges attract. (b) The function of a slit: selection of a narrow beam of particles.

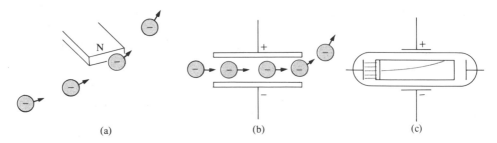

(a)   (b)   (c)

Fig. 3-4 (a) Negative particles are deflected upward by the north pole of a magnet. (b) Negative particles are attracted to the positively charged plate and repelled by the negatively charged plate. (c) The cathode ray beam (electrons) in a discharge tube is deflected by an electric field, as in (b).

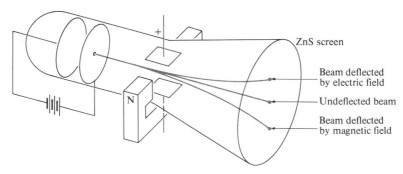

Fig. 3-5 J. J. Thomson's discharge tube. The electron beam is deflected by both magnetic and electric fields, allowing calculation of the electron charge-to-mass ratio, e/m.

negative to the positive electrode. (2) Cathode rays in a magnetic field are bent in the same fashion as a stream of negatively charged particles. (3) Cathode rays in an electric field are bent in the same fashion as a stream of negatively charged particles. (Opposite charges attract; like charges repel. A beam of negative particles is bent toward a positive electrode and away from a negative electrode.)

To explain his experimental observations, Thomson proposed that cathode rays were composed of small, negatively charged particles. These negative particles are called *electrons*.

Thomson carried his experiments further, made them more quantitative, and in 1897 developed a method for determining the *charge-to-mass ratio* of the electron, e/m, where e is the electron charge and m is the electron mass. He used a discharge tube like the one in Fig. 3–5, equipped with both magnetic and electric deflecting fields. In this way he was able to deflect the electron beam down with the magnetic field, note the distance of the deflection on the ZnS screen, and restore the beam to its original path by using an appropriate electric field. He could then calculate the charge-to-mass ratio of the electron* by knowing several of the dimensions of the discharge tube and the strengths of the electric and magnetic fields. Thomson found this very important ratio to be

$$e/m = -1.76 \times 10^8 \text{ coulombs/g.}$$

(The coulomb is a unit of electric charge, just as a gram is a unit of mass. The minus sign indicates that the electron has a negative charge.)

An application. You might recognize some similarity between Thomson's discharge tube (Fig. 3–5) and television tubes as we know them today. Indeed there is a striking similarity. Television tubes are evacuated discharge tubes; they use luminescent screens, like the ZnS screen used by Thomson; they use electron beams that strike the screen; and the electron beams are positioned and moved by magnetic fields. In fact, if Thomson and the other scientists involved had not done their very fundamental, original studies on electron deflection, we might not have television today.

It is not very probable that the scientists of 1900 envisioned that their experiments using discharge tubes would lay the foundations for such a gigantic television industry, just 50 years later. Their experiments were conducted for different reasons: to find out more about cathode rays, electricity, electrons, the atom. This is the type of experimentation that scientists often call *fundamental research*, i.e., it is not directed at an immediate technologically useful goal. Today it is the scientist conducting fundamental research who often must counter the question "What can we use your research for?" with the answer, "I don't know; the uses are still undiscovered." Sometimes this answer seems hard to justify. But when we realize that just a few fundamental discoveries can lead to many years of applied research and ultimately to tremendously useful and successful products, the answer, "The uses are still in the future," by the fundamental research scientist becomes more reasonable.

---

*   The calculation of e/m is somewhat involved, and will not be given here.

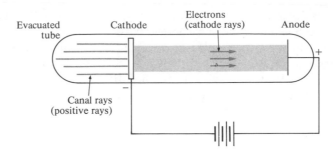

Fig. 3-6 A diagram of the discharge tube used by Goldstein to observe canal rays and protons.

Remember that, even after Thomson's work, it took about 50 years to perfect television. The advances made in applied research, engineering, and technology were great and were difficult, but they were necessary for the development of the industry. We must all realize the values of both fundamental and applied research and understand their true uses. Both are necessary in today's world.

The proton. Experiments with a discharge tube having a perforated cathode, like the one in Fig. 3-6, were conducted by Eugene Goldstein around 1896. He observed colored rays passing through the perforations in the cathode and called them *canal rays* because of their channel-like appearance. Upon further investigation he found that the color of the rays was determined by the type of gas that was present in the tube. But the canal rays appeared to be composed of positively charged particles in all the experiments, irrespective of the kind of gas in the discharge tube. Goldstein drew this conclusion because the canal rays appeared to be formed from the gas in the tube, to be attracted by the negatively charged cathode and to flow through the holes in the cathode. He obtained further proof of the positive nature of the rays by deflecting them with charged plates. The canal rays flowing out of the perforations in the cathode were attracted by a negatively charged plate and repelled by a positively charged plate.

The experimental techniques of deflecting the positively charged canal rays were improved by W. Wien, who was then able to measure the charge-to-mass (e/m) ratios for canal rays, just as Thomson had done for cathode rays. The observed results for the canal rays, however, were much different from those obtained by Thomson. For the canal rays, the e/m value depended on the type of gas present in the discharge tube; for cathode rays, the e/m did not depend on the gas present. The e/m values obtained for the positive canal rays were all very much smaller than the e/m value for the cathode ray (the electron). And the e/m values for the positive rays seemed to increase as the atomic weight of the gas in the tube decreased, with the lightest gas, hydrogen, giving the largest e/m.

These experimental observations of Goldstein and Wien were explained in the following way. Positive particles were formed from the gas in the tube by collision of

gas particles with the high-energy electrons, which were moving very rapidly from the cathode to the anode. This collision was sufficient to break the atoms or molecules of gas into charged fragments. With helium (He) gas, for example, interaction of a He atom and a high-energy electron (e^-) could produce a positively charged He atom, He^+, and a second electron. Charged atoms, such as He^+, or charged groups of atoms are called *ions*. The formation of the He^+ ion can be written in equation form as

$$He + e^- \xrightarrow{\text{high voltage}} He^+ + 2e^-.$$

The He^+ ions that are formed in the tube are then attracted by the negative electrode and flow toward it. Some of the He^+ ions pass through the holes in the cathode and produce the positively charged canal rays.

The proposed mechanism given in the equation above for the production of canal rays can also account for the variation in observed e/m values for different gases, and for the small e/m values compared with the electron charge-to-mass ratio. If the proposed mechanism is correct, then the charge on the He^+ ion should be equal to the electron charge (since both a He^+ ion and an e^- are produced simultaneously from a neutral He atom). Accepting this assumption, we have an explanation of the small e/m value for He^+ compared to e/m for an electron. The charge, e, is the same numerical value (but of opposite sign) in both cases, so the m-values must be very different. Since the e/m for an electron is much larger than that for He^+, the mass of He^+ must be much greater than the mass of an e^-.

▶ This is true because fractions are involved. To get a smaller value of e/m for a fixed e, the magnitude of m must be increased. For a larger e/m, m must be decreased. To see this more clearly, consider fractions such as $\frac{2}{4}$ and $\frac{2}{6}$. For a given numerator, the value of the fraction decreases as the denominator increases. ◀

The above relationship also explains why the e/m value for He canal rays is larger than the e/m value for Ar canal rays: The m-value (atomic weight) of Ar is larger than that of He. Since hydrogen gas gave the largest value of e/m observed for canal rays (but still smaller than e/m for the e^- by a factor of about 2000), it was postulated that H^+ is actually a fundamental subatomic particle with the same numerical charge as the electron (but of opposite sign) and with about 2000 times as much mass as the electron. This positively charged particle was called the *proton*.

3-3 THE OIL-DROP EXPERIMENT

From the experiments of J. J. Thomson, scientists learned the charge-to-mass ratio of the electron. If either the charge or the mass of the electron could be determined independently, the other property could be calculated using Thomson's e/m. In 1909 R. A. Millikan conducted an ingenious experiment that enabled him to determine the numerical value of the electron charge.

Figure 3–7 is a diagram of Millikan's experimental apparatus. Oil was atomized into a box and the droplets allowed to fall slowly between two parallel-plate elec-

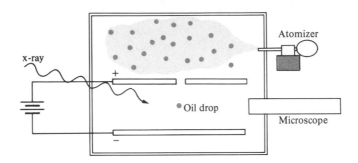

Fig. 3–7 Millikan's oil-drop experiment. The electron charge was found to be −1.60 × 10⁻¹⁹ coulomb.

trodes. The rate of fall of the very small droplets was observed and measured by means of a microscope. A charge was placed on the droplets by passing x-rays through the box. The x-rays knocked electrons off air particles; some of these electrons deposited on the oil droplets, giving them a negative charge. The rate of fall of the droplets could then be regulated by the parallel-plate electrodes. With the bottom plate charged negatively, the rate of fall could be decreased, stopped, or reversed by adjusting the strength of the electric field. Careful measurements of the rates of fall of many droplets and of the electric fields used (again the calculations are somewhat complex, involve other measurable variables, and are not presented here) enabled Millikan to calculate the value of the *electron charge*, $e = -1.60 \times 10^{-19}$ coulomb. Combining the electron charge with Thomson's value of the electron charge-to-mass ratio gave the electron mass, m:

$$\frac{e}{e/m} = \frac{-1.60 \times 10^{-19} \text{ coulomb}}{-1.76 \times 10^{8} \text{ coulombs/g}},$$

$$m = 9.0 \times 10^{-28} \text{ g.}$$

The proton mass can be calculated in a similar way, since its charge is equal in magnitude but of opposite sign to the electron. The proton charge-to-mass ratio was determined by Wien's experiments. The proton mass is much larger than that of the electron, being 1.67×10^{-24} g, or about 1837 times larger than the electron mass.

▶ The rays called x-rays are rays of electromagnetic radiation just like visible light rays, but they penetrate materials much more and are invisible to your eyes. Also x-rays have the same wave nature as visible light, except that the wave-crest separation is much smaller in x-rays, giving them a greater energy (see Section 5–2). ◀

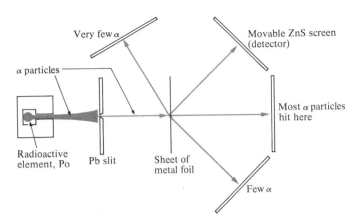

Fig. 3-8 Rutherford's α-particle-scattering experiment. A narrow beam of α-particles from a radioactive source was selected by a lead slit. A few of the α-particles were strongly deflected by the thin metal foil and detected by a movable ZnS screen.

3-4 ARRANGEMENT OF ELECTRONS AND PROTONS

The basic assumption of Dalton's atomic theory was that an atom was the smallest particle composing an element. But now at least two types of charged particles, electrons and protons, that are smaller than atoms [one atom of uranium (U) weighs 3.95×10^{-22} g] appear to exist. These particles, then, must be contained within atoms of the various elements. But how can they be arranged?

Rutherford's experiments. In 1911 Lord Rutherford and his students and associates conducted a series of experiments that were directed at this very question. Figure 3–8 is a diagram of the apparatus that Rutherford used. A narrow beam of α (the Greek letter alpha) particles was produced by means of a *radioactive* source such as the element polonium (Po). The beam was collimated by a lead (Pb) slit, and directed at a very thin sheet of material, such as silver (Ag) or gold (Au) metal foil. The α-particles passed through the thin metal sheet and struck a ZnS screen used as a detector. Most of the α-particles passed directly through the metal foil undeflected. A few, however, were scattered at rather large angles to the main beam, and some α-particles were even deflected backward.

▶ Note that α-particles are He atoms with two electrons removed, yielding a particle with a +2 charge, $\alpha = He^{2+}$. A radioactive substance spontaneously emits radiation, i.e., gives off energy (see Fig. 3–9). The emitted radiation can be of several forms, α-particles being associated with one form. Common examples of radioactive substances are the elements radium (Ra) and uranium (U). These substances decompose, sometimes very rapidly; their properties are often hard to determine and study. We shall consider this subject in more detail in Chapter 21. ◀

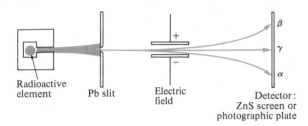

Fig. 3–9 Alpha, beta, and gamma rays are spontaneously given off by radioactive substances such as radium (Ra) or polonium (Po) as they decompose. The rays can be separated in an electric field, since gamma (γ) rays are uncharged, alpha (α) rays are composed of positively charged particles, and beta (β) rays of negatively charged electrons.

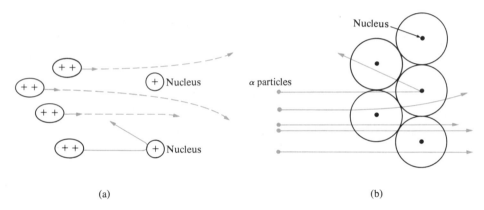

(a)   (b)

Fig. 3–10 (a) A schematic view of interactions of α-particles with nuclei of metal atoms in metal foil. (b) Rutherford's model of the atom, showing how it explained the α-particle-scattering experiments.   Large circles represent metal atoms with nuclei at their centers.

Observing particles being scattered at large angles was completely unexpected and could not be explained by any theories of the atom known at that time. Rutherford was so surprised by the result that he remarked, "It was quite the most incredible event that has ever happened to me in my life. It was almost as incredible as if you fired a 15-inch shell at a piece of tissue paper and it came back and hit you."

To account for these unexpected results Rutherford proposed a new model of the atom (see Fig. 3–10). In his model all the protons were located in a very small volume at the center of the atom. This high-mass, low-volume, positively charged atom center was called the *nucleus*. The electrons of the atom were distributed around the nucleus in a manner similar to the planets around the sun. The electrons took up essentially all the volume of the atom. Subsequently, it has been found that nuclear diameters are about 10^{-13} cm, while atom diameters are about 10^{-8} cm.

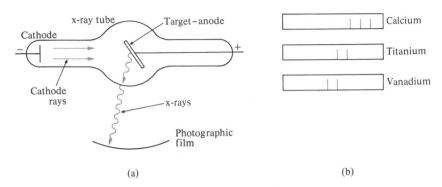

Fig. 3-11 An x-ray tube as used by Moseley. The x-rays produced from the anode material were detected by means of photographic film. (b) A sketch of Moseley's photographic data of x-rays produced by the anode materials Ca, Ti, and V.

Why was it necessary for Rutherford to propose such a model for the atom, i.e., how does the model account for Rutherford's experimental observations? One observation was that most of the α-particles passed through the thin metal foil undeflected. In order for this to be true, most of the atom must be free of obstructions to the travel of the α-particle. Rutherford therefore proposed that most of the atom volume was occupied by the electrons—particles too small to interfere with the travel of the larger particles. Some of the α-particles, however, were deflected by a large amount from a straight-line path through the metal foil, and some were even deflected backward. Rutherford reasoned (and calculated) that such large scattering of the α-particles would require close approach or collision with a massive, positively charged particle—and this could not happen very frequently. He therefore proposed that the protons, which contain most of the mass of the atom and are positively charged, are collected in a very small region at the center of the atom. Since these nuclei are so small, the α-particles would not pass near them very often and would therefore ordinarily pass undeflected through the metal foil.

Moseley's experiments. In 1913 Henry Moseley, a young scientist working in Rutherford's laboratory, conducted a series of experiments using various metallic elements as the anode in a cathode ray (Crookes) tube, as diagrammed in Fig. 3-11. Several years earlier Roentgen had observed that x-rays were produced from the anode under these conditions and that the character of the x-rays depended on the anode material. Moseley carefully photographed the x-rays that were produced with each different anode material. Detailed analysis of his data revealed a remarkably simple and systematic progression, illustrated in Table 3-1.

The data in Table 3-1 are plotted in Fig. 3-12. From these data we can see that the x-ray wavelength decreases as the atomic weight increases, except in the case of Co and Ni. Moseley noted these discrepancies, along with a few others, and found

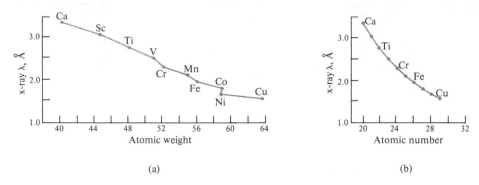

Fig. 3–12 A plot of Moseley's data of x-ray wavelengths versus (a) atomic weight and (b) atomic number. Note the discrepancy between Co and Ni in (a), and the smooth systematic change in (b).

that he could resolve the difficulties and explain his results if he assumed a nuclear charge of $+20$ for Ca, $+21$ for Sc, $+22$ for Ti, and so on. This high positive charge on the nucleus supported Rutherford's idea of a nucleus composed of protons. Then Ca has 20 protons, Sc 21 protons, Ti 22 protons, and so on. The number of protons that an element has in its nucleus was called the *atomic number*. As we can see in Table 3–1 and Fig. 3–12, Moseley's x-ray data show a smooth relationship to the atomic number of the elements.

Table 3–1. Characteristic x-ray wavelengths emitted by a series of elements and their relation to the atomic weights of the elements

Element	Atomic weight	Wavelength* of x-ray, Å	Atomic number
Ca	40.08	3.35	20
Sc	44.96	3.03	21
Ti	47.90	2.74	22
V	50.94	2.50	23
Cr	52.00	2.29	24
Mn	54.94	2.10	25
Fe	55.85	1.94	26
Co	58.93	1.79	27
Ni	58.71	1.66	28
Cu	63.55	1.54	29

* The wavelength λ of the x-rays gives the distance between crests of the x-ray wave. The unit of wavelength is the *angstrom*, Å. It is used for convenience, since $1 \text{ Å} = 1 \times 10^{-8}$ cm.

3–5 MORE ATOMIC MASS

Neutrons. In the preceding section we showed that the agreement between the experimentally determined atomic weight of an element and the number of protons contained in the nucleus of the element (the atomic number) was very poor (see Table 3–1). This was disturbing, since Rutherford's model of the atom attributed essentially all the atomic mass to the protons. For this reason Rutherford proposed that the nucleus contained another particle, called the *neutron*, that had no electrical charge, but had approximately the same mass as the proton. This particle was discovered experimentally by Chadwick, working in Rutherford's laboratory, but not until 1932.

The nucleus, then, is composed of protons and neutrons. The number of protons contained in the nucleus of any neutral atom is equal to the number of electrons outside the nucleus and is called the *atomic number*, often given the symbol Z. The atomic number is *the* unique characteristic of an element. Each atom of oxygen has 8 protons in its nucleus, regardless of how oxygen is combined chemically. In O atoms, O_2 molecules, O_3 molecules (ozone), the compound CaO, etc., the Z for oxygen is always 8. If the number of protons in the nucleus of a particular element were to be changed, the element would be transformed into a different element. This transformation is very difficult and cannot be accomplished under ordinary circumstances.

The neutrons in the nucleus, along with the protons, contribute essentially all the atomic weight of the elements. The sum of the number of protons and neutrons is called the *mass number*, A. For oxygen there are 8 protons and (usually) 8 neutrons. The mass number is $A = 8 + 8 = 16$. Standard notation for Z and A is $^{16}_{8}O$, that is, Z appears as a left-hand subscript and A as a left-hand superscript to the element symbol.

If we examine the atomic weights of the elements, we find that they are not whole numbers. This means that every atom of any given element cannot have the same number of neutrons in its nucleus. Lithium (Li), for example, has atomic number 3 and therefore has 3 protons (always), but its atomic weight is 6.941. There must be some Li atoms, therefore, with more than 3 neutrons, giving an A greater than 6 (3 protons + 3 neutrons). It is now known that some Li atoms have $A = 6$ ($^{6}_{3}Li$) and some have $A = 7$ ($^{7}_{3}Li$). These are called *isotopes,* atoms with the same atomic number but different mass number. *Isotopes* are therefore atoms of the same element containing different numbers of neutrons. Generally, the chemical properties of isotopes of the same element are very similar.

The mass spectrograph. An instrument has now been developed that can measure differences in the masses of isotopes. It is called the *mass spectrograph* or *mass spectrometer*, and is a very close relative of the discharge tubes of Thomson and Goldstein. In fact, just before World War I Thomson and F. W. Aston began experimenting with discharge tubes in order to compare the weights of atomic-sized particles directly. After the war the mass spectrograph (see Fig. 3–13) was developed by Aston, A. J. Dempster, and others.

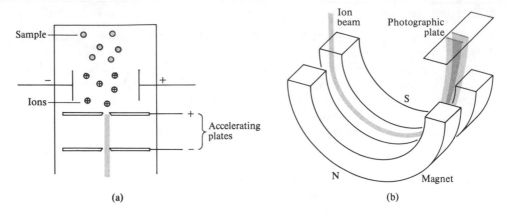

Fig. 3–13 A schematic view of a mass spectrograph, showing (a) the ionizing chamber and (b) the magnetic analyzer and detector. Note the film detector and record. An electronic detector is often used instead of film.

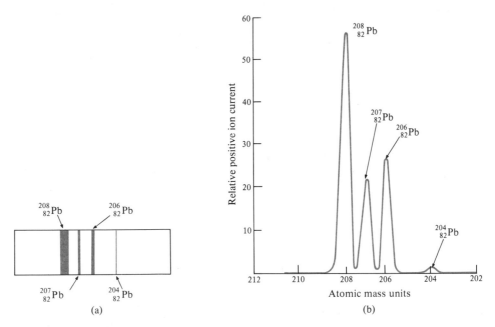

Fig. 3–14 A mass spectrum of lead (Pb) as it would appear (a) on a film, and (b) as detected electronically. The intensities of the lines are proportional to the number of ions of the particular Pb isotope that are present in the sample.

▶ The major difference between a spectrograph and a spectrometer is in the detecting system. A spectrograph uses a photographic film as a detector, while a spectrometer uses an electronic detector of some sort, usually in conjunction with a digital readout device and a recorder. ◀

The entire instrument is evacuated. The sample is then introduced into the ion source and is bombarded by cathode rays. Electrons are knocked off the sample by the rays and positively charged particles are formed. These charged particles are *ions* (positive ions are formed when electrons are lost; negative ions are formed when electrons are gained). If we chose the element lead (Pb) as our example, then electrons would be knocked off the Pb atoms and Pb ions (Pb^+) would be formed. (Actually other ions such as Pb^{2+} would also be formed. For the sake of simplicity, we shall consider only singly charged ions.)

The positive ions are accelerated by the charged parallel plates (attracted by the negative plate and repelled by the positive plate) into the curved chamber that passes between the poles of a powerful magnet (see Fig. 3–13a). The paths of the positive ions are bent in the magnetic field (see Section 3–3), with the paths of the lighter ions being curved the most, as shown in Fig. 3–13b. The ions strike the photographic film at the end of the curved chamber. The position of the line on the film is characteristic of the mass of the particle and the intensity of the line indicates the number of ions with that particular mass.

Table 3–2. Stable isotopes of various elements

Element	Mass number	Isotope weight, amu	Isotope abundance, %	Average atomic weight, amu
H, hydrogen	1	1.0078	99.985	1.0080
	2	2.0141	0.015	
C, carbon	12	12.00000	98.89	12.011
	13	13.00335	1.11	
O, oxygen	16	15.9949	99.759	15.9994
	17	16.9991	0.037	
	18	17.9992	0.204	
Al, aluminum	27	26.9815	100	26.9815
Cr, chromium	50	49.9461	4.31	51.996
	52	51.9405	83.76	
	53	52.9407	9.55	
	54	53.9389	2.38	
Tl, thallium	203	202.9723	29.50	204.37
	205	204.9745	70.50	

▶ Actually the line position is determined by the charge-to-mass ratio of the ion. The line intensity is measured in units of optical density. A line image is formed on the film because of the slit used to collimate the ion beam. ◀

With our sample of Pb, for example, three lines would be seen on the film (Fig. 3–14a). This same spectrum is shown in Fig. 3–14b, recorded electronically instead of photographically.

Precise mass-spectrographic measurements have provided the best method for determining atomic weights, isotopic weights, and the weight of the proton. Table 3–2 lists the isotopic weights, isotopic abundances, and average atomic weights of several elements.

Table 3–3 lists the values for the mass and charge of the three fundamental particles.

Table 3–3. Values of mass and charge for the electron, proton, and neutron

Particle	Symbol	Mass, g	Mass, amu	Charge*
Electron	e^-	9.1096×10^{-28}	0.00055	-1
Proton	p	1.6725×10^{-24}	1.00728	$+1$
Neutron	n	1.6748×10^{-24}	1.00867	0

* The charge given is the relative charge, with the electron assigned a -1 value. Its actual charge is -1.602×10^{-19} coulomb.

3–6 THE CHANGING ATOMIC MODEL

We can now see that the atomic model that Dalton so ingeniously developed is no longer adequate. We know that the atom is composed of yet smaller particles: electrons, protons, and neutrons. We know that every atom of any given element is not necessarily identical to all other atoms of the element, i.e., isotopes exist.

Rutherford's model of the atom takes us a bit farther. He assumed that the atom was composed of two parts: (1) the nucleus, containing the protons and neutrons, and (2) a region surrounding the nucleus, containing the electrons. The nucleus occupies a very small part of the atomic volume, but holds most of the atomic mass (this gives the nucleus a very high density). Since all the protons of the atom are located in the nucleus, it has positive charge. (Note that all these positive charges are in a very small volume. But like charges repel! What holds these protons together?) The region surrounding the nucleus contains the electrons. This region composes most of the volume of the atom and only a small amount of the atomic mass (this gives it a low density). In a neutral atom, the number of electrons is equal to the number of protons.

But Rutherford's model of the atom still does not give much insight into why atoms form molecules or why atoms react in varying ratios. In the next chapter we shall encounter some other difficulties with Rutherford's atomic model.

QUESTIONS

1. Define the following terms: Atom, cathode ray, canal ray, electron, proton, neutron, isotope, atomic number, mass number, nucleus, α-particle, radioactivitity, x-ray, nuclear charge, positive ion, negative ion, neutral atom.

2. a) What is the nuclear charge of the following atoms: Na, P, Ar, Ti, W, Eu, U?
   b) How many protons do each of the atoms in part (a) contain?
   c) What is the atomic number of each of the atoms in part (a)?

3. Give the atomic number of the following elements: carbon, fluorine, tungsten, barium, cadmium, arsenic, iron, indium, iodine, argon, cesium.

4. a) What is the mass number of each isotope of chromium (Cr) listed in Table 3–2?
   b) How many neutrons are located in the nucleus of each Cr isotope in part (a)?

5. Give the atomic number, mass number, and the number of neutrons for each of the following:
   $^{21}_{10}\text{Ne}$ $^{47}_{22}\text{Ti}$ $^{59}_{27}\text{Co}$ $^{75}_{33}\text{As}$ $^{95}_{42}\text{Mo}$ $^{109}_{47}\text{Ag}$ $^{116}_{50}\text{Sn}$
   $^{144}_{62}\text{Sm}$ $^{206}_{82}\text{Pb}$ $^{238}_{92}\text{U}$

6. a) Why are canal-ray e/m values smaller than e/m for the e^-?
   b) Describe the data needed and the procedure used to calculate the proton mass after Wein's experimental results were known.

7. In J. J. Thomson's discharge-tube experiments, what was the function of the magnetic field?

8. What differences would there have been in Thomson's experiments if protons instead of electrons had been investigated?

9. Stable (nonradioactive) isotopes with the indicated mass numbers exist for the following elements:

nitrogen	15	germanium	70
magnesium	26	silver	107
nickel	60	mercury	196

   Write the element symbol and indicate the atomic number and the mass number, in standard notation form.

10. How many electrons do neutral atoms of Li, Si, Ca, Mn, Zn, Y, Br, Xe, Ta, Pt, Pb, and Th contain?

11. In Millikan's oil-drop experiment, what purpose did the charged parallel plates serve?

12. Show how the electron mass can be calculated using the results of the experiments of Thomson and Millikan.

13. Why did Millikan need to use x-rays in his experiment?

14. Using Rutherford's model of the atom, how can you account for the formation of ions?

15. Why were some α-particles deflected by the metal foil in Rutherford's experiment?

16. What purpose did the following things serve in Rutherford's experiment: polonium, ZnS, metal foil, Pb?

17. Why did Rutherford postulate that most of the volume of the atom was occupied by electrons?

18. Explain how Moseley's results supported Rutherford's atomic model.

19. What purpose did the cathode rays serve in Moseley's experiments?

20. Describe the essential differences between electrons, protons, and neutrons.

21. Describe Rutherford's model of the atom. Draw a picture labeling the pertinent parts.

22. Consider the following chemical reaction:

$$^{108}_{48}Cd + \,^{120}_{52}Te \rightarrow CdTe.$$

a) Give the atomic number for all reactants.

b) What is the atomic number of Cd in the product, CdTe? What is the atomic number of Te in CdTe?

c) How many neutrons are contained in the nucleus of the Cd atom reactant?

d) How many electrons are there in the Te reactant?

23. Explain why Dalton's atomic theory cannot account for the results of Rutherford's experiment.

24. One electron weighs 9.1×10^{-28} g (0.00055 amu).

a) How many grams does 1.0 mole of electrons weigh?

b) How many grams does 1.0 mole of protons weigh?

c) How many grams does 1.0 mole of neutrons weigh?

d) How many grams do 6.02×10^{23} protons weigh?

e) How many neutrons are needed to give a total weight of 1.0 g?

25. A sample of $^{12}_{6}C$ powder weighs 24.0000 g.

a) How many electrons, how many protons, and how many neutrons does the sample contain?

b) How many grams each do the electrons, protons, and neutrons in part (a) weigh?

c) Compare the sum of the weights in part (b) to the sample weight. Can you suggest any reason for the discrepancy? (See Section 21–4.)

26. In the mass spectrograph, positive ions of elements are formed. Explain how this can happen.

27. What is the function of the magnet in the mass spectrograph? (Compare this with question 7.)

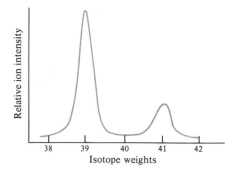

Isotope weights

Figure 3–15

28. The element silicon (Si) has three stable isotopes. Precise mass-spectrographic measurements indicate the following isotopic weights: 27.9769, 28.9765, and 29.9738 amu. The relative abundances of the isotopes are 92.21%, 4.70%, and 3.09%, respectively. Calculate the average atomic weight of Si.

29. Using the data in Table 3–2, sketch a probable mass spectrogram for the isotopes of Cr.

30. Chlorine (Cl) has two stable isotopes. Mass number 35 has an isotopic weight of 34.9689 amu and is 75.53% abundant. Mass number 37 is 24.47% abundant. What is the isotopic weight of $^{37}_{17}Cl$?

31. The mass spectrum given in Fig. 3–15 is for an element.
    a) What is the element?
    b) What are the mass numbers of the isotopes of the element?
    c) What is the atomic number of the element?
    d) How many neutrons are contained in each isotope?
    e) Estimate the abundance of each isotope.

4. PERIODIC PROPERTIES AND THE ATOM

4-1 ELEMENT PROPERTIES AND PERIODICITY

During the nineteenth century many elements were discovered, along with more detailed knowledge of their physical and chemical properties. The elements differed from one another in many of their properties: Some were hard and shiny, while others were soft and powdery; some were liquids and others gases at ordinary temperatures, and so on. The compounds formed by element combinations revealed an even greater variety of properties among the elements.

Nevertheless, several scientists, beginning with William Prout around 1815, tried to discover some form of order for the large array of elements that were apparently very different. Prout proposed that all elements were composed of combinations of different numbers of hydrogen atoms, which he thought were the only fundamental particles. With this hypothesis, all the elements would necessarily have atomic weights that are whole-number multiples of the atomic weight of hydrogen. Some of the early atomic-weight determinations appeared to support Prout's hypothesis, but as more accurate atomic-weight values were determined, by Jöns Jakob Berzelius and other chemists, it became evident that the atomic weights of the elements did not conform to the integral-multiple rule and that Prout's hypothesis was not correct. But no sooner had Prout's proposal been abandoned than the attempts of several other scientists to order and classify the elements appeared.

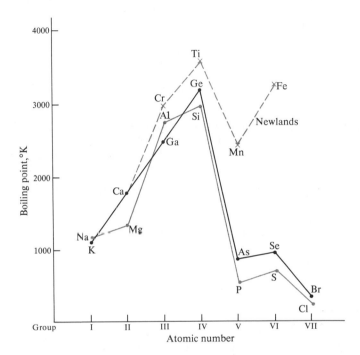

Fig. 4-1 Variation of boiling points of Group A elements of periods three and four of the periodic table. Boiling points of elements of the same group are similar and the variations along period three are similar to those along period four. The elements Cr, Ti, Mn, and Fe are included in the groups according to Newlands' arrangement. Note, particularly, the unsystematic changes of Mn and Fe, indicating that they are assigned to the wrong group.

Early Classifications. One of the first attempts at grouping elements was made by Johann Döbereiner during the 1820's. He noticed that the elements iron (Fe), cobalt (Co), and nickel (Ni) all had similar properties and formed compounds that are colored. He also studied another group of three elements, chlorine (Cl), bromine (Br), and iodine (I), which were all colored substances and were quite similar chemically. For example, they reacted vigorously with some metals, such as Na, to form simple compounds, such as NaCl.

Döbereiner observed similar behaviors for other groups of three elements. These groups became known as *Döbereiner's triads*, and stimulated more attempts at further classification of the elements.

Following Döbereiner's work and an increasing number of studies on the chemical and physical properties of all elements and their compounds, John Newlands, in 1865, proposed his *law of octaves*. He noticed that the properties of elements seemed to repeat when the elements were arranged in order of increasing atomic weight, in a series of rows of seven elements each. Newlands arranged the elements in the following way,

H	Li	Be	B	C	N	O
F	Na	Mg	Al	Si	P	S
Cl	K	Ca	Cr	Ti	Mn	Fe

indicating that the eighth element (F) was similar to the first (H), the ninth element (Na) similar to the second (Li), and so on. One difficulty was immediately apparent: The properties of Mn and Fe were not very similar to those of P and S. In fact, they were quite different. Manganese and iron are both hard, metallic elements that are good conductors of heat and electricity. Phosphorus and sulfur, on the other hand, are relatively soft, colored powders that are poor conductors of heat and electricity; they also form chemical compounds that are unlike those of Mn and Fe (see also Fig. 4–1). It was evident that Newlands' ideas were not completely right, but that they were headed in the proper direction.

4-2 DEVELOPMENT OF THE PERIODIC TABLE

In the late 1860's Dmitri Mendeleev and Lothar Meyer, working independently, discovered that they could arrange the elements known at that time in a way that would correlate their periodic properties. Since Mendeleev presented his arrangement in more detail, he is generally given the main credit for establishing the *periodic law*, which states that the properties of the elements and their compounds vary in a periodic way according to their atomic weight.

Mendeleev developed his ideas about the periodicity of the elements by comparing the chemical properties of the elements and their compounds, and by asserting that the atomic weight is the one thing about an element that might not change, even upon chemical combination.

▶ As people do when they make most great discoveries and advances, Mendeleev seemed to be guided to a considerable degree by his feelings and his intuition; in his case, chemical intuition. ◀

Mendeleev noticed that the elements formed an interesting and illuminating pattern if they were arranged in rows, according to increasing atomic weight (a pattern similar to that of Newlands, whose ideas were unknown to Mendeleev). He made an arrangement of the first few elements as follows (given with the atomic weights of 1872).

Li	Be	B	C	N	O	F
7	9.4	11	12	14	16	19
Na	Mg	Al	Si	P	S	Cl
23	24	27.4	28	31	32	35.3
K	Ca	–	Ti	V	Cr	Mn
39	40		50	51	52	55

▶ Note that Mendeleev did not place hydrogen in his series. It was known, even then, to have very distinctive properties, unlike those of any other element. The noble gases were unknown during Mendeleev's time and, consequently, are not included in his tables. ◀

The above table of elements illustrates Mendeleev's ideas about periodic properties and atomic weight. The atomic weight progressively increases, and all the elements in any particular vertical group have similar properties. The three elements Li, Na, and K occur in the first vertical column or group; they are all highly reactive, relatively soft metallic elements that form compounds with other elements, such as chlorine, that are also very similar in their properties. In general, this type of repetition of properties occurs in each of the other groups.

We can illustrate periodic occurrence of properties by considering the atomic volumes of the elements (the volume occupied by one mole of the element in the semi-solid state), as Lothar Meyer did. Figure 4–2 gives a plot of atomic volume versus atomic weight.

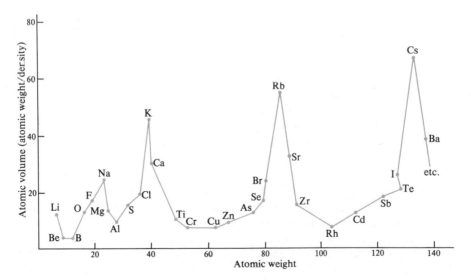

Fig. 4–2 Periodic variation of atomic volume (the atomic weight of an element divided by its density in the solid state) with increasing atomic weight. For the sake of clarity, the data points are connected, and not all the elements known to Mendeleev and Meyer are included.

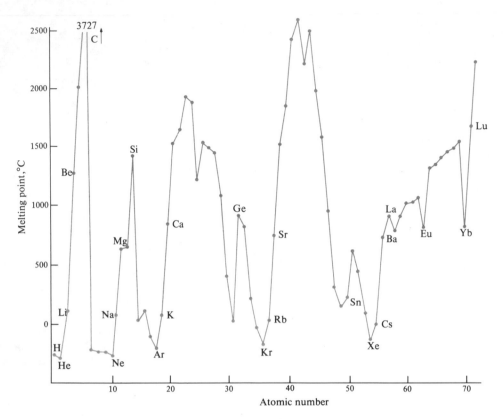

Fig. 4-3 Melting points of the elements (atomic numbers 1 through 72) show rough periodic variations.

Inspection of these data shows that the variation in atomic volume is cyclical. After Li there is a decrease, followed by an increase to Na; a second decrease is followed by another increase to a maximum at K; and so on. The maxima in the atomic-volume plot recur periodically throughout the sequence of elements, with the elements in the first group, that is, Li, Na, K, etc., always appearing at one of the maxima.

Figure 4-3 shows a similar periodic behavior of the melting points of the elements (some of the elements in Fig. 4-3 were unknown in Mendeleev's time); further examples of periodicity are given in Section 4-4.

Note that Mendeleev left a blank space between Ca and Ti in his table. He reasoned that Ti formed compounds with the same atom ratios and with properties similar to those of C and Si, but not to those of B and Al. Consequently, he placed Ti in the group with C and Si, rather than with B and Al, even though no element was then known that had an atomic weight between those of Ca and Ti. Mendeleev predicted that such an element would be discovered. This ability to predict was one

The Atomic Weights of the Elements

Distribution of the Elements in Periods

Groups	Oxide Formula	Typical or 1st small period	Large periods 1st	2nd	3rd	4th	5th
I.	R_2O	Li = 7	K 39	Rb 85	Cs 133	—	—
II.	RO	Be = 9	Ca 40	Sr 87	Ba 137	—	—
III.	R_2O_3	B = 11	Sc 44	Y 89	La 138	Yb 173	—
IV.	RO_2	C = 12	Ti 48	Zr 90	Ce 140	—	Th 232
V.	R_2O_5	N = 14	V 51	Nb 94	—	Ta 182	—
VI.	RO_3	O = 16	Cr 52	Mo 96	—	W 184	U 240
VII.	R_2O_7	F = 19	Mn 55	—	—	—	—
VIII.			Fe 56	Ru 103	—	Os 191	—
			Co 58.5	Rh 104	—	Ir 193	—
			Ni 59	Pd 106	—	Pt 196	—
I.	R_2O	H = 1, Na = 23	Cu 63	Ag 108	—	Au 198	—
II.	RO	Mg = 21	Zn 65	Cd 112	—	Hg 200	—
III.	R_2O_3	Al = 27	Ga 70	In 113	—	Tl 204	—
IV.	RO_2	Si = 28	Ge 72	Sn 118	—	Pb 206	—
V.	R_2O_5	P = 31	As 75	Sb 120	—	Bi 208	—
VI.	RO_3	S = 32	Se 79	Te 125	—	—	—
VII.	R_2O_7	Cl = 35.5	Br 80	I 127	—	—	—
		2nd small period	1st	2nd	3rd	4th	5th
				Large periods			

From Mendeleev's *Principles of Chemistry*, Vol. 1.

Fig. 4–4 An early version of Mendeleev's periodic table of the elements. The groups are arranged horizontally instead of vertically. The elements Sc, Ga, and Ge were predicted to have atomic weights 44, 70, and 72, as shown. Blanks were also left in the table between atomic weights 96 and 103, 184 and 191, 232 and 240. Subsequently, the elements Tc, Re, and Pa were discovered and filled these vacancies.

of the major contributions of the *periodic table of the elements* developed by Mendeleev.

Figure 4–4 shows a periodic table as published by Mendeleev about 1871. Just as in the short arrangement of three rows of elements given above, this periodic table contained rows of the elements, in order of increasing atomic weight, and

vertical columns or groups, with all elements of a given group having similar properties. Mendeleev predicted—and left blank spaces in his table for—four new elements: those with atomic weights 44, 68, 72, and 100. These elements have subsequently been discovered; they are scandium (Sc), gallium (Ga), germanium (Ge), and technetium (Tc).

▶ As a crude analogy to the development of the periodic law and the periodic table, we might consider the task of making some semblance of order out of a large number of differently colored marbles of varying sizes. After several false starts, we might try the idea of arranging the marbles according to increasing weight. Let us say that we find that the marble with lowest weight is red, the next pink, and then in the order orange, yellow, green, blue, and violet, according to increasing weight. Then, after searching through all the remaining marbles, we find that the next-heaviest marble is a second red one, and the next a second pink, then another orange, yellow, green, blue, and violet. When the next two marbles are a third red and a third pink, in that order, a definite pattern begins to emerge.

```
Light ─────────────────────────→ Heavy
  │      R   P   O   Y   G   B   V
  │      R   P   O   Y   G   B   V
  ↓
Heavy  R   P
```

A continued search for marbles of increasing weight yields a third orange, green, blue, and violet and a fourth red. When these are placed in the above sequence in the following way,

```
       R   P   O   Y   G   B   V
       R   P   O   Y   G   B   V
       R   P   O   G   B   V   R
```

it is evident that the orderly pattern has been interrupted. But, you might observe, the sequential order would be preserved if a spot were left for a third yellow marble which was not found or was overlooked during the first search. Then you might rearrange the order with a vacant position for the missing yellow marble:

```
       R   P   O   Y   G   B   V
       R   P   O   Y   G   B   V
       R   P   O   -   G   B   V
       R
```

Now the orderly arrangement is restored and can be continued until all the marbles are catalogued. After a more thorough search, you might even find the missing yellow marble. ◀

Not only was Mendeleev able to make predictions about finding new elements (using his periodic table), he also predicted their properties and the properties of their compounds. Table 4–1 gives an outstanding example of these predictions, for the case of the element he called "ekasilicon" (now called germanium, Ge) with atomic weight 72. From Table 4–1 we see how these predictions can be made. The

Table 4–1. Properties predicted by Mendeleev for the element ekasilicon (germanium)

Properties	Tin and its compounds	Silicon and its compounds	Mendeleev's predictions for ekasilicon	Germanium and its compounds
Atomic weight, amu	118.7	28.1	72	72.6
Melting point, °C	232	1410	High	937
Density, g/ml	7.3	2.3	5.5	5.3
Atomic volume	16.3	12.1	13	13.6
Oxide formula	SnO_2	SiO_2	EsO_2	GeO_2
Density of MO_2, g/ml	7.0	2.6	4.7	4.7
Sulfide formula	SnS_2	SiS_2	EsS_2	GeS_2
Chloride formula	$SnCl_4$	$SiCl_4$	$EsCl_4$	$GeCl_4$
Boiling point of MCl_4, °C	114	57.6	100	83

properties of the unknown element and its compounds should fall somewhere between the properties of the elements immediately above and below it on the table. The physical properties of Ge are between those of silicon (Si) and tin (Sn). The compounds of Ge are similar to those of Si and Sn and their properties lie between those of the compounds of Si and Sn.

The Periodic Law. The experiments stimulated by Mendeleev's periodic law and the periodic table, along with improved knowledge of the chemistry of the elements and their compounds and more advanced experimental techniques, uncovered a few discrepancies in Mendeleev's arrangement of the elements. It was found that nickel (Ni), although chemically similar to palladium (Pd) and platinum (Pt), has an atomic weight less than that of cobalt (Co). After the discovery of the noble gases, it was found that Argon (Ar) has a larger atomic weight than potassium (K), which clearly belongs in Group I, as assigned by Mendeleev. Finally, tellurium (Te), which is very similar to sulfur (S) and selenium (Se), has a higher atomic weight than iodine (I), a member of Group VII.

    These apparent exceptions to the periodic law as stated by Mendeleev were accounted for when Henry Moseley discovered that the energies of x-rays emitted by elements in a discharge tube changed in a slightly irregular fashion as the atomic weight of the element changed (see Section 3–4 for a description of Moseley's experiment). On the other hand, Moseley found that the variation of x-ray energies with changing atomic number was smooth and did not contain any irregularities (see Figs. 3–11 and 4–5). The irregularities noted by Moseley occurred with the same element pairs—Co and Ni, Ar and K, Te and I—in which there appeared to be variations in the atomic weights in the periodic table, as mentioned above. It seems, then, that a more consistent basis for the periodic law and the periodic table would be to arrange the elements in order of increasing atomic number rather than increasing atomic weight.

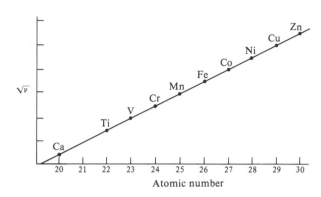

Fig. 4–5 A plot of Moseley's x-ray data for elements of period four. The square root of the emitted x-ray frequency (see Section 5–2 for the relation of energy, frequency, and wavelength) increases systematically as the atomic number increases.

Because of these few inconsistencies in the increasing atomic weight formulation of the periodic law, and because of Moseley's experiments, Mendeleev's periodic law has been changed slightly. The *modern periodic law* states: The properties of the elements vary in a periodic way according to their atomic number.

4–3 THE MODERN PERIODIC TABLE

The modern periodic table was developed from Mendeleev's ideas. Elements are listed in the order of increasing atomic number. Our modern table takes many forms; one of the most common—called the long form—is given in Table 4–2. In this table the elements are arranged in eight main vertical columns called *groups* and seven horizontal rows called *periods*. The elements in the long periods intervening between the main Groups IIA and IIIA are called *transition metals*, and are often designated as Group IIIB, IVB, and so on (see Table 4–2 on pages 82 and 83). The elements in the two long periods that are at the bottom, and separate from the table, are also transition metals and actually occupy positions between atomic numbers 57 and 72 and between 89 and 104. They are generally called the *lanthanide series* and the *actinide series*. Figure 4–6 gives other common group and family names.

Generalizations from the Table. We can use the periodic table to great advantage in systematizing and ordering our chemical knowledge and thinking. The main groups of the table contain elements of similar properties. They occur in vertical columns and are numbered IA, IIA, IIIA, IVA, etc. Then all elements in Group IA (the alkali metals Li, Na, K, Rb, Cs, and Fr, for example), should have similar properties. (Although the element hydrogen is placed in Group IA, its properties are unique and unlike those of the other Group IA members.) The groups designated IB,

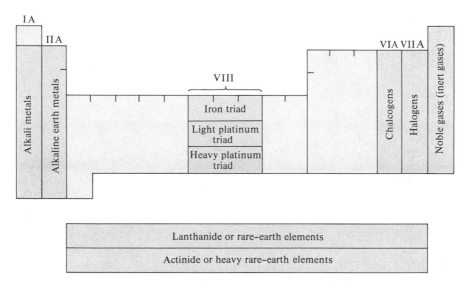

Fig. 4-6 Common names given to some of the element families.

IIB, IIIB, etc., generally have some properties in common with Groups IA, IIA, IIIA, etc., but also often show many properties different from those in the A groups.

▶ The designation of these groups as A or B is somewhat arbitrary. Some chemists prefer a different scheme of A and B subgroups. We must, therefore, be careful that these designations do not get confused in communications. The group at the far right of the table, Noble Gases, is often called Group 0 or VIIIA or VIIIB. ◀

The group called Noble Gases is given a different type of designation, since the members of this group exhibit distinctly different properties. These elements are all gases under ordinary conditions. Until recently it was thought that they did not form normal chemical compounds with any of the other elements. They were therefore often called *inert gases*. It was not until 1962 that the first true compounds with noble-gas elements, primarily between Xe and F, were isolated. Compounds of the noble gases, however, are quite rare; it is still true that the noble gases are relatively inert and do not react with the majority of the elements.

If we could conduct experiments on all the known elements, we would find that they can be divided into three separate categories: metals, metalloids (or semi-metals), and nonmetals. *Metals* are substances that are good conductors of heat and electricity; they also often have metallic luster and can be rolled into sheets (that is, they are *malleable*), or drawn into wire (that is, they are *ductile*). Our experiments show that the elements located on the left of the periodic table are metallic.

Table 4–2. A modern form of the periodic table

Simplified atomic weights for practice in chemical calculations

(*Not sufficiently accurate for quantitative work.*)

Values in parentheses indicate mass numbers of the most stable or best known isotope.

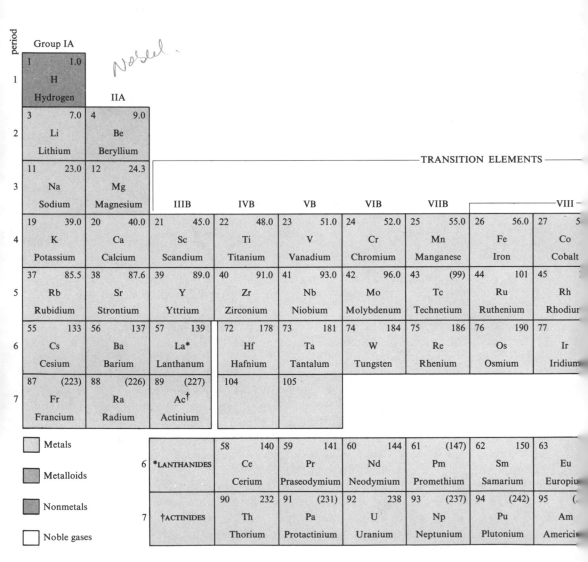

NOBLE GASES

						2 4.0 He Helium
		IIIA	**IVA**	**VA**	**VIA**	**VIIA**	

			IIIA	IVA	VA	VIA	VIIA	NOBLE GASES
			5 10.8 B Boron	6 12.0 C Carbon	7 14.0 N Nitrogen	8 16.0 O Oxygen	9 19.0 F Fluorine	10 20.0 Ne Neon
IB	**IIB**		13 27.0 Al Aluminum	14 28.0 Si Silicon	15 31.0 P Phosphorus	16 32.0 S Sulfur	17 35.5 Cl Chlorine	18 40.0 Ar Argon
58.7 Ni Nickel	29 63.5 Cu Copper	30 65.4 Zn Zinc	31 69.7 Ga Gallium	32 72.6 Ge Germanium	33 75.0 As Arsenic	34 79.0 Se Selenium	35 80.0 Br Bromine	36 84.0 Kr Krypton
106 Pd Palladium	47 108 Ag Silver	48 112 Cd Cadmium	49 115 In Indium	50 119 Sn Tin	51 122 Sb Antimony	52 128 Te Tellurium	53 127 I Iodine	54 131 Xe Xenon
195 Pt Platinum	79 197 Au Gold	80 201 Hg Mercury	81 204 Tl Thallium	82 207 Pb Lead	83 209 Bi Bismuth	84 (210) Po Polonium	85 (210) At Astatine	86 (222) Rn Radon

157 Gd Gadolinium	65 159 Tb Terbium	66 162 Dy Dysprosium	67 165 Ho Holmium	68 167 Er Erbium	69 169 Tm Thulium	70 173 Yb Ytterbium	71 175 Lu Lutetium
(247) Cm Curium	97 (249) Bk Berkelium	98 (251) Cf Californium	99 (254) Es Einsteinium	100 (253) Fm Fermium	101 (256) Md Mendelevium	102 (253) No Nobelium	103 (257) Lr Lawrencium

Nonmetals are substances that are poor conductors, i.e., they are good insulators and do not permit rapid flow of heat or electricity. Generally, they show no metallic properties and occur as powders or gases under normal conditions. The nonmetals are located on the right of the periodic table, and also include hydrogen (H).

Then there is a small group of elements, the *metalloids*, whose properties fall somewhere between those of metals and nonmetals. The boundary between metals and nonmetals is not sharp; it is along this boundary that the metalloids are found. There is some disagreement on how many elements should be classified as metalloids. Generally, the elements B, Si, Ge, As, Sb, and Te are considered metalloids. Under some experimental conditions P, Se, and Bi also show properties of metalloids. Note the diagonal, stepwise relationship of these metalloids. (Whether Po and At are true metalloids or not is not well established. Both elements are radioactive and consequently very difficult to study.)

In summary, we can say that:

1. Elements in the same group in the periodic table generally have similar properties.
2. The elements can be divided into three categories: metals, nonmetals and metalloids.
3. Metals occur at the left of the table; most of the elements are metals.
4. Nonmetals occur at the right of the table, and include hydrogen.
5. Metalloids occur at the metal–nonmetal element boundary in a diagonal, stepwise progression from B to Te (or At). Only a few elements show metalloid properties.

In the next section, the next chapter, and throughout the rest of the book, we shall see how useful the periodic table is in correlating the properties of the elements and their compounds, thereby helping us to understand these properties.

Changes in the Table. It is interesting to note that the periodic table is under almost constant revision and addition. Tables 4–3 and 4–4 (pages 86–89) show periodic tables for the years 1945 and 1970. We can see many changes in atomic weights since 1945 (In, Ta, Tb, Ho, Th, for example) and some changes in element symbols (elements number 18 and 41, for example). But the most striking change is the discovery of new elements 43, 85, 87, and 95 through 103. Elements number 104 and 105 have recently been reported by scientists from both the U.S. and Russia. Experiments designed to confirm these reports are now being conducted. Because of their expected positions in the periodic table (see Table 4–4), elements 104 and 105 should have properties similar to hafnium (Hf) and tantalum (Ta), respectively. Confirming experiments, then, might be based on the expected chemical behaviors of compounds of elements 104 and 105 using the known chemistry of Hf and Ta compounds as analogies.

Table 4–4 gives the predicted positions of elements 104 through 126. In 1945 G. T. Seaborg advanced the hypothesis that the actinide series of elements would be completed by producing *man-made* (i.e., not occurring naturally on earth) elements. In 1966, his prediction of the actinide series fulfilled, Seaborg envisioned production of elements with atomic numbers as high as 118 and possibly as high as 126.

4–4 PERIODIC PROPERTIES

We can get a better feeling for the uses of the periodic table (and also some of the limitations involved) if we consider some of the properties of the elements and their compounds and observe the systematic variations. We have chosen three groups to illustrate periodic variations: Groups IA, IIA, and VIIA. These groups exhibit many interesting and systematic variations, and will serve as an introduction to some of the chemistry that is in later chapters.

Group IA, the Alkali Metals. The *alkali metals* lithium (Li), sodium (Na), potassium (K), rubidium (Rb), cesium (Cs), and francium (Fr) compose the first group at the left of the periodic table. All these elements are metals; they all show similar properties. Many of the properties vary systematically down the group, i.e., from Li to Cs (since all of the isotopes of Fr are radioactive, it has not been as well studied as the other alkali metals and will not be included in the following discussion.)

Table 4–5. Physical properties of alkali metals

Alkali metal	Atomic number	Atomic weight, amu	Atomic radius, Å	Melting point, °C	Boiling point, °C	Electrical resistivity, ohm-cm	Density, g/cc
Li	3	6.941	1.225	179	1331	8.6	0.534
Na	11	22.990	1.572	97.8	890	4.4	0.97
K	19	39.102	2.025	63.2	766	6.6	0.86
Rb	37	85.468	2.16	39.0	701	12.5	1.52
Cs	55	132.906	2.35	28.6	685	19	1.87

Physical properties. Table 4–5 gives some properties of alkali-metal atoms. Inspection of the data shows that the properties of the elements follow certain general trends as the atomic number increases. For example, the atomic radius increases, while the melting and boiling points decrease. These trends are more easily seen when

Table 4–3. Periodic table of 1945

IA									
1 H 1.008	IIA								
3 Li 6.940	4 Be 9.02	IIIB	IVB	VB	VIB	VIIB		VIII	
11 Na 22.997	12 Mg 24.32								
19 K 39.096	20 Ca 40.08	21 Sc 45.10	22 Ti 47.90	23 V 50.95	24 Cr 52.01	25 Mn 54.93	26 Fe 55.85	27 Co 58.94	
37 Rb 85.48	38 Sr 87.63	39 Y 88.92	40 Zr 91.22	41 Cb 92.91	42 Mo 95.95	43	44 Ru 101.7	45 Rh 102.91	
55 Cs 132.91	56 Ba 137.36	57 La* 138.92	72 Hf 178.6	73 Ta 180.88	74 W 183.92	75 Re 186.31	76 Os 190.2	77 Ir 193.1	
87	88 Ra	89 Ac†							

Elements known

Elements predicted

	58 Ce 140.13	59 Pr 140.92	60 Nd 144.23	61	62 Sm 150.43	63 Eu 152.0
*LANTHANIDES						
†ACTINIDES	90 Th 232.12	91 Pa 231	92 U 238.67	93 Np 237	94 Pu	95

			IIIA	IVA	VA	VIA	VIIA	NOBLE GASES
								2 He 4.003
			5 B 10.82	6 C 12.010	7 N 14.008	8 O 6.000	9 F 19.00	10 Ne 20.183
	IB	IIB	13 Al 26.97	14 Si 28.06	15 P 30.98	16 S 32.06	17 Cl 35.457	18 A 39.944
29 Cu 63.57	30 Zn 65.38	31 Ga 69.72	32 Ge 72.60	33 As 74.91	34 Se 78.96	35 Br 79.916	36 Kr 83.7
47 Ag 107.880	48 Cd 112.41	49 In 114.76	50 Sn 118.70	51 Sb 121.76	52 Te 127.61	53 I 126.92	54 Xe 131.3
79 Au 197.2	80 Hg 200.61	81 Tl 204.39	82 Pb 207.21	83 Bi 209.00	84 Po	85	86 Rn 222

(Ni 58.69) (Pd 106.7) (Pt 195.23)

65 Gd 156.9	66 Tb 159.2	67 Dy 162.46	68 Ho 163.5	69 Er 167.2	70 Tm 169.4	71 Yb 173.04	Lu 174.99

Table 4-4. Periodic table of 1970, showing predicted positions of elements 106–126. Note some of the changes in element symbols since 1945, and particularly the numbers of elements that were discovered between 1945 and 1970.

IA	IIA	IIIB	IVB	VB	VIB	VIIB		VIII	
1 H 1.0080									
3 Li 6.941	4 Be 9.01218								
11 Na 22.9898	12 Mg 24.305								
19 K 39.102	20 Ca 40.08	21 Sc 44.9559	22 Ti 47.90	23 V 50.9414	24 Cr 51.996	25 Mn 54.9380	26 Fe 55.847	27 Co 58.9332	
37 Rb 85.4678	38 Sr 87.62	39 Y 88.9059	40 Zr 91.22	41 Nb 92.9064	42 Mo 95.94	43 Tc 98.9062	44 Ru 101.07	45 Rh 102.9055	
55 Cs 132.9055	56 Ba 137.34	57 La* 138.9055	72 Hf 178.49	73 Ta 180.9479	74 W 183.85	75 Re 186.2	76 Os 190.2	77 Ir 192.22	
87 Fr (223)	88 Ra 226.0254	89 Ac† (227)	104	105	106	107	108	109	
119	120	121							

Known in 1945

Discovered after 1945

Predicted

*LANTHANIDES	58 Ce 140.12	59 Pr 140.9077	60 Nd 144.24	61 Pm (147)	62 Sm 150.4	63 Eu 151.96
†ACTINIDES	90 Th 232.0331	91 Pa 231.0359	92 U 238.029	93 Np 237.0482	94 Pu (242)	95 Am (243)
	122	123	124	125	126	

	IB	IIB	IIIA	IVA	VA	VIA	VIIA	NOBLE GASES
								2 He 4.00260
			5 B 10.81	6 C 12.011	7 N 14.0067	8 O 15.9994	9 F 18.9984	10 Ne 20.179
			13 Al 26.9815	14 Si 28.086	15 P 30.9738	16 S 32.06	17 Cl 35.453	18 Ar 39.948
Ni 58.71	29 Cu 63.546	30 Zn 65.37	31 Ga 69.72	32 Ge 72.59	33 As 74.9216	34 Se 78.96	35 Br 79.904	36 Kr 83.80
Pd 106.4	47 Ag 107.868	48 Cd 112.40	49 In 114.82	50 Sn 118.69	51 Sb 121.75	52 Te 127.60	53 I 126.9045	54 Xe 131.30
Pt 195.09	79 Au 196.9665	80 Hg 200.59	81 Tl 204.37	82 Pb 207.2	83 Bi 208.9806	84 Po (210)	85 At (210)	86 Rn (222)
	111	112	113	114	115	116	117	118

Gd 157.25	65 Tb 158.9254	66 Dy 162.50	67 Ho 164.9303	68 Er 167.26	69 Tm 168.9342	70 Yb 173.04	71 Lu 174.97
Cm (247)	97 Bk (247)	98 Cf (249)	99 Es (254)	100 Fm (253)	101 Md (256)	102 No (254)	103 Lr (259)

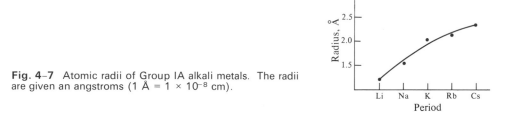

Fig. 4–7 Atomic radii of Group IA alkali metals. The radii are given an angstroms (1 Å = 1 × 10⁻⁸ cm).

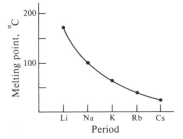

Fig. 4–8 Melting points of Group IA alkali metals.

the data are plotted, as shown in Figs. 4–7 and 4–8 for the atomic radii and the melting points. From these two figures it is clear that the radii systematically increase from Li to Cs and the melting points systematically decrease from Li to Cs. Now trends such as these help us a great deal when they are found, but they also raise an immediate question. Why do the trends occur in this particular way?

We can try to explain these observations using Rutherford's atomic model. In the case of the atomic radii we might expect the atoms to get larger as the number of electrons increases—i.e., with increasing atomic number—since the electrons are all negatively charged and like charges repel. This is the observed trend; Rutherford's model seems to give at least the proper qualitative result.

The trend observed for the melting points, on the other hand, is impossible to explain using Rutherford's model. This is true since the model does not provide any idea of what might be involved when a solid alkali metal melts, producing a liquid. The applicability of Rutherford's atomic model to one set of data but not to another is an example of one of the problems that is constantly with us in science. We think that we understand some observed behavior because it is in accord with an existing theory. Then the very next set of observations cannot be explained using the theory; our thinking has to be revised and a new model devised to fit the experimental data.

We can further illustrate the dilemma presented by Rutherford's model by considering the data on electrical resistivity and density given in Table 4–5 for the alkali metals. It is apparent from Figs. 4–9 and 4–10 that, since the trends shown by the resistivity and the density contain irregularities, they are going to be much more difficult to explain than were the data on the atomic radius and the melting point.

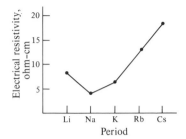

Fig. 4–9 Electrical resistivity of Group IA alkali metals. Resistivity and resistance are directly related. A unit of resistance is called an *ohm.*

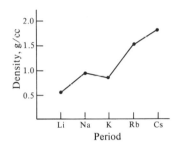

Fig. 4–10 Density of Group IA alkali metals. Potassium seems to have too low a density compared with the rest of the group.

Figures 4–9 and 4–10 indicate two things that are often observed. First, there are exceptions in the systematic trends of the elements. Potassium, for example, seems to have too low a density and lithium too high a resistivity. Second, lithium often exhibits properties that do not follow the trends established by the other alkali metals. This anomalous behavior is not limited to lithium. All the second-period elements, Li through F, seem to behave quite differently in many respects from the other members of their own group. Any model that we have of the atom should, of course, account for these observations; Rutherford's atomic model fails completely.

Chemical properties. The alkali metals are the most active metals known; that is, they react readily with virtually every nonmetallic element. For this reason they are never found in the elemental form in nature. All the alkali metals also react rather violently with water (H_2O), producing hydrogen gas in the process. Preparation of the Group IA elements is generally accomplished by the electrolysis of a molten ore containing a compound of the desired alkali metal.

    The alkali metals form familiar and well-characterized compounds with the Group VIIA elements (the halogens). Common table salt, sodium chloride, is one example; its chemical formula is NaCl. The alkali-metal halides have the general formula MX, where M stands for the alkali metal and X for the halogen that composes the particular compound. Some examples are: LiF (lithium fluoride), NaI (sodium iodide), KBr (potassium bromide), RbCl (rubidium chloride), CsI (cesium iodide), etc. Note that all the compounds contain one alkali-metal atom combined with one

halogen atom. If we would prepare the alkali halides in the following way,

$$2M + X_2 \rightarrow 2MX,$$

we would obtain two moles of alkali-metal halide from two moles of alkali-metal atoms and one mole of halogen molecules.

Table 4–6 gives some properties of the alkali-metal chlorides.

Table 4–6. Properties of alkali-metal chlorides

Property	Color	Melting point, °C	Boiling point, °C	Density, g/cc
LiCl	White	610	1382	2.07
NaCl	White	808	1465	2.16
KCl	White	772	1407	1.99
RbCl	White	717	1381	2.76
CsCl	White	645	1300	3.97

Here, just as with the alkali-metal atoms, some property trends are evident. The melting points of the various halides are plotted in Fig. 4–11. Again we find that Li behaves in an anomalous way; the melting point of LiCl is much lower than might be expected. Examine the rest of the data in Table 4–6. Can you find other trends? How about some exceptions? It would be interesting to try to look into these trends, and any exceptions to them, in more detail.

Group IIA, the Alkaline-Earth Metals. The *alkaline-earth metals*—beryllium (Be), magnesium (Mg), calcium (Ca), strontium (Sr), barium (Ba), and radium (Ra)—are all very active metallic elements. They resemble the Group IA alkali metals in many of their properties and reactions. The solid elements have a high metallic luster and are good conductors. They are harder than the alkali metals and, as shown in Table 4–7, have higher melting and boiling points than their Group I neighbors. (Radium is radioactive. Data for it are not included.)

Table 4–7. Physical properties of the alkaline-earth metals

Property	Atomic number	Atomic weight, amu	Atomic radius, Å	Melting point, °C	Boiling point, °C	Density, g/cc
Be	4	9.01	0.889	1285	2477	1.86
Mg	12	24.30	1.364	650	1120	1.74
Ca	20	40.08	1.736	845	1492	1.54
Sr	38	87.62	1.914	757	1370	2.60
Ba	56	137.34	1.981	710	1638	3.74

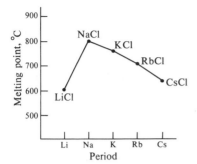

Fig. 4–11 Melting points of alkali-metal chlorides.

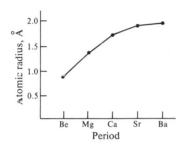

Fig. 4–12 Atomic radii of Group IIA alkaline-earth metals (1 Å = 1 × 10⁻⁸ cm).

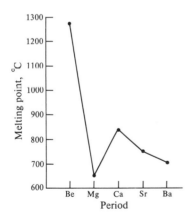

Fig. 4–13 Melting points of Group IIA alkaline-earth metals. Melting points of Be and Mg seem to be too high and too low, respectively.

From Table 4–8 and plots of the atomic radii and the melting points of the alkaline-earth metals in Figs. 4–12 and 4–13, respectively, we can see that the trends in the atomic properties of Group II are less general than in Group I. Again there are interesting exceptions to smoothly sysematic variations. Note the very high melting point of Be and the relatively low melting point of Mg.

Chemical properties. The alkaline-earth metals, just as do the alkali metals, react readily with most nonmetallic elements and also with H_2O, although not with the rapidity or violence of their Group I neighbors. Although essentially all alkali-metal compounds dissolve well in water, some common alkaline-earth compounds—such as MgO (magnesium oxide), CaF_2 (calcium fluoride), $SrCO_3$ (strontium carbonate), and $BaSO_4$ (barium sulfate)—do not.

The oxides of the alkaline-earth metals can be prepared by heating the metal in a stream of oxygen (O_2) gas,

$$2M + O_2 \rightarrow 2MO,$$

or by decomposing the metal carbonates (MCO_3),

$$MCO_3 \xrightarrow{\text{heat}} MO + CO_2.$$

Some examples of alkaline-earth oxides are: BeO (beryllium oxide), CaO (calcium oxide), BaO (barium oxide), etc. Note the 1:1 alkaline-earth-to-oxygen atomic ratio. All the oxides have high melting points (they are refractory), as given in Table 4–8, which also lists other physical properties. Figure 4–14a shows the dependence of the oxide melting points on atomic number.

Table 4–8. Properties of oxides of alkaline-earth metals

Alkaline-earth metal oxide	Color	Melting point, °C	Density, g/cc
BeO	White	2820	3.01
MgO	White	3075	3.58
CaO	White	2860	3.35
SrO	White	2730	4.7
BaO	White	2196	5.72

Table 4–9. Properties of alkaline-earth metal chlorides

Alkaline-earth chloride	Color	Melting point, °C	Boiling point, °C	Density, g/cc
$BeCl_2$	White	410	492	1.899
$MgCl_2$	White	714	1418	2.316
$CaCl_2$	White	782	2000?	2.512
$SrCl_2$	White	873	2000?	3.052
$BaCl_2$	White	960	1830?	3.856

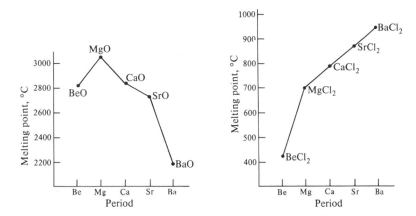

Fig. 4-14 Melting points of (a) alkaline-earth metal oxides and (b) alkaline-earth chlorides. In both cases the Be compound exhibits unsystematic behavior.

The halides of the Group IIA metals have the general formula MX_2, CaF_2 (calcium fluoride), $SrCl_2$ (strontium chloride), BaI_2 (barium iodide), etc. Table 4–9 lists some properties of the alkaline-earth metal chlorides, and Fig. 4–14b shows the variation of their melting points.

If we compare the variations in the melting points of the Group IA chlorides (Fig. 4–11) and the Group IIA oxides and chlorides (Fig. 4–14a and b), some interesting variations are evident. As we said before, the properties of lithium and beryllium, which belong to the second period, are often exceptions to the group trends. This is shown by the relatively low melting points of LiCl, BeO, and $BeCl_2$. We find an unusual behavior if we compare all three figures, since the melting points of NaCl through CsCl and MgO through BaO decrease, while the melting points of $MgCl_2$ through $BaCl_2$ increase.

We can also make another interesting observation. An alkali-metal halide, such as NaCl, is found experimentally to have an atom ratio of 1 : 1, whereas experiments show that alkaline-earth metal halides, such as $MgCl_2$, have atom ratios of 1 : 2.

Again we have the possibility of explaining some very interesting variations of properties. Why should the behavior of Li and Be be so different from that of the other members of their groups? Why are there such unusual differences in the variations of the melting points of the Group IA and Group IIA chlorides? What determines the combining ratio of the atoms? Why NaCl and not $NaCl_2$, or why $MgCl_2$ and not $MgCl_3$? We need an atomic model that will answer these questions; Rutherford's model is inadequate.

Group VIIA, the Halogens. The *halogens*—fluorine (F), chlorine (Cl), bromine (Br), iodine (I), and astatine (At)—are all active nonmetals, reacting with most of the metallic elements. At room temperature and pressure, fluorine and chlorine are gases, bromine

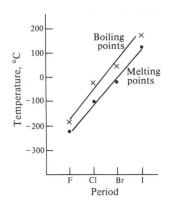

Fig. 4-15 Melting points and boiling points of Group VIIA halogens.

is a liquid, and iodine is a solid (which can easily be melted and vaporized). All the halogens form colored, diatomic gases of the general formula X_2 (Cl_2, I_2, etc.). Table 4–10 gives some properties of the elements, and Fig. 4–15 shows the variations in melting and boiling points. (Astatine is radioactive; its properties are not included.)

Note some of the systematic trends in the properties, and especially the increase in melting point from F to I (compare with the variations in melting point shown for the Group IA and IIA metals in Figs. 4–10 and 4–15).

Table 4-10. Properties of the halogens

Halogen	Atomic number	Atomic weight, amu	Atomic radius, Å
F	9	18.998	0.72
Cl	17	35.453	1.00
Br	35	79.904	1.14
I	53	126.904	1.35

Halogen	Melting point, °C	Boiling point, °C	Density, g/cc	Gas color
F	−219.6	−187.9	1.11	F_2, pale yellow
Cl	−102.4	−34.0	1.57	Cl_2, yellow-green
Br	−7.2	58.2	3.14	Br_2, brown-red
I	113.6	184.5	4.94	I_2, violet

Chemical properties. The halogens exhibit many useful and interesting chemical properties. We shall mention only a few here; others will be considered later. Hydrogen gas (H_2) reacts with the halogens to form the hydrogen halides according

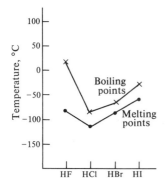

Fig. 4-16 Melting points and boiling points of hydrogen halides.

to the following general equation:

$$H_2 + X_2 \rightarrow 2HX.$$

Hydrogen fluoride (HF), hydrogen chloride (HCl), hydrogen bromide (HBr), and hydrogen iodide (HI) are all colorless gases that have sharp, pungent odors. They all dissolve readily in water, forming a series of acids; for example, HCl dissolved in water is called hydrochloric acid. Note the variations of the melting and boiling points of the hydrogen halides given in Fig. 4-16. Again the second-period element, F, shows behavior that is unusual for the group.

Since the halogens form compounds with so many other elements, we can also consider a variation of properties across a period (rather than within or between groups). Table 4-11 gives the melting points of the compounds formed by the reaction of fluorine with the elements of period three, sodium through chlorine.

Table 4-11. Melting points of period-three fluorides

Compound	NaF	MgF$_2$	AlF$_3$	SiF$_4$	PF$_5$	SF$_6$	ClF
Melting point, °C	995	1263	1290	−90.3	−83	−51	−155.6

Table 4-11 shows two very interesting things. One is the systematic change in the combining ratio of the atoms. Progressively, the compounds contain 1, 2, 3, 4, 5, 6, and 1 fluorine atoms. Second, there is a dramatic change in the melting points of the compounds between the Group IIIA compound, AlF$_3$, and the Group IVA compound, SiF$_4$. The fluorides of Groups IA, IIA, and IIIA have quite high melting points; the fluorides of Groups IVA, VA, VIA, and VIIA have very low melting points.

As with all of the other trends that we have encountered, these with the halogens and their compounds are helpful and interesting. And again they raise the question: Why do they occur?

4–5 PERIODIC TRENDS AND THE ATOM

We have seen that the periodic table is very useful in systematizing our chemical knowledge. Metallic elements are found on the left of the table, nonmetallic elements on the right, and metalloids along their boundary. The elements in any one of the main groups (IA, IIA, IIIA, IVA, VA, VIA, VIIA) all have similar properties and form similar chemical compounds. The properties of the elements and their compounds generally vary in a systematic way within a group. For example, the atomic radii increase down a group, with increasing atomic number. But there are often exceptions in these systematic trends, particularly with the second-period elements (Li through F).

As illustrated by the properties shown in Table 4–11, compounds formed between metallic and nonmetallic elements are much more stable (have higher melting points, as well as other markedly different properties) than those formed between metalloids and nonmetals or between two nonmetals. Also, there appears to be a systematic change in the combining ratios of F with the period-three elements: one atom of F in the Group IA compound, two atoms of F in the Group IIA compound, and so on through Group VIA.

Not only do these periodic trends help us in coordinating our chemical knowledge and our thinking, they also make predictions possible. For example, if we were trying to determine the melting points of the radioactive elements Fr and At, we could predict from Figs. 4–8 and 4–15, respectively, that the melting point of Fr would probably be about 25 °C and that of At about 240 °C. We could also safely expect that Fr would form a halide FrX (1 : 1 atom ratio) and, from Fig. 4–11, we could predict that FrCl would have a melting point somewhere around 580 °C. We can make many more predictions and draw many more correlations just from the data in this chapter. For example, what might be the melting point of element 119 or the atomic radius of Ra or the melting point of the oxide of element 120 or the formula of the chloride of Ge, and so on? Try a few predictions. Then check and see (if the experimental data are known) how close your estimates were.

Our real goal, though, is to understand the reasons behind these systematic trends and their exceptions. Rutherford's model of the atom offers very little, if any, help in understanding group trends, differences between or exceptions to group trends, the reasons for metallic and nonmetallic type behavior of the elements, the differences in combining ratios between various elements, the apparent systematic changes in combining ratios, etc. In short, Rutherford's atomic model fails to explain the observed periodic properties of the elements and their compounds. In the next chapter the need for a new atomic model will become even more apparent.

QUESTIONS

1. Define the following terms: Group, period, transition metal, noble gas, alkali metal, alkaline-earth metal, halogen, metal, nonmetal, metalloid, diatomic gas, atomic number, active metal.

2. Define periodic law. How has the definition changed since the time of Mendeleev?

3. How many elements belong to Group IIIA? Write the symbols and the corresponding names for all these elements.

4. a) How many periods make up the periodic table?
   b) How many elements are contained in period four? Name the elements and give their corresponding symbols.

5. Discuss the relationships, if any, between Döbereiner's triads, Newlands' octaves, and Mendeleev's periodic table.

6. Alkaline-earth metals form with sulfur, compounds with 1:1 atomic ratios.

   a) Write an equation for the reaction between magnesium and sulfur atoms that produces magnesium sulfide.
   b) How many moles of magnesium are needed to make 0.10 mole of the product in part (a)?
   c) How many grams of magnesium sulfide are formed by reacting 0.20 mole magnesium and 6.4 g sulfur?
   d) How many grams of magnesium sulfide are formed by reacting 0.20 mole magnesium and 12.8 g sulfur?
   e) Explain why the answers to parts (c) and (d) are the same or why they are different.

7. a) Write the equation for the reaction of each alkali metal with chlorine gas.
   b) Write the equation for the reaction of each alkaline-earth metal with oxygen gas.
   c) Write the equations for the reaction of aluminium metal with oxygen gas and with chlorine gas.
   d) Write the equations for the reaction of indium metal with argon gas and with sulfur.

8. Examine Mendeleev's first periodic table, given in Fig. 4–4. Which elements appear to fall in the wrong groups? Suggest some rearrangements that would eliminate these difficulties. Justify your suggestions.

9. a) Predict some properties of the element radium (Ra); be as quantitative as possible.
   b) Predict the melting points of $RaCl_2$ and RaO.
   c) Would the melting point of RaI_2 be higher or lower than that of $RaCl_2$?

10. Which of the following compounds would you expect to have high melting points (considerably above room temperature) and which low melting points (below room temperature)?

    a) NaI b) CaO c) SO_2
    d) NO e) $ScCl_3$ f) CO_2
    g) HF h) LiF i) NH_3
    j) H_2S k) $BrCl$ l) WO_3
    m) SrF_2 n) Sm_2O_3 o) SO_3

11. Identify the following elements as metals, nonmetals or metalloids.

    a) N   b) Sb   c) Cd
    d) Re   e) Eu   f) Ca
    g) Br   h) Ne   i) C
    j) Al   k) As   l) Fe
    m) Cs   n) Th   o) Fm

12. An unknown element is shiny and is a good conductor of electricity. It reacts readily with Cl_2 to form a compound MCl that has a melting point of 610 °C and a density of 2.07 g/cc.

    a) From the data given in this chapter, predict the identity of element M and compound MCl.
    b) Write an equation for the reaction producing MCl.

c) How many grams of MCl could be obtained from 0.40 mole of the unknown element?

d) What would be the volume of the MCl produced in part (c)?

13. Balance the following equations.

a) $Na + Br_2 \rightarrow NaBr$ b) $Ca + Br_2 \rightarrow CaBr_2$ c) $Al + Br_2 \rightarrow AlBr_3$

d) $Cu + Cl_2 \rightarrow CuCl$ e) $Cd + Cl_2 \rightarrow CdCl_2$ f) $Y + Cl_2 \rightarrow YCl_3$

g) Assign the metals in parts (a) through (f) to their appropriate groups in the periodic table. Note the similarities and the systematic variations.

14. Balance the following equations.

a) $K + O_2 \rightarrow K_2O$ b) $Ba + O_2 \rightarrow BaO$ c) $La + O_2 \rightarrow La_2O_3$

d) $Ga + O_2 \rightarrow Ga_2O_3$ e) $Si + O_2 \rightarrow SiO_2$ f) $P + O_2 \rightarrow P_2O_5$

g) $S + O_2 \rightarrow SO_3$ h) $Ti + O_2 \rightarrow TiO_2$ i) $Zn + O_2 \rightarrow ZnO$

j) $W + O_2 \rightarrow WO_3$

k) Arrange the elements in parts (a) through (j) according to groups in the periodic table. Note any similarities and systematic variations. Compare with question 13.

15. In what groups would the unknown elements, M or X, in the following compounds be located (indicate all reasonable possibilities).:

a) M_2O_3 b) MCl c) BaX_2

d) Al_2X_3 e) MO f) MF_2

g) MO_2 h) SiX_4 i) HX

j) $MClF$

16. The atomic radii of some of the fourth-period elements are as follows.

Element	K	Ca	Sc	Ti	V	Cr	Mn	Fe	Co	Ni
Atomic number	19	20	21	22	23	24	25	26	27	28
Radius, Å	2.35	1.97	1.62	1.47	1.34	1.27	1.26	1.26	1.25	1.24

Identify a trend in these atomic radii and then discuss the trend, using Dalton's atomic theory and Rutherford's model of the atom. Include any appropriate explanations of the trend.

17. From their positions on the periodic table, which of the following elements would you expect to be good conductors of heat and electricity? Ti, Ne, N, Au, Tb, Br, Mo

18. Which of the following elements are gases under ordinary conditions? Ta, Ba, Pt, Fe, H, Ge, Sn, N

19. Discuss the effect that Henry Moseley's work had on the periodic law established by Mendeleev.

20. Explain why K is placed in Group IA rather than Ar, even though Ar has the larger atomic weight.

21. a) What properties might be envisioned for the elements with atomic numbers 117 and 119?

b) Indicate some compounds that these elements might form. Predict some of their properties.

22. Write an equation for the reactions between Al and S, In and O_2, Cd and S, La and O_2.

23. Which element exists as a diatomic gas under normal conditions, reacts with Mg to form a compound MgX, and reacts with Ga to form a compound Ga_2X_3?

24. Complete and balance the following equations:

a) $Be + O_2 \rightarrow$ b) $Mg +$ $\rightarrow MgS$ c) $Zn + O_2 \rightarrow$

d) $Cd + S \rightarrow$ e) $Li + Cl_2 \rightarrow$ f) $+ Cl_2 \rightarrow CuCl_2$

g) $Cd +$ $\rightarrow CdF_2$ h) $Sr + Br_2 \rightarrow$

i) Why does equation (f) illustrate an exception to the generally observed group reactions?

25. Using the periodic table and the data given below, predict the values that are missing from the tables by plotting the appropriate graphs.

a)

Noble gas	He	Ne	Ar	Kr	Xe	Rn
Melting point, °K	1.0	24.6	83.8	—	161.7	202
Boiling point, °K	4.23	—	87.3	121	166.1	208

b)

Substance	Melting point, °C	Boiling point, °C
H_2O	0	——
H_2S	−83	−60.8
H_2Se	——	−41.5
H_2Te	−51	−1.8

c)

Substance	Melting point, °K	Boiling point, °K
ClF	118	173
Cl_2	170	—
BrCl	—	278
ICl	300	373

5. ATOMIC PROPERTIES AND THEORIES

5-1 EXPERIMENTS AND THEORIES

In Chapter 4 we considered the periodic properties of the elements. We saw that these observed periodic trends could not be explained by any of the atomic theories existing during Rutherford's time. Another set of experimental results, concerning the way that light is absorbed and emitted by atoms, was equally perplexing to scientists during the late 1800's and early 1900's. They found that under the proper experimental conditions, atoms could be observed to absorb or give off light in a discontinuous fashion. When an atom emitted light, it was generally not white light, but light of a particular color. Burning sodium (Na), for example, emitted yellow light. Neon (Ne) produced a reddish glow in a discharge tube. Such optical properties of atoms could not be explained by Rutherford's atomic model, or by any other theory of that time. However, when Niels Bohr proposed a new model for the atom, the foundations for understanding both the periodic and the optical properties of atoms were laid. Bohr's ideas proved to be the basis for our current theories and understanding of the atom.

5-2 LIGHT AND ITS INTERACTION WITH MATTER

We are all familiar with visible light, and we also know that white light, such as that emitted by the sun, is composed of various colors (red, orange, yellow, green, blue, etc.). One can produce these colors, as in the case of a rainbow, by passing white light

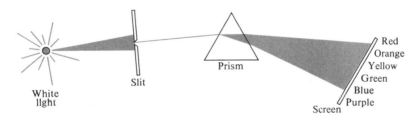

Fig. 5–1 A continuous spectrum of white light from a light bulb or from the sun produces a band of colors. There are no breaks observed in the spectrum.

Fig. 5–2 Representations of a light wave (with wavelength λ) and a water wave.

through a prism. As Fig. 5–1 shows, this produces a *continuous* spectrum, i.e., one in which there is a gradual change from one color to another. The various colors that are obtained can be characterized by their *wavelengths*, λ (the Greek letter lambda). Prior to 1900 scientists had already described light as oscillating electric and magnetic fields that have characteristic wavelengths (and travel with a constant velocity, $c = 2.9979 \times 10^{10}$ cm/sec).

Figure 5–2 is a representation of a light wave, which is similar in some respects to waves in a pool of water. If a pebble is thrown into the pool, small water waves emanate from the "plunk point" of the pebble. Both water waves and light waves have alternating crests and valleys and, consequently, characteristic wavelengths.

Just as the water waves in the pool can have various wavelengths (different distances between crests and valleys), light is composed of waves with continuously varying wavelengths. In fact, *visible light*—i.e., light that we can see with our eyes—is only a very small segment of the entire collection of wavelengths (generally called the *electromagnetic spectrum*) that are possible.

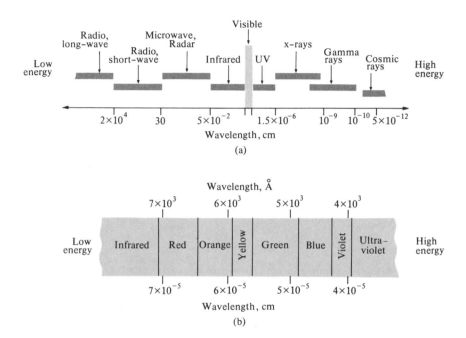

Fig. 5-3 (a) The electromagnetic spectrum. There is a continuous variation in wavelength from the long-wavelength, low-energy radio waves to the short-wavelength, high-energy cosmic rays. (b) An expanded picture of the visible region of the electromagnetic spectrum. Note how small the visible section is compared with the entire spectrum.

Figure 5-3 presents the electromagnetic spectrum, plus an expanded diagram of the visible portion of it. We can see that the various colors of visible light have different wavelengths and that x-*rays* and *infrared* rays are simply light rays whose wavelengths are different from those of visible light; x-rays have shorter wavelengths and greater energy than visible light.

An important relationship between the energy and the wavelength of electromagnetic radiation was developed by Max Planck in 1900 and Albert Einstein in 1905. Their work led to the proposal that light is composed of small bundles of energy, called *photons*, that have discrete wavelengths and obey the equations

$$E = \frac{hc}{\lambda} \quad \text{or} \quad E = h\nu,$$

where c is the velocity of light, λ the wavelength of light, ν (the Greek letter nu) the frequency of light, and h is a universal constant (i.e., it is a fixed, unchanging number) called *Planck's constant*.

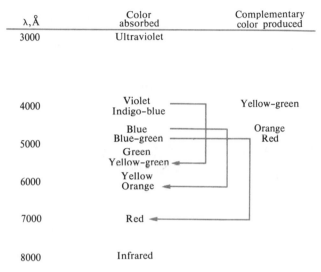

λ,Å	Color absorbed	Complementary color produced
3000	Ultraviolet	
4000	Violet Indigo–blue	Yellow–green
5000	Blue Blue–green	Orange Red
	Green Yellow–green	
6000	Yellow Orange	
7000	Red	
8000	Infrared	

Fig. 5–4 The absorption (removal) of one color from white light produces the complementary color.

▶ The frequency v of the radiation is related to its wavelength by the relation $v\lambda = c$. Note that v and λ are inversely proportional to one another. With λ measured in cm and the velocity of light in cm/sec, the units of v are Hertz (Hz) or cycles per second. ◀

Absorption of Light. We know that things differ vastly in color: Diamonds are clear and colorless, while rubies are deep red; glass can be tinted with many colors; water solutions can be made in any desired color. We are now in a position to begin to understand *why* these various colors exist. Different wavelengths—or colors—of light are absorbed by different substances. The preferential absorption of certain wavelengths of light, with the remaining light being transmitted, creates a colored material. Red light passes through a ruby, while blue-green light is absorbed; a deep blue water solution transmits blue light and absorbs yellow and orange light (see Fig. 5–4). We can conduct quantitative experiments to determine precisely which wavelengths of light are absorbed by a given sample. The record of these wavelengths, often a trace on a recorder graph or a photograph, is called an *absorption spectrum.*

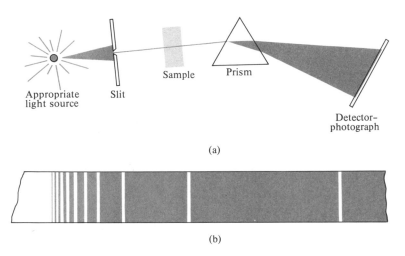

(a)

(b)

Fig. 5–5 (a) The apparatus needed to determine an absorption spectrum. (b) The absorption spectrum of hydrogen atoms as it would appear if a photographic detector were used.

The apparatus for experimentally determining absorption spectra is diagrammed in Fig. 5–5. A narrow beam of light passes through the sample and then onto a detector. Figure 5–5 also gives a representation of an absorption spectrum obtained for a hydrogen atom sample. Note that the spectrum obtained was not continuous. Instead, a series of lines was observed that had progressively smaller and smaller spacings until they merged into a continuous band. This type of spectrum is called a *line spectrum*, since it is discontinuous and is composed of lines.

Similar experiments have been conducted with other elements such as He, Li, Na, Mg, etc. Each element has its own characteristic absorption spectrum. The absorption lines recorded for H are different from those found for He or Li; the lines of He are different from those of Li or Na, and so on.

Emission of Light. We also know that under certain conditions substances emit light as well as absorb it. The zinc sulfide (ZnS) detectors that J. J. Thomson and Lord Rutherford used in their experiments, for example, emitted bright yellow-green light when struck by high-energy particles. To study these emission phenomena, scientists conducted experiments with various elements to determine the wavelengths of the light that they emitted. As in the case with absorption spectra, the emission of light was found to occur in discrete lines—that is, at certain definite wavelengths—

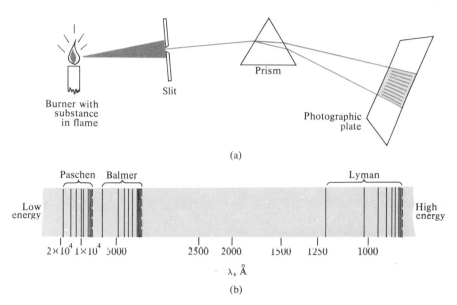

(a)

(b)

Fig. 5–6 (a) Apparatus for determining an emission spectrum. A substance such as an alkali metal is burned in a flame. The emitted light is passed through a prism—which further separates the emitted wavelengths of light—and onto a detector. (b) The emission spectrum of hydrogen atoms as detected by a photographic plate. The emission lines appear dark against a bright background. Note the three series of lines. The high-energy (long-wavelength) limit for each series is indicated by the dashed line. The region between the last line drawn and the dashed line is composed of a series of very closely spaced lines (not shown).

rather than in continuously varying bands. A record of these emission lines is called an *emission spectrum*.

Materials can be made to emit light in several different ways. Probably the most common are the absorption of high-energy radiation, such as ultraviolet light, or the heating of the substance to a high temperature. Figure 5–6a shows the apparatus necessary to determine the emission spectrum of a sample by burning it in a flame. This is a good method for the alkali metals, for example. Sodium emits bright yellow light, potassium pale violet; lithium, calcium, and strontium are all red emitters, while barium produces green light.

Hydrogen atoms have the simplest emission spectrum. Around 1900 two sets of hydrogen emission lines were known; they had been studied by J. J. Balmer and F. Paschen. Figure 5–6b is a representation of an emission spectrum for hydrogen, containing these sets of lines. The Balmer series lies in the visible portion of the electromagnetic spectrum and was therefore the first series investigated in detail. The Paschen series lies in the infrared.

5-3 ATOMIC MODELS AND OPTICAL SPECTRA

Not only must a model of the atom explain the observed periodic properties of the elements (as discussed in Chapter 4), but it must also account for the observed line spectra of the elements. Why do hydrogen atoms absorb energy at only a few discrete wavelengths, instead of continuously over all energy ranges? Why should elements emit radiation of characteristic, discrete wavelengths instead of emitting a continuous band of radiation? Why do the observed emission lines for hydrogen atoms occur in several groups? Why is the visible emission of lithium in the red, that of sodium in the yellow, and that of potassium in the violet?

We need to answer these questions, and many more. But our model of the atom is inadequate. The Rutherford atom is too simple; it needs to be revised, added to, or changed.

▶ There was another difficulty with the Rutherford atom. Experimentally, and theoretically, it could be shown that a system consisting of a charged particle orbiting around an oppositely charged particle was unstable. Ultimately the two opposite charges unite. In the case of an atom, the electrons would circle closer and closer to the nucleus until they were attracted into it (see Fig. 5-7). An atomic model that predicts gradual decay and final destruction for the atom is obviously not completely correct. ◀

The man who bridged this gap between the known experimental observations and Rutherford's model of the atom was Niels Bohr. In 1913 Bohr used Rutherford's basic picture of the atom, ingeniously combined it with Planck's theory of radiation, and produced a beautifully simple picture of the atom that solved many of the difficulties of the time and formed the basis for our current theories of the atom and its interactions.

5-4 THE BOHR ATOM

Basically Bohr's model of the atom was similar to that of Rutherford. Bohr retained the picture of a very small, positively charged nucleus that contained all the protons and neutrons of the element and hence most of the mass of the atom. He also placed the negatively charged electrons outside the nucleus, composing most of the volume of the atom. He envisioned the electrons as being similar to the planets rotating around the sun.

It was at this point that Bohr took a gaint step forward by applying the ideas concerning electromagnetic radiation developed by Planck and Einstein to the structure of the atom. It had been proposed that energy was radiated in discrete bundles of energy, $E = h\nu$. These bundles were characterized by their frequency ν, and called *quanta*. Bohr assumed that the electrons in an atom could travel around the nucleus in circular paths, or *orbits*, having only certain discrete energies. The energy of an electron located in one of these orbits is *quantized*, i.e., it has a specific energy, and we

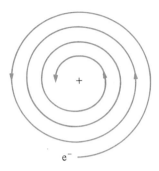

Fig. 5-7 The collapse of the Rutherford atom. As it travels around the nucleus, the electron radiates energy and approaches the nucleus in a spiral path.

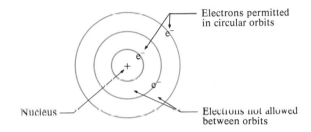

Fig. 5-8 The Bohr atom. The nucleus contained the protons and neutrons. Electrons circled the nucleus only in shells of certain definite radii and were not allowed between these shells.

call these electron orbits *energy levels*. Note that electrons could exist only in these orbits with definite energies.

As Fig. 5-8 shows, there were gaps between the orbits where no electrons were, or could be, located. It was this condition that made possible Bohr's explanation of line spectra. He assumed that electron transitions between orbits involved bundles of energy of characteristic frequency v. The electron either absorbs this definite amount of energy from some external source (which causes it to jump to a higher level of energy) or it emits a definite amount of energy (which corresponds to an electron falling to a lower level of energy).

The Hydrogen Atom. Bohr first applied his theory to the simplest atom known, the hydrogen atom. Let us consider Bohr's postulates, using the hydrogen atom as an example.

Bohr's atomic model involves the following postulates:

1. The electron travels around the nucleus in circular orbits that have certain definite energies.

2. The energy of these quantized orbits is characterized by an integer n, called the *principal quantum number*, which can have any integral value from 1 to infinity: $n = 1, 2, 3$, etc.

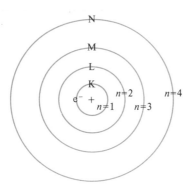

Fig. 5–9 Bohr's model of the hydrogen atom.

3. In order to change energies, an electron must move from one energy level to another. No gradual changes are allowed.

4. Upon changing energy levels, an electron absorbs or emits energy equal to the difference in energy between the levels.

Figure 5–9 is a diagram of the hydrogen atom. The circular electron energy levels should be thought of as spheres or shells, with the nucleus at the center. For this reason they are often designated as the K, L, M, N, O, etc., *shells*. Since the energy of an electron in any given level or shell is determined by the principal quantum number n, the energy levels are also designated as *principal quantum level* 1 (or 2, 3, 4, etc.) corresponding to $n = 1, 2, 3, 4$, etc.

We can represent the various energy levels in a more convenient way. Instead of drawing each entire orbit, we use only a short segment of the orbit. This is called an *energy level diagram*, and is pictured in Fig. 5–10. The energy level closest to the nucleus is taken to be the level with the lowest energy. An electron in this energy level is as close to the nucleus as possible, and is therefore attracted most strongly by the positive nucleus charge. This is the most stable arrangement possible, i.e., the hardest to disturb; it is assigned the lowest energy, and often called the *ground state* of the atom. Electrons located in energy levels that are farther from the nucleus have increasingly greater energies.

Using Figs. 5–9 and 5–10 for reference, let us consider Bohr's postulates and the hydrogen atom, which has only one electron and one proton. The electron travels around the proton (nucleus) in any of the circular orbits or energy levels. It is never found anywhere but in one of the orbits. When the electron is in the energy level closest to the nucleus—the K-shell with $n = 1$—opposite charge attraction and atom stability are greatest. The atom is then in its lowest energy (or ground) state.

But the hydrogen atom can also exist in higher energy states, because the electron can be located in a higher energy level (one farther from the nucleus). After a hydrogen atom in its ground state absorbs just the right amount of energy to transfer its

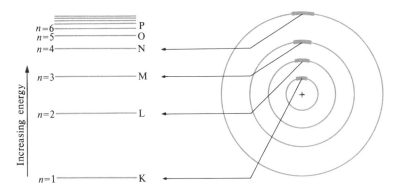

Fig. 5–10 An energy-level diagram and the corresponding Bohr orbits.

electron from the first shell (K) to the second shell (L), it exists in a higher energy state (an *excited state*), with the electron located at a distance farther from the nucleus than in the lower energy state. The transfer of an electron to a higher energy level requires that the electron absorb energy from some external source; this transfer creates an atom with higher energy, and leaves an *electron hole* or *vacancy* in the lower energy level. If the electron falls back into this vacant position from a location in a higher energy level, energy is emitted from the hydrogen atom to its surroundings. The energy of the atom is decreased; the atom becomes more stable.

▶ A crude analogy to electron energy levels in an atom is an ordinary stepladder. You can stand on the steps of the ladder, but not in between the steps, just as an electron exists in energy levels but not between them. You feel most secure and stable on the lowest step, as an electron does in its lowest energy level or ground state. When you climb the ladder, you leave a vacant step behind, one to which you can return, as an electron does when it absorbs energy and jumps to a higher energy level. Climbing the ladder also increases your energy, which you can demonstrate by jumping from successive steps. The higher the step, the harder is your impact with the ground. Similarly with an electron, the farther it is from the nucleus, the higher its energy and the greater the energy released when it falls to a lower level. ◀

Although Bohr's postulates originally concerned only the hydrogen atom, their application is now generalized to include the atoms of all the elements. The same physical pictures that were developed for the hydrogen atom (see Figs. 5–9 and 5–10, for example) apply to atoms that have many electrons, protons, and neutrons.

We should also note that Bohr's theory of the atom removes, at least partially, some of the objections to—and difficulties with—Rutherford's model of the atom. When we use Bohr's theory, we have at least a basis for understanding the optical emission and absorption line spectra of the elements and the periodic properties of the elements.

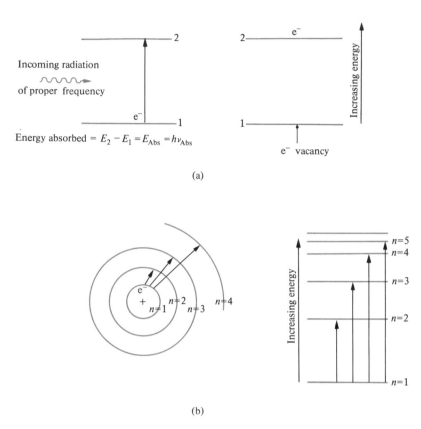

Fig. 5–11 (a) Absorption spectra. An electron at level 1 absorbs energy from incoming radiation (one form would be light) and is transferred to level 2, at a higher energy, leaving level 1 unoccupied. The energy absorbed is $E_2 - E_1$. The energy of the incident radiation must match this energy difference, giving the radiation a characteristic wavelength or frequency ($E_{\text{absorbed}} = E_2 - E_1 = h\nu_{\text{absorbed}}$). (b) Bohr's explanation of absorption spectra. Electrons are excited to higher energy levels. Absorption lines of characteristic wavelengths would be found for electron jumps between energy levels $1 \to 2$, $1 \to 3$, $1 \to 4$, $1 \to 5$, etc.

5–5 THE BOHR ATOM AND OPTICAL SPECTRA

Absorption and Emission Spectra. A hydrogen atom in its lowest energy state (or any other energy state) can absorb energy from its surroundings; this absorption transfers (or excites) the electron into a higher energy level. Figure 5–11 is a diagram of this process.

Once the electron has jumped to this upper level or excited state, there is an electron vacancy at the lower energy level. This is an unstable situation. There is a great attraction for the electron to return "home." This transition from an upper to a lower

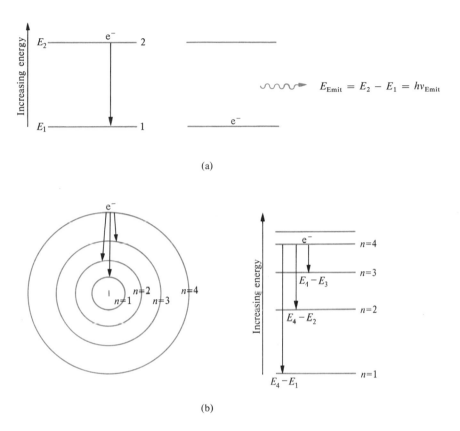

Fig. 5–12 (a) Emission spectra. An electron at E_2 falls to an unoccupied position in E_1. The energy released (E_2 is greater than E_1) has a characterististic wavelength or frequency. (b) Emission spectra and the Bohr model. Emission lines would correspond to electron transitions between levels 4 and 1, 4 and 2, 4 and 3.

energy level generally occurs quite rapidly and produces energy equal to the difference in energy between the levels. Figure 5–12 shows these changes, illustrating postulates 3 and 4 of the Bohr theory.

Using this idea of electrons absorbing and emitting radiation of characteristic energies, and consequently characteristic wavelengths or frequencies, Bohr was able to explain the line spectra observed for the hydrogen atom. He could calculate from his theory the wavelengths of the absorption lines and of the emission lines that should be observed. When his calculations were compared with the experimentally observed data of Balmer and Paschen, shown in Figs. 5–5 and 5–6, the agreement was amazingly good.

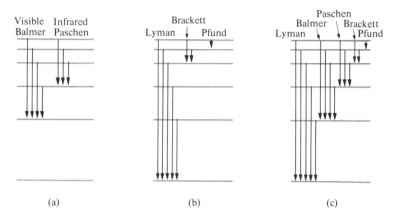

Fig. 5-13 (a) Emission lines for the hydrogen atom. Wavelength values calculated by Bohr were in close agreement with those observed experimentally by
Balmer and Paschen. (b) Emission lines for the hydrogen atom, predicted by
Bohr and later observed experimentally. (c) A comparison of the five series of
emission lines of the hydrogen atom.

But Bohr predicted many more lines than had been detected experimentally at
that time. He predicted that there should be two other sets of lines in the infrared, and
another set—at much higher energy—in the ultraviolet part of the spectrum (see
Figs. 5-6 and 5-13). These predictions stimulated further experimental investigation,
which led to the discovery of three new groups of lines, at wavelengths very close to
those predicted by Bohr. The three sets of lines—the Brackett and Pfund in the infrared and the Lyman in the ultraviolet—are all named after their discoverers.

5-6 THE BOHR ATOM AND PERIODIC PROPERTIES

We have seen how Bohr's theory explains optical spectra; it was also a first step in the
understanding of the periodic properties of the elements.

Energy-Level Populations. To understand the periodic properties using Bohr's
atomic model, we must consider the ground-state electron populations of the atomic
energy levels for each element. Hydrogen (atomic number 1) has one proton in its
nucleus and one electron in its electron energy levels. The model requires that this one
electron be in the first principal quantum level—$n = 1$ or the K-shell—as shown in
Table 5-1, since this is the energy level of greatest stability (closest to nucleus).

Table 5–1. Electron populations for elements of periods 1, 2, and 3

								Noble gases	
Period 1 Number e⁻ in K-shell, $n = 1$,	H (1)							He (2)	
2 e⁻ max.	1							2	Complete shell
Period 2	Li (3)	Be (4)	B (5)	C (6)	N (7)	O (8)	F (9)	Ne (10)	
K-shell	2	2	2	2	2	2	2	2	
Number e⁻ in L-shell, $n = 2$,									
8 e⁻ max.	1	2	3	4	5	6	7	8	Complete shell
Period 3	Na (11)	Mg (12)	Al (13)	Si (14)	P (15)	S (16)	Cl (17)	Ar (18)	
K-shell	2	2	2	2	2	2	2	2	
L-shell	8	8	8	8	8	8	8	8	
Number e⁻ in M-shell, $n = 3$,									
18 e⁻ max.	1	2	3	4	5	6	7	8	Incomplete shell

Helium (atomic number 2) has two protons and two electrons. In the model, then, both electrons are placed in the first quantum level.

The element lithium begins the second period of the periodic table. We can account for this by placing the third electron in the second principal quantum level— $n = 2$ or the L-shell—as shown in Table 5–1. Then for each of the next seven elements that compose period 2—Be through Ne—one electron is added to the L-shell. For neon there are two electrons in the $n = 1$ energy level and eight electrons in the $n = 2$ energy level. This completes the second period in the periodic table, and also completes the $n = 2$ energy level. That is, the $n = 2$ level (L-shell) can hold a maximum of 8 electrons. It has subsequently been found that the maximum electron population of any given principal quantum level is equal to twice the principal quantum number squared, or $2(n^2)$. These maximum electron populations are given in Table 5–2 on page 116.

The next element, sodium, begins period 3. We continue to build our model of the atom by placing electrons in principal quantum level 3, $n = 3$ or the M-shell. Note that Na has one electron, Mg two, Al three, etc., in the third quantum level. Now compare these electron populations with those for the elements of period 2. Lithium has one electron, Be has two, B three, etc., in the second quantum level. We now have an idea why elements in the same group in the periodic table have similar properties. *They have the same number of electrons in their outermost energy level.*

Group IA elements have one electron, Group IIA elements two electrons, Group IIIA elements three electrons, etc., in their outer energy level. Tables 5–1 and 5–3 give these similar electron configurations.

Table 5–2. Electron populations and principal quantum levels

Principal quantum level	Shell	Maximum electron population	$2(n^2)$
$n = 1$	K	2	2
$n = 2$	L	8	8
$n = 3$	M	18	18
$n = 4$	N	32	32
$n = 5$	O	50	50
$n = 6$	P	72	72

Some Periodic Properties. The electron configurations of the elements of the first three periods can furnish a few more ideas about observed periodic properties of the elements. We see that the noble gases Ne and Ar both have 8 electrons in their outermost energy level (the noble gas He can have only two electrons, which completes the K-shell). If we determined the electron configurations for the other noble gases (Kr, Xe, and Rn), we would find in each case 8 electrons in the outermost energy level. Experimentally we know that the noble gases form very few chemical compounds with the other elements. This indicates that there is an unusually high degree of stability associated with having 8 electrons in the outermost energy level.

Another property that we have studied seems to be related to the high degree of stability associated with 8 outer electrons. The combining ratios of atoms in simple chemical compounds that we considered in Chapter 4 (recall Table 4–11) also indicate a special significance to this "rule of eight." We see that one atom of a Group IA element, containing one electron in its outer energy level, combines with one atom of a Group VIIA element, containing seven electrons in its outer energy level (see Table 5–4). One atom of a Group IIA element combines with one atom of a Group VIA element, but with two atoms of a Group VIIA element (2 electrons from Group IIA plus 14 electrons from two Group VIIA atoms gives 16 or 2 × 8 electrons). Check the other compounds listed in Table 4–11. Do they follow this "rule?" Would you predict that all compounds will follow the rule? Or, knowing the fate of other general rules that we have considered, would you look for some exceptions?

Table 5–3. Elements, groups, and populations of electrons in outermost levels

Element (period 2)	Li	Be	B	C	N	O	F	Ne
Element (period 3)	Na	Mg	Al	Si	P	S	Cl	Ar
Group	IA	IIA	IIIA	IVA	VA	VIA	VIIA	0
Number electrons in outermost energy level	1	2	3	4	5	6	7	8

Table 5–4. Combining ratios

Group	IA	IIA	VIA	VIIA
Element	Na	Ca	O	Cl
Number electrons in outermost energy level	1	2	6	7
Formation of common compounds	Na 1 e⁻	+		Cl → Na Cl 7 e⁻ 1 + 7 = 8 e⁻

$$Na \quad 1\,e^- \qquad Cl \rightarrow Na\,Cl \qquad 7\,e^- \quad 1 + 7 = 8\,e^-$$

$$Ca + O \rightarrow Ca\,O$$
$$2\,e^- \quad 6\,e^- \qquad 2 + 6 = 8\,e^-$$

$$Ca \qquad + \qquad 2\,Cl \rightarrow Ca\,Cl_2$$
$$2\,e^- \qquad\qquad 2\,(7\,e^-)\,2 + 14 = 16\,e^- = (2 \times 8\,e^-)$$

The Noble Gases. There is at least one more thing that we should consider in Table 5–1. We have already seen that the outer electron configurations for the noble gases He, Ne, and Ar are 2, 8, and 8, respectively. For both He and Ne the outermost electron energy level is filled with its maximum number of electrons, i.e., the level is complete. But in the case of Ar, the M-shell contains only 8 electrons, and its maximum population is 18. Now if we were to conclude from the elements He and Ne that noble gases have filled outer energy levels (this would seem to be a reasonable idea, since it would give the noble gases a very distinct characteristic to fit their very distinct behavior), then we would not expect Ar to be a noble gas. In fact, we would predict that the element with 10 more electrons, filling the M-shell with 18, would be a noble gas. This element, which has a total of 28 electrons and 28 protons, is atomic number 28, the element nickel (Ni). As you can see from the position of Ni in the periodic table, it is a transition metal, exhibits metallic properties, and is certainly not a noble gas.

5-7 DIFFICULTIES WITH THE BOHR ATOM

This difficulty with Ar—the fact that it was a noble gas and yet had an unfilled outer energy level—was not the only problem that the model of the atom built on Bohr's theory encountered. As a result of more accurate and precise experiments in optical absorption and emission, using improved techniques and apparatus, scientists observed many more absorption and emission lines for the hydrogen atom than could be explained by Bohr's theory, as shown in Fig. 5–14. These lines occurred near the major lines that Bohr predicted and were less intense, like a fine structure grouped around the major lines. They also observed that even *more* lines were present when the spectrum was studied while the sample was in a magnetic field. This became known as the *Zeeman effect*, and could not be explained by means of Bohr's model.

The most serious difficulty with Bohr's theory, however, was its complete failure when its predicted optical absorption and emission lines for atoms with more than one electron—He, Li, Be, etc.—were compared with the experimentally observed spectra. The theory simply did not work. The observed spectral lines of elements could not be explained in any quantitative fashion, except for those of hydrogen. Even in the case of helium the theory failed miserably. It was found that Bohr's theory was useful only for one-electron systems, i.e., for the hydrogen atom or for He^+ or Li^{2+}, etc. Naturally, a theory that can handle only one-electron systems cannot be completely satisfactory; a new development was needed.

5-8 THE NEW QUANTUM THEORY

Development of the Wave Equation. In 1926, an Austrian physicist named Erwin Schrödinger advanced a new concept of the atom. His description was in the form of a mathematical equation, which has become known as the *Schrödinger wave equation*. He developed this equation using the ideas of Max Planck (blackbody radiation), Albert Einstein (the photoelectric effect), and Louis de Broglie (all matter possesses wave properties). The ideas of these three scientists, and the experiments upon which they were based, seemed to indicate that electrons exhibited both wavelike and particle-like character. In some experiments electrons acted like small particles; in other experiments they acted like light waves (see Fig. 5–15). This double behavior has become known as the *wave-particle duality*.

We now believe that all matter exhibits this dual behavior. For large objects, those that we can easily see, the particle nature is emphasized. For very small objects, those that are difficult to observe, the wave nature is emphasized. Schrödinger's wave equation embodies this wave behavior of particles of matter.

Note that for an electron to behave as a particle, we must consider it as a small, discrete unit. For an electron to behave as a wave, we must consider it as continuous, i.e., not divided up into separate parts. Then to consider an electron *both* as a wave and a particle appears to be inconsistent. Indeed it leads to indeterminacy, a result that was expressed by the German physicist Werner Heisenberg, and is called *Heisenberg's uncertainty principle*. According to this principle, we cannot know simultaneously the exact energy and the exact position of an electron.

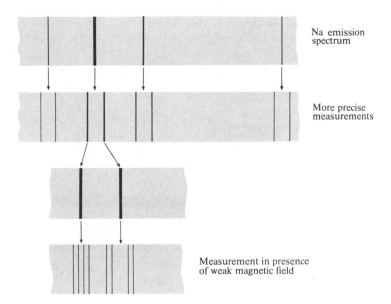

Na emission spectrum

More precise measurements

Measurement in presence of weak magnetic field

Fig. 5-14 A representation of the photographic record of the sodium emission spectrum. More emission lines appear under better or different experimental techniques.

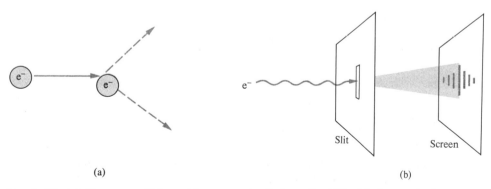

(a)

(b)

Fig. 5-15 (a) Electrons colliding with one another behave like billiard balls, transferring energy from one to another. (b) Electrons also behave like waves, and produce a diffraction pattern just as light rays do.

This means that when we calculate the exact energy of an electron by means of Schrödinger's wave equation, which relates both wave and particle behavior of the electron, we can't determine the exact position of the electron in space. We can say only that the electron will probably be contained in a certain region of space; i.e., its exact position is indeterminable. This is an idea that is significantly different from that of Bohr.

We can see the difference by considering the first principal quantum level ($n = 1$ or K-shell) of the hydrogen atom, using both theories. In the Bohr atom, the electron traveled around the nucleus in a definite, circular orbit, as shown in Fig. 5–16a. The energy of the electron was determined exactly by the principal quantum number n. We therefore knew *both* the exact energy and the exact position of the electron.

According to the quantum theory (i.e., Schrödinger's equation), we can calculate the exact energy of the electron, but then we know only a probable electron location. This probability of position, often called the *electron* or *probability density*, is diagrammed in Fig. 5–16b. The darker shading indicates that the electron has a greater likelihood of being found at that particular position. In the case of the Bohr atom, we can imagine taking a stop-action photograph of the electron that would locate it at a given point in its circular orbit. But with the quantum electron, an imaginary stop-action photograph would only show a blur like that in Fig. 5–16b. Figure 5–17 is a graphic representation of the location of the electron according to the two theories.

This difference between the theories is a significant one, but by no means the only one. There are several other benefits that we derive from the quantum theory, though sometimes at the expense of simplicity.

Quantum Numbers. Schrödinger's wave equation is a rather complicated mathematical equation. We shall not present its formulation and solution here, but we can enjoy its benefits in a qualitative way. The equation can be set up and solved exactly for the hydrogen atom. We can then generalize this information to atoms of the other elements.

▶ Actually the Schrödinger equation can be set up for any atom, but it can be solved exactly only for one-electron systems. Various approximations must be used to obtain solutions for more complex atoms, but generally the results are good qualitative pictures for multi-electron atoms. ◀

Solving the wave equation for the hydrogen atom yields three quantum numbers. It is important for us to realize that these quantum numbers are direct results of solving the equation. In the case of the Bohr theory, only one quantum number was found; it arose because of a restriction arbitrarily placed on the behavior of the atom, since its observed behavior could not be explained by any existing atomic model. Bohr assumed that the energy of an electron takes on certain definite values determined by a number n, because this assumption allowed him to explain some experimentally observed data. But solutions of Schrödinger's equation provide three quantum numbers without any arbitrary assumptions or requirements to fit known experimental data.

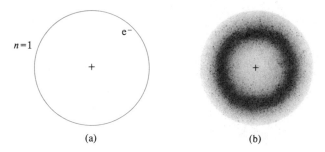

(a)   (b)

Fig. 5-16 The hydrogen atoms as viewed by (a) Bohr and (b) the new quantum theory.

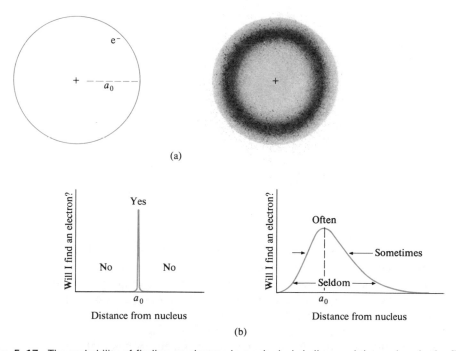

Fig. 5-17 The probability of finding an electron in a spherical shell around the nucleus in the first principal quantum level according to: (a) The Bohr theory. An electron is located only in a shell of radius a_0 around the nucleus. (b) The new quantum theory. An electron is most often found in a shell with radius a_0; but there is also a probability of finding the electron at other distances from the nucleus.

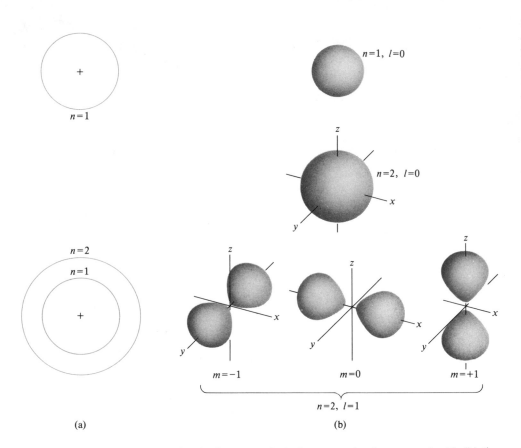

Fig. 5–18 (a) The Bohr orbits for the first two principal quantum levels compared with (b) the quantum-theory representation of the first principal quantum level and the four possible electron regions of the second principal quantum level.

Solving the Schrödinger wave equation for the hydrogen atom yields only certain allowed values for the *principal quantum number n*. This quantum number determines the energy of the electron and, roughly, the volume of the atom. A second quantum number *l*, called the *orbital quantum number*, has only certain allowed values, depending on the value of *n*; it determines the shape of the allowed regions for the electron in the atom. As we see when we compare Figs. 5–18a and 5–18b, Bohr's orbits and the electron regions of the quantum theory are not the same. The electron regions as shown have definite shapes: spherical and dumbbell. The shading inside these regions indicates that there is a high probability of finding an electron in that particular region in space. We shall call these regions of particular shapes and high electron probability density *electron* (or *atomic*) *orbitals*.

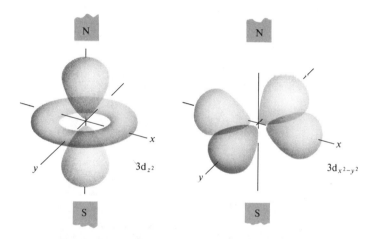

Fig. 5–19 Differences between two of the 3d orbitals in their orientations to a magnet directed along the z-axis. Interactions of electrons with the magnetic field are different in these two 3d orbitals.

▶ These regions have the highest probability of electron population, but there is a small chance of finding an electron outside the figures that are drawn. ◀

The third quantum number, m, is called the *magnetic quantum number*; it takes on discrete values that depend on the value of l. The m quantum number describes the different interactions of the electron orbitals with external magnetic fields. As we see in Fig. 5–19, orbitals with different orientations with respect to a magnetic field are affected in different ways.

There is a fourth quantum number, called the *spin quantum number s*, that has values of $\pm\frac{1}{2}$ and arises from the consideration of an electron spinning on its axis. A spinning electron generates its own magnetic field, which can interact with an external

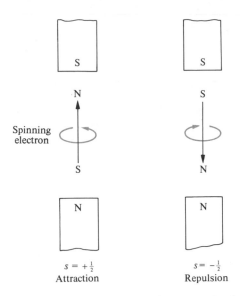

Fig. 5-20 The two values of the spin quantum number, $s = \pm\frac{1}{2}$. An electron imagined as a small, charged particle spinning on its own axis acts like a small bar magnet and interacts in two different ways with an external magnetic field, depending on the direction of spin.

Table 5-5. Quantum numbers

Symbol	Designation	Value
n	Principal	Integral values from 1 to infinity; 1, 2, 3, etc., to ∞
l	Orbital	Integral values from 0 to $(n-1)$, for any given value of n; 0, 1, 2, etc., to $(n-1)$
m	Magnetic	Integral values from $-l$ to $+l$, for any given value of l
s	Spin	$\pm\frac{1}{2}$

magnetic field in two different ways, similar to a tiny bar magnet, as shown in Fig. 5-20. Table 5-5 summarizes the four quantum numbers.

Atomic Energy Levels. There are now many more energy levels available to electrons in atoms than were available in the original Bohr theory. Not only are there energy levels associated with the principal quantum number, but also with the other three

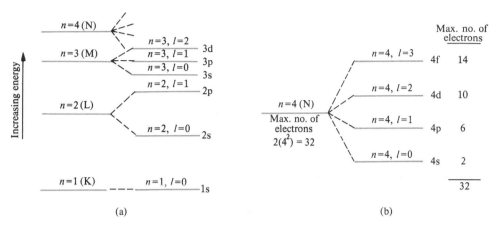

Fig. 5–21 The electron energy levels calculated for the hydrogen atom when no external magnetic or electric fields are present. The energy-level separations are not drawn to actual scale.

quantum numbers, *l*, *m*, and *s*. For the hydrogen atom these levels can be calculated exactly. Figure 5–21 gives a qualitative picture. We shall generalize this energy-level scheme to all other atoms and make corrections when we find that they are needed.

Note that new divisions of the principal quantum levels have been introduced. We see that the principal levels, instead of being single energy levels, are composed of a number of separate energy levels. The number of separate levels composing any given principal quantum level is equal to the principal quantum number of the level; that is, *n* = 3 has 3 divisions, *n* = 4 has 4 divisions, etc. We call these sets of smaller divisions *sublevels* or *subshells*. They are generally designated s, p, d, and f, for historical reasons. These designations correspond to values of 0, 1, 2, and 3, respectively, for the orbital quantum number *l*.

▶ Spectroscopists assigned the symbols s, p, d, f after the *sharp, principal, diffuse*, and *fundamental* series of optical emission lines that were observed experimentally. The sharp series was found to correspond to electron transitions between s-sublevels— from 3s to 2s, for example—the principal series between p-sublevels, etc. If other sublevel designations are needed beyond f, they are listed alphabetically beginning with g, h, etc. ◀

In Fig. 5–21 we see that the order of increasing energy of the sublevels is s, p, d, f for principal quantum level four. This same order of energy for the subshells is observed within every principal quantum level; that is, s is lower than p, p lower than d (where it is present), d lower than f (where it is present). Note that this says nothing

Table 5–6. Sublevels

Symbol	*l*	*m*	Energy increases	Total no. orbitals	Maximum no. of electrons
s	0	0		1	2
p	1	$-1, 0, 1$		3	6
d	2	$-2, -1, 0, 1, 2$		5	10
f	3	$-3, -2, -1, 0, 1, 2, 3$		7	14

about the relationship *between* principal quantum levels. Table 5–6 summarizes information about the sublevels.

With the energy-level scheme pictured in Fig. 5–21, we can see that many more optical absorption and emission lines are possible than was the case in Bohr's original atomic model. But this energy-level picture does not yet explain the Zeeman effect, i.e., optical spectra observed within a magnetic field. When an external magnetic field is applied to an atom, the sublevels split into still more divisions, determined by the magnetic quantum number *m*. This splitting of the levels makes possible even more electron transitions and consequently more absorption and emission lines.

▶ When several electron energy levels have exactly the same energy, we say that they are *degenerate*. When these levels are split apart and no longer have identical energies, their degeneracy has been removed. ◀

Refer back to Fig. 5–18b, which shows diagrams of the three possible 2p-orbitals. All the orbitals have principal quantum number $n = 2$ and orbital quantum number $l = 1$. Since the Schrödinger equation tells us that for $l = 1$ there are three possible values for the magnetic quantum number $m = -1, 0, 1$, there are three possible p-orbitals. Note that they differ only in their spatial orientation. The $2p_x$ orbital is directed along the x-direction in space, the $2p_y$ along the y-direction, and the $2p_z$ along the z-direction. All p-orbitals have the same external geometry, as is shown for the 2p-orbitals in Fig. 5–18b.

When we consider the spin quantum number *s*, we find that the orbitals can be subdivided still further. Each electron can have a spin of $+\frac{1}{2}$ or $-\frac{1}{2}$. That means that each s-orbital is composed of two energy levels (one for an electron spin of $+\frac{1}{2}$ and one for a spin of $-\frac{1}{2}$); each of the three p-orbitals is composed of two levels (giving a total of six sublevels); each of the five d-orbitals is composed of two levels (10 total sublevels); etc.

5–9 THE AUFBAU PRINCIPLE

We have seen that the new quantum theory gives us a basis for explaining complex optical absorption and emission spectra. But the new electron-energy-level structure that has been developed also enables us to make another stab at explaining periodic properties.

What we actually do is fill up the energy levels that have been derived based on the hydrogen atom with the proper number of electrons for the succeeding atoms. This is called the *aufbau* (or building-up) *principle*. We start by placing electrons in the orbitals of lowest energy first. Then, with the help of two rules, we should be able to construct the electron structure for every element:

1. The first rule, called the *Pauli exclusion principle*, states that no two electrons can have the same four quantum numbers. In the 3s-orbital, for example, one electron has the quantum numbers (or lies in an energy level determined by the quantum numbers) $n = 3$, $l = 0$, $m = 0$ and $s = +\frac{1}{2}$. There is "room" in the 3s-orbital for a second electron with the quantum numbers $n = 3$, $l = 0$, $m = 0$, but $s = -\frac{1}{2}$. Another electron in the 3s-subshell would have to have quantum numbers identical to one of the first two electrons. The maximum electron population of the 3s-orbital, and of all s-orbitals, is therefore two.

2. The second rule, called *Hund's rule*, states that when electrons are filling a set of orbitals of the same energy, they will have the same spin quantum number s as long as they do not violate the Pauli exclusion principle (Rule 1). When two electrons have the same s-value, we say that they have *parallel spins* or that they are *unpaired*. When they have different s-values (one electron with $s = +\frac{1}{2}$ and the second with $s = -\frac{1}{2}$), then we say they have *opposed spins* or are *paired*.

As an example of Hund's rule, we might consider filling the 2p-orbitals. The first three electrons enter the $2p_x$, $2p_y$, and $2p_z$ orbitals (all having identical energies) with parallel spins; say, all have $s = +\frac{1}{2}$. In order not to violate the Pauli exclusion principle, the next three electrons must pair with the first three, i.e., their s-values must be $-\frac{1}{2}$. When the three 2p-orbitals contain three electron pairs—or six electrons total—they have the maximum number of electrons possible.

We now see why Table 5–6 listed the maximum electron populations of the s, p, d, and f sublevels as 2, 6, 10, and 14. These electron populations are natural consequences of the electron-energy-level system that was developed from the quantum theory.

Periodic Properties. Now we can build up the periodic table of the elements, proceeding just as we did in Section 5–5, except that we have more energy levels to consider. Helium (atomic number 2) has two paired electrons in the 1s-orbital, filling the sublevel. This electron configuration is written $1s^2$, indicating that there are 2 electrons (the right-hand superscript) in the s-orbital of principal quantum level $n = 1$. The addition of one more electron (and proton) gives the element lithium an electron configuration $1s^2 2s^1$; one more electron yields atomic number 4, Be, with 4 electrons, or $1s^2 2s^2$. Compare this electron configuration with that of magnesium (Mg), $1s^2 2s^2 2p^6 3s^2$, the second element in Group IIA. Again we find the same number of electrons in the outermost energy level. Compare elements of the other major groups to convince yourself that it is a valid generalization to say that elements of the same group have similar outer-electron configurations.

When we consider the electron configuration of the noble gas argon, we find another interesting result. The eight electrons in the outer principal quantum level of Ar have the configuration $3s^2 3p^6$. We see that the noble gas neon has a similar outer-electron configuration, $2s^2 2p^6$. In fact, all the noble gas elements (except He) have this characteristic $ns^2 np^6$ electron configuration. Recall the difficulty with argon presented in Section 5–5; it is a noble gas, but does not have a filled principal quantum level. Now we see that the distinguishing characteristic of noble gases is their $ns^2 np^6$ electron configuration, which is apparently a very stable arrangement of electrons. Argon fits in this group very nicely.

With the next element, however, our energy-level scheme runs into trouble. Element 19, potassium, follows argon and should be an alkali metal. The two previous Group IA elements, Li and Na, have $1s^2 2s^1$ and $1s^2 2s^2 2p^6 3s^1$ electron configurations. Both elements have one electron in an outer s-orbital. If we use the energy-level structure shown in Fig. 5–22, then potassium is not going to have the correct outer-electron configuration. Figure 5–23 shows the solution to the problem. The 4s-orbital is lower in energy than the 3d-orbital, and therefore is filled with electrons (two) before any electrons enter the 3d-levels.

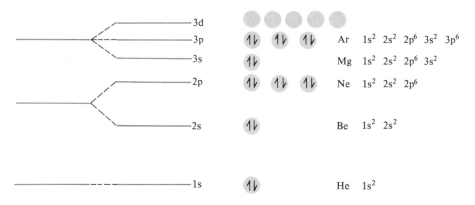

Fig. 5–22 An energy-level diagram, showing the electron populations of He, Be, Ne, Mg, and Ar.

After the 4s-sublevel is filled with two electrons in the element calcium, electrons enter the 3d-energy levels, obeying Hund's rule. Figure 5–23 shows this for the elements manganese, nickel, and zinc. Note that the problem previously mentioned for nickel (see Section 5–5) no longer exists. Nickel has metallic properties just like the other 9 transition metal elements, Sc through Zn, that are filling the 3d-sublevel. It should not, and does not, have any noble-gas properties, since it does not have the characteristic noble-gas electron configuration.

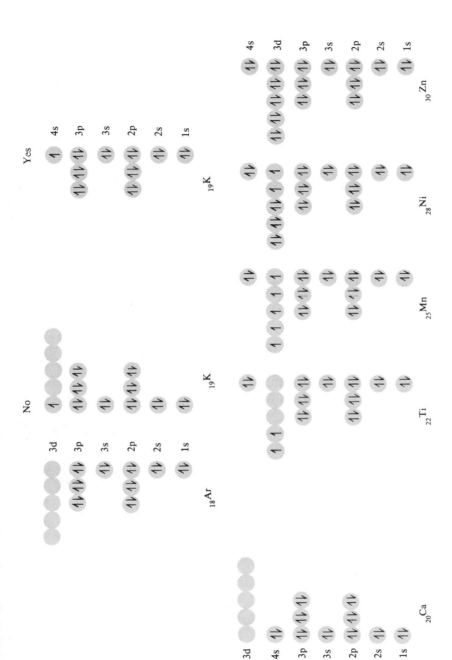

Fig. 5-23 Electrons fill the 4s sublevel before entering the 3d sublevel.

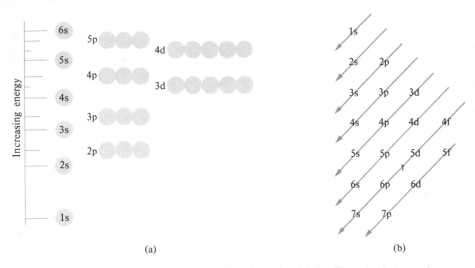

(a)   (b)

Fig. 5-24 (a) Approximate relative energies of atomic orbitals. Though this is not drawn to scale, note that the separations at low energy are much larger than those at higher energies. (b) An easy method for remembering the order of filling of energy levels.

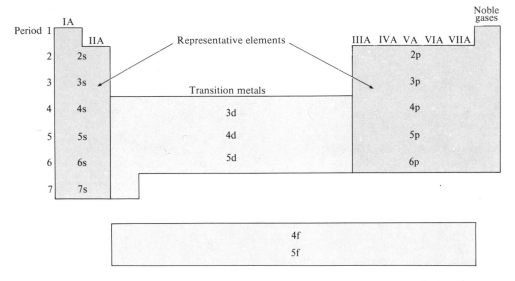

Fig. 5-25 Representative elements are those that are filling s- or p-sublevels; i.e., the last electrons added are in s- or p-orbitals. Transition elements are filling d- or f-sublevels.

The next level to be filled, starting with the element gallium (Ga), is the 4p, and the one following that is the 5s. Figure 5–24 shows the relative energies of the first several sublevels, along with an easy method for remembering the electron filling order. Placing the proper number of electrons in the orbitals in the order indicated in Fig. 5–24 gives the electron configuration of any atom.

▶ Actually the relative atomic-orbital energies change as the atomic number changes. The diagram in Fig. 5–24 applies to elements with atomic numbers of less than about 30. For our purposes Fig. 5–24 is a sufficiently good approximation. ◀

By building up the periodic table in this fashion, we can see that in certain sections of the table a particular kind of sublevel is being filled. Figure 5–25 shows these sections of the periodic table. Elements that are often called *representative elements* are found in the sections that are filling the s- and p-orbitals. The *transition metals* are found in the sections in which the d- and f-orbitals are being filled.

5-10 ATOMIC THEORIES AND CHEMICAL COMPOUNDS

The atomic model that Niels Bohr proposed in 1913 was a major advance for science, helping us understand many experimental observations that were unexplained, predicting new behaviors of atoms, and laying the foundations for other equally dramatic changes. Bohr's theory enabled scientists to make the first real interpretations of optical absorption and emission spectra. It explained, quantitatively, the spectra observed for the hydrogen atom and other one-electron systems. And it gave us the first ideas about the cause of the periodic properties of the elements. It was, at least in its qualitative form as we have considered it, a simple and easily pictured theory. Perhaps for these very reasons Bohr's theory ran into problems. It failed to explain the properties, particularly the optical spectra, of more complex atoms, i.e., atoms with more than one electron.

Another giant step forward was made in 1926, when Schrödinger proposed his wave equation and the new quantum theory was born. This theory provides the basis for the all-important quantum numbers, whereas Bohr could only postulate their existence. It leads us to a more detailed understanding of the atom, but at the expense of the easily visualized physical pictures possible with Bohr's theory. It makes possible better explanations of atomic properties: optical spectra of complex atoms, periodic properties, and many others. And it enables us to go one step further, to answer the questions that we had posed earlier: Why do atoms "stick" together, i.e., why are compounds formed? And why do atoms combine in different ratios?

Actually, we can again use Bohr's theory to begin to explain the reasons for chemical bonds. But the basis for most of our current understanding and theories of chemical bonding is the new quantum theory. In Chapter 6 we shall see how these theories of the atom account for the formation of chemical bonds and the properties of the compounds that are formed.

QUESTIONS

1. Define or discuss the following terms: Light, wavelength, electromagnetic spectrum, Balmer series, hydrogen atom, Bohr's postulates, frequency, visible light, x-rays, emission of radiation, Lyman series, absorption spectrum, quantum, quantized orbit, ground state, principal quantum level, the quantum numbers n, l, m, and s, sublevel, L-shell, noble gas, wave-particle duality, electron probability density, $2p_y$ orbital, electron spin, degenerate energy levels, paired electrons, parallel spins, electron filling order.

2. Sketch a continuous spectrum, including only the visible region of the electromagnetic spectrum. Label the spectrum with the various colors and with the appropriate values of frequency (v) in Hertz and of wavelength (λ) in centimeters and in angstroms.

3. Indicate the essential difference(s) between absorption and emission spectra.

4. If a transparent solid material absorbs the following wavelengths of electromagnetic radiation, what color will the material be in ordinary light?

   a) Green
   b) Infrared
   c) 4000 Å
   d) Orange
   e) Ultraviolet
   f) x-ray
   g) 3000 Å
   h) Red
   i) 8500 Å
   j) 6×10^{-5} cm

5. Make the following conversions:

   a) 6×10^{-5} cm to Å
   b) 8000 Å to cm
   c) 5.0×10^{14} Hz to cm
   d) 6000 Å to Hz
   e) 4×10^{-5} cm to Hz
   f) 7.5×10^{14} Hz to Å.

6. a) Explain the difference between a continuous emission spectrum and a line emission spectrum.
   b) Why does the element lithium (Li) emit a line spectrum?

7. Figure 5–26 represents the emission spectrum of $^4_2$He. The wavelengths are given in angstroms.

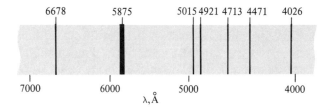

Figure 5–26

   a) Indicate the emission lines of highest and lowest energy.
   b) Calculate the frequency of the two lines in part (a).
   c) Give the approximate color of each of the lines in the figure.
   d) Assume that the figure shown is a photographic record of the spectrum. Would the lines be bright against a dark background or dark against a light background? Why?

8. Would you expect the emission spectrum of sodium (Na) or of uranium (U) to be more complex, i.e., composed of more lines? Why?

9. Using Bohr's model of the atom, give a qualitative explanation of the absorption spectrum of a Li atom.

10. Discuss the similarities and differences between the Rutherford and Bohr pictures of the atom. What major contribution(s) did Bohr make?

11. Question 16 in Chapter 4 gives a table of atomic radii of the fourth-period elements. Explain this trend, using Bohr's model of the atom.

12. Write the chemical equation for the reaction that would be expected between the following elements.

   a) K and Br_2 b) Aluminum and S
   c) Zn and oxygen molecules   d) Ne and sodium
   e) Sr and O_2 f) Yttrium and O_2

13. Assign the elements with the following electron configurations to their periodic groups.

   a) $1s^2 2s^2 2p^6 3s^2 3p^5$ b) $1s^2 2s^2 2p^6 3s^2 3p^6 3d^{10} 4s^2$
   c) $1s^2 2s^2 2p^6$ d) $1s^2 2s^2 2p^6 3s^2 3p^4$
   e) $1s^2 2s^2 2p^6 3s^1$ f) $1s^2 2s^2 2p^6 3s^2 3p^6 3d^{10} 4s^2 4p^2$
   g) $1s^2 2s^2 2p^6 3s^2$ h) $1s^2 2s^2 2p^6 3s^2 3p^3$

14. Assign electrons to subshells for:

   a) Sodium   b) Atomic number 20
   c) The element containing 30 protons   d) Cr
   e) Cu

15. Describe the necessary components of an experimental apparatus to measure the emission spectrum of an element.

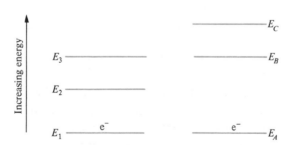

Figure 5–27

16. Consider two different systems with energy level diagrams as shown in Fig. 5–27 and electrons located at E_1 and E_A. Describe what will happen if the two systems are exposed to incident radiation of the following energies:

   a) Incident radiation energy less than $E_2 - E_1$.
   b) Incident radiation energy equal to $E_2 - E_1$.
   c) Incident radiation energy less than $E_3 - E_1$ but greater than $E_2 - E_1$.
   d) Incident radiation energy equal to $E_3 - E_1$ and $E_B - E_A$.
   e) Incident radiation energy greater than $E_B - E_A$ but less than $E_C - E_A$.
   f) Incident radiation energy equal to $E_C - E_A$.
   g) Incident radiation energy greater than $E_C - E_A$.

17. Using energy-level diagrams, indicate plausible explanations for the experimental observations that Li, Na, and K emit red, yellow, and violet light, respectively, when they are burned in flames.

18. Indicate what must be done to the following atoms to give them a noble-gas electron configuration.

a) F
b) Ca
c) Na
d) S
e) O
f) Al
g) Mg
h) C
i) Cl
j) N

19. Consider the electron configurations of five elements.

1) $1s^2 2s^2 2p^5$
2) $1s^2 2s^2 2p^6 3s^2 3p^6 4s^1$
3) $1s^2 2s^1$
4) $1s^2 2s^2 2p^6 3s^2 3p^2$
5) $1s^2 2s^2 2p^6 3s^2 3p^6 4s^2 3d^5$

a) For each electron configuration, indicate which sublevel(s) and which principal quantum level(s) have their maximum electron populations.
b) Identify the periods in which each of the five elements are found.
c) Identify the five elements.

20. We call atoms that have an electrical charge *ions*. Assume that one electron has been removed from each of the atoms Li, Na, and K to form Li^+, Na^+, K^+ (*cations*) and that one electron has been added to F and Cl atoms to form F^- and Cl^- (*anions*). Write the complete electron configurations for all these atoms and ions. Compare your results with the electron configurations of the noble gases He, Ne, and Ar.

21. a) Calculate the frequency of red light of 7000 Å.
b) Calculate the energy of the light in part (a).
c) Is blue light of higher or lower energy than red light? Longer or shorter wavelength?

22. Explain why more precise experimental measurements often show that optical emission lines are actually several individual but closely spaced lines.

23. Why does the electron in the 1s-sublevel of the hydrogen atom have two possible energies in a weak magnetic field?

24. Sketch what we believe to be the shape of the 2s and $2p_x$ orbitals.

25. How do Bohr's theory and the new quantum theory differ in their representation of an electron in the first principal quantum level, $n = 1$ or K-shell?

26. Indicate the possible values for the n and l quantum numbers in the K-, L-, and M-shells.

27. Sketch the orbitals associated with the following quantum numbers:

a) $n = 3, l = 0, m = 0$
b) $n = 3, l = 1, m = 0$
c) $n = 2, l = 1, m = 0$
d) $n = 2, l = 0, m = 0$
e) $n = 1, l = 0, m = 0$

28. How many sublevels are there in each of the following shells?

a) $n = 1$
b) $n = 2$
c) $n = 3$
d) $n = 4$
e) K
f) L
g) M
h) N

29. How many orbitals are contained in each of the following sublevels?

a) $l = 1$
b) $l = 2$
c) 1s
d) 2p
e) 3d
f) 3p
g) $l = 3$
h) $l = 0$
i) 2s
j) 4f

30. What is the spatial orientation of the 2s-orbital? Of the $2p_z$ orbital?

31. Using the M-shell as an example, draw an energy-level diagram showing the progressive splittings of the shell into more complex energy levels, labeling each level with the appropriate quantum numbers, n, l, m, and s.

32. Explain why the maximum electron population of the M-shell is 18.

33. Explain why the maximum electron population of a d-sublevel is 10.

34. Write the entire electron configuration of P, Ca, Ti, and Mn, and indicate the number of unpaired electrons for each atom.

35. What evidence leads us to believe that in some atoms the 4s-sublevel has a lower energy than the 3d?

36. In the absence of a magnetic field, which of the 3p-orbitals will an electron enter first?

37. a) Given that a free atom (no external field) has an outer electron configuration of $3d^5$, how many unpaired electrons does it have?
    b) Assign four quantum numbers to each of the 3d-electrons in part (a).
    c) Given that another electron is added to the atom in part (a) so that the configuration is $3d^6$, how many unpaired electrons are there?

38. Discuss some of the differences between Bohr's theory and the new quantum theory.

6. CHEMICAL BONDING

6-1 DALTON TO BOHR AND BEYOND

In 1803 Dalton proposed that atoms united with one another in forming chemical compounds. This assumption proved to be correct, and was a major advance for science. But it also posed a difficult question: Why do atoms unite with one another, and why in varying ratios?

Many experiments related to this question were performed. Electrons and protons were discovered and their charges measured. Rutherford's experiments led him to propose that atoms were made up of electrons orbiting around a central nucleus composed of protons and neutrons. Rutherford's idea, just as Dalton's before it, was ingenious and a great help to science, but the question of why atoms unite remained unanswered. It was not until 1913, when Niels Bohr proposed his atomic model, that the theoretical foundations for understanding the formation of chemical compounds were laid. With the advent of the quantum theory, understanding increased and the current theories of chemical bonds were developed. We shall now study chemical bonds and compound formation, using as a basis the ideas of Bohr and the quantum theory.

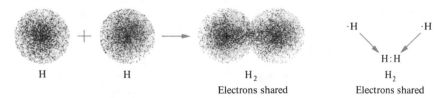

H   H H_2 H_2
                              Electrons shared   Electrons shared

Fig. 6–1 The formation of a molecule of hydrogen, H_2, from two separate hydrogen atoms. The shaded electron clouds give a more realistic picture of the hydrogen atoms and the hydrogen molecule than the electron-dot representations.

6–2 COVALENT BONDS

Hydrogen Atoms and the Hydrogen Molecule. Experimentally we know that hydrogen gas exists as diatomic molecules, H_2. When one hydrogen atom comes close to another hydrogen atom, the two combine to form a molecule,

$$H + H \rightarrow H_2.$$

The diatomic hydrogen molecule is more stable than two separate hydrogen atoms. Why should this be?

We can understand the "desire" of two hydrogen atoms to react with one another by considering the electron configurations of hydrogen, H $1s^1$, and helium, He $1s^2$. Experimentally we know that helium (a noble gas) does not enter into ordinary chemical combinations or reactions. If we postulate that: (1) The reason for the very high stability of the helium atom is its electron configuration, and (2) the chemical bond between two hydrogen atoms is related to the electron configuration of the helium atom, then we might expect that this very stable electron configuration of helium would be formed when the hydrogen atoms unite.

Does this model fit the case of hydrogen? Figure 6–1 shows the formation of one hydrogen molecule, H_2, from two separate hydrogen atoms. The two separate H atoms both have $1s^1$ electron configurations, but the H_2 molecule has two electrons in the region around and between the two hydrogen nuclei. That is, the two hydrogen atoms share the two electrons. This gives, in effect, both hydrogen atoms the electron configuration of helium, $1s^2$, at least part of the time.

This does not mean that H_2 and He are the same, or even similar. There are major differences between the atom helium (He) and the molecule hydrogen (H_2). Helium has only one nucleus, containing two protons and two neutrons, with two electrons in the 1s atomic orbital. A molecule of H_2, on the other hand, has two nuclei separated by 0.74 Å with each nucleus containing one proton.* The two

---

* For the sake of clarity, we shall consider in this discussion only the most abundant isotopes of helium, ^4_2He, and of hydrogen, ^1_1H.

electrons are now no longer in the separate 1s atomic orbitals of the two hydrogen atoms, but belong equally to both atoms. The electrons are in a *molecular orbital*, i.e., an orbital belonging to the entire molecule rather than to separate atoms.

Hydrogen does seem to fit our proposed model. Two separate atoms of hydrogen combine to form a diatomic molecule in which two electrons are shared and the very stable electron configuration of helium is, in effect, formed.

We say that a chemical bond has been formed between the two hydrogen atoms. The distance between the two atomic nuclei is called the *bond length* or *bond distance*. The attraction between two atoms in a chemical compound is called a *chemical bond*. When the attraction between two atoms is caused by the sharing of a pair of electrons, the bond is a *covalent bond*. Compounds that contain only covalent bonds are called *covalent compounds*.

In the case of the hydrogen molecule, the bonding electron pair is shared equally by both hydrogen atoms. Since both atoms are identical, they have the same attraction for the electrons; the electrons spend the same amount of time around each hydrogen atom. This type of chemical bond is called a *nonpolar covalent bond*. It is a nonpolar bond because there is no separation of charge between the atoms. Generally, this happens only when a bond is formed between two identical atoms.

Lewis Structures. We should be able to derive a general model for chemical bonds from the postulated model that accounted for the formation of the molecule H_2. A reasonable generalization would be: (1) A chemical bond is formed when an electron pair is shared by two atoms. (2) In forming chemical bonds, atoms strive to attain a noble-gas electron configuration. Since this noble-gas electron configuration is ordinarily ns^2np^6 (except for helium), atoms "desire" to be associated with four electron pairs, or a total of eight electrons. This has become known as the *octet rule* or the *rule of eight*.

Proposals (1) and (2) were actually made by G. N. Lewis in 1916. They have been replaced by more sophisticated bonding theories, but they still remain very useful for many compounds and make possible easily visualized reasons for chemical bond formation. We shall, therefore, use these ideas extensively, indicating some of the places in which more advanced theories are necessary.

We can visualize the application of electron configuration to chemical bond formation more easily by using *Lewis structures* or *Lewis electron-dot formulas*. In these structures, illustrated in Table 6-1, an electron is represented by a dot. Only the electrons involved in bonding are shown; these are called *valence electrons*, and are generally the electrons in the outer principal quantum level (Na·, Mg:, ·Al:, ·Si:). Electrons that are paired are represented by two adjacent dots or a short line, while those that are unpaired are shown as single dots. The element phosphorus, for example, has an electron configuration $1s^22s^22p^63s^23p^3$. In the third, and outer, principal quantum level there are two 3s-electrons paired and three 3p-electrons unpaired. Hence the proper Lewis structure.

$$·\overset{..}{P}·   \text{or}   ·\overline{P}·$$

Table 6–1. Typical Lewis structures

Formula	Lewis structure
Li	Li·
Rb	Rb·
Sr	Sr:
Ga	·Ga:
P	·P̈·
S	·S̈:
Br	:B̈r:
Ar	:Är:
NaF	Na⁺:F̈:⁻
BaI₂	⁻:Ï: Ba²⁺ :Ï:⁻
HBr	H:B̈r: or H—B̈i:
H₂Se	H:S̈e:H or H—S̈e—H
AsH₃	H:Äs:H or H—Äs—H
	H H
CHCl₃	H H
	:C̈l:C:C̈l: or Cl—C—Cl
	:C̈l: Cl

Other Diatomic Molecules. Experimentally we know that chlorine gas is composed of diatomic molecules, Cl_2. According to the above postulates of the proposed model for chemical bonds we should be able to understand why Cl_2 is formed by considering the electron configuration of chlorine in conjunction with the electron configuration of the noble gas argon. The electron configuration for Cl is $1s^2 2s^2 2p^6 3s^2 3p^5$ and that for Ar is $1s^2 2s^2 2p^6 3s^2 3p^6$. We can see from the Lewis structures for Cl and Ar,

$$:\ddot{C}l\cdot \qquad :\ddot{A}r:$$

that Cl needs one more electron to have the Ar noble-gas electron configuration. As in the case of the hydrogen molecule, H_2, the noble-gas electron configuration can be formed if two chlorine atoms share one pair of electrons. Then, in effect, both Cl atoms have eight electrons, completing the stable $3s^2 3p^6$ electron configuration,

$$:\ddot{C}l\cdot + \cdot\ddot{C}l: \rightarrow :\ddot{C}l:\ddot{C}l: \qquad \text{or} \qquad :\ddot{C}l—\ddot{C}l:$$

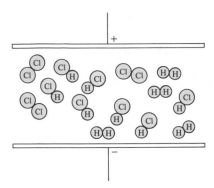

Fig. 6-2 The orientation of dipoles between two charged plates. The negatively charged end of the HCl molecule is attracted by the positive plate and the positively charged end by the negative plate. Since H_2 and Cl_2 are nonpolar molecules, they are not affected by the charged plates and are randomly oriented.

A nonpolar covalent bond is formed between the two chlorine atoms. The atoms, being identical, share the electron pair equally. (In the language of our new bonding theories based on the quantum theory, we would say that the electrons are no longer located in the separated atomic orbitals of the two chlorine atoms, but belong to a molecular orbital that encompasses the entire Cl_2 molecule.)

The situation is only slightly different when we consider a molecule formed from two unlike atoms, H and Cl, for example. Experiments show that the molecule HCl has a 1 : 1 atom ratio. We know from our previous discussions of H_2 and Cl_2 that both hydrogen and chlorine atoms need one electron to complete a noble-gas electron configuration. Both atoms can complete this configuration by sharing an electron pair between them:

$$H\cdot \; + \; \cdot \ddot{\underset{..}{Cl}}\colon \; \rightarrow \; H\colon\ddot{\underset{..}{Cl}}\colon \qquad \text{or} \qquad H-\ddot{\underset{..}{Cl}}\colon$$

In this case the electron pair is shared by two unlike atoms. There is no reason to expect that the sharing will be equal. In fact, we find experimentally that the sharing is unequal. We determine this by checking to see whether the ends of the HCl molecule have opposite charges. We could, for example, place some HCl molecules between two charged parallel plates, as shown in Fig. 6-2. From experiments such as this, we observe that some of the plus and minus charges on the plates are neutralized. This neutralization can occur if the HCl molecules contain separated charges and are thereby oriented when they are between the plates. Then the attraction between the positively charged plate and the negative end of the molecule and the negatively charged plate and the positive end of the molecule would neutralize some of the charge present on the plates.

From experimental observations such as these (in conjunction with other types of experimental data), it is inferred that the H end of the HCl molecule has a small positive charge (usually indicated by $\delta+$) and the Cl end a small negative charge

H : H

: C̈l : C̈l :

δ^+ H : C̈l : $^{\delta^-}$

Fig. 6–3 The H_2, Cl_2, and HCl molecules. The HCl molecule is slightly polar, with the Cl end having a partial negative charge, $\delta-$, and the H end a partial positive charge, $\delta+$.

($\delta-$). (The δ is the lower-case Greek *delta*.) We can account for this separation of charge by assuming that chlorine attracts the shared electron pair more strongly than hydrogen does. A bond in which the electron pair is shared unequally is a *polar covalent bond*.

Figure 6–3 shows a diagram of the electron density in the HCl molecule (as pictured by the quantum theory) and the corresponding Lewis structure. Molecules such as HCl, in which there is a separation of the centers of positive and negative charge (i.e., there is a plus charge and an equal minus charge separated by some distance) are called *dipoles*. Dipoles can be described by their *dipole moment*, which is equal to the charge times the distance between the charges. (Whether molecules containing more than two atoms have dipole moments depends on their structure. We shall discuss this later in this section.) The magnitude of the dipole moment, then, is an indication of the degree of unequal sharing of bonding electron pairs. Table 6–2 gives some representative values for diatomic molecules.

Table 6–2. Representative values of dipole moments for diatomic molecules

Substance	Dipole moment, Debyes*	Substance	Dipole moment, Debyes*
HF	1.9	ClF	0.9
HCl	1.1	Cl_2	0.0
HBr	0.8	BrCl	0.6
HI	0.4	ICl	0.5
H_2	0.0		
NO	0.2	CsF	7.9
CO	0.1	CsCl	10.4

*   Data taken from the *Handbook of Chemistry and Physics*, The Chemical Rubber Publishing Co., Cleveland, Ohio. The dipole unit, the *Debye*, is named for Peter Debye, 1936 Nobel laureate in chemistry.

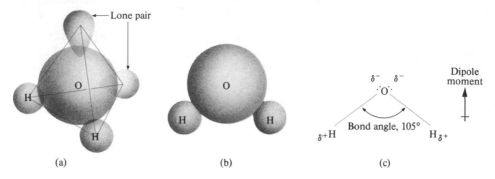

Fig. 6-4 The water molecule drawn three ways. Because of bond polarity and the bent structure of the H_2O molecule, water has a rather high dipole moment (see Table 6-3).

Polyatomic Molecules. Covalent bonds are formed between atoms in many compounds that contain more than two atoms. Common examples are: water (H_2O), carbon dioxide (CO_2), sulfur dioxide (SO_2), ammonia (NH_3), methane (CH_4), carbon tetrachloride (CCl_4), chloroform ($CHCl_3$), and so on. Let us consider some of these polyatomic molecules, using the model for bond formation that seemed to work for H_2, Cl_2, and HCl.

Experiments have shown that two atoms of hydrogen are combined with one atom of oxygen in one molecule of H_2O. The electron configurations for H and O are $1s^1$ and $1s^22s^22p^4$, respectively. Hydrogen needs one electron per atom to complete the 1s energy level and oxygen needs two electrons per atom to form the stable $2s^22p^6$ outer electron configuration of the noble gas neon. The following Lewis structures show that both these conditions can be fulfilled if two hydrogen atoms share two pairs of electrons with one oxygen atom:

$$\text{H·} + \text{·}\ddot{\text{O}}\text{:} + \text{·H} \rightarrow \text{H}\underset{\text{H}}{\overset{}{\ddot{\text{O}}\text{:}}} \quad \text{or} \quad \text{H}{-}\underset{|}{\overset{}{\ddot{\text{O}}\text{:}}}\ \ \text{H}$$

Note that there are a total of eight electrons associated with one H_2O molecule, two pairs of bonding electrons (four electrons total) and two *unshared pairs*, often called *lone pairs*, on the oxygen atom (four electrons total). Eight electrons per H_2O molecule are necessary, since it is composed of one oxygen atom with six valence electrons and two hydrogen atoms, each having one electron.

Two polar covalent bonds are formed between the two hydrogen atoms and the oxygen atom in H_2O. Experimentally we find that the water molecule has an angular shape (see Fig. 6-4) and that the oxygen end of each hydrogen-oxygen bond carries a small negative charge. These two factors combine to give H_2O a relatively high dipole moment.

A second example of a polyatomic molecule taken from those listed above is methane. Composed of one carbon atom and four hydrogen atoms, its chemical

formula is CH_4. The electron configuration of C is $1s^2 2s^2 2p^2$ and $\cdot\overset{\cdot}{C}\cdot$ is its corresponding electron-dot formula. In order to attain the $2s^2 2p^6$ noble-gas electron configuration, carbon must share four pairs of electrons. It might seem, then, that the formula CH_4 would be predicted by our proposed bonding model, since the four hydrogen atoms can share four pairs of electrons with the carbon atom. But if we consider the carbon electron configuration in more detail, we shall see that it needs to be changed slightly before the four hydrogen atoms can be accommodated. One of the two 2s electrons needs to be moved up (to a higher energy) into a 2p orbital. This changes the outer electron configuration of C from $2s^2 2p^2$ to $2s^1 2p^3$ and the corresponding Lewis structure from $\cdot\overset{\cdot}{C}\cdot$ to $\cdot\overset{\cdot}{\underset{\cdot}{C}}\cdot$. Then we can visualize what happens in the formation of methane, CH_4:*

$$2H\cdot \;+\; \cdot\overset{\cdot}{\underset{\cdot}{C}}\cdot \;+\; 2H\cdot \;\rightarrow\; H\!:\!\overset{\cdot\cdot}{\underset{\cdot\cdot}{C}}\!:\!H \qquad \text{or} \qquad \begin{array}{c} H \\ | \\ C \\ {\diagup}{|}{\diagdown} \\ H\;\;H\;\;H \end{array}$$

Even though some changes had to be made in the atomic electron structure of carbon, the general rule about completing noble-gas electron configurations when forming a compound still seems to apply. We must assume that the energy needed to promote the one 2s electron into a 2p orbital is furnished by the formation of two more bonds. The necessity for changing the electron configuration of carbon to comply with experimental data requires a slight bending of, or addition to, our bond model. This is not an unusual circumstance. We must constantly revise existing ideas to meet experimental observations.

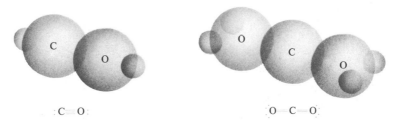

:C≡O: :O=C=O:

Fig. 6-5 The carbon monoxide (CO) and carbon dioxide (CO_2) molecules.

Multiple Bonds. Carbon dioxide (CO_2) and carbon monoxide (CO) are common substances that have been well characterized experimentally. It is known that they are both linear molecules and that the C—O bond length is shorter in CO than in CO_2 (see Fig. 6-5). We should be able to explain this difference in bond length by using our proposed bond model. We could write a Lewis structure for CO_2

---

* It has been found experimentally that CH_4 has a tetrahedral geometry. We shall consider this in Section 6-5.

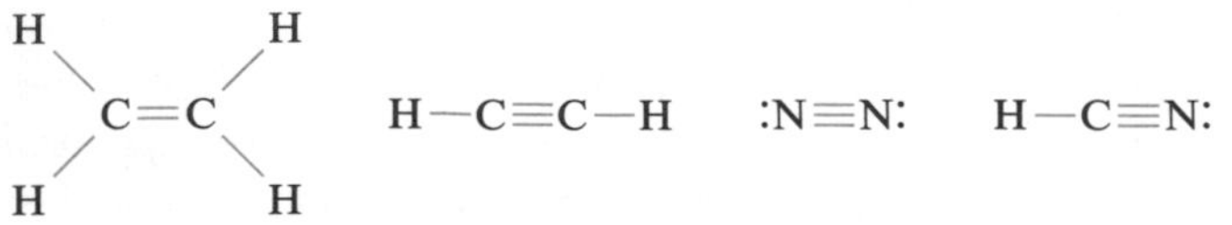

$$\begin{matrix} H & & H \\ & \diagdown & \diagup & \\ & C=C & \\ & \diagup & \diagdown & \\ H & & H \end{matrix} \qquad H-C\equiv C-H \qquad :N\equiv N: \qquad H-C\equiv N:$$

Fig. 6–6 Examples of single, double, and triple bonds, using Lewis structures. Ethylene (C_2H_4) contains a C=C double bond with all atoms in one plane. Acetylene (C_2H_2), nitrogen (N_2), and hydrogen cyanide (HCN) are linear molecules containing triple bonds.

such as

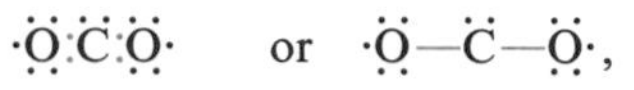

$$\cdot \ddot{O}\!:\!\ddot{C}\!:\!\ddot{O}\cdot \quad \text{or} \quad \cdot\ddot{O}-\ddot{C}-\ddot{O}\cdot,$$

but this structure does not obey the octet rule, since the oxygen atoms have only seven electrons associated with them and the carbon atom only six. If the octets of the carbon and oxygen atoms are to be completed, more electrons must be shared between them. If four electrons are shared between the carbon atom and each oxygen atom (i.e., four electrons in each bond), we can satisfy the octet rule,

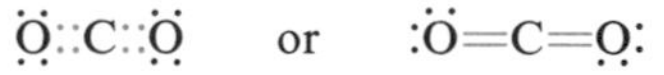

$$\ddot{O}\!::\!C\!::\!\ddot{O} \quad \text{or} \quad :\ddot{O}=C=\ddot{O}:$$

With this Lewis structure we see that both oxygen atoms and the carbon atom are associated with eight electrons. When two electron pairs (four electrons total) are shared between two atoms, a *double bond* is formed; when only one electron pair is shared, there is a *single bond*.

The situation with carbon monoxide is somewhat similar, except that, to fulfill the octet rule, three electron pairs must be shared:

$$:C\!:::\!O: \quad \text{or} \quad :C\equiv O:$$

When three electron pairs are shared between two atoms, a *triple bond* is formed.

We now have a qualitative idea of why the C—O bond length is less in CO than it is in CO_2. There are more bonds between the C and O atoms in CO. The greater number of bonds "pull" the atoms closer together. Crudely, we can think of the bonds as small springs. As the number of springs is increased, the strength of the bond is increased and the atoms are moved closer together.

In order to break the bond between two atoms, the atoms must be separated until they no longer interact. The energy required to break a bond is called the *bond energy*.* We can see that as the number of bonds formed between two atoms increases, the bond strength also increases (the spring becomes stronger). Therefore we can expect that in bonds between the same two atoms, triple bonds will be stronger than double bonds and double bonds stronger than single bonds. More Lewis structures illustrating various bonding arrangements are shown in Fig. 6–6.

---

*   Common units for bond energies are *kilocalories per mole* and *electron volts per molecule*; 1 eV = 23.07 kcal/mole.

The C—O bond lengths for CO and CO_2 have been experimentally determined to be 1.13 and 1.16 Å, respectively. The "average" length of the double bonds between carbon and oxygen in other chemical compounds has been found to be 1.22 Å. This indicates that there is some triple-bond character to the C—O bonds in CO_2; the bonds are actually shorter than a double bond should be. This is not predicted by the proposed bonding model. In order to account for this behavior of CO_2 and for similar experimental observations with many other molecules, such as CO and SO_2, we must revise our bonding model once again.

Resonance. If we try to write Lewis structures for a molecule such as sulfur dioxide (SO_2), we find that no single structure can correctly represent the SO_2 molecule and still fulfill the octet rule. A structure such as

comes pretty close, but note that we have arbitrarily placed the double bond between the sulfur atom and the oxygen atom on the left. There is actually no reason to select this particular oxygen atom over the one on the right. We could just as easily choose

as the proper Lewis structure.

Experiments show that both S—O bonds in SO_2 are identical and have properties that lie somewhere between those expected for single and double bonds. For example, when the bond length is measured or the bond energy is determined, it is found that the experimental values lie between those expected for an average single bond and an average double bond between sulfur and oxygen atoms.

We can account for this by making another slight alteration or addition to our bonding model. Both Lewis structures can be written and the actual structure visualized as some intermediate or hybrid structure between the two. Lewis structures, such as those written for SO_2, that approximate the true molecular structure and lead to the formation of a more realistic hybrid structure are called *resonance structures*. The resonance structures for SO_2 are

leading to

an intermediate structure or *resonance hybrid*.

▶ Remember that the individual resonance structures that are written for any compound do not actually exist; and they are not rapidly changing from one resonance structure to another, i.e., resonating. The structure of the compound that we believe actually exists is represented by the resonance *hybrid*. ◀

When we use the proposed model that chemical bonds are formed by electron pair sharing to complete noble-gas electron configurations, there are many compounds that require resonance structures for the proper description of their bonding. According to this approach to chemical bonding, the actual structures for these compounds need not contain single, double, or triple bonds, but may involve some intermediate type bond. Try writing some resonance structures that obey the octet rule for molecules such as sulfur trioxide (SO_3) and ozone (O_3) and for the nitrite ion (NO_2^-).

There are several common and seemingly simple molecules that appear to violate our bond model unless rather exotic resonance structures are proposed. No ordinary Lewis structures that adequately describe the properties of the substances can be written for oxygen (O_2) and nitrogen II oxide (nitric oxide), NO, gases. It has been found experimentally that O_2 has two unpaired electrons and NO one unpaired electron. The structure

$$\ddot{O}=\ddot{O}$$

can be written for O_2, but all electrons are paired. This structure, then, predicts the wrong properties for O_2. Try writing some structures for O_2 and NO that will account for the proper number of unpaired electrons. Do they satisfy the octet rule? Actually, this complete failure to explain the O_2 molecule is one of the most serious problems with the bond model that we have been considering. We can readily overcome the difficulty, however, when we use more recent bond theories.

Shapes of Molecules. Another major problem with the model of chemical bonding that we have considered is that it gives very little help in predicting or explaining the geometric forms in which molecules occur. We can consider the three triatomic molecules H_2O, CO_2, and SO_2 as examples. Water has a bent structure with a bond angle of about 105°, close to the "ideal" tetrahedral angle of 109°28'. Carbon dioxide is a linear molecule (all three atoms lie in a straight line), while sulfur dioxide is angular, but with a bond angle of about 120°, close to the "ideal" trigonal angle of 120°.

Figure 6–7 gives the shapes of these triatomic molecules. In order to understand the reasons for the varying shapes, we must resort to a different theory of chemical bonding, or at least make appropriate additions to our present model (see Section 6–5 for further discussion).

▶ We can envision the ideal trigonal bond angle by observing the bent SO_2 structure in Fig. 6–7. Place the S atom at the center of an equilateral triangle; the two O atoms

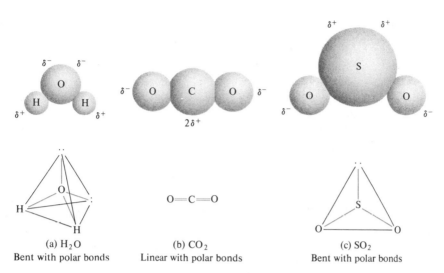

(a) H_2O (b) CO_2 (c) SO_2

Bent with polar bonds   Linear with polar bonds   Bent with polar bonds

Fig. 6–7 The different shapes of the three triatomic molecules, H_2O, CO_2, and SO_2. Note that H_2O and SO_2 have dipole moments, but that CO_2 does not. Since CO_2 is linear, the centers of positive and negative charge coincide.

then fall at two of the vertices of the triangle, with the third vertex above the S. The angle formed by the O—S—O atoms is 120°, called the *ideal trigonal angle.*

The ideal tetrahedral bond angle is shown in Fig. 6–15 for CH_4. With the C atom at the center of a tetrahedron (e.g., a tetrapack containing cream or milk), the angle formed between the H—C—H atoms is 109.28°, called the *ideal tetrahedral angle.* ◄

Carbon compounds present similar difficulties. It has been found experimentally that the carbon atom usually forms four bonds that are directed in a tetrahedral fashion, as shown in Fig. 6–7 for the H_2O molecule and in Fig. 6–15 for H_2O, NH_3, CH_4, and $CHCl_3$. The bond model that we have been using provides no reasons for, or understanding of, this tetrahedral structure of carbon. This structure is, however, explained by more sophisticated ideas (see Section 6–5).

Note that the shapes of the molecules, along with the bond polarities, determine the size of the molecule's dipole moments. In Fig. 6–7 we can see that CO_2 has no net dipole moment, even though the C—O bonds are polar. In the CO_2 molecule, the centers of positive and negative charge coincide. This is not the case for H_2O and SO_2, both of which are bent.

Methane (CH_4) and carbon tetrachloride (CCl_4) present a situation similar to CO_2. Even though each bond is polar, there is no net dipole moment for the molecules, since the plus and minus charge centers are the same. Table 6–3 lists dipole moments for some polyatomic molecules, plus their general shapes. Why should chloroform have a dipole moment when CCl_4 does not? Which "end" of the molecule would be positively charged?

6-3 IONIC BONDS

When we inspect the data on dipole moments in Tables 6–2 and 6–3, one large difference is clearly evident. The dipole moments of CsCl and CsF are much larger than those of any of the other compounds. This indicates that the separation into plus and minus charges is much greater in CsCl and CsF than in the other compounds. In fact, experimentally it is found that the Cs atom has essentially a full $+1$ charge and the Cl or F atom a full -1 charge, not just partial charges ($\pm \delta$), as in the compounds considered previously.

▶ Dipole moments can be measured for gaseous CsCl and CsF molecules, as indicated previously for HCl. In solid CsCl and CsF, infrared spectroscopic studies indicate the presence of a charge on the Cs that is essentially $+1$ and on the Cl or F that is essentially -1. ◀

Table 6–3. Dipole moments of polyatomic molecules

Molecule	Shape	Dipole moment, Debyes
CO_2	Linear	0
CS_2	Linear	0
COS	Linear	0.71
HCN	Linear	2.95
BF_3	Trigonal planar	0
SO_3	Trigonal planar	0
CH_4	Tetrahedral	0
CCl_4	Tetrahedral	0
NH_3	Trigonal pyramidal	1.47
PH_3	Trigonal pyramidal	0.55
AsH_3	Trigonal pyramidal	0.16
H_2O	Angular	1.84
H_2S	Angular	0.92
SO_2	Angular	1.63
NO_2	Angular	0.39
O_3	Angular	0.52

In order for the Cs atom to have a $+1$ charge in the compounds, it must have lost one electron (remember that electrons are negatively charged and that a loss of negative charge leaves a positive charge behind). The Cl or F atoms, therefore, must have gained one electron and become negatively charged. Charged atoms (or groups of atoms) that have either lost or gained electrons are called *ions*. Positively charged ions are *cations*; negatively charged ions are *anions*. An *ionic bond* is a chemical bond

in which electron transfer has occurred, and the attractive force is the opposite charge attraction between the cation and anion formed. Compounds that contain ionic bonds are called *ionic* or *electrovalent compounds*.

There are many common examples of ionic compounds: sodium chloride (common table salt), NaCl; sodium hydroxide (lye), NaOH; calcium carbonate (limestone), $CaCO_3$; aluminum oxide (sapphire or ruby), Al_2O_3; and so on. Figure 6–8 gives some other examples, together with their appropriate Lewis structures. Note that it is customary to write the chemical formulas for ionic compounds without including the plus and minus charges. The charges are indicated when the Lewis structures are drawn. To maintain charge balance, it is always necessary that the number of electrons lost be equal to the number of electrons gained.

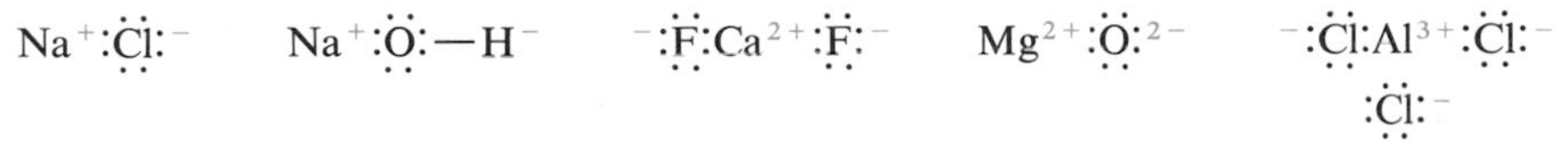

$$Na^+ :\!\ddot{Cl}\!:^- \qquad Na^+ :\!\ddot{O}\!: -H^- \qquad {}^- :\!\ddot{F}\!: Ca^{2+} :\!\ddot{F}\!:^- \qquad Mg^{2+} :\!\ddot{O}\!:^{2-} \qquad {}^- :\!\ddot{Cl}\!: Al^{3+} :\!\ddot{Cl}\!:^-$$

$$:\!\ddot{Cl}\!:^-$$

Fig. 6–8 Lewis structures for the ionic compounds sodium chloride (NaCl), sodium hydroxide (NaOH), calcium fluoride (CaF_2), magnesium oxide (MgO), and aluminum chloride ($AlCl_3$).

Factors Involved in Ionic Bonding. The formation of ionic bonds can be considered as occurring in three separate steps:

1. The loss of electrons by one element.
2. The gain of electrons by another element.
3. The attraction of the positive and negative ions formed.

Actually we cannot detect these separate processes when an ionic bond forms. Again we are proposing a simplified and easily visualized model that enables us to think about the problem in more concrete terms.

To illustrate the three steps involved, let us consider the ionic compound sodium fluoride, NaF. It is known from experiments that Na^+ cations and F^- anions make up the compound. Then the hypothetical steps in forming the compound would be:

1. $Na\cdot \rightarrow Na^+ + e^-$
2. $:\!\ddot{F}\!\cdot + e^- \rightarrow :\!\ddot{F}\!:^-$
3. $Na^+ \cdots F^-$ opposite charge attraction,
   $Na^+ + :\!\ddot{F}\!:^- \rightarrow Na^+ :\!\ddot{F}\!:^-$

In the first step an electron is removed from the neutral sodium atom. This process takes energy. Atoms do not eject electrons spontaneously; they hold onto them. The energy that is required to remove an electron from a gaseous atom is called the *ionization potential* (IP). Ionization potentials are generally expressed in *electron volts* (eV), which are units of energy. To pull an electron from the neutral sodium atom requires an expenditure of energy equal to the ionization potential:

$$\text{Na·} + (\text{IP} = 5 \text{ eV}) \rightarrow \text{Na}^+ + e^-.$$

All atoms except hydrogen (see Fig. 6–9) have more than one ionization potential. The *first ionization potential* (IP_1) is the energy required to remove an electron from the neutral atom; the *second ionization potential* (IP_2) is the energy required to remove an electron from the singly charged cation; and so on. For example, the first three ionization potentials for calcium are:

$$\text{Ca:} + (\text{IP}_1 = 6 \text{ eV}) \rightarrow (\text{Ca·})^+ + e^-$$
$$(\text{Ca·})^+ + (\text{IP}_2 = 12 \text{ eV}) \rightarrow \text{Ca}^{2+} + e^-$$
$$\text{Ca}^{2+} + (\text{IP}_3 = 51 \text{ eV}) \rightarrow \text{Ca}^{3+} + e^-$$

Note that the ionization potentials increase in energy from IP_1 to IP_2 to IP_3. This is true for all atoms; IP_1 is always the lowest in energy. Can you guess the reason for this? Why is IP_3 so much larger than IP_2 for Ca? (See the discussion in Section 6–4.) Table 6–4 lists the first ionization potential (where known) and the electron configurations for all the elements.

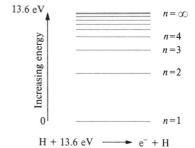

Fig. 6–9 The ionization potential of the hydrogen atom. It takes 13.6 eV of energy to remove the electron from the 1s orbital of a hydrogen atom. Removal of this one electron leaves a bare hydrogen nucleus, which is a single proton.

The second (imagined) step in the formation of the sodium fluoride ionic bond involves the addition of an electron to the neutral fluorine atom. In the case of fluorine, this process releases energy, i.e., it is a favorable process. With some of the other elements, the addition of an electron to the neutral atom takes energy. The change in energy when an electron is added to a neutral gaseous atom is called the *electron affinity* (EA) of the atom. For fluorine this process can be diagrammed as follows:

$$\text{:F·} + e^- \rightarrow \text{:F:}^- + (\text{EA} = 3.45 \text{ eV}).$$

Table 6–4. Electron configuration and first ionization potential for the elements

Atomic number	Element	Electronic configuration	First ionization potential, eV
1	H	1s	13.60
2	He	$1s^2$	24.48
3	Li	(He)2s	5.39
4	Be	$(He)2s^2$	9.32
5	B	$(He)2s^22p$	8.30
6	C	$(He)2s^22p^2$	11.26
7	N	$(He)2s^22p^3$	14.53
8	O	$(He)2s^22p^4$	13.61
9	F	$(He)2s^22p^5$	17.42
10	Ne	$(He)2s^22p^6$	21.56
11	Na	(Ne)3s	5.14
12	Mg	$(Ne)3s^2$	7.64
13	Al	$(Ne)3s^23p$	5.98
14	Si	$(Ne)3s^23p^2$	8.15
15	P	$(Ne)3s^23p^3$	10.48
16	S	$(Ne)3s^23p^4$	10.36
17	Cl	$(Ne)3s^23p^5$	13.01
18	Ar	$(Ne)3s^23p^6$	15.76
19	K	(Ar)4s	4.34
20	Ca	$(Ar)4s^2$	6.11
21	Sc	$(Ar)4s^23d$	6.54
22	Ti	$(Ar)4s^23d^2$	6.82
23	V	$(Ar)4s^23d^3$	6.74
24	Cr	$(Ar)4s3d^5$	6.76
25	Mn	$(Ar)4s^23d^5$	7.43
26	Fe	$(Ar)4s^23d^6$	7.87
27	Co	$(Ar)4s^23d^7$	7.86
28	Ni	$(Ar)4s^23d^8$	7.63
29	Cu	$(Ar)4s3d^{10}$	7.72
30	Zn	$(Ar)4s^23d^{10}$	9.39
31	Ga	$(Ar)4s^23d^{10}4p$	6.00
32	Ge	$(Ar)4s^23d^{10}4p^2$	7.88
33	As	$(Ar)4s^23d^{10}4p^3$	9.81
34	Se	$(Ar)4s^23d^{10}4p^4$	9.75
35	Br	$(Ar)4s^23d^{10}4p^5$	11.84
36	Kr	$(Ar)4s^23d^{10}4p^6$	14.00
37	Rb	(Kr)5s	4.18
38	Sr	$(Kr)5s^2$	5.69
39	Y	$(Kr)5s^24d$	6.38
40	Zr	$(Kr)5s^24d^2$	6.84

(Continued)

Table 6–4 (*continued*)

Atomic number	Element	Electronic configuration	First ionization potential, eV
41	Nb	$(Kr)5s4d^4$	6.88
42	Mo	$(Kr)5s4d^5$	7.10
43	Tc	$(Kr)5s^24d^5$	7.28
44	Ru	$(Kr)5s4d^7$	7.36
45	Rh	$(Kr)5s4d^8$	7.46
46	Pd	$(Kr)4d^{10}$	8.33
47	Ag	$(Kr)5s4d^{10}$	7.57
48	Cd	$(Kr)5s^24d^{10}$	8.99
49	In	$(Kr)5s^24d^{10}5p$	5.78
50	Sn	$(Kr)5s^24d^{10}5p^2$	7.34
51	Sb	$(Kr)5s^24d^{10}5p^3$	8.64
52	Te	$(Kr)5s^24d^{10}5p^4$	9.01
53	I	$(Kr)5s^24d^{10}5p^5$	10.45
54	Xe	$(Kr)5s^24d^{10}5p^6$	12.13
55	Cs	$(Xe)6s$	3.89
56	Ba	$(Xe)6s^2$	5.21
57	La	$(Xe)6s^25d$	5.61
58	Ce	$(Xe)6s^24f5d$	5.6
59	Pr	$(Xe)6s^24f^3$	5.46
60	Nd	$(Xe)6s^24f^4$	5.51
61	Pm	$(Xe)6s^24f^5$	—
62	Sm	$(Xe)6s^24f^6$	5.6
63	Eu	$(Xe)6s^24f^7$	5.67
64	Gd	$(Xe)6s^24f^75d$	6.16
65	Tb	$(Xe)6s^24f^9?$	5.78
66	Dy	$(Xe)6s^24f^{10}$	6.8
67	Ho	$(Xe)6s^24f^{11}$	—
68	Er	$(Xe)6s^24f^{12}$	6.08
69	Tm	$(Xe)6s^24f^{13}$	5.81
70	Yb	$(Xe)6s^24f^{14}$	6.2
71	Lu	$(Xe)6s^24f^{14}5d$	5.0
72	Hf	$(Xe)6s^24f^{14}5d^2$	7
73	Ta	$(Xe)6s^24f^{14}5d^3$	7.88

(*Continued*)

Table 6–4 (*continued*)

Atomic number	Element	Electronic configuration	First ionization potential, eV
74	W	$(Xe)6s^24f^{14}5d^4$	7.98
75	Re	$(Xe)6s^24f^{14}5d^5$	7.87
76	Os	$(Xe)6s^24f^{14}5d^6$	8.5
77	Ir	$(Xe)6s^24f^{14}5d^7$	9
78	Pt	$(Xe)6s4f^{14}5d^9$	9.0
79	Au	$(Xe)6s4f^{14}5d^{10}$	9.22
80	Hg	$(Xe)6s^24f^{14}5d^{10}$	10.43
81	Tl	$(Xe)6s^24f^{14}5d^{10}6p$	6.11
82	Pb	$(Xe)6s^24f^{14}5d^{10}6p^2$	7.42
83	Bi	$(Xe)6s^24f^{14}5d^{10}6p^3$	7.29
84	Po	$(Xe)6s^24f^{14}5d^{10}6p^4$	8.43
85	At	$(Xe)6s^24f^{14}5d^{10}6p^5$	9.5
86	Rn	$(Xe)6s^24f^{14}5d^{10}6p^6$	10.75
87	Fr	$(Rn)7s$	4
88	Ra	$(Rn)7s^2$	5.28
89	Ac	$(Rn)7s^26d$	6.9
90	Th	$(Rn)7s^26d^2$	6.95
91	Pa	$(Rn)7s^25f^26d$	—
92	U	$(Rn)7s^25f^36d$	6.08
93	Np	$(Rn)7s^25f^46d$	—
94	Pu	$(Rn)7s^25f^6$	5.1
95	Am	$(Rn)7s^25f^7$	6.0
96	Cm	$(Rn)7s^25f^76d$	—
97	Bk	$(Rn)7s^25f^9$	—
98	Cf	$(Rn)7s^25f^{10}$	—
99	Es	$(Rn)7s^25f^{11}$	—
100	Fm	$(Rn)7s^25f^{12}$	—
101	Me	$(Rn)7s^25f^{13}$	—
102	No	$(Rn)7s^25f^{14}$	—
103	Lr	$(Rn)7s^25f^{14}6d$	—

Table 6–5. Electron affinities of some atoms and their relation to electron configurations*

Element	Group	Electron affinity, eV	Electron configuration of: Atom	Anion
H		0.75	$1s^1$	$1s^2$
Li	IA	(0.54)	(He) $2s^1$	(He) $2s^2$
Na		(0.74)	(Ne) $3s^1$	(Ne) $3s^2$
Be	IIA	(−0.6)	(He) $2s^2$	(He) $2s^2 2p^1$
Mg		(−0.3)	(Ne) $3s^2$	(Ne) $3s^2 3p^1$
B	IIIA	(0.2)	(He) $2s^2 2p^1$	(He) $2s^2 2p^2$
Al		(0.6)	(Ne) $3s^2 3p^1$	(Ne) $3s^2 3p^2$
Ga		(0.2)	(Ar) $3d^{10} 4s^2 4p^1$	(Ar) $3d^{10} 4s^2 4p^2$
In		(0.2)	(Kr) $4d^{10} 5s^2 5p^1$	(Kr) $4d^{10} 5s^2 5p^2$
C	IVA	1.25	(He) $2s^2 2p^2$	(He) $2s^2 2p^3$
Si		(1.6)	(Ne) $3s^2 3p^2$	(Ne) $3s^2 3p^3$
Ge		(1.2)	(Ar) $3d^{10} 4s^2 4p^2$	(Ar) $3d^{10} 4s^2 4p^3$
O	VIA	1.47	(He) $2s^2 2p^4$	(He) $2s^2 2p^5$
S		2.07	(Ne) $3s^2 3p^4$	(Ne) $3s^2 3p^5$
Se		(1.7)	(Ar) $3d^{10} 4s^2 4p^4$	(Ar) $3d^{10} 4s^2 4p^5$
Te		(2.2)	(Kr) $4d^{10} 5s^2 5p^4$	(Kr) $4d^{10} 5s^2 5p^5$
F	VIIA	3.45	(He) $2s^2 2p^5$	(He) $2s^2 2p^6$
Cl		3.61	(Ne) $3s^2 3p^5$	(Ne) $3s^2 3p^6$
Br		3.36	(Ar) $3d^{10} 4s^2 4p^5$	(Ar) $3d^{10} 4s^2 4p^6$
I		3.06	(Kr) $4d^{10} 5s^2 5p^5$	(Kr) $4d^{10} 5s^2 5p^6$

*   Values of electron affinities which are in parentheses have been estimated by calculation methods and have not been verified experimentally.

The electron affinity is positive when energy is released and negative when energy is absorbed. Table 6–5 gives some representative values. Note that Group VIIA elements have much higher electron affinities than the other elements. Why should this be true?

▶ Electron affinities are difficult to measure experimentally. They are often calculated using other atomic property data that are more easily obtained experimentally. Ionization potentials, on the other hand, can be determined relatively easily and quite accurately using spectroscopic methods. ◀

In the third (imagined) step, the cation and anion that have been formed attract one another. This process releases energy. We can see then that, in order for a stable ionic bond to form, it is necessary for the energy released in steps 2 and 3 to be larger than the energy required for step 1. In the case of NaF, step 1 requires the IP_1 of the sodium atom, 5.14 eV; step 2 releases the EA of the fluorine atom, 3.45 eV; then, for NaF to be a stable compound, step 3 must release more than $5.14 - 3.45 = 1.69$ eV.

Electronegativity. We have studied nonpolar covalent bonds such as those in H_2, Cl_2, O_2, N_2, etc., and ionic bonds such as those in NaF, CsF, CsCl, CaF_2, etc. We have also considered compounds with polar covalent bonds, which can be considered as covalent bonds with some ionic character. One question suggests itself immediately: How do we know when a compound has polar covalent bonding and when it has ionic bonding? What distinguishes these two types of chemical bonds from one another? This is a rather difficult question. From experiments we know that no chemical compounds have 100% ionic bonding, although some of the alkali halides come very close. We must resort to saying that those compounds that are predominantly ionic will be called ionic compounds, while those that are only somewhat ionic will be called polar covalent compounds.

In order to try to define "predominantly" and "somewhat" a little better, Linus Pauling* introduced a new idea in the 1930's. He called the attraction that an atom has for an electron in a chemical bond the *electronegativity* of the atom. The electronegativity is closely related to the ionization potential and electron affinity of an atom. Robert S. Mulliken, 1966 Nobel laureate in chemistry, proposed that the electronegativity be equal to the average of the IP and EA:

$$\frac{IP + EA}{2}.$$

Although easy in concept, this proposal is difficult in practice, since so few reliable values of electron affinities are known. Generally, therefore, we use Pauling's scale of electronegativities, which he developed by considering the known bond energies of many different substances. Table 6–6 gives values of Pauling's electronegativities for the elements. The element fluorine has the highest electronegativity and has been assigned the value of 4.0 (electronegativities have no units; they are pure numbers). All other elements have assigned electronegativity values that are relative to the value for fluorine.

▶ Remember that the IP and EA are actual, physically measurable quantities of atoms, while electronegativities are assigned values that apply when chemical bonds are being considered. ◀

---

*　Linus Pauling won the Nobel prize in chemistry in 1954 and the Nobel prize for peace in 1962. He and Madame Curie are the only persons to have received two Nobel prizes.

Table 6–6. Electronegativities

H 2.1																
Li 1.0	Be 1.5											B 2.0	C 2.5	N 3.0	O 3.5	F 4.0
Na 0.9	Mg 1.2											Al 1.5	Si 1.8	P 2.1	S 2.5	Cl 3.0
K 0.8	Ca 1.0	Sc 1.3	Ti 1.4	V 1.6	Cr 1.6	Mn 1.5	Fe 1.8	Co 1.8	Ni 1.8	Cu 1.9	Zn 1.6	Ga 1.6	Ge 1.8	As 2.0	Se 2.4	Br 2.8
Rb 0.8	Sr 1.0	Y 1.2	Zr 1.4	Nb 1.6	Mo 1.8	Tc 1.9	Ru 2.2	Rh 2.2	Pd 2.2	Ag 1.9	Cd 1.7	In 1.7	Sn 1.8	Sb 1.9	Te 2.1	I 2.5
Cs 0.7	Ba 0.9	*	Hf 1.3	Ta 1.5	W 1.7	Re 1.9	Os 2.2	Ir 2.2	Pt 2.2	Au 2.4	Hg 1.9	Tl 1.8	Pb 1.8	Bi 1.9	Po 2.0	At 2.2
Fr 0.7	Ra 0.9															

*   The lanthanides appear to have electronegativities of 1.1–1.2.

†   The actinides appear to have electronegativities of 1.3 except: Ac Pa U
                                                                      1.1 1.5 1.7

We can see that the nonmetallic elements have high electronegativity values, while the metallic elements—particularly Groups IA and IIA—have relatively low ones. When a nonmetal such as fluorine is combined with a metal such as sodium, there is a large difference in electronegativity values between the atoms. The atom with the higher electronegativity attracts the electrons in the chemical bond more strongly. The fluorine atom, therefore, attracts bonding electrons more strongly than the sodium atom. In this case the large difference in electronegativity values reflects the fact that there is an electron transferred from the sodium to the fluorine atom, i.e., the attraction of the fluorine for electrons in a bond is so much greater than that of sodium that an electron actually moves from the sodium to the fluorine atom. The result is an ionic bond; there is a large difference in electronegativity values between the bonded atoms. When the difference in electronegativities of the atoms is small, a polar covalent bond is formed; when the difference is zero, a nonpolar covalent bond is formed.

▶ Note that a nonpolar bond could be formed between two unlike atoms (as well as between like atoms) if they happened to have identical electronegativity values. ◀

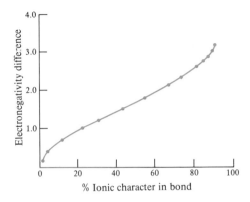

Fig. 6-10 Percent ionic character in a chemical bond is related to the electronegativity difference between the two atoms in the bond.

Figure 6–10 shows a more quantitative relationship between electronegativity differences and the degree of ionicity of a bond. Again the extreme cases are rather easy to categorize. Above about 80% ionic character we tend to speak of ionic bonds, and at 0.0% ionic character we speak of nonpolar covalent bonds. Between these two extremes lie the polar covalent bonds (or covalent bonds with some ionic character). The boundary between polar covalent and ionic bonds is neither sharp nor well defined.

The concept of electronegativity is useful in many other ways. The polarity of a chemical bond is indicated by the difference in electronegativity values of the two atoms. For example, we could predict that a F—Cl bond should be more polar than a Cl—Br bond, since the electronegativity differences are $4.0 - 3.0 = 1.0$ and $3.0 - 2.8 = 0.2$, respectively. Experimentally we find that the dipole moments of ClF and BrCl support this prediction (see Table 6–2), with the larger dipole moment of ClF signifying greater polarity. Table 6–7 gives similar data for the hydrogen halides.

Electronegativity differences can also be related to bond strengths. The strength of a chemical bond increases as the electronegativity difference between the two atoms increases. Table 6–7 shows this for the hydrogen–halide series.

Table 6–7. Relation of electronegativity difference to bond polarity and bond strength

Hydrogen-halide bond	Electronegativity of:		Electronegativity difference	% ionic character	Dipole moment, Debyes	Bond strength, eV
	H	Halogen				
H—F	2.1	4.0	1.9	59	1.9	5.87
H—Cl	2.1	3.0	0.9	19	1.1	4.48
H—Br	2.1	2.8	0.7	12	0.8	3.82
H—I	2.1	2.5	0.4	4	0.4	3.08

▶ Bond strengths can be conveniently measured experimentally by determining the amount of heat liberated or absorbed when elements react or when bonds are broken. Pauling used this fact in establishing his electronegativity series for the elements. ◀

Hydrogen and fluorine atoms have the largest difference in electronegativity in the series. The separation of plus and minus charge is therefore greater in HF than in the other hydrogen halides; i.e., fluorine attracts the shared electron pair most strongly. This is shown by the large dipole moment of HF. The separated charges assist in binding the two atoms together. The partial negative charge on the fluorine atom attracts the partial positive charge on the hydrogen atom, thereby increasing the strength of the bond between the two atoms. This opposite charge attraction (or partial ionic character) in a covalent bond becomes greater as the partial charges become larger. Hence HF has a greater bond strength than the other hydrogen halides, and there is a periodic trend in the series paralleling the trend in bond polarity indicated by the dipole moments.

6-4 VARIATIONS IN BONDING

We have seen that the type of chemical bond that is formed between two atoms depends on the ionization potentials and electron affinities of the atoms or, more concisely, on their electronegativity difference. These differences in bonding and also therefore in ionization potential, electron affinity, and electronegativity should be understandable if we recall the ideas of atomic structure and chemical bonding that we have been discussing. If these ideas are at all sound, we should be able to explain —and predict—any periodic variations of the properties, and give reasons why some atoms have high ionization potentials and electron affinities while others have relatively low ones, and why certain atoms tend to form ions and others do not. When we know and understand these property variations, we can make some helpful generalizations about the type of chemical bonds that will be formed between two atoms.

Ionization Potentials.   There are two interesting periodic variations of the ionization potentials of elements. One is the variation within a group. Generally ionization potentials decrease from top to bottom in a group. Table 6-8 shows this for the Group IA alkali metals.

We can understand this trend if we think about Bohr's model and the concepts of the quantum theory. There are three things to consider:

1.   The positive charge on the nucleus attracts the electrons in the atom. The larger the nuclear charge, the greater the attraction for the electrons. Lithium, for example, has a nuclear charge of $+3$, while that of potassium is $+19$.

Table 6–8. Ionization potentials for the Group IA alkali metals

Group IA element	Atomic number	IP_1, eV	Electron population of principal quantum level					
			1	2	3	4	5	6
Li	3	5.4	2	1				
Na	11	5.1	2	8	1			
K	19	4.3	2	8	8	1		
Rb	37	4.2	2	8	18	8	1	
Cs	55	3.9	2	8	18	18	8	1

2. The amount of attraction between the electrons and the nucleus decreases rapidly as the distance between them is increased. In lithium the first electron that is ionized is in principal quantum level $n = 2$, but in potassium it is in principal quantum level $n = 4$. Remember that, according to our theories of the atom, the higher the principal quantum number, the greater the distance of the electrons from the nucleus. This increase in separation between the electrons and the nucleus decreases the attractive force between them, making it easier to remove the outer electron.

3. The electrons that intervene between the nucleus and the outer electron (the one ionized) *shield* or *screen* part of the nuclear charge from the outer electron. In lithium, for example, the two electrons in the 1s sublevel shield part of the $+3$ nuclear charge from the 2s electron. This reduces the nuclear charge that is attracting the 2s electron from $+3$ to a value close to $+1$. Applying this same reasoning to the other alkali metals indicates that the *effective nuclear charge* that is attracting the electron in the outer s sublevel is close to $+1$ for all the alkali metals. The $+19$ nuclear charge for potassium is effectively shielded from the 4s electron by the 18

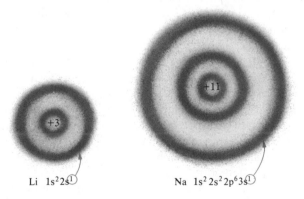

Li $1s^2 2s^1$ Na $1s^2 2s^2 2p^6 3s^1$

Fig. 6–11 The screening effect and the greater distance from the nucleus of the 3s electron of sodium give it a lower ionization potential than the 2s electron in lithium.

negatively charged electrons in the 1s, 2s, 2p, 3s, and 3p sublevels that are located between the nucleus and the 4s sublevel.

Considering all three of these factors for the alkali metals, we can see that as the atomic number increases the outer electron that is ionized is increasingly distant from the nucleus, which has an effective charge close to $+1$ for all the alkali metals, due to the screening effect of intervening electrons. The predominant factor, then, is the increase in the size of the alkali metal atoms as the atomic number increases. As the distance between the outer electron and the nucleus becomes greater (see Fig. 6–11), less energy is needed to remove the electron. In a given group of the periodic table, therefore, the ionization potential decreases as the atomic number increases (going down the group).

Table 6–9. Ionization potentials for the period 2 elements

Period 2 element	Li	Be	B	C	N	O	F	Ne
Atomic number	3	4	5	6	7	8	9	10
Ionization potential, eV	5.4	9.3	8.3	11.3	14.5	13.6	17.4	21.6
Electron configuration	$1s^2 2s^1$	$1s^2 2s^2$	$1s^2 2s^2 2p^1$	$1s^2 2s^2 2p^2$	$1s^2 2s^2 2p^3$	$1s^2 2s^2 2p^4$	$1s^2 2s^2 2p^5$	$1s^2 2s^2 2p^6$

The second interesting trend of ionization potentials is that found in periods in the periodic table. Table 6–9 gives the first ionization potentials for the period 2 elements, along with their electron configurations.

There is a noticeable increase from left to right in the period; lithium has an IP_1 of 5.4 eV, while neon has an IP_1 of 21.6 eV. We can explain this general trend by considering the same three things that were important in the group trend—the nuclear charge, the separation between the electrons and the nucleus, and the screening of nuclear charge by inner electrons—in conjunction with our atomic models. Note that electrons are being added to the same principal quantum level as the atomic number increases. This is a situation entirely different from the one that is true for the group trend. It means that the distance of the outer electrons from the nucleus does not increase with increasing atomic number. Furthermore, electrons in the same principal quantum level do not screen the nuclear charge from one another effectively. Therefore, when electrons are added to the same principal quantum level, as they are in going across period 2, they are attracted by an increasingly greater nuclear charge. Since these outer electrons are not more distant from the increased nuclear charge, they are attracted more strongly. Hence the ionization potential increases with increasing atomic number within a given period.

Figure 6–12 is a chart showing the relative sizes of the atoms and common ions.

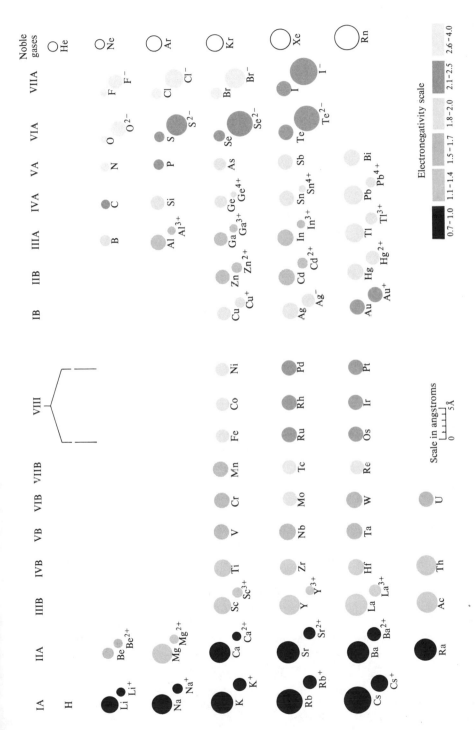

Fig. 6-12 The relative sizes of the atoms and ions of most of the elements. Electronegativities are indicated by the color code in lower right corner. (Based on arrangement by J. A. Campbell, *J. Chem. Educ.* **23**, 525, 1946.)

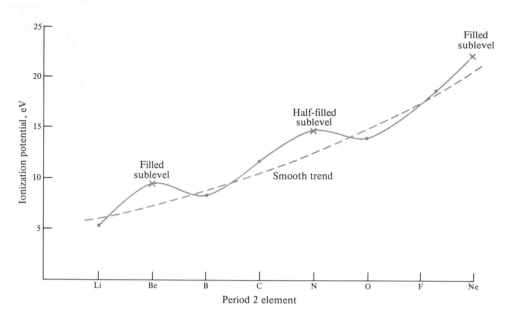

Fig. 6-13 The variation of the first ionization potential of the period 2 elements. The elements beryllium, nitrogen, and neon (with filled, half-filled, and filled electron sublevels, respectively) fall above a curve that shows a smoothly increasing trend.

Note that cations are smaller in size than atoms, while anions are larger.* Why should this be true? The same arguments used for variations in ionization potential can explain the trends observed for atom sizes.

If we examine the trend of ionization potentials of the period 2 elements (Table 6-9) more closely, we can find some variations in the trend. As shown in Fig. 6-13, the general increase from left to right across period 2 is apparent. But the trend is not smooth. The elements beryllium, nitrogen, and neon seem to have unsystematically high ionization potentials. These experimental data indicate another generalization that we can make when we note the electron configuration of these three elements: Beryllium and neon both have complete sublevels (the sublevels contain their maximum number of electrons), while nitrogen has a half-filled sublevel. In general we can say that filled and half-filled sublevels have extra stability. It is thus harder to remove an electron from one of these three elements.

In summary, ionization potentials decrease down a group and increase across a period. The metallic elements on the left of the periodic table have low ionization potentials; the nonmetallic elements on the right of the table have high ionization potentials. Then the elements with the lowest ionization potentials would be the metals in the lower-left-hand corner of the table, and the elements with the highest

---

*   The difference in size between anions and their corresponding atoms is probably much less than that indicated in Fig. 6-12. The discrepancy is a result of different methods being used to measure the atom sizes and the anion sizes.

ionization potentials would be the nonmetals in the upper-right-hand corner of the
table.

Electron Affinity. The electron affinities of elements exhibit periodic trends similar
to those of their ionization potentials, and for similar reasons. From top to bottom
in a group, the electron affinity decreases; from left to right in a period, the electron
affinity increases. Table 6–10 illustrates the variation within a group.

Table 6–10. Electron affinities of Group VIIA elements

Group VIIA element	Atomic number	Electron affinity, eV	Electron population in principal quantum level				
			1	2	3	4	5
F	9	3.45	2	7			
Cl	17	3.61	2	8	7		
Br	35	3.36	2	8	18	7	
I	53	3.06	2	8	18	18	7

There is one exception to the similar trends for ionization potentials and electron
affinities. The electron affinities of the noble gases are very low, and probably negative
(they have not yet been determined experimentally). They do not follow the general
trend of increasing electron affinities across a period, as the ionization potentials of
the noble gases do. Can you suggest a reason for the difference in periodic behavior
of the electron affinities and ionization potentials of the noble gases?

Electron affinities exhibit periodic variations for the same reasons that ionization
potentials do. From top to bottom in a group there is an increase in atomic size and
a concurrent screening of nuclear charge. Therefore an extra electron is less attracted
by the larger atoms at the bottom of the group. (Note the electron affinities of F and
Cl. Can you suggest a reason for this unsystematic behavior?) Across a period, however, the size of the atom does not increase and shielding of the increasing nuclear
charge is ineffective. This gives the atoms at the right of the periodic table (excluding
the noble gases) a greater attraction for an added electron.

Electronegativity. The periodic variations of the electronegativities of the elements
are very similar to those of the electron affinities. As shown in Table 6–6, electronegativities are lowest for the metals in the lower left corner of the periodic table and
highest for the nonmetals in the upper right corner. Again the noble gases are unsystematic in their behavior; they have electronegativities of zero.

Thus we can see how the electronegativity of an element is related to its ionization
potential and its electron affinity. An atom of the alkali metal sodium has a low IP
and a low EA, that is, it gives up an electron rather easily and has little attraction for
an extra electron. Its "attractive power" for an electron in a chemical bond is quite
low. Fluorine, on the other hand, has a high IP and EA, and consequently a high
electronegativity value. It has a strong attraction for an electron in a chemical bond.

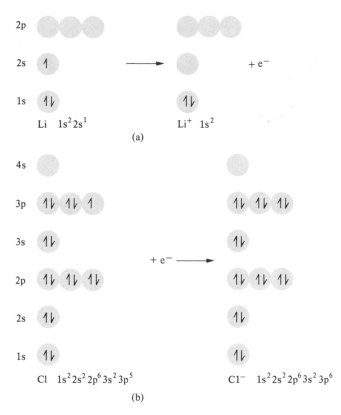

Fig. 6–14 A diagram showing that: (a) The ionization potential of lithium should be relatively low, since the loss of one electron gives Li^+, which has the same stable electron configuration that helium does. (b) The electron affinity of chlorine should be relatively high, since the addition of one electron gives Cl^-, which has the stable electron configuration that argon does.

Bond Differences. We are now in a position to understand why certain elements generally form ionic bonds with one another, while other elements favor covalent bonds. We can make some helpful generalizations that allow predictions of bond types, and we can justify many of the generalizations and periodic trends that we have encountered by using our ideas about atomic structure.

It has been found experimentally that the metallic elements at the left of the periodic table—Groups IA and IIA, for example—have low ionization potentials and electron affinities (giving them low—assigned—electronegativity values), while the nonmetals at the right of the table—Groups VIA and VIIA, for example—have high ionization potentials and electron affinities (giving them high—assigned—electronegativity values). This corresponds quite nicely with the idea that atoms strive to attain a noble-gas electron configuration. The metallic elements in Group IA should lose one electron relatively easily; that completes their noble-gas electron

configuration. (For an example, see Fig. 6–14a.) Their ionization potential should
therefore be relatively low. In the process the electron configuration of Na is changed
from $1s^2 2s^2 2p^6 3s^1$ for Na· to the neon configuration of $1s^2 2s^2 2p^6$ for Na$^+$:

$$Na· + (IP_1 = 5.1 \text{ eV}) \rightarrow Na^+ + e^-.$$

The electron affinity for sodium should be low, since the process

$$Na· + e^- \rightarrow Na: + (EA = 0.74 \text{ eV})$$

changes the electron configuration from $1s^2 2s^2 2p^6 3s^1$ to $1s^2 2s^2 2p^6 3s^2$, which is not
a stable noble-gas electron configuration.

For the Group VIIA elements the situation is different. The ionization potentials
are high, since removal of an electron does not give a noble-gas electron configuration:

$$:\ddot{F}: + (IP_1 = 17.4 \text{ eV}) \rightarrow :\dot{F}\cdot^+ + e^-$$
$$1s^2 2s^2 2p^5 \qquad\qquad\qquad 1s^2 2s^2 2p^4$$

The electron affinities are also high, since addition of one more electron does complete the stable octet, as shown in Fig. 6–14b:

$$:\ddot{Cl}: + e^- \rightarrow :\ddot{Cl}:^- + (EA = 3.61 \text{ eV})$$
$$1s^2 2s^2 2p^6 3s^2 3p^5 \quad 1s^2 2s^2 2p^6 3s^2 3p^6$$

Energy is released in the process.

Then when a Group IA element combines with a Group VIIA element to form a
chemical compound, we would expect the Group IA element to give up one electron
to attain a noble-gas electron configuration (it has a low ionization potential and a
low electron affinity). The Group VIIA element should accept one electron to complete its octet (it has a high ionization potential and a high electron affinity). The
combination of elements from these two groups favors an electron transfer between
the two atoms; an ionic bond is formed.

In general we can say that elements with low ionization potentials and low
electron affinities—and therefore low electronegativities—form ionic bonds with
elements that have high ionization potentials and electron affinities—and therefore
high electronegativities. This means that the metals at the left of the periodic table,
and in particular the Group IA and IIA elements, form ionic bonds (in general) with
the nonmetals at the right of the table (excluding the noble gases), and in particular
with the Group VIA and VIIA elements. Note that these elements are widely separated on the table. Elements that are close to one another on the periodic table tend
to form covalent bonds, while those that are far apart tend to form ionic bonds.

Remember that the combining ratios of the atoms are also consistent with our
ideas of atomic structure and bond formation. An alkali metal gives up one electron
to form an ionic bond with a halogen, which accepts the electron. But a compound
between an alkali metal and a Group VIA element, such as Na_2S, should contain two
alkali metal atoms to one atom of the Group VIA element, since two electrons are
needed by the VIA element to attain a noble-gas electron configuration. A similar
situation exists with a compound such as $CaCl_2$. Two Cl atoms are needed to accept

the two electrons released by the Ca atom in attaining the argon electron configuration.

We have already considered several other examples of bond formation in covalent compounds. Note that we have used the same reasoning process for both the covalent and ionic cases. The only difference involves the degree of sharing of electrons between the bonded atoms. Elements such as C, O, N, S, etc., tend to form covalent bonds with one another (they are close together on the periodic table), while elements such as Na, K, Mg, Ca, etc., tend to form ionic bonds with elements such as O, S, F, Cl, etc. (they are far apart on the periodic table).

6–5 MOLECULAR GEOMETRY

Even when we include resonance, one of the deficiencies of the model of chemical bonding that we have considered is that it does not predict or explain the geometrical shapes of molecules. We saw in Fig. 6–7 that the relatively simple compounds H_2O, SO_2, and CO_2 have quite different geometrical structures and yet each is composed of three atoms. Experiments also show that the compounds H_2O, NH_3, and CH_4 have similar geometrical arrangements, even though they contain different numbers of atoms (see Fig. 6–15).

Currently, there are several different bonding theories that can be used to explain (and in some cases predict) the geometries of molecules. We shall introduce only two here, and apply them to the molecules mentioned above. These examples will present ideas often used in studying organic chemistry, and should supply some feeling for the theories involved.

An Extension of the Lewis Concept. Actually chemists have made considerable progress toward understanding molecular geometries by adding to, and further refining, the bonding model that we have been using. If we consider the number of bonds between the atoms and the number of lone pairs of electrons in the molecule, and arrange the bonds and electron pairs in the most probable pattern around the central atom, i.e., the pattern in which the groups are as far apart as possible (see Table 6–11), we can obtain some idea of molecular geometry.

Table 6–11. The most probable configuration of a number of groups distributed around a central atom

No. groups	Geometry	Examples
2	Linear	CO_2, $HgCl_2$
3	Trigonal planar	SO_3, BF_3
4	Tetrahedral	CH_4, SiF_4
6	Octahedral	WF_6, $[Co(NH_3)_6]^{3+}$

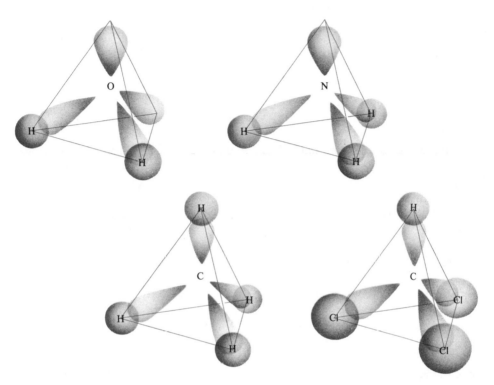

Fig. 6-15 Four molecules—water, ammonia, methane, and chloroform—that have structures based on tetrahedral symmetry.

We can consider H_2O, SO_2, and CO_2 in the light of these ideas. The Lewis structures for these three examples are:

$$\ddot{O}=C=\ddot{O}$$

(We write the structure for SO_2 in this fashion, in violation of the octet rule, to make electron counting more convenient. We could also use any of the resonance structures for SO_2 shown in Section 6-2.) In the case of H_2O, the oxygen has two bonds to hydrogen atoms and two lone electron pairs. We therefore have four groups to arrange around the central oxygen atom. A tetrahedron is the most probable arrangement for a group of four things distributed about a central point. In H_2O, the two hydrogen atoms and the two lone pairs are located at the points of a tetrahedron surrounding the central oxygen atom (see Fig. 6-15).

The situation is different for sulfur dioxide. In SO_2 there are two bonds between the S atom and the two O atoms (regardless of the number of electrons that compose each bond, they are located in the same general region of space around the central atom and are considered as one group) and one lone pair located on the central S atom. The most probable arrangement of these three groups around the S atom locates the two oxygen atoms and the lone electron pair at the vertices of a triangle, with the sulfur atom at the center (see Fig. 6–7).

Similar reasoning applies to carbon dioxide. Here there are only two groups to distribute around the central C atom, the two carbon-oxygen bonds. The CO_2 molecule is therefore linear, since this is the most probable distribution for two groups around a central point.

We can consider the ammonia (NH_3) and methane (CH_4) molecules in the same way. In each case there are four groups to be placed around the central atom. With NH_3, there are three bonds and one lone pair of electrons; with CH_4, there are four bonds. In each case the arrangement is tetrahedral (see Fig. 6–15).

This idea of placing bonds and electron pairs around a central atom in the most probable symmetry explains (and in many cases predicts) the experimentally observed molecular geometry of many chemical compounds. It has, of course, been developed in more detail than we have given. One refinement that is necessary is evident from our examples. The bond angle in SO_2—that is, the angle between the two oxygen atoms and the sulfur atom—has been found experimentally to be slightly less than the 120° trigonal angle that we would expect when the lone pair and two oxygen atoms are spread equally around the sulfur atom. A similar situation is noted for H_2O and NH_3. The H—O—H bond angle of about 105° in water and the H—N—H bond angle of 107° in ammonia are both less than the standard tetrahedral angle of 109°, while the H—C—H bond angle in methane is equal to the tetrahedral angle.

An explanation of these bond angle variations requires an addition or refinement of the model, which involves the unusual behavior of lone electron pairs on the central atom. Note that in CH_4 there are no lone pairs and that the bond angle is the one expected. But in the other compounds the lone pairs appear to "squeeze" the bond angles of the atoms together slightly. Since there are two lone pairs in H_2O, they affect the H—O—H bond angle more strongly than does the single lone pair in SO_2 and NH_3.

We can justify this idea of lone pairs "expanding" to occupy more than their share of space by noting that they are attracted by the nuclear charge of only one atom, while bonded electron pairs are influenced by two nuclei. Hence the larger charge clouds of the lone electron pairs interact with the bonding electron pair charge distributions more strongly than the bonding pairs interact with one another. This decreases the bond angle by forcing the atoms bonded to the central atom to move slightly closer together.

Hybridization. Another theory or approach to molecular geometries that is a great help in organic as well as inorganic chemistry involves the mixing-up of the s, p, and d atomic orbitals to produce composite orbitals that we call *hybrids*. The process is generally called the *hybridization* of atomic orbitals.

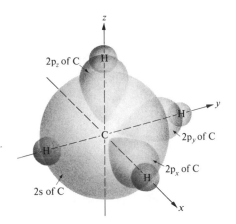

Fig. 6–16 A possible formation of bonds between one carbon and four hydrogen atoms which fails to account for the observed tetrahedral structure of CH_4.

We can take methane as a common example in which hybridization is needed to give the proper molecular shape. From experiments we know that CH_4 has the shape of a regular tetrahedron; all four of the C—H bonds are equivalent. The Lewis structure gives little help, since the hydrogens could be arranged in either a square plane about the carbon atom or tetrahedrally (or in any other peculiar fashion, for that matter).

We obtain a better picture if we consider the atomic orbitals actually involved in the bonding. In the case of carbon, remember that it is necessary to promote an electron from the 2s to the 2p sublevel to accommodate four bonding groups:

$$C \quad 1s^2 2s^2 2p^2 \quad \rightarrow \quad C \quad 1s^2 2s^1 2p^3$$

Then the four hydrogen atoms can each share one electron, as in Fig. 6–16.

But note what happens. Three hydrogens are associated with the three 2p orbitals of the carbon atom, and one hydrogen with the 2s carbon orbital. The hydrogen cannot be arranged tetrahedrally about the carbon atom, since the 2p orbitals are perpendicular to one another. Also, the 2s and 2p orbitals are not equivalent and their bonding properties are different. There should therefore be three identical C—H bonds and one C—H bond that is different, and there should not be a tetrahedral arrangement of bonds around the carbon. This picture of bonding, then, is contrary to experimental observations.

To circumvent these difficulties, we can mix the 2s and the three 2p atomic orbitals of carbon (or other elements with s and p bonding atomic orbitals) to give four hybrid orbitals called the sp^3 (pronounced s-p-three) orbitals. The charge distributions of the sp^3 orbitals are directed tetrahedrally instead of at right angles, as are the three p orbitals. This accommodates the tetrahedral bonding characteristics of the carbon atom in the formation of methane, carbon tetrachloride, and its many other compounds, as well as other atoms that form tetrahedrally shaped molecules.

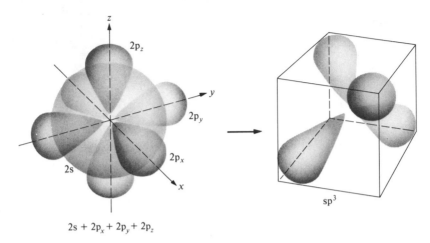

$$2s + 2p_x + 2p_y + 2p_z$$

Fig. 6–17 Four hybrid sp³ orbitals can be formed by proper combination of one s and three p atomic orbitals. The regions of maximum electron density in the four sp³ orbitals are directed to the four corners of a tetrahedron.

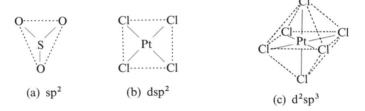

(a) sp²   (b) dsp²   (c) d²sp³

Fig. 6–18 (a) Sulfur uses three sp² hybrid orbitals formed by proper combination of one s and two p atomic orbitals to form the one planar SO_3 molecule. Bonds are directed at the vertices of an equilateral triangle. (b) In the complex ion tetrachloroplatinate (II), $PtCl_4^{2-}$, the central Pt atom uses dsp² hybrid orbitals to form bonds with the four Cl atoms, producing this experimentally observed square-planar structure. (c) In the hexachloroplatinate (III), $PtCl_6^{3-}$, complex ion, the central Pt atom uses d²sp³ hybrid orbitals to produce the experimentally observed octahedral symmetry.

Figure 6–17 diagrams the formation of sp³ hybrid orbitals from an s and three p orbitals. The fact that this particular mixture of atomic orbitals produces four sp³ hybrid orbitals that are oriented tetrahedrally is not obvious. Actually, the shapes of hybrid orbitals are calculated mathematically, using the equations of the quantum theory. Other hybrid orbitals are formed by combining s and p and d atomic orbitals in various ways. Some of these hybrids are pictured in Fig. 6–18 and listed in Table 6–12. (Note the similarities between Tables 6–11 and 6–12.)

Table 6-12. Hybrid orbitals formed from s, p, and d atomic orbitals

Hybrid	Number of groups	Geometry	Examples
sp	2	Linear	CO_2, BeH_2
sp^2	3	Planar, $120°$	SO_3, BH_3
sp^3	4	Tetrahedral	CH_4, VCl_4
dsp^2	4	Square planar	$[PtCl_4]^{2-}$, $[Ni(CN)_4]^{2-}$
d^2sp^3	6	Octahedral	$[Co(NH_3)_6]^{3+}$, $[Ti(H_2O)_6]^{3+}$

6-6 DIFFERENCES IN BONDING THEORIES

Since the geometries of the molecules just considered seem to be explained by two somewhat different ideas, we might ask the question: Which theory is actually correct? Our answer would have to be: neither one. In fact, even if we considered all the presently accepted bonding theories, the answer would still be the same. These theories, just like all others proposed by human beings, enable us to think about pertinent situations and often to solve some of the associated problems, which usually uncovers new problems that require new considerations and ideas. The best we can say for our present ideas is that they help us, at least sometimes, with our present problems and often lead us to increased understanding and improved ideas. We use our chemical bonding theories in situations in which they give reasonable results or agree with experiment. When they cannot be applied to the problem or are contrary to experimental fact, we must use a different theory or devise a new one.

In the following chapter we shall see how the ideas of atomic structure and chemical bonding are directly applicable to a study of chemical reactions.

QUESTIONS

1. Define or discuss the following terms: Diatomic molecule, atomic orbital, molecular orbital, bond length, chemical bond, covalent bond, polar covalent bond, ionic bond, octet rule, valence electrons, dipole moment, dipole, lone pair, double bond, bond energy, resonance, resonance hybrid, cation, anion, ionization potential, electron affinity, electronegativity, tetrahedral, percent ionic character, screening effect, effective nuclear charge, half-filled level, hybridization, sp^3.

2. Classify the following compounds as nonpolar covalent compounds and as polar covalent (covalent with ionic character) compounds:
   a) HCl   b) Br_2 c) H_2O d) CO   e) H_2
   f) BrCl   g) N_2 h) HBr   i) I_2 j) NO

3. Indicate which of the compounds in Question 2 would have a dipole moment of zero.

4. Draw Lewis structures for elements with atomic numbers 1 through 20.

5. Explain why the element fluorine ordinarily exists as a diatomic molecule F_2.

6. In chemical compounds, the elements fluorine and chlorine often exist as the anions F^- and Cl^-. Explain this observation.

7. Draw diagrams showing the electron density in the series of compounds HF, HCl, HBr, and HI.

8. Why does CO_2 have a zero net dipole moment while CO has a dipole moment of 0.1 debye?

9. a) Why does H_2O have a dipole moment while CO_2 does not?
   b) Would you expect H_2S to have a dipole moment? Why?
   c) Would you expect SO_3 to have a dipole moment? Why?

10. Refer to Table 6–2 and rank the diatomic molecules listed in order of increasing polarity.

11. a) Draw the Lewis structures of N_2, Cl_2, H_2 and S_2.
    b) Explain why the elements in part (a) form diatomic molecules readily, while elements such as Li, Be, Al, and Ne do not.
    c) Experimental evidence shows that the S_2 molecule contains two unpaired electrons. Modify the Lewis structure drawn in part (a) to account for this observation.

12. Explain why molecules such as CCl_4, CO_2, and SO_3 can contain polar bonds and yet have no dipole moment associated with the molecule.

13. Draw the Lewis structures for:

    a) NaF   b) $CaCl_2$ c) CsCl
    d) AlF_3 e) KBr   f) BaO

14. Predict the approximate percent ionic character in bonds formed between:

    a) Li and Cl   b) K and Cl   c) Cs and Cl   d) Na and O   e) Mg and O
    f) Si and O   g) S and O   h) Cl and O   i) Ar and O   j) O and O

15. Draw a graph showing the variation of electronegativity values in Group VIA. Explain any observed trend.

16. Select the Lewis structures that do not obey the octet rule and, where possible, draw new structures that do obey the rule.

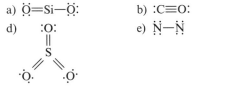

17. a) Explain what is meant by a single bond, a double bond, and a triple bond.
    b) If single, double, and triple bonds can be formed between atoms X and Z, what general statements can be made about the bond energies and bond lengths between X and Z?

18. Write as many resonance structures as you can that obey the octet rule for O_3, CO_3^{2-}, NO_2^-, and N_3^-.

19. Predict whether the following elements tend to form cations or anions.

    a) Sc   b) Br   c) Se   d) Ca   e) Be
    f) I   g) Ar   h) Al   i) Li   j) Ba

20. a) Explain why the first ionization potential of As is higher than that of Ge (see Table 6–4 for numerical values).

   b) Explain why the first ionization potential of As is higher than that of Se.

21. Why do the electron affinities of atoms generally decrease from top to bottom (going down) in a given group?

22. Justify the statement that in Group IA the electronegativities of the elements decrease as the atomic number increases.

23. a) Predict what ion charge would most probably be expected for Na, Ca, Al, O, and Cl.

   b) Justify the predictions in part (a).

24. Using barium fluoride (BaF_2) as an example, illustrate the properties that must be considered in forming a stable ionic compound.

25. Suppose that elements Q, X, and R are found in Groups IA, IIA, and VIIA, respectively.

   a) Will the elements be metals, nonmetals, or metalloids?

   b) Will the elements have high, low, or intermediate first ionization potentials?

   c) Will the elements have high, low, or intermediate electron affinities?

   d) Will the elements have high, low, or intermediate electronegativities?

   e) Will the elements form cations or anions readily?

   f) Will the ionic charge on X most probably be $1+$, $2+$, or $3+$?

   g) Will the compound formed between X and R contain ionic, polar covalent, or nonpolar covalent bonds?

   h) Will the compound formed between X and R have the formula XR, X_2R, XR_2?

26. Make predictions about the relative C—C bond lengths and bond strengths in the series: ethane (H_3C—CH_3), ethylene (H_2C=CH_2), and acetylene (HC≡CH).

27. Make a graph showing the variation of the first and second ionization potentials of the Group IIA alkaline earth metals.

   a) Explain any observed trend.

   b) The third ionization potentials for Be, Mg, and Ca are about 153, 80, and 51 eV, respectively. Why should IP_3 be so much greater than IP_2 for the Group IIA elements?

28. Explain why a general decrease is observed in the atomic radii of the elements K through Ni in period 4.

29. By drawing an appropriate graph, fill in the following tables.

Group IA element	M^+ ionic radius, Å	Group IIA element	M^{2+} ionic radius, Å
Li	0.60	Be	0.31
Na	0.95	Mg	0.65
K	1.33	Ca	0.99
Rb	–	Sr	1.13
Cs	1.67	Ba	–
Fr	1.76	Ra	1.40

30. It has been found experimentally that the molecule CF_4 is tetrahedral with no dipole moment and four equivalent C—F bonds. Explain these experimental observations.

31. The ionic radii of Na^+, Ca^{2+}, and F^- are 0.95, 0.99, and 1.36 Å, respectively. Calculate the bond lengths in NaF and CaF_2.

32. Write appropriate electron-dot formulas for stable compounds formed between Ge and O; Si and H; P and H; Na and H; Ca and H; S and H; Rb and Br; Sr and Cl; I and Cl.

33. In the series of compounds H_2O, NH_3, CH_4, why does the bond angle vary in the following way: H—O—H angle $= 105°$, H—N—H angle $= 107°$, H—C—H angle $= 109°$?

34. Why are the ionization potentials of the noble gases high, while their electron affinities and electronegativities are low?

35. The electron configurations of some elements are: (He) $2s^22p^1$; (Ne) $3s^23p^2$; (Ne) $3s^2$; (Ne) $3s^23p^5$, where (He) and (Ne) represent the $1s^2$ and $1s^22s^22p^6$ electron configuration of He and Ne, respectively. Without specifically identifying the elements:

   a) Place them in their proper periodic groups.
   b) Rank their values of electron affinity from low to high.

36. Explain why the H—N—H bond angle in NH_3 is 107°, while in the ammonium ion (NH_4^+) it is 109°.

37. Using H_2O and HF as examples, show what a lone pair is.

7. CHEMICAL REACTIONS

7-1 THE CHEMICAL BOND

Actually, in the study of chemical bonds and bond formation, we have been studying chemical reactions. We have learned that two hydrogen atoms bond together to form a diatomic hydrogen molecule and that chlorine atoms also form diatomic molecules. When these molecules are formed, a chemical reaction occurs between the atoms, i.e., a chemical change takes place. We can write these reactions in equation form,

$$H + H \rightarrow H_2 \quad \text{and} \quad Cl + Cl \rightarrow Cl_2.$$

Another chemical reaction that has been considered is the formation of sodium chloride, common table salt. We have seen that NaCl is an ionic compound since an electron is transferred from the Na atom to the Cl atom when the compound is formed. We can write a simple equation describing this chemical reaction:

$$2Na_{(s)} + Cl_{2(g)} \rightarrow 2NaCl_{(s)}.$$

▶ In many of the chemical equations throughout the book, we have used subscript s, g, ℓ, and aq to emphasize that the species involved are solid (s), gas (g), liquid (ℓ), or aqueous (aq). An aqueous designation indicates that the species is dissolved in water (see Chapter 10). ◀

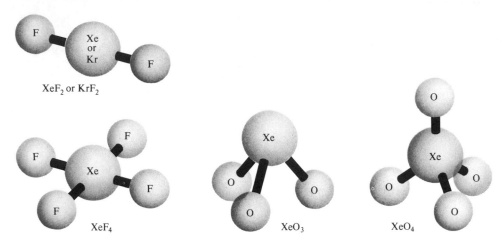

Fig. 7–1 The geometries of several noble-gas compounds. (See Table 7–1.)

We shall use some typical reactions of the noble gases, of hydrogen, and of oxygen to illustrate other common types of chemical reactions. A study of these relatively simple reactions combined with the fundamentals of chemical bonding will lead to some important generalizations concerning reaction types and also to the consideration of more complex systems.

7–2 NOBLE GAS REACTIONS

Prior to 1962 no stable chemical compounds of the noble gases were known. The firm belief, held by almost all chemists, was that a noble-gas element simply would not combine with any other element (hence the widely accepted designation of *inert gases*). Since 1962 several reactions of the noble gases, primarily of xenon (Xe), have been discovered. Compounds such as XeF_2, XeF_4, XeF_6, XeO_3, XeO_4, and KrF_2 have now been prepared, and many of their properties have been determined (see Table 7–1 and Fig. 7–1).

The xenon fluorides can be prepared by direct combination of the elements under the proper reaction conditions:

$$Xe + F_2 \rightarrow XeF_2, \qquad Xe + 2F_2 \rightarrow XeF_4, \qquad Xe + 3F_2 \rightarrow XeF_6.$$

Each of these reactions requires special conditions in order for the product to be formed. For example, XeF_4 has to be prepared at 400 °C with a fluorine pressure of 8 atm and a xenon pressure of 1.7 atm.* Or XeF_2 and XeF_4 can be prepared at room temperature by irradiating the Xe + F_2 gas mixture with ultraviolet light of wavelength 2200–3400 Å. In general, the formation of XeF_6 requires rather high pressures (60–100 atm), a high proportion of F_2 (1-to-20 mixture of Xe to F_2) and a high reac-

---

* A gas pressure of one atmosphere (1 atm) corresponds approximately to the pressure exerted by the air around the earth. See Section 8–2.

tion temperature (300 °C). Note that all these preparations involve the direct reaction of elements. They illustrate a general group of chemical reactions, often called *direct-combination reactions*.

The only known method of preparing the xenon oxides involves the use of the xenon fluorides,

$$XeF_6 + 3H_2O \rightarrow XeO_3 + 6HF.$$

This reaction of XeF_6 with water is one of a group of reactions called *hydrolysis reactions*. Neither XeO_3 nor XeO_4 can be prepared by direction combination of Xe and O_2. In fact, both these compounds are rather unstable and have been known to explode unexpectedly, even at room temperature or below (see Table 7–1).

The explosion of XeO_3 (or XeO_4) results in the very rapid decomposition of the compound,

$$2XeO_3 \rightarrow 2Xe + 3O_2,$$

and the release of heat, 96 kcal/mole XeO_3 (reactions which release heat are called *exothermic*). This equation describes a group of reactions, often called *decomposition reactions*. Another example is the thermal decomposition of the xenon fluorides; for example,

$$XeF_4 + heat \rightarrow Xe + 2F_2.$$

This thermal instability of the xenon fluorides is responsible for one of the major uses of noble-gas compounds: They provide F_2 free of unwanted impurities for use in other chemical reactions, for example, in the synthesis of certain fluorine-substituted organic compounds that are valuable as pharmaceuticals.

Table 7–1 Properties of noble-gas compounds

Compound	Oxidation state of Xe or Kr	Melting point, °C	Structure	Color
XeF_2	2	129.0	Linear	Colorless solid and vapor
XeF_4	4	117.0	Square planar	Colorless solid and vapor
XeF_6	6	49.5	Distorted octahedron (?)	Colorless solid, yellow melt and vapor
XeO_3	6	Explodes, often at 25 °C	Pyramidal	Colorless
XeO_4	8	Explodes, even below 0 °C	Tetrahedral	Colorless
KrF_2	2	Sublimes at −40 °C	Linear	Colorless

The fact that XeO_3 decomposes so easily also means that it can be used to supply O_2 gas free of unwanted impurities, and may be responsible for its very strong oxidizing capability, represented by the reaction

$$XeO_3 + 6HCl \rightarrow 3Cl_2 + Xe + 3H_2O.$$

Here XeO_3 is involved in a reaction that is one of a general group called *oxidation-reduction reactions*, which we shall cover in detail in Sections 7–6, 7–7, and 7–8.

7-3 HYDROGEN: SOME TYPICAL REACTIONS

Unlike the noble gases, the element *hydrogen* reacts chemically with almost all the other elements. The best-known compound that hydrogen forms is water, which is the product of a direct-combination reaction between the gaseous elements hydrogen and oxygen:

$$2H_{2(g)} + O_{2(g)} \rightarrow 2H_2O_{(\ell)} + heat.$$

This reaction can be reversed—that is, H_2O can be decomposed—by a process called *electrolysis* (using electrical energy to bring about a chemical change):

$$2H_2O_{(\ell)} \xrightarrow{\text{Electrolysis}} 2H_{2(g)} + O_{2(g)}.$$

Since the H_2 and O_2 gases produced can be conveniently separated (see Fig. 7–2), this is a good method for preparing reasonably pure H_2 gas (and O_2 as well), although it is relatively expensive.

We can prepare hydrogen by using several other types of chemical reactions. For example, a mixture of carbon monoxide and hydrogen gases (often called water gas) is obtained by passing steam over coke (carbon) at about 1000 °C:

$$C_{(s)} + H_2O_{(g)} \xrightarrow{\text{Heat}} CO_{(g)} + H_{2(g)}.$$

We can separate the CO and H_2 gases by converting carbon monoxide to carbon dioxide by reacting the water–gas mixture with more steam at about 500 °C:

$$CO_{(g)} + H_2O_{(g)} \rightarrow CO_{2(g)} + H_{2(g)}.$$

▶ This reaction requires the presence of a *catalyst*, which is a substance that speeds up a chemical reaction, but does not appear to be changed by the reaction. ◀

Both these reactions can be classified as *single-displacement reactions*, since one element in one of the reactants appears to be displaced (O from H_2O in the two above reactions).

Another displacement reaction is often used to prepare hydrogen in the laboratory: the reaction of an active metal such as zinc with an acid such as hydrochloric acid (HCl gas dissolved in H_2O; see Chapter 12). A similar, but quite violent, reaction occurs between sodium metal and water:

$$Na_{(s)} + H_2O_{(\ell)} \rightarrow Na^+_{(aq)} + OH^-_{(aq)} + H_{2(g)}.$$

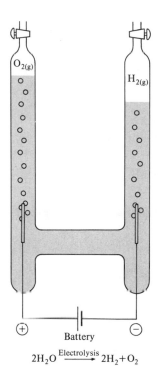

$$2H_2O \xrightarrow{\text{Electrolysis}} 2H_2 + O_2$$

Fig. 7–2 Apparatus for preparing hydrogen (and oxygen) by electrolyzing water. Two metal electrodes are connected to a battery. The volume of H_2 gas produced is twice as large as the volume of O_2 gas.

Decomposition reactions are also used to prepare H_2 gas. Methane gas (CH_4) is a product of the petroleum industry and can be decomposed by using the proper catalyst:

$$CH_{4(g)} \xrightarrow{\text{Catalyst}} C_{(s)} + 2H_{2(g)}.$$

The hydrides of the Group IIA alkaline-earth metals can also be decomposed to produce hydrogen:

$$CaH_{2(s)} \xrightarrow{\text{Heat}} Ca_{(s)} + H_{2(g)}.$$

Substances such as CaH_2, SrH_2, BaH_2, etc., are apparently ionic compounds composed of, for example, Ca^{2+} and H^- ions. They can be prepared by the reverse of the above reaction:

$$Ca_{(s)} + H_{2(g)} \rightarrow CaH_{2(s)}.$$

Hydrogen also undergoes important direct-combination reactions with the Group VIIA halogens. The reactions with fluorine (F_2) and chlorine (Cl_2) occur at room temperature, while the reactions with bromine (Br_2) and iodine (I_2) require temperatures of 400–600 °C,

$$H_{2(g)} + Cl_{2(g)} \rightarrow 2HCl_{(g)}.$$

A very important direct-combination reaction occurs between hydrogen and nitrogen gases under high pressures (300 to 1000 atm), at temperatures between 400 and 600 °C, and in the presence of a catalyst:

$$N_{2(g)} + 3H_{2(g)} \xrightarrow[\text{pressure}]{\text{Catalyst,}\atop\text{heat,}} 2NH_{3(g)}.$$

This reaction is called the *Haber process* for the preparation of ammonia (NH_3).

Hydrogen is also often used (in single-replacement reactions) in the production of free metals from their metal-oxide ores, as illustrated by the reactions with copper (II) oxide and tungsten (VI) oxide:

$$CuO_{(s)} + H_{2(g)} \rightarrow Cu_{(s)} + H_2O_{(g)},$$
$$WO_{3(s)} + 3H_{2(g)} \rightarrow W_{(s)} + 3H_2O_{(g)}.$$

7-4 OXYGEN: SOME TYPICAL REACTIONS

Oxygen, like hydrogen, is a diatomic gas (at room temperature and pressure) that reacts chemically with almost all other elements. It is by far the most abundant element in the earth's crust and also composes about 21% of the earth's atmosphere. It can be prepared by electrolyzing water or by the thermal decomposition of potassium chlorate, $KClO_3$, according to the reaction

$$KClO_{3(s)} \xrightarrow{\text{Heat}} 2KCl_{(s)} + 3O_{2(g)},$$

which is catalyzed by manganese dioxide, MnO_2 (see Fig. 7-3). Several metal oxides also decompose to produce oxygen; mercuric oxide, for example:

$$2HgO_{(s)} \xrightarrow{\text{Heat}} 2Hg_{(\ell)} + O_{2(g)}.$$

Oxygen forms many compounds by direct-combination reactions. Some of these reactions require high temperatures, or pressures, or catalysts:

$$S_{(s)} + O_{2(g)} \rightarrow SO_{2(g)}$$
$$2SO_{2(g)} + O_{2(g)} \rightarrow 2SO_{3(g)}$$

and

$$2C_{(s)} + O_{2(g)} \rightarrow 2CO_{(g)}$$
$$2CO_{(g)} + O_{2(g)} \rightarrow 2CO_{2(g)}$$

and

$$N_{2(g)} + O_{2(g)} \rightarrow 2NO_{(g)}$$
$$2NO_{(g)} + O_{2(g)} \rightarrow 2NO_{2(g)}$$

All these reactions produce *molecular oxides*; that is, SO_2, CO, NO_2, etc., are molecules and contain covalent bonds. Ionic oxides—such as MgO and Al_2O_3—are also produced via direct combination with oxygen:

$$2Mg_{(s)} + O_{2(g)} \rightarrow 2MgO_{(s)}, \qquad 4Al_{(s)} + 3O_{2(g)} \rightarrow 2Al_2O_{3(s)}.$$

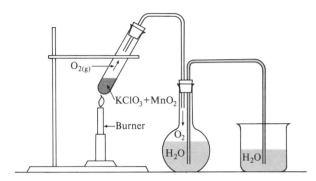

Fig. 7–3 The preparation and collection of O_2 gas via the thermal decomposition of potassium chlorate ($KClO_3$) in the presence of the catalyst manganese dioxide (MnO_2). The O_2 gas generated displaces an equal volume of water.

Some typical oxides of both types are shown in Table 7–2.

Table 7–2 Some properties of typical oxides

Oxide	Predominant bonding	State and color at 25 °C and 1 atm		Melting point, °C
CaO			White	2614
Al_2O_3			White	2070
FeO			White	1370
Fe_3O_4			Black	1590
MnO			Green	1780
Cu_2O	Ionic	Solid	Red	1236
CuO		(powder)	Black	1326
Nb_2O_5			White	1485
WO_3			Yel-grn	1470
SnO_2			White	1630
In_2O_3			Yellow	1910
UO_2			Br-blk	2878
H_2O	Covalent	Liquid	Colorless	0
CO			Colorless	−199
CO_2			Colorless	−57.5
N_2O	Covalent	Gas	Colorless	−90.9
NO			Colorless	−163.6
NO_2			Brown	−11.2
SO_2			Colorless	−73.2

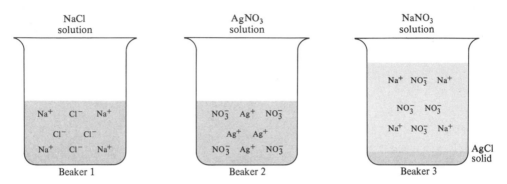

Fig. 7-4 When the two solutions in beakers 1 and 2 are mixed in beaker 3, the ions trade places. Solid silver chloride (AgCl) forms, with Ag^+ and Cl^- ions packed close together; Na^+ and NO_3^- ions are left in the water above the solid AgCl precipitate.

Direct reaction with oxygen is used in the production of nitric acid (HNO_3 dissolved in water, and more accurately represented by $H^+ + NO_3^-$), a common and very important acid, in the *Ostwald process*:

1) $4NH_{3(g)} + 5O_{2(g)} \xrightarrow[\text{Pt catalyst}]{750\text{-}900\ °C} 4NO_{(g)} + 6H_2O_{(g)}$

2) $2NO_{(g)} + O_{2(g)} \longrightarrow 2NO_{2(g)}$

3) $3NO_{2(g)} + H_2O_{(\ell)} \longrightarrow 2H^+_{(aq)} + 2NO_{3(aq)}^- + NO_{(g)}$

Other nitrate compounds, such as ammonium nitrate (NH_4NO_3), are often used as fertilizers and also as explosives. The decomposition*

$$NH_4NO_{3(s)} \rightarrow N_2O_{(g)} + 2H_2O,$$

for example, can occur very rapidly, producing an explosion.

The reaction that occurs between silver nitrate ($AgNO_3$) and sodium chloride when both compounds are dissolved in water illustrates another common type of reaction. We can observe this reaction by mixing the contents of two separate beakers, as shown in Fig. 7-4, with beaker 1 containing NaCl dissolved in water and beaker 2 containing $AgNO_3$ dissolved in water.

Remember that NaCl is an ionic compound. The particles that compose the solid NaCl or that are in the salt water solution are ions, a cation Na^+ and an anion Cl^-. Since $AgNO_3$ is also an ionic compound, beaker 2 contains Ag^+ cations and NO_3^- anions.

---

*   Nitrous oxide (N_2O) is a colorless and almost odorless gas that has anesthetic properties and is often called *laughing gas*.

▶ NO_3^- is called the nitrate ion. It is a *complex ion*, since it is charged and is composed of more than one atom. It is a group of four atoms—three oxygen and one nitrogen—bonded together, with the whole unit having a single negative charge, i.e., one more electron than the sum of 21 valence electrons from the four atoms. Try writing a Lewis structure for NO_3^- that obeys the octet rule. ◀

If we pour the contents of both beakers into a third beaker and mix thoroughly, we can make some interesting observations. The most obvious thing that happens is that a white solid is formed. In the third beaker, then, there is a clear, colorless water solution and, at the bottom of the beaker, a white solid. We can determine the composition of the solid and also determine which ions still remain in the clear water solution by conducting a series of appropriate experiments designed to analyze for the specific components that are present (this is called a *chemical analysis*). These experiments show that Na^+ and NO_3^- ions are present in the water and that the solid is composed of Ag^+ and Cl^- ions, i.e., it is silver chloride, $AgCl$. Note what happens when we mix the contents of beakers 1 and 2. Before the reaction, the water in beaker 1 contains Na^+ and Cl^- ions, and beaker 2 contains Ag^+ and NO_3^- ions. After we mix them, we can see that the ions have switched places. Now Na^+ and NO_3^- ions remain in the water, while Ag^+ and Cl^- are bonded together, forming the $AgCl$ solid. Hence reactions of this type are often called *double-displacement reactions*.

7–5 THE ATMOSPHERE

Many of the reactions just considered for hydrogen and oxygen—together with the compounds that are formed—are involved in one way or another with the earth's atmosphere and with problems concerning the atmosphere and their effects on our lives. Table 7–3 gives the approximate composition of clean, dry air near sea level. This composition can be significantly changed in particular localities, and even over wide areas of the earth, by automobile and jet airplane exhaust, industrial effluents, discharges from the furnaces of homes, smoke from burning trash, etc.

Table 7–3. Approximate composition of clean, dry air near sea level

Component	% by volume	Component	% by volume
N_2	78.09	H_2	0.00005
O_2	20.94	N_2O	0.000025
Ar	0.93	CO	0.00001
CO_2	0.0318	Xe	0.000008
Ne	0.0018	O_3	0.000002
He	0.00052	NH_3	0.000001
CH_4	0.00015	NO_2	0.0000001
Kr	0.0001	SO_2	0.00000002

In urban areas, for example, where large amounts of sulfur-containing fuels are used for industrial and home heating, the SO_2 (sulfur dioxide) and SO_3 (sulfur trioxide) content of the air is so much higher than in pure air that it could possibly be dangerous to health.

▶ In 1965 the average SO_2 content of the air in Chicago was $1.30 \times 10^{-5}\%$ versus the normal content of $2 \times 10^{-8}\%$. ◀

These gaseous oxides of sulfur are produced by reactions such as

$$S_{(s)} + O_{2(g)} \rightarrow SO_{2(g)}$$
$$2SO_{2(g)} + O_{2(g)} \rightarrow 2SO_{3(g)}$$
$$2H_2S_{(g)} + 3O_{2(g)} \rightarrow 2SO_{2(g)} + 2H_2O_{(g)}.$$

Either SO_2 or SO_3 can combine with water to form acids. Reactions of this kind in the atmosphere cause rainfall in some localities to have much higher acid content than normal rain. Similar reactions occurring in streams can also alter their properties.

▶ An increase in acidity is generally measured on the pH scale. Normal, pure H_2O has a pH value of 7.0 (see Chapter 13); higher acid content is indicated by a lowering of the pH. In 1958 the pH of rainwater in Europe showed values of only about 5.0 over limited areas in the Netherlands. By 1962 the Netherlands reported rainfall with pH less than 4.0, and sections of central Europe had rain of pH 5.0. In 1966 the area of precipitation with pH values between 4 and 5 had reached central Sweden.

Similar effects were measured in streams and lakes in Sweden, where pH decreases from 7.3 to 6.8 and 6.8 to 6.2 were measured between 1965 and mid-1967. Some fish, such as salmon, are killed in water with pH values below 5.5, and a pH less than 4.0 threatens all aquatic life. ◀

Another possibly dangerous air pollutant is carbon monoxide (CO), which can be produced both by incomplete combustion in automobiles and by burning common fuels:

$$2C_{(s)} + O_{2(g)} \rightarrow 2CO_{(g)}.$$

Carbon monoxide is a poisonous gas, and when its concentration in the atmosphere becomes too large, as it might in congested traffic, people who breathe it may suffer impairment of their judgments and reaction times.

▶ The average CO content of the air in Chicago in 1965 was $1.7 \times 10^{-2}\%$, compared with the normal $1 \times 10^{-5}\%$. ◀

Carbon dioxide (CO_2) is generally not considered an air pollutant, since it is a normal constituent of the air. However, its global concentration is rising above the natural level because of the enormous quantities of it produced by burning fuels such as coal, oil, and natural gas. The increase in the amount of CO_2 in the atmosphere

might affect global temperatures via the "greenhouse effect," which arises from the preferential absorption of certain wavelengths of electromagnetic radiation by CO_2 molecules. Energy from the sun arrives as short-wavelength radiation and passes through the carbon dioxide in the air. On the other hand, the heat energy emitted by the earth is of relatively long wavelength and is absorbed and radiated back to earth by the CO_2 molecules. An increase in the level of carbon dioxide in the air consequently permits less heat from the earth to escape into outer space, and thereby increases the mean global temperature.

The formation of photochemical smog in many areas of the United States, particularly in the major urban centers, is a very complex and poorly understood process. It is known, however, that the nitrogen oxides mentioned above, oxygen, and ozone (another form of the element oxygen with the formula O_3) along with hydrocarbons (organic chemical compounds described in Chapter 14) are often involved in smog formation. Table 7–4 presents a simplified scheme proposed for the formation of photochemical smog.

Table 7–4. Proposed steps in a simplified reaction scheme for the production of photochemical smog

---

1) NO_2 + light → NO + O

2) $O + O_2 \rightarrow O_3$

3) $O_3 + NO \rightarrow NO_2 + O_2$

4) O + Hc → HcO·
   (Hydrocarbon)   (Activated substance I)

5) $HcO\cdot + O_2 \rightarrow HcO_3\cdot$
   (Activated substance III)

6) $HcO_3\cdot$ + Hc → Aldehydes, ketones, etc.
   (Organic compounds)

7) $HcO_3\cdot + NO \rightarrow HcO_2\cdot + NO_2$
   (Activated substance II)

8) $HcO_3\cdot + O_2 \rightarrow O_3 + HcO_2\cdot$

9) $HcO_x\cdot + NO_2 \rightarrow$ Peroxyacyl nitrates
   (Activated substance x)   (Smog)

7–6 OXIDATION-REDUCTION REACTIONS

Sections 7–2, 7–3, and 7–4 described the reactions of hydrogen, oxygen, and the noble gases in terms of particular reaction types, i.e., direct combination, single replacement, etc. These reactions, and all chemical reactions, can be grouped into two very broad classifications: (1) reactions in which there is no electron transfer, and (2) reactions in which there is electron transfer. The double replacement reaction be-

tween $AgNO_3$ and $NaCl$ described above is an example of Group 1 reactions. No electrons are transferred between the reactants during the reaction; ions simply switch places.

Most of the other reactions described earlier in this chapter, on the other hand, are examples of Group 2 reactions; electrons are transferred between reactants during the course of the reaction. This large group of reactions is generally called *oxidation-reduction* (or *redox*) *reactions*.

The reaction between sodium metal and chlorine—forming sodium chloride—is an example of an oxidation-reduction reaction. Electrons are transferred from the sodium atoms to the chlorine atoms. We can represent this exchange schematically as follows:

$$Na\cdot + \cdot \ddot{\underset{..}{Cl}}: \rightarrow Na^+ + :\ddot{\underset{..}{Cl}}:^-$$

The sodium atom, with one valence electron and an electron configuration of $1s^2 2s^2 2p^6 3s^1$, gives up one electron. The cation Na^+ is formed; it has the $1s^2 2s^2 2p^6$ electron configuration of the noble gas neon. The chlorine atom, with seven valence electrons, accepts one electron, forming the Cl^- anion, which has the stable argon electron configuration of $1s^2 2s^2 2p^6 3s^2 3p^6$. This reaction between sodium and chlorine is typical of oxidation-reduction reactions; electrons are lost and gained, shifted between the reactants involved.

Oxidation Numbers. For oxidation-reduction reactions it is helpful to develop a method of following the electron changes. We need to determine when electrons are transferred and when they are not; and if they are transferred, how many are involved.

To help with this electron bookkeeping, chemists have devised a rather arbitrary system of assigning electrons to the various atoms in chemical compounds. In this system the basic rule is that electrons involved in chemical bonds are assigned to the more electronegative atom. This (arbitrary) shifting of electrons forms apparent charges on the atoms, since electrons are either given to or taken from an atom, depending on its electronegativity value. These assignments are made irrespective of what the real situation in the chemical bond is. For example, a bond may be only slightly polar, with the bonding electrons being shared, but in assigning the electrons to determine the oxidation numbers of the atoms, we give the electrons completely to the more electronegative atom. The charge which an atom appears to have when electrons are assigned to it according to a set of rules is called an *oxidation number* or an *oxidation state*. The oxidation number of an atom can be positive, negative, or zero, as shown by the examples in Table 7–5.

General Rules for Oxidation Numbers. To simplify the job of assigning oxidation numbers, the following rules have been developed (common oxidation states for the elements are given in Table 7–6 on pages 188–189).

1. The oxidation number of a free element is zero. This applies to each atom in the formula unit. If there is more than one atom present in the formula unit, the oxidation number of each atom is zero.

Table 7–5. Oxidation numbers: some examples

Rule for oxidation number for:	Examples of species	Assigned oxidation number
Elements	Na	0 for Na
	I_2	0 for each I
	S_8	0 for each S
Simple ions	Li^+	+1 for Li^+
	Al^{3+}	+3 for Al^{3+}
	S^{2-}	−2 for S^{2-}
IA elements in compounds	Na^+	+1 for Na^+
	KCl	+1 for K
	Rb_2CO_3	+1 for each Rb
IIA elements in compounds	Mg^{2+}	+2 for Mg^{2+}
	$Sr(NO_3)_2$	+2 for Sr
	BaO	+2 for Ba
F in compounds	F^-	−1 for F^-
	LiF	−1 for F
	CdF_2	−1 for each F
O in compounds	MgO	−2 for O
	Al_2O_3	−2 for each O
	Fe_3O_4	−2 for each O
H in compounds	OH^-	+1 for H
	H_2O	+1 for each H
	NH_3	+1 for each H

2. The oxidation number of a simple ion (a single atom with a charge) is equal to the charge on the ion.

3. In compounds, the oxidation number of Group IA alkali metals is always +1.

4. In compounds, the oxidation number of Group IIA alkaline-earth metals is always +2.

5. In compounds, the oxidation number of fluorine is always −1. A −1 oxidation number is common for the other Group VIIA halogens, but all except fluorine also exhibit other oxidation numbers.

6. In compounds, the oxidation number of oxygen is always −2, with two exceptions. In peroxides, such as hydrogen peroxide (H_2O_2), barium peroxide (BaO_2), etc., the oxidation number of oxygen is −1. In compounds with fluorine, oxygen takes on a positive oxidation number, such as +2 in OF_2. Neither of these exceptions is very common.

Table 7–6

+1

+2

IA	IIA	IIIB	IVB	VB	VIB	VIIB		VIII
1 +1 **H**								
3 +1 **Li**	4 +2 **Be**							
11 +1 **Na**	12 +2 **Mg**							
19 +1 **K**	20 +2 **Ca**	21 +3 **Sc**	22 +4 +3 **Ti**	23 +5 +4 +3 +2 **V**	24 +6 +3 +2 **Cr**	25 +7 +6 +4 +3 +2 **Mn**	26 +3 +2 **Fe**	27 **Co**
37 +1 **Rb**	38 +2 **Sr**	39 +3 **Y**	40 +4 **Zr**	41 +5 +3 **Nb**	42 +6 +5 +4 +3 +2 **Mo**	43 +7 **Tc**	44 +8 +6 +4 +3 +2 **Ru**	45 **Rh**
55 +1 **Cs**	56 +2 **Ba**	57 +3 **La** *	72 +4 **Hf**	73 +5 **Ta**	74 +6 +5 +4 +3 +2 **W**	75 +7 +6 +4 +2 −1 **Re**	76 +8 +6 +4 +3 +2 **Os**	77 **Ir**
87 +1 **Fr**	88 +2 **Ra**	89 +3 **Ac** †						

☐ Metals

☐ Metalloids

■ Nonmetals

☐ Noble gases

*LANTHANIDES	58 +4 +3 **Ce**	59 +4 +3 **Pr**	60 +3 **Nd**	61 +3 **Pm**	62 +3 +2 **Sm**	63 **Eu**
†ACTINIDES	90 +4 **Th**	91 +5 +4 **Pa**	92 +6 +5 +4 +3 **U**	93 +6 +5 +4 +3 **Np**	94 +6 +5 +4 +3 **Pu**	95 **Am**

(Handwritten annotations across top: "O", "+3", "+4", "-3", "-2", "-1")

	IIIA	IVA	VA	VIA	VIIA	NOBLE GASES
						2 — He
	5 +3 B	6 +4, +2, −4 C	7 +5, +4, +3, +2, −3 N	8 −2 O	9 −1 F	10 — Ne
	13 +3 Al	14 +4 Si	15 +5, +4, +3, −3 P	16 +6, +4, +2, −2 S	17 +7, +5, +3, +1, −1 Cl	18 — Ar

	IB	IIB	IIIA	IVA	VA	VIA	VIIA	NOBLE GASES
+3, +2 Ni	29 +2, +1 Cu	30 +2 Zn	31 +3 Ga	32 +4 Ge	33 +5, +3, −3 As	34 +6, +4, −2 Se	35 +5, +1, −1 Br	36 — Kr
+4, +2 Pd	47 +1 Ag	48 +2 Cd	49 +3 In	50 +4, +2 Sn	51 +5, −3 Sb	52 +6, +4, 2 Te	53 +7, +1, −1 I	54 — Xe
+4, +2 Pt	79 +3, +1 Au	80 +2, +1 Hg	81 +3, +1 Tl	82 +4, +2 Pb	83 +5, +3 Bi	84 +4, +2 Po	85 +7, +5, +3, +1, −1 At	86 — Rn

+3 Gd	65 +4, +3 Tb	66 +3 Dy	67 +3 Ho	68 +3 Er	69 +3 Tm	70 +3, +2 Yb	71 +3 Lu
+3 Cm	97 +4, +3 Bk	98 +3 Cf	99 — Es	100 — Fm	101 — Md	102 — No	103 — Lr

7. In compounds, the oxidation number of hydrogen is always $+1$, except in metal hydrides, such as NaH, CaH_2, etc., in which hydrogen is bonded to a metal with low electronegativity, where the oxidation number of hydrogen is -1.

8. The sum of the oxidation numbers of the atoms in the formula for a compound or a complex ion (a charged group of atoms) must be equal to the experimentally observed charge on the species. In a neutral compound the observed charge is zero. The sum of the oxidation numbers of the elements composing the compound must therefore equal zero. In the case of complex ions, the sum of the oxidation numbers must be equal to the charge on the ion.

For example, in the compound potassium fluoride (KF), the oxidation numbers of K and F are $+1$ and -1, respectively (rules 3 and 5 above). The sum of these two oxidation numbers, $(+1) + (-1) = 0$, must be zero, since KF is uncharged. A common complex ion is the hydronium ion (H_3O^+). Here the ion has a single positive charge (this charge can be experimentally detected and determined). The sum of the oxidation numbers of hydrogen and oxygen must therefore yield $+1$. The oxidation number of oxygen is -2 and that of hydrogen $+1$ (rules 6 and 7). Since there are three hydrogen atoms contained in one H_3O^+ formula unit, the sum of the oxidation numbers of the atoms for H_3O^+ is

$$(+1) + (+1) + (+1) + (-2) = +1.$$

This sum is equal to the observed charge on the complex ion.

▶ The big block of elements including Groups IVB, VB, VIB, VIIB, and VIII are not covered by the general oxidation-number rules, because many of these elements exhibit several oxidation numbers, depending on the other elements with which they are combined. Titanium, for example, can form three different compounds with oxygen: TiO, Ti_2O_3, and TiO_2. The oxidation states of titanium in these three compounds can be calculated, since the oxidation number of oxygen is -2 in all three compounds. The oxidation number of Ti is $+2$ in TiO, $+3$ in Ti_2O_3, and $+4$ in TiO_2. ◀

Example 7–1. What is the oxidation number of chromium in the complex ion $Cr_2O_7^{2-}$?

The oxidation number of each of the seven oxygen atoms is -2. The total assigned negative charge contributed by these seven oxygen atoms is $7(-2) = -14$. Since the charge on the $Cr_2O_7^{2-}$ ion is two minus, the sum of positive and negative charges must be -2.

Atoms	Oxidation number	Total charge	
2Cr	?	$2(?)$	$= +12$
7O	-2	$7(-2)$	$= -14$
		$Cr_2O_7^{2-}$	$= \quad -2$

The two Cr atoms therefore must be assigned a positive charge of $+12$, since $(+12) + (-14)$ $= -2$. Each Cr atom has an oxidation number of $+6$, since the total $+12$ charge is supplied by two atoms: $+12/2 = +6$.

Example 7–2. What is the formula of the compound barium hydroxide?

Barium is a Group IIA alkaline-earth metal. It therefore has an oxidation state of $+2$ in the compound barium hydroxide. The hydroxide ion is an ion with the formula OH^-. For the compound to be uncharged, two hydroxide ions need to combine with one barium ion,

$$\begin{array}{ccc} & ^-OH & Ba^{2+} & OH^- \\ \text{Oxidation} & & & \\ \text{number} & (-2) + (+1) + (+2) + (-2) + (+1) = 0, \end{array}$$

giving $Ba^{2+}(OH^-)(OH^-)$ or $Ba(OH)_2$ as the formula for barium hydroxide. Note that we drop the ion charges in writing the formula (as with NaCl) and enclose the hydroxide ion in parentheses. The subscript indicates that two OH^- ions are present in the compound (as with $CaCl_2$).

Oxidation Numbers and Definitions. Although oxidation numbers are arbitrary (and usually fictitious), we can conveniently use them in determining oxidation-reduction processes and in defining the related concepts. We can illustrate these uses by considering the reaction between sodium metal and chlorine gas as represented by the equation

$$\begin{array}{cc} & 2Na + Cl_2 \rightarrow 2NaCl \\ \text{Oxidation} & 0 \quad\quad 0 \quad\quad +1 -1 \\ \text{number} & \end{array}$$

The oxidation numbers can be assigned according to the general rules given above. The reactants are both free elements; their oxidation numbers are both zero. In the NaCl product, the Group IA element Na is in a compound; its oxidation number is $+1$. The oxidation number of Cl must be -1, so that NaCl is uncharged (the -1 oxidation state is the most common one for the Group VIIA halogens).

Now we can see that during the course of the reaction, the oxidation number of sodium has gone up, from 0 to $+1$, and the oxidation number of chlorine has gone down, from 0 to -1. To explain an increase in oxidation number, we propose that the atom must lose electrons, which are negatively charged. A decrease in oxidation number, then, indicates that the atom has gained electrons. Each sodium atom is said to lose one electron (its oxidation number increases from 0 to $+1$), while each chlorine atom accepts one electron (its oxidation number decreases from 0 to -1). Note that the total number of electrons gained is equal to the total number of electrons lost. The equation tells us that two Na atoms react. These two atoms furnish two electrons that are picked up by the Cl_2 molecule, which contains two Cl atoms. To preserve charge neutrality, the number of electrons lost and gained must be equal.

Since changes in oxidation number are readily determined, it is convenient to use them to define oxidation and reduction, as well as other associated terms. There is an *oxidation* in any chemical reaction when there is an increase in oxidation number.

In the example above, sodium metal has undergone oxidation, since its oxidation number increased. We say that sodium metal (Na) is the *substance oxidized*. *Reduction* is associated with a decrease in oxidation number. In the reaction with sodium metal, chlorine gas undergoes reduction, since the oxidation number of each chlorine atom decreases. We say that chlorine gas (Cl_2) is the *substance reduced*. Note that during oxidation electrons are lost and that during reduction electrons are gained. The substance oxidized, therefore, *loses* electrons, while the substance reduced *gains* electrons. Remember that there must be both an increase and a decrease in oxidation number in an oxidation-reduction reaction. Oxidation and reduction occur simultaneously.

There are two other common terms associated with oxidation-reduction reactions: oxidizing agent and reducing agent. The *oxidizing agent* is the substance that brings about the oxidation; it oxidizes another substance. The *reducing agent* is the substance that carries out the reduction; it reduces another substance. Note that the oxidizing agent oxidizes something else. That means that there will be an increase in oxidation number in the other substance. Then the oxidizing agent itself must be reduced; it will show a decrease in oxidation number. Similarly with the reducing agent. It will be oxidized and show an increase in oxidation number, since it is reducing another substance. Both an oxidizing agent and a reducing agent must be present in order for an oxidation-reduction reaction to occur (in certain special cases the same substance can serve as both oxidizing and reducing agent). Table 7–7 gives a summary of all these terms.

Table 7–7. Terms associated with oxidation-reduction reactions

Term	Oxidation number change	Electron change
Oxidation	Increase	Loss
Reduction	Decrease	Gain
Oxidizing agent	Decrease	Accepts
Reducing agent	Increase	Releases
Substance oxidized	Increase	Releases
Substance reduced	Decrease	Accepts

Example 7–3. Determine the oxidizing agent, the substance reduced, and the substance that gains electrons in the reaction between sodium metal and water:

$$2Na + 2H_2O \rightarrow 2Na^+ + 2OH^- + H_2.$$

To find changes in the oxidation number, we must assign an oxidation number to each atom. The elements Na and H_2 are in the zero oxidation state. The simple ion Na^+ has an oxidation number of $+1$, equal to its charge. The atoms hydrogen and oxygen have oxidation numbers of

+1 and −2 in both H_2O and OH^-, as do almost all H and O atoms in compounds. The oxidation numbers are:

$$\begin{array}{l}
\text{Oxidation} \\
\text{number}
\end{array} \qquad
\begin{array}{ccccc}
2Na & + & 2H_2O & \rightarrow & 2Na^+ & + & 2OH^- & + & H_2 \\
0 & & +1\;-2 & & +1 & & -2\;+1 & & 0
\end{array}$$

Inspection shows that Na increases in oxidation number from 0 to +1 and H decreases from +1 to 0. Actually, we can now see that all the oxidation numbers need not be assigned. We are interested only in the ones that are changing. Only the following oxidation numbers are necessary:

$$\begin{array}{l}
\text{Oxidation} \\
\text{number}
\end{array} \qquad
\begin{array}{ccccc}
2Na & + & 2H_2O & \rightarrow & 2Na^+ & + & 2OH^- & + & H_2 \\
0 & & +1 & & +1 & & & & 0
\end{array}$$

With sufficient experience and practice, you will be able to inspect the equations rapidly and write down only the necessary oxidation numbers. Until that time we suggest that you determine and write down an oxidation number for each atom in the reaction.

The oxidizing agent carries out an oxidation on another substance and is simultaneously reduced itself. Therefore the oxidizing agent contains an element decreasing in oxidation number, hydrogen in H_2O forming H_2 (a change from +1 to zero in oxidation number) in this reaction. Then H_2O is the oxidizing agent, and is also the substance reduced. Since the oxidation number of H in H_2O decreases, the H_2O molecule is also the substance gaining electrons (negative charge).

Example 7–4. Determine how many electrons are shifted per atom, what element increases in oxidation number, and which substance is oxidized in the reaction of hydrogen bromide with chlorine:

$$2HBr + Cl_2 \rightarrow 2HCl + Br_2.$$

The oxidation numbers can all be assigned according to the general rules,

$$\begin{array}{l}
\text{Oxidation} \\
\text{number}
\end{array} \qquad
\begin{array}{ccccc}
2HBr & + & Cl_2 & \rightarrow & 2HCl & + & Br_2 \\
+1\;-1 & & 0 & & +1\;-1 & & 0
\end{array}$$

Note that the oxidation number of hydrogen does not change and is not needed in answering the questions.

To find the number of electrons gained or lost per atom, we must find the changes in oxidation number. In the HBr reactant, the Br atom is in the −1 oxidation state, while in the Br_2 product, the oxidation state of a Br atom is zero. Then each Br atom has an oxidation-number increase of 1; each Br atom loses one electron. Each atom of Cl in the Cl_2 reactant has an oxidation number of zero and in the HCl product an oxidation number of −1. Each Cl atom therefore gains one electron, since there is an oxidation-number decrease of 1 for each Cl atom.

The element that increases in oxidation number is bromine (−1 to zero); it is contained in the substance HBr. Since Br has an increase in oxidation number, the substance HBr is the substance oxidized.

Note that throughout the examples we have considered substances oxidized and reduced, substances acting as oxidizing and reducing agents, but elements or atoms (generally contained in a substance) changing in oxidation number. We make this

distinction because it conforms more closely to the situation that is observed experimentally. We can actually observe HBr being changed by Cl_2. But we do not really know whether electrons are being released by the Br atoms in HBr or by the HBr molecule as a whole. Remember that the assigned oxidation numbers usually do not reflect the actual charges on the atom; we use them only for our convenience in handling the problem. Therefore—when we speak of a substance that is oxidized or reduced or speak of it as an oxidizing or reducing agent—we generally refer to an entire molecule or formula unit, not to one specific atom within the unit, to conform to the experimentally observed situation. Sometimes it is convenient to consider an ion as the oxidizing or reducing agent. In this case it is still the entire formula unit that is either oxidized or reduced.

7-7 BALANCING CHEMICAL EQUATIONS

In order for a chemical equation to be a proper and complete equation, it must satisfy a set of conditions.

1. A chemical equation must be consistent with experimental facts. We cannot write down any arbitrary equation we please. Things simply might not react in the way that we propose. The ultimate test is the experiment. Chemical equations must reflect the results observed in experiments.

2. A chemical equation must be consistent with the laws of conservation of mass and conservation of electrical charge. Neither mass nor electrical charge can be created or destroyed in any ordinary chemical reaction. These last two conditions are associated with the numerical coefficients that appear in the equations. When these two conditions are fulfilled and the proper coefficients are placed, the equation is said to be *balanced*.

There are two methods of balancing equations. The first is by *inspection of the equation* (see Section 2-7). Coefficients are inserted in the equation by trial and error until mass and charge are found to be conserved. This method can be used for any equation, but it actually works well only for relatively simple equations. For the more complex equations, we need to use the second method, which is called *matching electron transfer*. As we might gather from its name, this second method is applicable to oxidation-reduction reactions.

With more complex oxidation-reduction reactions, it is often necessary to use a set method to balance the equation describing the reaction. Consider, for example, the oxidation-reduction reaction represented by the equation

$$K_2Cr_2O_7 + S + H_2O \rightarrow SO_2 + KOH + Cr_2O_3.$$

The proper coefficients have been omitted. See if you can balance the equation by inspection. Then, using the example given below as a guide, balance it by using the following rules for the matching-electron-transfer method:

1. Assign oxidation numbers to those atoms that change in oxidation number.
2. Determine the number of electrons shifted per atom.

3. Determine the number of electrons shifted per formula unit.

4. Equalize electron loss and gain.

5. Conserve mass and electrical charge.

When potassium dichromate ($K_2Cr_2O_7$) and potassium iodide (KI) are mixed in water containing sulfuric acid (H_2SO_4), it is found that the following equation describes the reaction:

$$Cr_2O_7^{2-} + I^- + H^+ \rightarrow Cr^{3+} + I_2 + H_2O.$$

Only the chemical species that change or are formed are included in the equation. This is called a *net ionic equation*. Experimental measurements indicate that $K_2Cr_2O_7$ and KI dissolved in water form K^+ plus $Cr_2O_7^{2-}$ ions and K^+ plus I^- ions, and, furthermore, that the K^+ ion does not enter into the reaction. It is therefore not included in the reaction, nor is the anion of sulfuric acid.

The first step in balancing the equation is to assign oxidation numbers to those atoms that change in oxidation number. (We shall assign all oxidation numbers and then select those that change.) Recalling from Example 7–1 that chromium in $Cr_2O_7^{2-}$ has an oxidation number of $+6$, we can make all the assignments as follows:

$$
\begin{array}{l}
\text{Oxidation} \\
\text{number}
\end{array}
\quad
\begin{array}{cccccc}
Cr_2O_7^{2-} & + \ I^- & + \ H^+ & \rightarrow Cr^{3+} & + \ I_2 & + \ H_2O \\
+6-2 & -1 & +1 & +3 & 0 & +1-2
\end{array}
$$

The second step is to determine the number of electrons gained or lost per atom. We can do this by finding the difference in oxidation number between the atom on the reactant side and the same atom on the product side of the equation. One pair of atoms will show an increase in oxidation number and a second pair a decrease:

$$
\begin{array}{l}
\text{Oxidation} \\
\text{number}
\end{array}
\quad
\begin{array}{cccccc}
Cr_2O_7^{2-} & + \ I^- & + \ H^+ & \rightarrow Cr^{3+} & + \ I_2 & + \ H_2O \\
+6 & -1 & & +3 & 0 &
\end{array}
$$

Each Cr atom decreases in oxidation number from $+6$ to $+3$ during the course of the reaction. Thus each Cr atom gains three electrons. Each I atom increases in oxidation number from -1 to 0. Therefore each I atom loses one electron. We can diagram these changes as follows:

$$
\begin{array}{l}
\text{Oxidation} \\
\text{number}
\end{array}
\quad
\begin{array}{cccccc}
Cr_2O_7^{2-} & + \ I^- & + \ H^+ & \rightarrow Cr^{3+} & + \ I_2 & + \ H_2O \\
+6 & -1 & & +3 & 0 &
\end{array}
$$

$$\underset{3\,e^-/Cr\text{ atom}}{\uparrow} \quad \underset{1\,e^-/I\text{ atom}}{\downarrow}$$

The third step in balancing an oxidation-reduction equation is to determine the number of electrons shifted per formula unit. To do this you must account for the number of electrons shifted per atom and the number of atoms of the element changing in oxidation state that are contained in the formula unit. For example, each Cr atom in $Cr_2O_7^{2-}$ accepts three electrons; but there are two Cr atoms in each $Cr_2O_7^{2-}$ formula unit (indicated by the subscript after the chromium symbol). Then each

$Cr_2O_7^{2-}$ unit gains a total of six electrons:

$$3 \text{ e}^- \text{ gained/}\cancel{Cr \text{ atom}} \times 2\cancel{Cr \text{ atoms}}/Cr_2O_7^{2-} \text{ unit} = 6 \text{ e}^- \text{ gained/}Cr_2O_7^{2-} \text{ unit.}$$

In the case of I^-, there is only one iodine atom in the I^- formula unit. The electron change per atom is therefore equal to the electron change per formula unit:

$$1 \text{ e}^- \text{ lost/}\cancel{I \text{ atom}} \times 1\cancel{I \text{ atom}}/I^- \text{ unit} = 1 \text{ e}^- \text{ lost/}I^- \text{ unit.}$$

We can diagram these changes by using the equation:

$$
\begin{array}{ccccccccc}
Cr_2O_7^{2-} & + & I^- & + & H^+ & \rightarrow & Cr^{3+} & + & I_2 & + & H_2O \\
+6 & & -1 & & & & +3 & & 0 \\
\uparrow & & \downarrow \\
\end{array}
$$

2(3 e⁻/Cr atom)   1(1 e⁻/I atom)
6 e⁻/Cr₂O₇²⁻ unit   1 e⁻/I⁻ unit

The fourth step is to equalize electron loss and gain. We can do this by multiplying the number of electrons lost and the number gained by the proper coefficients to make the loss and gain equal. In our example, six electrons are picked up by each $Cr_2O_7^{2-}$ unit and only one electron is released by each I^- unit. To make this electron loss equal to the electron gain, we must multiply the electron loss by six. This means that six I^- ions are needed to react with one $Cr_2O_7^{2-}$ ion, since it takes six I^- ions to supply the six electrons needed by the $Cr_2O_7^{2-}$ ion for reaction to occur. Then 1 and 6 are the proper coefficients for $Cr_2O_7^{2-}$ and I^-, respectively, in the equation:

$$
\begin{array}{ccccccccc}
Cr_2O_7^{2-} & + & 6I^- & + & H^+ & \rightarrow & Cr^{3+} & + & I_2 & + & H_2O \\
+6 & & -1 & & & & +3 & & 0 \\
\uparrow & & \downarrow \\
\end{array}
$$

1(6 e⁻/Cr₂O₇²⁻)   6(1 e⁻/I⁻)
6 e⁻/1Cr₂O₇²⁻   6 e⁻/6I⁻

Six electrons are supplied by six I^- formula units and picked up by one $Cr_2O_7^{2-}$ formula unit. Note that in the equation the coefficient 1 before the $Cr_2O_7^{2-}$ is not written, but understood. Whenever there is no specific coefficient indicated in an equation, a coefficient of 1 is understood.

An easy way of ensuring equal electron loss and gain is to multiply the electron loss per formula unit by the electron gain per formula unit, and vice versa. In our example, 6 e⁻/$Cr_2O_7^{2-}$ would be multiplied by 1 and 1 e⁻/I^- would be multiplied by 6. The multiplying coefficients are then transferred to the equation: 1 (understood) before $Cr_2O_7^{2-}$ and 6 before I^-.

▶ Sometimes, when we do this, we fail to obtain numbers of the lowest common denominator. For example, suppose that 6 and 2 were the multiplying coefficients to be transferred into the equation; they are both divisible by 2, yielding 3 and 1. The numbers 6 and 2 have the common denominator 2; the numbers 3 and 1 have the

lowest common denominator of 1. If the coefficients can both be divided by the same number, this should be done before they are transferred into the equation. ◄

The fifth and final step in balancing the equation is to conserve mass and electrical charge. We do this by inspecting the equation and inserting any required coefficients. In our example, we have

$$Cr_2O_7^{2-} + 6I^- + H^+ \rightarrow Cr^{3+} + I_2 + H_2O.$$

The coefficients for $Cr_2O_7^{2-}$ and I^- are set as 1 and 6. On the product side, then, the same number of Cr, O, and I atoms must appear as in one $Cr_2O_7^{2-}$ unit and six I^- ions. This can be accomplished by having 2 Cr^{3+} ions, 3 I_2 molecules and 7 H_2O molecules as products, since there would then be two chromium atoms, six iodine atoms (3 I_2 molecules × 2 I atoms/I_2 molecule = 6 I atoms), and seven oxygen atoms (1 oxygen atom in each of 7 H_2O molecules):

$$Cr_2O_7^{2-} + 6I^- + H^+ \rightarrow 2Cr^{3+} + 3I_2 + 7H_2O.$$

Only the H^+ coefficient remains; it must be 14, since there are 14 hydrogen atoms included in the 7 H_2O molecules on the product side of the equation.
The balanced equation is

$$Cr_2O_7^{2-} + 6I^- + 14H^+ \rightarrow 2Cr^{3+} + 3I_2 + 7H_2O.$$

But we also have a check; the charges must balance. The total charge on the reactant side of the equation must be equal to the total charge on the product side. The charge on the reactant side is:

one $Cr_2O_7^{2-}$ unit,	$1(2-) = -2$ charge,	
six I^- units,	$6(1-) = -6$ charge,	
fourteen H^+ units,	$14(1+) = +14$ charge,	

$$\text{Total charge} = (-2) + (-6) + (+14) = +6.$$

The charge on the product side is:

two Cr^{3+} units, $2(3+) = +6$ charge,
I_2 and H_2O uncharged.
          Total charge reactants $= +6 =$ Total charge products.

Since these charges are equal, the equation is balanced correctly.

Example 7-5. Balance the equation

$$Cu + H^+ + NO_3^- \rightarrow Cu^{2+} + NO + H_2O.$$

First we assign oxidation numbers:

Oxidation number

$$\begin{array}{ccccccc} Cu & + & H^+ & + & NO_3^- & \rightarrow & Cu^{2+} & + & NO & + & H_2O \\ 0 & & +1 & & +5-2 & & +2 & & +2-2 & & +1-2 \end{array}$$

Then we find the number of electrons gained and lost, first per atom and then per formula unit:

$$Cu \quad + \quad H^+ \quad + \quad NO_3^- \quad \rightarrow \quad Cu^{2+} \quad + \quad NO \quad + \quad H_2O$$

$$0 \qquad\qquad\qquad +5 \qquad\qquad +2 \qquad\qquad +2$$

2 e$^-$/Cu atom   3 e$^-$/N atom
2 e$^-$/Cu unit   3 e$^-$/NO$_3^-$ unit

The change in oxidation number for copper is 0 to $+2$, or two electrons lost per Cu atom; for nitrogen the change is $+5$ to $+2$, or three electrons gained per N atom. Since there is one atom of Cu and one atom of N per formula unit in both Cu and NO_3^-, the electron change per formula unit is equal to the change per atom.

Electron loss and gain can be equalized by taking 3 Cu atoms and 2 NO_3^- ions, since:

$$3Cu \quad + \quad H^+ \quad + \quad 2NO_3^- \quad \rightarrow \quad Cu^{2+} \quad + \quad NO \quad + \quad H_2O$$

$$0 \qquad\qquad\qquad\qquad +5 \qquad\qquad +2 \qquad\qquad +2$$

3(2 e$^-$/Cu unit)   2(3 e$^-$/NO$_3^-$ unit)
      or   or
6 e$^-$/3 Cu units   6 e$^-$/2 NO$_3^-$ units
              6 e$^-$ lost = 6 e$^-$ gained

Mass is balanced by having: 3 Cu on the left and 3 Cu^{2+} on the right; 2 nitrogen atoms in 2 NO_3^- ions on the left and 2 nitrogen atoms in 2 NO molecules on the right; 6 oxygen atoms in 2 NO_3^- ions $(2 \times 3 = 6)$ on the left and 6 oxygen atoms in 2 NO plus 4 H_2O molecules $(2 \times 1 + 4 \times 1 = 2 + 4 = 6)$ on the right; 8 H^+ ions are then needed to balance the 8 hydrogen atoms in the 4 H_2O molecules on the right:

$$3Cu + 8H^+ + 2NO_3^- \rightarrow 3Cu^{2+} + 2NO + 4H_2O.$$

To check the equation, let us see whether the charges on both sides are equal: On the left side, $+8$ from the 8 H^+ plus -2 from the 2 NO_3^- $= +6$. On the right side, $+6$ from the 3 Cu^{2+} and zero from 2 NO and 4 H_2O. The charges are equal $(+6 = +6)$ and therefore the equation is balanced.

It is important to remember that balanced equations show relative relationships between formula units or moles of formula units. Weight relations are not given directly. They must be calculated using the formula unit relationships from the equation in conjunction with the gram-formula weights.

Example 7–6. Determine how many grams of water can be produced from 2.0 moles of ammonia, according to the reaction:

$$NH_3 + O_2 \rightarrow NO + H_2O.$$

The balanced equation will show the number of moles of H_2O that can be produced from 2.0 moles of NH_3. We can then calculate that weight of H_2O by knowing the gram-formula

weight of H_2O. The first step, then, is to balance the equation. We assign oxidation numbers:

$$NH_3 + O_2 \rightarrow NO + H_2O$$

Oxidation number
$$-3+1 \quad 0 \quad +2-2 \quad +1-2$$

We determine the electron shifts per formula unit and equalize them:

$$4NH_3 \quad + \quad 5O_2 \quad \rightarrow \quad NO \quad + \quad H_2O$$

$$-3 \qquad\qquad 0 \qquad\qquad +2-2 \qquad\qquad -2$$

5 e⁻/N atom   2 e⁻/O atom
4(5 e⁻/NH₃ unit)   5(4 e⁻/O₂ unit)

20 e⁻ lost = 20 e⁻ gained

Mass is conserved:

$$4NH_3 + 5O_2 \rightarrow 4NO + 6H_2O,$$

and the equation is balanced. The balanced equation indicates that four moles of NH_3 will produce six moles of H_2O. Since we have only 2.0 moles of NH_3, we can obtain 3.0 moles of H_2O:

4 moles NH_3 produce 6 moles H_2O,
1 mole NH_3 produces $\frac{6}{4}$ moles H_2O,
2.0 moles NH_3 produce $\frac{6}{4}(2.0)$ moles H_2O,
2.0 moles NH_3 produce 3.0 moles H_2O.

To find the number of grams of H_2O contained in 3.0 moles of H_2O, we need to know the number of grams of H_2O in 1.0 mole of H_2O. The number of grams in 3.0 moles is just 3 times the number of grams in 1.0 mole. Since 1.0 mole H_2O weighs 18 g, 3.0 moles H_2O weigh $3 \times 18 = 54$ g. According to the given reaction, 54 g of H_2O can be produced from 2.0 moles of NH_3.

7–8 THE GRAM-EQUIVALENT

A New Concept. Now that we have spent considerable time on oxidation-reduction equations, learning how they are balanced and how the balanced equations are used, it may be somewhat disheartening to find that we can solve many of the problems without balancing the equations involved. Actually, this is the case. But this short-cut is certainly not a cure-all for equation-related problems. It involves the introduction of a new idea and approach. Just as is the case with other new ideas, understanding the idea and being able to use it are often not simultaneous. We recommend, therefore, that you study both methods of problem-solving, the first based on moles and balanced equations and the second based on gram equivalents and unbalanced equations. When you have the ideas of both methods well in hand, then you can use the method that is easier for a given problem.

Definition and Use. One *gram-equivalent of an oxidizing agent* is the number of grams of the oxidizing agent that gain one mole of electrons. One *gram-equivalent of reducing agent* is the number of grams of reducing agent that lose one mole of electrons.

▶ We have used the term *gram-equivalent*. Actually, *gram-equivalent weight* is the more precise term, but it is less common and more complex. Sometimes the term is shortened to *equivalent*. We shall stick with gram-equivalent, since it is commonly used and, we hope, leads to less confusion. ◀

Note that our definition of gram-equivalent requires that one gram-equivalent of reducing agent react with exactly one gram-equivalent of oxidizing agent. This is true because one mole of electrons is released by one gram-equivalent of reducing agent, and one mole of electrons is picked up by one gram-equivalent of oxidizing agent. This is the useful feature of gram-equivalents. We can see how this feature makes possible problem-solving if we consider the following case and the succeeding examples.

Consider the unbalanced, oxidation-reduction equation

$$CuO + NH_3 \rightarrow Cu + N_2.$$

There is a third product, H_2O, which is not included and not necessary for the solution of the problem. When we are working with gram-equivalents, the only substances necessary are the ones involved in the oxidation-reduction process, those containing the elements that show changes in oxidation number. We know that one gram-equivalent of CuO will react with one gram-equivalent of NH_3. But how do we find the number of moles or the number of grams in one gram-equivalent of CuO or NH_3? To do this we need to know how many electrons are gained or lost per formula unit. As before, we assign oxidation numbers, determine the number of electrons shifted per atom, and then the number shifted per formula unit:

$$
\begin{array}{ccccccc}
& CuO & + & NH_3 & \rightarrow & Cu & + & N_2 \\
\text{Oxidation} & +2-2 & & -3+1 & & 0 & & 0 \\
\text{number} & \uparrow & & \downarrow & & & & \\
& 2\,e^-/\text{Cu atom} & & 3\,e^-/\text{N atom} & & & & \\
& 2\,e^-/\text{CuO unit} & & 3\,e^-/NH_3\ \text{unit} & & & &
\end{array}
$$

Since CuO and NH_3 have one Cu and one N atom, respectively, per formula unit, the number of electrons shifted per atom is equal to the number shifted per formula unit.

Now to find the number of gram-equivalents of CuO in one mole of CuO and of NH_3 in one mole of NH_3, we must use the definition of the gram-equivalent: One gram-equivalent is the amount of a substance that loses or gains one mole of electrons. We found that one formula unit of CuO gains two electrons. Therefore one mole of CuO formula units gains two moles of electrons, and it takes only one-half mole of CuO to accept one mole of electrons. One gram-equivalent of CuO is therefore equal to one-half mole of CuO, since both these quantities gain one mole of electrons. With the reducing agent, NH_3, there are three gram-equivalents of NH_3 contained in one mole of NH_3, since one mole of NH_3 formula units loses three moles of electrons. Then one gram-equivalent of NH_3 is equal to one-third mole of NH_3.

To find the number of grams of CuO and NH_3 in one gram-equivalent of each, we need to consider the weight of one mole of each compound, i.e., its formula weight, in conjunction with the number of gram-equivalents contained in one mole. One

mole of CuO weighs $63.5 + 16.0 = 79.5$ grams. Since one gram-equivalent of CuO is equal to one-half mole (or there are two gram-equivalents in one mole), there are $\frac{1}{2} \times 79.5 = 39.8$ grams of CuO in one gram-equivalent of CuO. The number of grams of substance contained in one gram-equivalent is called the *gram-equivalent weight* of the substance. In the case of NH_3, the gram-equivalent weight is equal to the formula weight divided by three, $\frac{17}{3} = 5.7$ g/g-equivalent, since three gram-equivalents compose one mole.

From the previous definitions, we know that one gram-equivalent of oxidizing agent (CuO) reacts with one gram-equivalent of reducing agent (NH_3). We can say, then, that one-half mole of CuO reacts with one-third mole of NH_3, or that three $(6 \times \frac{1}{2})$ moles of CuO react with two $(6 \times \frac{1}{3})$ moles NH_3, etc. Since there are 39.8 g CuO and 5.7 g NH_3 in one gram-equivalent, 39.8 g CuO react exactly with 5.7 g NH_3.

In summary:

1. Gram-equivalents enable us to solve problems with oxidation-reduction reactions without balancing equations.

2. One gram-equivalent of oxidizing agent reacts with one gram-equivalent of reducing agent.

3. The number of gram-equivalents contained in one mole of formula units is equal to the number of electrons shifted per formula unit.

4. The gram-equivalent weight is equal to the gram-formula weight divided by the electron change per formula unit.

Example 7–7. In the following reaction, calculate the number of gram-equivalents in one mole of Al atoms and the gram-equivalent weight of Fe_2O_3:

$$Al + Fe_2O_3 \rightarrow Fe + Al_2O_3.$$

To determine the number of electrons changed per formula unit, the changes in oxidation-number are needed.

	Al	+	Fe_2O_3	$\rightarrow$	Fe	+	Al_2O_3
Oxidation number	0		+3		0		+3

$3\,e^-/\text{Al atom} \quad 3\,e^-/\text{Fe atom}$
$3\,e^-/\text{Al unit} \quad 6\,e^-/Fe_2O_3 \text{ unit}$

One gram-equivalent of reducing agent loses one mole of electrons. In this reaction, one mole of Al atoms loses three moles of electrons. Therefore there are three gram-equivalents of Al in one mole of Al (the number of gram-equivalents per mole is equal to the number of electrons changed per formula unit).

In Fe_2O_3 there are six gram-equivalents in one mole (six electrons gained per formula unit). One gram-equivalent therefore weighs one-sixth as much as one mole. The gram-equivalent weight of Fe_2O_3 *in this reaction* is

$$\frac{\text{gram-formula weight } Fe_2O_3}{6} = \frac{159.7 \text{ g } Fe_2O_3/\text{mole } Fe_2O_3}{6 \text{ g-equivalents}/\text{mole } Fe_2O_3} = 26.6 \text{ g } Fe_2O_3/\text{g-equivalent.}$$

Example 7–8. Determine how many grams of NH_3 react with 7.0 moles of O_2 according to the equation:

$$NH_3 + O_2 \rightarrow NO_2 + H_2O.$$

We need not balance the equation if we use gram-equivalents. We can determine the number of gram-equivalents per mole of NH_3 and O_2 from the changes in oxidation number obtained from the equation. This enables us to calculate the number of gram-equivalents of O_2 in 7.0 moles of O_2. The same number of gram-equivalents of NH_3 are necessary. Then, after we calculate the gram-equivalent-weight of NH_3, we can find the number of grams of NH_3.

	NH_3	+	O_2	$\rightarrow$	NO_2	+	H_2O
Oxidation	-3		0		$+4-2$		-2
number							

$$\downarrow \qquad\qquad \uparrow$$

$7\,e^-/N$ atom $2\,e^-/O$ atom
$7\,e^-/NH_3$ unit $4\,e^-/O_2$ unit

There are four gram-equivalents of O_2 in each mole of O_2, since each mole of O_2 accepts four moles of electrons. Then there are 28 gram-equivalents of O_2 in the 7.0 moles of O_2 available for reaction:

$$7.0 \text{ moles } O_2 \times 4.0 \text{ g-equivalents } O_2/\text{mole } O_2 = 28 \text{ g-equivalents } O_2.$$

One gram-equivalent of O_2 reacts with one gram-equivalent of NH_3. Therefore 28 gram-equivalents of NH_3 are needed. The weight of one gram-equivalent (the gram-equivalent weight) of NH_3 is its gram-formula weight divided by the electron change per formula unit,

$$\tfrac{17}{7} = 2.4 \text{ g } NH_3/\text{g-equivalent } NH_3.$$

Then 28 gram-equivalents of NH_3 will weigh 28 times as much as one gram-equivalent,

$$28 \text{ g-equivalents } NH_3 \times 2.4 \text{ g } NH_3/\text{g-equivalent } NH_3 = 67 \text{ g } NH_3.$$

We find that 67 g NH_3 will react with 7.0 moles O_2.

QUESTIONS

1. Define or discuss the following: Direct-combination reaction, noble-gas fluorides, decomposition reactions, double-replacement reactions, catalyst, the preparation of oxygen, photochemical smog, oxidation-reduction, oxidation number, oxidizing agent, substance oxidized, gram equivalent, gram-equivalent weight.

2. Classify the following reactions as direct combination, decomposition, single replacement, or double replacement, and balance each equation.

   a) $NH_4Cl \rightarrow NH_3 + HCl$ b) $P + O_2 \rightarrow P_4O_{10}$
   c) $Fe + Cu^{2+} \rightarrow Fe^{2+} + Cu$ d) $Ag + S \rightarrow Ag_2S$
   e) $HNO_2 \rightarrow H_2O + NO + O_2$
   f) $Ba^{2+} + Cl^- + Na^+ + SO_4^{2-} \rightarrow BaSO_{4(s)} + Na^+ + Cl^-$
   g) $Hg^{2+} + NO_3^- + H_2S \rightarrow HgS + H^+ + NO_3^-$
   h) $Cu(NO_3)_2 \rightarrow CuO + NO_2 + O_2$

3. What is the distinguishing feature about an oxidation-reduction reaction?

4. a) What is an oxidation number?

   b) When do oxidation numbers correspond to actual atomic charges?

5. Assign oxidation numbers to each of the atoms in:

   a) $KMnO_4$ b) CrF_2 c) $NaIO_3$ d) $S_2O_3^{2-}$ e) NO_3^-

   f) $CaMoO_4$ g) Sc_2O_3 h) $Gd_2(SO_4)_3$ i) V_2O_5 j) CrO_4^{2-}

6. Explain why oxygen generally has an oxidation number of -2 in compounds.

7. Balance the following equations:

   a) $Ba^{2+} + HSO_4^- \rightarrow BaSO_4 + H^+$ b) $Na + H_2 \rightarrow NaH$

   c) $Cl_2 + HI \rightarrow I_2 + HCl$ d) $CaO + HF \rightarrow CaF_2 + H_2O$

   e) $Al + O_2 \rightarrow Al_2O_3$ f) $HCN + NH_4OH \rightarrow H_2O + NH_4^+ + CN^-$

   g) $Ca^{2+} + OH^- \rightarrow Ca(OH)_2$ h) $WO_3 + S \rightarrow WS_2 + SO_2$

8. Assign oxidation numbers to all the atoms in:

   a) CO_2 b) $NaBr$ c) I_2

   d) PH_3 e) CaH_2 (ionic hydride)   f) BaO

9. Refer to Question 7:

   a) Which reactions involve oxidation-reduction?

   b) In each oxidation-reduction reaction, list the oxidizing agent, the reducing agent, the substance oxidized, the substance reduced, the element increasing in oxidation number, the element decreasing in oxidation number, the substance gaining electrons, and the substance losing electrons.

10. Assign oxidation numbers to all the atoms in:

    a) SO_4^{2-} b) NH_4NO_3 c) $Cr_2O_7^{2-}$

    d) CO_3^{2-} e) PO_4^{3-} f) Fe_3O_4

11. Balance the following oxidation-reduction reactions:

    a) $Na + Zn^{2+} \rightarrow Na^+ + Zn$

    b) $PbF_2 + H_2 \rightarrow Pb + HF$

    c) $Fe^{2+} + Ce^{4+} \rightarrow Fe^{3+} + Ce^{3+}$

    d) $MnO_4^{2-} + Fe^{3+} \rightarrow Fe^{2+} + MnO_4^-$

    e) $Al + Pb^{2+} \rightarrow Al^{3+} + Pb$

    f) $Pb + NO_3^- + H^+ \rightarrow Pb^{2+} + NO + H_2O$

    g) $Cr^{3+} + MnO_4^- + H_2O \rightarrow Cr_2O_7^{2-} + Mn^{2+} + H^+$

    h) $Mn^{2+} + MnO_4^- + H_2O \rightarrow MnO_2 + H^+$

    i) $MnO_4^{2-} + H_2O \rightarrow MnO_2 + MnO_4^- + H^+$

    j) $H_2O + F_2 \rightarrow O_2 + HF$

12. Consider the following balanced equation:

$$5Cu + 2MnO_4^- + 16H^+ \rightarrow 2Mn^{2+} + 5Cu^{2+} + 8H_2O.$$

    a) What is shown by the subscripts in MnO_4^- and H_2O?

    b) What is learned from the equation?

    c) What is learned from the coefficients in the equation?

    d) How many moles of water can be produced from 10 moles of copper metal, according to this reaction?

13. For the equation $N_2 + 3H_2 \rightarrow 2NH_3$, give the reactants, product(s), oxidizing agent, reducing agent, substance oxidized, substance reduced, element decreasing in oxidation number, and element increasing in oxidation number.

14. Show the step-by-step process in balancing the following oxidation-reduction reaction:

$$Mn^{2+} + Cr_2O_7^{2-} + H^+ \rightarrow MnO_2 + Cr^{3+} + H_2O.$$

15. After assigning oxidation numbers, predict the formulas of the compounds formed between:

a) $La^{3+} + OH^- \rightarrow$

b) $Al^{3+} + NO_3^- \rightarrow$

c) $Al^{3+} + SO_4^{2-} \rightarrow$

d) $Na^+ + PO_4^{3-} \rightarrow$

e) $Y^{3+} + PO_4^{3-} \rightarrow$

f) $In^{3+} + CO_3^{2-} \rightarrow$

g) $Ba^{2+} + PO_4^{3-} \rightarrow$

h) $Ca^{2+} + H_2PO_4^- \rightarrow$

i) $K^+ + HCO_3^- \rightarrow$

j) $H^+ + C_2H_3O_2^- \rightarrow$

16. The following couples include only one-half of an oxidation-reduction process. In each case: 1) Determine whether the couple as written is an oxidation or a reduction. 2) Determine the change in oxidation number. 3) Determine the number of electrons shifted per formula unit.

a) $Fe^{2+} \rightarrow Fe^{3+}$

b) $F_2 \rightarrow F^-$

c) $NO_3^- + 4H^+ \rightarrow NO + 2H_2O$

d) $H_2SO_3 + H_2O \rightarrow SO_4^{2-} + 4H^+$

e) $2Hg \rightarrow Hg_2^{2+}$

f) $O_2 + 4H^+ \rightarrow 2H_2O$

g) $I_3^- \rightarrow 3I^-$

h) $AuCl_4^- \rightarrow Au + 4Cl^-$

17. In the reaction between calcium metal and hydrogen gas, calcium hydride is formed.

a) Recalling the oxidation number of hydrogen in ionic hydrides, write the equation for the reaction.

b) How many atoms of calcium react with one mole of hydrogen gas?

c) How many moles of calcium hydride are produced by reacting one mole of calcium and 1.0 gram of hydrogen?

18. Balance the equation:

$$Sn^{2+} + MnO_4^- + H^+ \rightarrow MnO_2 + Sn^{4+} + H_2O.$$

a) How many moles of H^+ react with 1.5 moles of Sn^{2+}?

b) How many moles of Sn^{4+} can be produced from 4.0 moles of MnO_4^-?

c) How many moles of MnO_2 can be obtained from 5.0 moles of MnO_4^- and 3.0 moles of Sn^{2+}?

d) How many grams of H_2O can be obtained from a mixture of 10 moles Sn^{2+}, 10 moles of MnO_4^-, and 16 moles of H^+?

19. According to the equation $2Fe^{3+} + 2I^- \rightarrow I_2 + 2Fe^{2+}$, how many grams of iodine can be produced by reacting 6.5 moles Fe^{3+} and 6.4 moles I^-?

20. If hydrogen and iodine gases are mixed, the product is hydrogen iodide.

a) Write the balanced equation.

b) How many grams of hydrogen react with 4.0 moles of iodine?

c) How many moles of product can be obtained from 10.0 grams of hydrogen and 254 grams of iodine?

d) How many moles of hydrogen are necessary to produce 12.8 grams of hydrogen iodide?

21. In the reaction $Zn + I_2 \rightarrow Zn^{2+} + 2I^-$:

a) How many grams of Zn react with 12.7 g of I_2?

b) How many atoms of Zn react with 6.02×10^{22} molecules of I_2?

c) How many I^- ions can be produced from 2.0 moles of I_2?

d) How many moles of Zn^{2+} and I^- can be obtained from 6.5 g Zn and 12.0×10^{22} molecules of I_2?

22. a) What information is given by the equation $H_2SeO_3 + H_2S \rightarrow Se + S + H_2O$?

b) Balance the equation in part (a). What information does this provide?

c) How many moles of H_2S lose one mole of electrons?

d) How many gram-equivalents of H_2S are contained in one mole of H_2S?

23. For the following reactants and products, determine: (1) the number of electrons shifted per formula unit, (2) the number of gram-equivalents per mole for each reactant, and (3) the gram-equivalent weight for each reactant.

a) $CuO + H_2 \rightarrow Cu + H_2O$ b) $Pb + F_2 \rightarrow PbF_2$

c) $H_2O + F_2 \rightarrow O_2 + HF$ d) $Cr^{3+} + Cd \rightarrow Cr^{2+} + Cd^{2+}$

e) $Al + Fe_2O_3 \rightarrow Al_2O_3 + Fe$

24. In the reaction $Br_2 + I^- \rightarrow Br^- + I_2$:

a) How many moles of Br_2 and I^- combine exactly?

b) How many moles of I^- combine with 1.0 gram-equivalent of Br_2?

c) How many gram-equivalents of I^- combine with 7.91 gram-equivalents of Br_2?

d) What are the molecular and gram-equivalent weights of Br_2?

25. It is found experimentally that 8.9 gram-equivalents of an oxidizing agent X are needed to produce a specified amount of product.

a) How many gram-equivalents of reducing agent Z are needed to carry out the reaction?

b) What information is needed to calculate the number of grams of X and Z that are necessary for the reaction?

26. Calculate the formula weights for all reactants and products and the gram-equivalent weights for the oxidizing and reducing agents for the reaction

$$H_2SO_3 + I_2 + H_2O \rightarrow H_2SO_4 + HI.$$

27. By listing the appropriate weights and checking the correct spaces, complete the following table for the reaction

$$H_2SO_3 + H_2S \rightarrow S + H_2O.$$

	H_2SO_3	H_2S	S	H_2O
Formula weight				
Weight of 1.0 mole				
Oxidizing agent				
Reducing agent				
Substance oxidized				
Substance reduced				
Substance gaining e^-				
Substance losing e^-				
Number e^- shifted per formula unit				
Gram-equivalent weight				

28. a) Balance the equation in Question 27.
    b) How many grams of reducing agent are needed to produce 9.6 g S?

29. Consider the equation in Question 27:
    a) How many gram-equivalents of H_2S react with 0.40 gram-equivalent of H_2SO_3?
    b) How many gram-equivalents of H_2S react with 0.10 mole of H_2SO_3?
    c) How many gram-equivalents of H_2S react with 8.2 grams of H_2SO_3?
    d) How many gram-equivalents of H_2S are needed to produce 0.30 mole of S?

8. GASES

8-1 INTRODUCTION

We have seen several cases in which many experimental observations coupled with the ingenious insight of one man led to a greatly increased understanding of our physical world. When scientists make more observations and couple them with the ideas of other scientists, mankind achieves a more detailed understanding of the behavior of matter.

What further types of observations can be made? Consider the behavior of a sample of water as its temperature is lowered below, and raised above, room temperature. When the water temperature reaches 0 °C, ice begins to form, i.e., the liquid water changes into a solid. When the temperature reaches 100 °C, the water begins to boil and is changed into water vapor, i.e., the liquid water changes into a gas. The characteristics and behaviors of a substance existing as a gas are quite different from those of the solid substance, or of the liquid substance; and the properties of the liquid are different from those of the solid. The characteristics, behaviors, and differences of these three states of matter, and their interconversions, are interesting and important topics. We begin our study with gases, since their behaviors are somewhat less complicated and more amenable to generalizations and simplifying assumptions.

8-2 THREE IMPORTANT VARIABLES: VOLUME, TEMPERATURE, AND PRESSURE

A *gas* is a substance that takes the shape and volume of its container. The properties and behaviors of all gases depend on the volume of the gas, the temperature of the gas, and the pressure of the gas. For any gas these three variables—volume, temperature, and pressure—are interrelated; the value of one depends on the values of the other two.

Volume. The *volume* of a substance is the space occupied by the substance (recall the units: ℓ; ml = cc = cm^3). A gas occupies the entire volume of its container. Solids and liquids, on the other hand, have their own definite volume, which is not influenced greatly by changes of temperature or pressure. The volume of a gas, however, is very dependent on its temperature and pressure. We can see that this is true by considering what happens to a balloon when it is squeezed (its pressure is changed) or placed in a refrigerator (its temperature is changed). We know that when we squeeze a balloon its size decreases, i.e., the volume of the air in the balloon decreases. If a balloon is inflated at room temperature and then placed in a refrigerator for a short time, again the volume of the balloon decreases. Qualitatively, then, it seems that the volume of a gas decreases with increasing pressure (an inverse relationship) and decreases with decreasing temperature (a direct relationship). What other experiments might we perform to investigate these relations? How could the measurements be made more quantitative?

Temperature and Pressure. We have already seen that the *temperature* of a substance indicates its degree of hotness (and is measured in °C or °K). Heat energy flows from hot regions of a substance to cold regions.

Pressure is probably a less familiar concept than temperature. *Pressure* can be defined as force per unit area, but is best thought of as a push. When pressure is exerted on a substance, the substance receives a push and may move in the direction of the push. Mass flows from regions of high pressure to regions of low pressure. In an automobile tire the air pressure inside the tire is greater than the air pressure outside the tire. If the tire is punctured, air rushes from the high-pressure area inside to the lower-pressure area outside the tire.

In all *fluids* (gases and liquids) the pressure at any given point in the fluid is the same in all directions. We can visualize this by considering a person swimming underwater, as in Fig. 8-1. The pressure on the swimmer is the same in all directions. The water is pushing with equal force from above and below and from all sides of the swimmer.

We can also note one more thing. The deeper the swimmer is in the water, the greater the pressure on him. This seems reasonable if we think of the swimmer as supporting a column of water above his body equal to the depth of the water. As the swimmer goes farther below the surface of the water, the amount of water that he supports increases. He feels this as an increase in pressure on his body. (Remember that he feels this same pressure or push equally from every direction.)

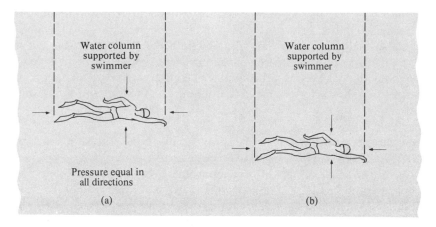

Fig. 8-1 The pressure exerted on a swimmer. The water column supported by the swimmer in (a) is smaller than that in (b). The pressure on the swimmer is less in (a) than in (b).

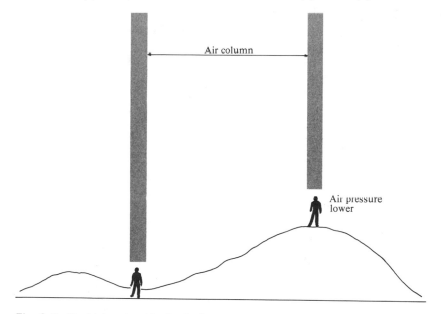

Fig. 8-2 The higher the altitude, the lower the atmospheric pressure.

We can use this same reasoning process when we consider the pressure exerted by gases. As an example, consider the pressure exerted by the air that surrounds the earth. It is fairly evident that the air pressure on our bodies is the same in all directions. If this were not the case, then we would constantly feel a push from the direction of higher pressure. This happens, for example, in a strong wind. We actually feel a push on our body coming from the direction of high pressure and trying to move us toward lower pressure.

It is also true that the pressure exerted by air is different at different distances above the surface of the earth. Just as an underwater swimmer feels less pressure as he nears the surface of the water, a mountain climber (see Fig. 8–2) feels less pressure (from the air) as he nears the top of a mountain (he is supporting a smaller column of air above his body). We can test this qualitative observation by making quantitative measurements of the air (or atmospheric) pressure at the foot and at the summit of the mountain.

Pressure Measurement. We can measure atmospheric pressure—the pressure exerted by the air surrounding the earth—conveniently by using a *barometer*. We can understand the principle of operation of a barometer by considering the preparation of a simple working model, as shown in Fig. 8–3. A long glass tube, closed at one end, is filled with mercury. (Actually liquids other than mercury could also be used, but because of its high density, low vapor pressure, and its existence as a liquid over a wide temperature range, mercury is very convenient.) The open end of the tube is then closed, say, with a finger (encased in a rubber glove!). The tube is inverted and placed in a pool of mercury. When you take away your finger while the open end of the tube is below the surface of the mercury pool, the mercury level in the tube drops a short distance below the upper closed end, forming a vacuum in this top section of the tube. The mercury does not flow completely out of the tube because the pressure of the air pushing down on the surface of the mercury pool is holding the mercury column up in the tube. At the surface of the mercury pool, the pressure exerted by the column of mercury in the tube is equal to the air pressure.

When we are making pressure measurements, we need to specify the distance above the surface of the earth, since we have seen that the pressure of the air changes with the distance above the earth's surface. Variations in gas pressure also occur for changes in gas temperature. As we do with most other measurements,

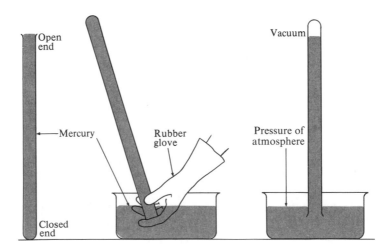

Fig. 8–3 Preparation of a mercury barometer.

therefore, we must choose some standard point of reference for measuring gas pressures. (Almost all measurements require a standard of some kind; for example, 0 °C and 100 °C are the selected temperatures for the freezing and boiling points of water; 1 meter is a certain defined length; ^{12}C has an atomic weight of 12.0000 amu; etc.). This standard condition for pressure measurement is: One *atmosphere* of air pressure at sea level and 0 °C supports a column of mercury (Hg) 760 mm (76.0 cm) high. One standard atmosphere of gas pressure is therefore equal to the pressure exerted by a column of mercury 760 mm high. We say that 1 atm = 760 mm Hg = 76.0 cm Hg. These are the units of pressure for gases: atmospheres, mm Hg, and cm Hg. The word *torr* (named after Torricelli, an Italian physicist) is now replacing the term mm Hg. One torr = one mm Hg and standard pressure is 760 torr = 760 mm Hg = 1 atm. The conditions of temperature and pressure 0 °C (273 °K) and 1 atm (760 torr) are called the conditions of *standard temperature and pressure (STP)* for gases.

We can also measure gas pressures by using a manometer. Figure 8–4 shows a simple manometer, which consists essentially of a U-tube partially filled with mercury. Manometers are generally used in conjunction with a system that holds a gas at pressures below (and sometimes above) atmospheric pressure. One end of the U-tube is connected to the gas handling system; the other end is open to atmospheric pressure. When the gas pressure is equal to the pressure of the atmosphere, the columns of mercury in the two arms of the U-tube are at the same

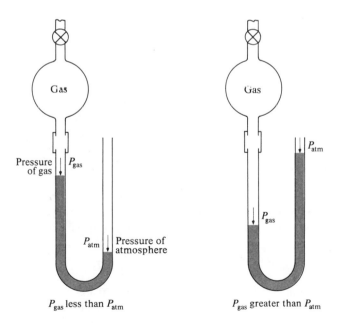

P_{gas} less than P_{atm} P_{gas} greater than P_{atm}

Fig. 8–4 A mercury manometer. Note the differences in the mercury levels under different gas pressures.

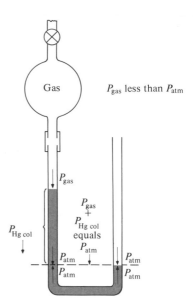

Fig. 8–5 Measurement of an unknown gas pressure using a mercury manometer.

level. When a pressure imbalance occurs between the gas system and the atmosphere, the mercury levels shift and are no longer equal. If the gas pressure is less than atmospheric pressure, the mercury column will rise in the arm connected to the gas system. If the gas pressure is higher than that of the atmosphere, the reverse will happen.

If we can obtain the prevailing atmospheric pressure from a barometer, we can make quantitative measurements of the pressure of an unknown gas by using a simple manometer, as shown in Fig. 8–5. The pressure of the gas in the sample bulb is less than the pressure of the atmosphere. The pressure is equal in both arms of the manometer at the level indicated by the dashed line, since the pressure in all fluids is the same at the same depth (or level) in the fluid. Then in the left arm of the manometer, the pressure being exerted up and the pressure being exerted down at the dashed line are equal to one another, and they are both equal to atmospheric pressure. The downward pressure at this point is composed of the pressure exerted by the liquid mercury column above the level of the dashed line and by the pressure of the gas in the space above the mercury column. Since the downward pressure at the dashed line is equal to P_{atm}, the sum of the gas pressure and the pressure exerted by the mercury column must also be equal to atmospheric pressure:

$$P_{atm} = P_{gas} + P_{Hg\ col}.$$

We can then determine the pressure of the unknown gas, P_{gas}, by measuring the height of the mercury column (say, in mm Hg) and reading P_{atm} from a barometer:

$$P_{gas} = P_{atm} - P_{Hg\ col}.$$

(Note that $P_{Hg\ col}$ will be a negative number if P_{gas} is greater than P_{atm}.)

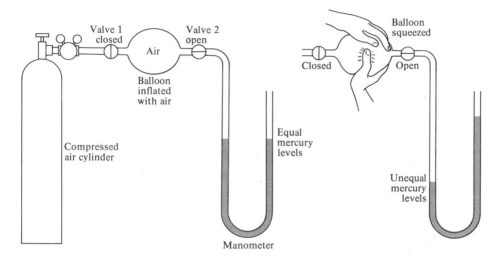

Fig. 8–6 Relationship between gas volume and gas pressure. A decrease in the volume of gas produces an increase in the gas pressure, as indicated by the manometer.

8–3 VOLUME-PRESSURE RELATIONS IN GASES

Qualitative Experiments. We can illustrate the relationship between the volume and the pressure of a gas by conducting a simple experiment using a balloon and a manometer to measure the gas pressure in the balloon. In the experimental setup in Fig. 8–6, a balloon is attached on one end to a cylinder containing air at high pressure and on the other end to a mercury manometer. Valves at each end of the balloon can connect or disconnect the air cylinder or the manometer from the balloon. If valve 2 is closed and then valve 1 opened, the balloon will inflate. If valve 1 is closed and valve 2 opened, the mercury columns in both arms of the manometer will be essentially equal, indicating little difference in pressure between the gas in the balloon and in the atmosphere.

Now with valve 1 closed and valve 2 open, we decrease the volume of the balloon by squeezing it. When we do so, the mercury columns in the manometer immediately indicate a pressure difference between the gas in the balloon (being squeezed) and the atmosphere. The mercury is pushed out of the manometer arm connected to the balloon (i.e., the mercury's level drops) and into the arm open to the atmosphere (its level rises). The gas in the squeezed balloon is at a higher pressure than the gas was when the balloon had its normal—and larger—volume.

This experiment shows that the volume and pressure of a gas have an inverse relationship. As the pressure of the gas increases, the volume of the gas decreases. As the external pressure on the balloon was increased, causing an increase in the air pressure inside the balloon, the volume of the balloon decreased, corresponding to a decrease in the volume of the air in the balloon.

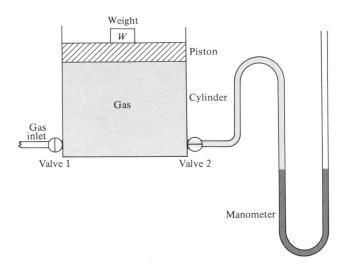

Fig. 8–7 Apparatus for quantitative experiments on relationships between gas volume and gas pressure.

Quantitative Experiments. An understanding of qualitative relationships is very important, but, where possible, a more detailed, quantitative understanding is even more helpful. We can make the relation between gas volumes and gas pressures more quantitative if we conduct the experiments in a more exact manner. We need only to vary the above experimental apparatus and procedures slightly.

Figure 8–7 shows a cylinder fitted with a piston in place of the balloon shown in Fig. 8–6. The cylinder and piston (similar to those found in an automobile engine) make possible the easy measurement of gas volumes and changes of external pressures.

We can conduct a series of experiments in which the gas in the cylinder occupies various volumes under various pressures. We can measure the volumes and pressures precisely and develop a quantitative relationship between them. In these experiments we shall follow certain procedures: A given amount of gas will be admitted to the cylinder through valve 1, which is then closed. The cylinder will then contain a fixed weight of gas. We shall always assume that no gas escapes during any part of the experiment. (The amount of gas in the cylinder is not, for the moment, important. Experimentally, we would find that a greater amount or weight of gas would occupy a larger volume. As more gas entered the cylinder, the piston would move up.) We can determine the volume of the gas by measuring the volume of the cylinder enclosing the gas (for this, we need to know the height of the piston and the diameter of the cylinder), and we can measure the gas pressure by using the attached manometer. One other thing needs mentioning. During the series of experiments, the temperature of the gas must not change. The value of the constant temperature can be determined and monitored with a thermometer.

We can conduct the series of experiments in the following way. We determine the volume of a given weight of gas under a certain constant pressure. This

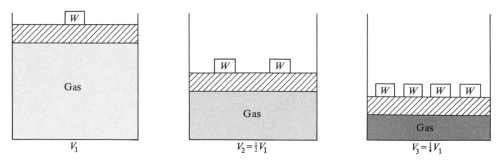

Fig. 8–8 Increasing the pressure exerted on a gas decreases the volume of the gas.

constant pressure can be atmospheric pressure or atmospheric pressure plus an added weight placed on top of—and pushing down on—the piston and enclosed gas. Then we can apply a greater pressure by increasing the weight on the piston by a known amount and redetermining the volume and pressure of the gas.

Table 8–1 records data from experiments such as these. The experimental conditions were selected (the proper weight of gas at 300 °K) so that the initial pressure was 1 atm and the initial volume of the gas was 1 ℓ. Then the pressure on the gas was increased by adding weights, each weight corresponding to 1 atm pressure, to the piston (see Fig. 8–8) and the volume and pressure of the gas were remeasured.

Table 8–1. Pressure–volume data from four experiments on a gas

Experiment	Weight on piston	Pressure on gas, atm	Volume of gas, ℓ
1	W	1	1
2	$2W$	2	$\frac{1}{2}$
3	$3W$	3	$\frac{1}{3}$
4	$4W$	4	$\frac{1}{4}$

We can see from Table 8–1 that, as the pressure was doubled, the volume was halved; at three times the original pressure, the volume was one-third of the original volume, and so on.

The experimental data in Table 8–1 can be handled in at least two ways: 1) The data can be inspected and conclusions drawn. 2) The data can be plotted in graphical form to reveal any characteristic behavior or trend. Scientists generally do both in seeking solutions to their problems. The first procedure is just common sense; the second can summarize the data in an easily visualized manner and often leads to an important conclusion.

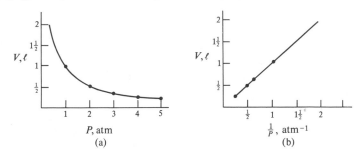

Fig. 8-9 (a) A plot of the pressure–volume data for a gas given in Table 8-1. (b) The same data as in (a), except that the volume is plotted versus 1 over the pressure.

Let us illustrate the graphical procedure using the data in Table 8-1. When we plot the volume of the gas V versus its pressure P, as in Fig. 8-9a, the resulting curve is a hyperbola. But we find an even simpler relationship—a straight line— when we plot the volume V versus 1 over the pressure, $1/P$. This graph, shown in Fig. 8-9b, indicates that there is a direct relationship between V and $1/P$, and the equation $V = 1/P$ can be obtained from the graph.

▶ The equation of a straight line is $y = mx + b$. In this case, the variable y is the gas volume V and x is 1 over the gas pressure $(1/P)$. The slope of the graph, m, is equal to 1 and the intercept on the V-axis, b, is $b = 0$; that is, the line passes through the origin.

To find a simple relationship between two quantities, such as a straight line, it is often necessary to make several graphs, each one plotting the same two quantities, but in different forms. It might be that we obtain a straight line if we plot A versus $1/B$, or if we plot log A versus B or $1/B$, or log A versus log B, and so on. This procedure of selecting the proper graph to give a simple relation between A and B is often a matter of trial and error, sometimes combined with educated guessing. ◀

Table 8-2. Pressure–volume data for a given weight of gas at 27.0 °C

Experiment	P, atm	$1/P$, atm^{-1}	V, ℓ
1	0.554	1.81	44.6
2	0.720	1.39	34.2
3	0.872	1.15	28.3
4	1.30	0.770	19.0
5	1.63	0.614	15.1
6	2.90	0.346	8.50
7	5.20	0.192	4.74
8	10.1	0.0991	2.44

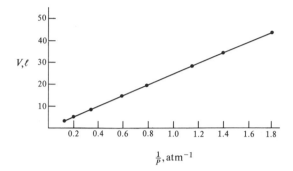

Fig. 8-10 The pressure–volume data given in Table 8–2.

Boyle's Law. The data in Table 8–1 were chosen to illustrate the relationship between gas volume and gas pressure. Generally, experimental data are not so neat, nor do they lend themselves to such easy interpretation. Consider Table 8–2, for example. A simple gas volume–gas pressure relationship is not evident just from an inspection of the data. But from the plot of V versus $1/P$ in Fig. 8–10, we can see that the relationship of these data is similar to the relationship that holds for the data in Table 8–1. The only difference in the two plots is in the slope of the straight line.

Experiments similar to those described above, but using different apparatus, were carried out by Robert Boyle around 1662. He experimented with several gases and found that for a fixed weight of gas at a given temperature, the volume of the gas was always equal to some constant number divided by the gas pressure. He formulated what we know as *Boyle's law*: At constant temperature, a given weight of gas occupies a volume that is inversely proportional to the pressure exerted on it.

In Figs. 8–9b and 8–10 we can see that the data follow Boyle's law. As the pressure decreases (and $1/P$ increases), the volume increases. The lines in the two figures have different slopes, indicating different constants of proportionality between V and $1/P$ in the two cases. In Fig. 8–9b the slope of the line is 1; the equation of the line is

$$V = (1)\frac{1}{P} = \frac{1}{P}.$$

In Fig. 8–10 the slope of the line is 24.6; the equation of the line is

$$V = (24.6)\frac{1}{P} = \frac{24.6}{P}.$$

Mathematically, the general equation for Boyle's law is

$$V = \frac{b}{P} \quad \text{or} \quad P = \frac{b}{V} \quad \text{or} \quad PV = b,$$

where b is a constant number ($b = 1$ and $b = 24.6$ in the two above cases).

▶ The Boyle's law proportionality constant is determined by the weight *and* the temperature of the gas present. If experiments are conducted on a gas using the same weight of gas each time, then the constant b will change if the experiments are run at different temperatures. Another b would be obtained under conditions of constant temperature and changing weights of gas. Check Section 8-7 and see if you can derive these relationships between b and the temperature and weight (number of moles) of a gas being studied. ◀

Using Boyle's law, we can see that if the pressure on a gas is doubled ($P_2 = 2P_1$), the volume of the gas is cut in half ($V_2 = \frac{1}{2}V_1$); if $P_2 = \frac{1}{4}P_1$, then $V_2 = 4V_1$; if $P_1 = 747$ torr and $P_2 = 763$ torr, then

$$P_2 = \frac{763}{747} P_1 \quad \text{and} \quad V_2 = \frac{747}{763} V_1.$$

The first two examples are easily seen, since simple numbers are involved. When the numbers become more complicated, the proper relations and procedures can sometimes be obscured. This might be the case in the third example. But note that the thought processes and operations performed are identical in all three cases, regardless of whether simple or complicated numbers are used. The ratio of the final gas pressure to the initial gas pressure is some fraction (it can be less than or greater than 1); the ratio of the final volume to the initial volume is just the inverse of the pressure ratio.

These pressure–volume relationships, at conditions of constant temperature and unchanging amount of gas, can be summarized mathematically as follows:
Since $V = b\,(1/P)$, then

$$(1)\ V_1 = b\left(\frac{1}{P_1}\right) \quad \text{and} \quad (2)\ V_2 = b\left(\frac{1}{P_2}\right).$$

Dividing (1) by (2), we obtain

$$\frac{V_1}{V_2} = \frac{\cancel{b}\,(1/P_1)}{\cancel{b}\,(1/P_2)} = \frac{1/P_1}{1/P_2}$$

and

$$\frac{V_1}{V_2} = \frac{P_2}{P_1}.$$

Example 8–1. The pressure of 1.0 ℓ of a given weight of gas held at constant temperature is changed from 1.0 atm to 380 torr. What is the final volume of the gas?

At constant temperature and with a fixed weight of gas, Boyle's law applies. The pressure changes from $P_1 = 1.0$ atm to $P_2 = 380$ torr. To determine the pressure change, we need to use the same units of pressure for both P_1 and P_2. Either P_1 or P_2 may be changed; it makes no

difference. Converting P_2 to atm gives

$$P_2 = \frac{380 \text{ torr}}{760 \text{ torr/atm}} = 0.5 \text{ atm.}$$

The pressure decreases by one-half, from 1.0 to 0.5 atm:

$$P_2 = \frac{0.5}{1.0} P_1 = \tfrac{1}{2} P_1.$$

Therefore the volume must increase by a factor of two, since the volume change is inverse to the pressure change:

$$V_2 = \frac{1.0}{0.5} V_1 = \frac{2}{1} V_1.$$

Since the initial volume is $V_1 = 1.0\ \ell$,

$$V_2 = 2V_1 = 2(1.0\ \ell) = 2.0\ \ell.$$

Example 8–2. A balloon is inflated with air to a volume of 1.5 ℓ at 752 torr and 300 °K. The balloon is connected to an evacuated flask which has a volume of 2/3 ℓ. All the air is forced from the balloon into the flask. What is the pressure of the gas in the flask at 300 °K?

Boyle's law applies, since a certain amount of gas is being studied at constant temperature. The volume of the gas is decreased from $V_1 = 1.5\ \ell$ to $V_2 = 2/3\ \ell$. The volume ratio is

$$\frac{V_2}{V_1} = \frac{2/3}{3/2} = \frac{4}{9}.$$

The pressure ratio is the inverse of the volume ratio, therefore

$$\frac{P_2}{P_1} = \frac{9}{4}$$

and

$$P_2 = \frac{9}{4} P_1.$$

Since $P_1 = 752$ torr,

$$P_2 = \frac{9}{4}(752 \text{ torr}) = 1692 \text{ torr.}$$

The final gas pressure in the flask must be greater than the initial pressure, since the volume of the gas was decreased.

8–4 VOLUME–TEMPERATURE RELATIONS IN GASES

Charles's Law. Experiments similar to those described above can be conducted to determine the relationship between the volume and the temperature of a gas. In these experiments the pressure of a given weight of gas will be held constant, the temperature of the gas varied and the resulting gas volumes (and volume changes)

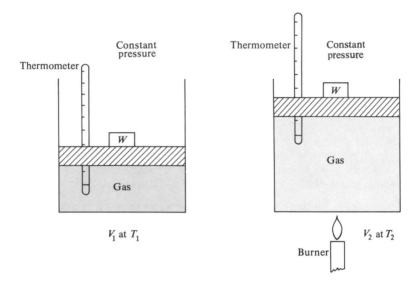

Fig. 8–11 Apparatus for quantitative experiments on relationships between gas volume and gas temperature.

determined. Figure 8–11 shows a diagram of the experimental apparatus. The cylinder contains a fixed weight of gas under a constant pressure, exerted by the weight on top of the piston. The temperature of the gas can be changed by heating (or cooling) the cylinder and monitored by means of the thermometer.

Experimentally it is found that as the cylinder is heated and the temperature of the gas increases, the gas expands, pushing the piston higher in the cylinder, and fills a larger volume. Conversely, if the cylinder is cooled and the temperature of the gas decreases, the volume of the gas also decreases.

Table 8–3 gives data from experiments such as these. In all the experiments a constant weight of gas was used, and the gas was kept at a constant pressure. We can see that the volume of the gas decreased as its temperature decreased, and the volume increased as the temperature increased. Closer inspection of Table 8–3 reveals a more quantitative relation. When the temperature (on the Kelvin scale) was doubled, the volume was doubled; when the Kelvin temperature decreased by half, the volume decreased by half, and so on. These observations indicate that the

Table 8–3. Temperature–volume data for a given weight of gas at constant pressure

Experiment	Temperature °C	°K	Volume, ℓ
1	-123	150	0.50
2	27	300	1.0
3	327	600	2.0
4	627	900	3.0
5	927	1200	4.0

volume and temperature of a gas are directly proportional, $V \propto T$ (the symbol $\propto$ means *is proportional to*). This behavior is different from that found for the volume and pressure of a gas: $V \propto 1/P$.

About 1787 experiments similar to those above led to *Charles's Law*: At constant pressure, the volume of a fixed weight of gas is directly proportional to its temperature, expressed in °K. Charles's law can be written in a simple mathematical equation,

$$V = cT \, (^\circ K),$$

where c is a constant number (a proportionality constant), and T is the Kelvin temperature.

▶ The value of c depends on the weight (the number of moles) of gas present and on the pressure of the gas. You should understand this after consulting Section 8–7. Also note that an equation such as $V_1/V_2 = T_1/T_2$ can easily be derived using Charles's law, since the constant c is a fixed number irrespective of the numerical values of V and T (remember that T must be expressed in °K). Consequently,

$$(1) \; V_1 = cT_1, \qquad (2) \; V_2 = cT_2.$$

Dividing (1) by (2) gives us

$$\frac{V_1}{V_2} = \frac{cT_1}{cT_2} = \frac{T_1}{T_2} \qquad \text{or} \qquad V_2 = V_1 \left(\frac{T_2}{T_1}\right). \; \blacktriangleleft$$

Note that T must be expressed in °K and not °C, since negative temperatures exist on the Celsius scale. If c is a positive number and T, in °C, is negative, then the gas volume V must also be negative. But this situation is physically impossible. It arises from our arbitrary definition of 0° on the Celsius temperature scale. The difficulty can be removed by redefining 0°, that is, by using the Kelvin temperature scale. In fact, studying the volume–temperature relationships of gases gives us one way of establishing absolute zero on the Kelvin scale, and its equivalent temperature on the Celsius scale.

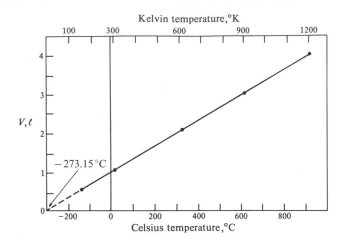

Fig. 8–12 A plot of the temperature–volume data for a gas given in Table 8–3, giving both Celsius and Kelvin temperatures. Absolute zero, 0 °K and −273.15 °C, is found by extending the curve to $V = 0$ for the gas.

Figure 8–12 plots the Charles's law data in Table 8–3. A straight line is obtained which can be extrapolated (extended where no actual data exist) to $V = 0$, and therefore $T = 0°$, since $V = cT$. This temperature is called 0 °K (absolute zero) and is found to be equal to −273.15 °C.

▶ This presents an interesting dilemma. Presumably we could make a gas vanish, i.e., have its volume go to zero, simply by cooling the gas far enough. Can you think of a circumstance that will relieve this difficulty? ◀

In using Charles's law, we are interested in finding changes in the volume of the gas brought about by temperature changes or changes in the temperature of the gas brought about by volume changes. Typical problems are illustrated in the following examples. Remember that Charles's law applies only when the gas pressure remains constant and no gas is added or lost.

Example 8–3. A balloon contains 596 ml of air at 25 °C and 755 torr. What will the volume of the balloon be if the temperature is changed to 15 °C?

The pressure remains constant at 755 torr, and there is no change in the weight of air in the balloon; Charles's law applies. The temperature must be expressed in °K. Therefore

$$T_1 = 25 + 273 = 298 \text{ °K} \qquad \text{and} \qquad T_2 = 15 + 273 = 288 \text{ °K}.$$

The temperature decreases from 298 to 288 °K; the volume must also decrease. The temperature ratio is $\dfrac{T_2}{T_1} = \dfrac{288}{298}$; the volume must change (decrease) by the same ratio, $\dfrac{V_2}{V_1} = \dfrac{288}{298}$. Therefore

the final volume is

$$V_2 = \frac{288}{298} (596 \text{ ml}) = 576 \text{ ml}.$$

Note that V_2 must be less than V_1, since $T_2/T_1 \left(\dfrac{288}{298}\right)$, is less than 1.

Example 8–4. Suppose that 3.2 g O_2 occupy 3.0 ℓ at 750 torr and 380 °K. What is the temperature of the O_2 at 750 torr if the volume is changed to 4.5 ℓ?

Charles's law applies, since a given weight of gas is at constant pressure. The volume of O_2 is increased from 3.0 to 4.5 ℓ:

$$\frac{V_2}{V_1} = \frac{4.5}{3.0}.$$

The temperature must increase by the same ratio,

$$\frac{T_2}{T_1} = \frac{4.5}{3.0}.$$

Therefore

$$T_2 = \frac{4.5}{3.0} T_1 = \frac{4.5}{3.0} (380 \text{ °K}) = 570 \text{ °K}.$$

The increase in temperature is directly proportional to the increase in volume.

8–5 GAS MIXTURES

Dalton's Law of Partial Pressures. The laws of Boyle and Charles apply to a given sample of gas without specifying, in general, what the gas is or whether more than one gas is present in the sample. In fact, many of the gas samples that we encounter are mixtures of two or of several gases. Air is probably the most common example. It is composed of N_2, O_2, CO_2, Ar, and water vapor, along with several other gases in small quantities.

An interesting observation about mixtures of gases, made by Dalton, has become known as *Dalton's law of partial pressures*. It states that the total pressure exerted by a mixture of gases is equal to the sum of the partial pressures of the gases. The *partial pressure* of a gas is the pressure that the gas would exert if it were alone in the container. Dalton's law can be expressed mathematically as

$$P_{\text{total}} = P_1 + P_2 + P_3 + \ldots$$

where P_1, P_2, P_3, etc., are the partial pressures of the gases in the mixture. (The three dots after P_3 indicate that as many partial pressures should be added as there are gases in the mixture. If just two gases are present, then $P_{\text{total}} = P_1 + P_2$, but if five gases are present, then

$$P_{\text{total}} = P_1 + P_2 + P_3 + P_4 + P_{5.})$$

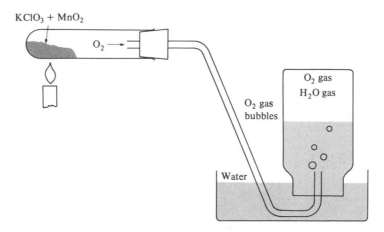

Fig. 8-13 Collecting oxygen gas by displacing water.

We can illustrate the use of Dalton's law of partial pressures by considering experiments in which gas samples such as O_2 are collected by displacement of water, using an experimental apparatus as diagrammed in Fig. 8-13. You can produce oxygen gas readily by heating a test tube containing potassium chlorate ($KClO_3$) and a very small amount of the catalyst manganese dioxide (MnO_2). The reaction proceeds according to the equation

$$2KClO_3 \xrightarrow[\text{Heat}]{MnO_2} 2KCl + 3O_2.$$

The O_2 gas produced by the reaction is bubbled into a bottle, displacing the water that initially filled the bottle. But if an experiment were performed to determine the composition of the gas collected in the bottle, it would show that water vapor was present along with the O_2 gas. As the bottle is filled with gas, some liquid H_2O evaporates, making the gas sample a mixture of the gases O_2 and $H_2O_{(g)}$. Then the total pressure of the gas inside the bottle is made up of the pressures of both gases. And according to Dalton's law, the total pressure inside the bottle is equal to the sum of the partial pressures of the O_2 and H_2O gases:

$$P_{in} = P_{O_2} + P_{H_2O}.$$

Now we can calculate the pressure of the oxygen, P_{O_2}, in the bottle, if we know the total pressure inside the bottle, P_{in}, and the pressure of the water vapor, P_{H_2O}. These two quantities can be easily obtained. We can read the value of P_{in} from a barometer; that is, P_{in} is equal to the prevailing atmospheric pressure, when the surface of the liquid H_2O remaining in the bottle is at the same level as the H_2O in the trough, as shown in Fig. 8-14. (Note that these H_2O levels inside and outside the bottle must be equal. If they are not, P_{in} will not equal P_{atm}. Why is this true?) You can find the value of P_{H_2O} in a standard table, such as the one in Appendix D (or Table 9-2).

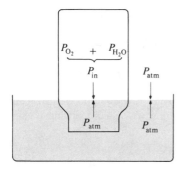

Fig. 8–14 Determining an unknown oxygen gas pressure after collection, as in Fig. 8–13.

These H_2O vapor pressure data are the result of many careful experiments. You can see that the numerical value of the P_{H_2O} depends on the temperature of the liquid H_2O (we shall consider this in more detail in Section 9–3). Then we can calculate the pressure of the O_2, using Dalton's law:

$$P_{in} = P_{O_2} + P_{H_2O}$$
$$P_{atm} = P_{in} \ (H_2O \text{ levels equal})$$
$$P_{atm} = P_{in} = P_{O_2} + P_{H_2O}$$
$$P_{atm} = P_{O_2} + P_{H_2O}$$
$$P_{O_2} = P_{atm} - P_{H_2O}$$
$$\uparrow \qquad\quad \uparrow$$

Read from   Standard
barometer   table

The following example illustrates the use of Dalton's law of partial pressures, in combination with Boyle's law and Charles's law.

Example 8–5. Suppose that 5.0 ℓ of nitrogen gas (N_2) is collected over water at 27 °C and at an atmospheric pressure of 747 torr. What would be the volume of dry N_2 gas at STP?

Note that the problem involves all three laws: Boyle's, Charles's, and Dalton's. A mixture of gases is present, since N_2 is collected over H_2O (Dalton's law applies, involving gases of N_2 and H_2O); the temperature is changed, producing a change in the volume of the gas (Charles's law); the pressure is changed, producing a change in the volume of the gas (Boyle's law). We must consider each one of these laws in determining the final volume of dry N_2 at STP.

▶ We can apply these gas laws to a particular problem in a stepwise fashion. To illustrate that this calculation procedure is valid, consider the N_2 gas in Example 8–5 to be collected in a (perfectly elastic) balloon instead of a bottle. Since we want to know the volume of dry N_2, we need to subtract the effects of H_2O vapor. We can do this by squeezing out the gas in the balloon and passing it through a substance that removes the H_2O but not the N_2. The N_2 can then be re-collected in the balloon, still at 27 °C and 747 torr. Now Dalton's law has effectively been satisfied. We have subtracted the P_{H_2O} (by removing the H_2O) from the total pressure, P_{atm}. Note that the temperature and atmospheric pressure were unchanged. What can be said about the volume of the balloon before and after removal of the H_2O vapor?

Now we can experiment on the balloon, first using Charles's law and then, independently, Boyle's law. The temperature of the balloon can be changed from 27 °C to 0 °C by placing it in a cold box, such as a refrigerator. In this experiment the pressure remains constant at 747 torr, but the balloon volume will change (will it increase or decrease?) because of the temperature change. This experiment satisfies the Charles's law step of the procedure. Then the pressure in the refrigerator can be increased from 747 to 760 torr by pumping air into the refrigerator, while the temperature remains constant at 0 °C. This experiment satisfies Boyle's law. We can then determine the volume of the dry N_2 at STP by measuring the volume of the balloon while it is in the refrigerator at 0 °C and 1 atm. ◀

Our problem is to calculate the final volume of dry N_2 gas. We know the initial volume of the N_2 and also that the temperature and pressure of the N_2 sample are changing. We must therefore determine the exact temperature and pressure changes of the dry N_2 and then use the laws of Boyle and Charles to calculate the final volume.

Conditions of N_2 gas	Initial	Final
Volume	5.0 ℓ	?
Temperature	27 °C = 300 °K	0 °C = 273 °K
Pressure	747 − 27 = 720 torr	760 torr

We can calculate the initial pressure of dry N_2 by using Dalton's law of partial pressures. The total pressure (P_{atm}) is the sum of the pressures of the N_2 (P_{N_2}) and the H_2O vapor (P_{H_2O}),

$$P_{atm} = P_{N_2} + P_{H_2O}.$$

Atmospheric pressure is given as 747 torr, and P_{H_2O}, as obtained from the standard table in Appendix D, is 27 torr at 27 °C. Then

$$P_{N_2} = P_{atm} - P_{H_2O},$$
$$P_{N_2} = 747 - 27 = 720 \text{ torr.}$$

The initial pressure of dry N_2 is therefore 720 torr.

We can then use Boyle's law to correct for the volume change caused by the pressure changes of the dry N_2. The pressure increases from $P_1 = 720$ torr to $P_2 = 760$ torr:

$$\frac{P_1}{P_2} = \frac{720}{760} \quad \text{or} \quad P_2 = \frac{760}{720} P_1.$$

Since the pressure increases, the volume must decrease, according to Boyle's law:

$$\frac{V_2}{V_1} = \frac{P_1}{P_2}; \quad V_2 = \frac{P_1}{P_2} V_1; \quad V_2 = \frac{720}{760} V_1.$$

Note that the fraction $\frac{720}{760}$ is the inverse of the fraction in the pressure change; it is, and must be, less than 1, since the final volume V_2 is less than the initial volume V_1. This fraction, derived from Boyle's law, is called the *pressure correction factor*.

If the pressure remains constant, the change in the volume of the gas caused by a change in the temperature can be calculated according to Charles's law. Remember that temperatures must be expressed in °K. The initial temperature is 27° + 273° = 300 °K and the final temperature is

standard temperature, or 273 °K. The temperature decreases from $T_1 = 300$ °K to $T_2 = 273$ °K:

$$\frac{T_1}{T_2} = \frac{300}{273} \quad \text{or} \quad T_2 = \frac{273}{300} T_1.$$

According to Charles's law, the volume change is in direct proportion to the temperature change. Therefore the volume changes by the same fraction as the temperature changes:

$$\frac{V_2}{V_1} = \frac{T_2}{T_1}; \quad V_2 = \frac{T_2}{T_1} V_1; \quad V_2 = \frac{273}{300} V_1.$$

This fraction, derived from Charles's law, is called the *temperature correction factor*.

Now we can combine the two correction factors, so that the final volume of gas is

$$V_2 = V_1 \text{ (temperature correction factor) (pressure correction factor)},$$

$$V_2 = V_1 \left(\frac{T_2}{T_1}\right)\left(\frac{P_1}{P_2}\right),$$

$$V_2 = 5.0\,\ell \left(\frac{273 \text{ °K}}{300 \text{ °K}}\right)\left(\frac{720 \text{ torr}}{760 \text{ torr}}\right),$$

$$V_2 = 4.3\,\ell.$$

The final volume of the dry N_2 gas at STP is 4.3 ℓ. It must be less than the initial volume of 5.0 ℓ, since the temperature decreased, causing a volume decrease, and the pressure increased, causing a volume decrease. This qualitative reasoning gives a rough check of the answer and also of the temperature and pressure correction factors.

8-6 VOLUME–WEIGHT RELATIONS IN GASES

Avogadro's Law. We have seen that the volume of a gas depends on both its temperature and its pressure, given a fixed weight or amount of gas. What happens when the amount of gas is changed? This is actually quite familiar and easily observed. When a balloon is inflated, the amount of air in the balloon increases and the volume of the balloon increases. If air escapes, the volume decreases. Both these experiments can be done under conditions of constant temperature and pressure, i.e., the temperature and pressure in the room remain unchanged.

Observations similar to these, combined with tremendous scientific insight, led Avogadro to propose what has become known as *Avogadro's law*: At the same temperature and pressure, if the volumes of gases are equal, then the gas samples contain the same number of moles of gas. This says that, at room temperature and pressure, 1.0 ℓ of O_2 and 1.0 ℓ of N_2 contain the same number of moles of O_2 and N_2, respectively. Since the number of moles of O_2 and N_2 are equal, the number of molecules of O_2 and N_2 are also equal. But note that the weights of the O_2 and N_2 samples are not the same (since O_2 and N_2 have different molecular weights).

Experimentally it has been shown that *one mole of any ideal gas at STP occupies 22.4 ℓ.* For example, 1.0 mole $O_2 = 32.0$ g O_2 occupies 22.4 ℓ at 0 °C and 1 atm;

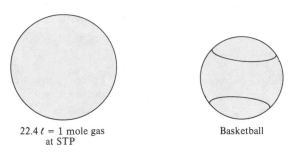

22.4 ℓ = 1 mole gas   Basketball
at STP

Fig. 8-15 The relative sizes of 1 mole of ideal gas at STP, 22.4 ℓ, and an ordinary basketball.

0.5 mole Ar gas = 20.0 g Ar occupies 11.2 ℓ at STP. (To get an idea of the size of 22.4 ℓ, see Fig. 8-15.)

▶ For our purposes all gases will be considered as ideal gases, i.e., the gases follow all the gas laws. There are some gases, however, that do deviate from the behaviors that are expected according to the general gas laws, especially at low temperatures, or at high pressures, or with large amounts of gas present. The deviations from ideal behavior are often small. Sometimes the gas laws can be modified so that they do account for nonideal gas behaviors. ◀

Avogadro's law and the corollary finding that one mole of gas occupies 22.4 ℓ at STP are helpful in determining molecular weights and in solving many other types of problems, as illustrated in the following examples.

Example 8-6. Two balloons are at 26 °C and 752 torr and have equal volumes. Balloon 1 contains 1.6 g O_2. How many moles of gas are present in balloon 2?

Both balloons are at the same temperature and pressure and have equal volumes. Therefore, according to Avogadro's law, both balloons contain the same number of moles of gas. There are 1.6 g O_2 in balloon 1; the molecular weight of O_2 is 32 g O_2/mole O_2. Then in balloon 1 there is

$$\frac{1.6 \text{ g } O_2}{32 \text{ g } O_2/\text{mole } O_2} = 0.050 \text{ mole } O_2.$$

Balloon 2 also contains 0.050 mole of gas.

Example 8-7. How many grams of helium (He) are needed to inflate a balloon to 6.72 ℓ at STP?

If we can determine the number of moles of He needed to give 6.72 ℓ at STP, we can calculate the number of grams of He, using its atomic weight. We can determine the number of moles of He required to inflate the balloon to 6.72 ℓ, since 1.0 mole of He at STP occupies 22.4 ℓ:

$$\frac{6.72 \text{ } \ell}{22.4 \text{ } \ell/\text{mole}} = 0.30 \text{ mole He at STP.}$$

One mole of He weighs 4.0 g; 0.30 mole of He weighs

$$(0.30 \text{ mole He}) (4.0 \text{ g He/mole He}) = 1.2 \text{ g He}.$$

Then 1.2 g He are needed to inflate a balloon to 6.72 ℓ at STP.
What would be the volume of the balloon at 25 °C and 740 torr?

Example 8–8. At STP it is observed experimentally that one liter of H_2 gas weighs 0.900 g and that one liter of Cl_2 gas weighs 31.6 g. The atomic weight of H = 1.01 amu. Calculate the molecular weight of Cl_2.

Since the volumes of H_2 and Cl_2 are the same at STP, the number of moles of H_2 and Cl_2 are also equal; and the same number of H_2 and Cl_2 molecules are present. Since the Cl_2 sample weighs more than the H_2 sample and there are equal numbers of molecules of each, the weight of one molecule of Cl_2 must be greater than the weight of one H_2 molecule by the fraction

$$\frac{1 \text{ } Cl_2 \text{ molecule}}{1 \text{ } H_2 \text{ molecule}} = \frac{31.6}{0.900}$$

or

$$1 \text{ } Cl_2 \text{ molecule} = \frac{31.6}{0.900} = 35.1 \text{ times as heavy as each } H_2 \text{ molecule.}$$

The weight of one H_2 molecule is 2 times the atomic weight of H = 2(1.01 amu) = 2.02 amu. The molecular weight of Cl_2 is 35.1 times the molecular weight of H_2, or (35.1) (2.02 amu) = 71.0 amu.

▶ Scientists now know, after considerable experimentation and thought, that H_2 and Cl_2 are diatomic gases. They did not know the number of atoms composing a particular gas during Avogadro's time. A wrong guess about the number of atoms in a molecule led to gross errors in the calculation of molecular weights by the methods used in this example. ◀

8–7 A UNIFYING EQUATION

The Ideal Gas Law. The results of the laws of Boyle, Charles, and Avogadro can be combined to give one general equation relating the pressure P, temperature T, volume V, and number of moles n of any ideal gas. According to these three laws:

$$\text{Boyle:} \quad V \propto \frac{1}{P} \qquad \text{Constant } T \text{ and } n$$

$$\text{Charles:} \quad V \propto T \text{ (°K)} \qquad \text{Constant } P \text{ and } n$$

$$\text{Avogadro:} \; V \propto n \qquad \text{Constant } T \text{ and } P$$

Since the volume of an ideal gas is proportional to $1/P$, T, and n separately, we can use reasoning similar to that given in the insert on pages 225–226 to show that

$$V \propto \left(\frac{1}{P}\right)(T)(n),$$

where T is expressed in °K. Then

$$V = R\left(\frac{1}{P}\right)(T)\,(n),$$

where R is a constant of proportionality, and is called the *ideal gas constant*. Rearranging the equation slightly gives

$$PV = nRT,$$

which is called the *ideal gas law*. All ideal gases obey this equation.

The ideal gas law is useful in calculating one of the four variables P, V, n, or T when we know data for three of the variables. For example, we can calculate the volume of an ideal gas sample if we can determine the numerical values for P, n, and T. But this is only true after a value for the ideal gas constant, R, has been established, since a calculation for V,

$$V = \frac{nRT}{P},$$

requires knowledge of not only P, n, and T, but also R. We can determine a value for R by using information that we have already acquired, in conjunction with the rearranged ideal gas equation,

$$R = \frac{PV}{nT}.$$

To calculate R, we need experimental values for P, V, n, and T for a given ideal gas sample. It has been found experimentally that at STP one mole of any ideal gas occupies a volume of 22.4 ℓ. This gives, for one mole of ideal gas:

$$P = 1.0 \text{ atm}, \qquad n = 1.0 \text{ mole}, \qquad V = 22.4\,\ell, \qquad T = 273 \text{ °K}.$$

Substituting these values for P, V, n, and T into the ideal gas equation enables us to calculate a value for R,

$$R = \frac{(1.0 \text{ atm})(22.4\,\ell)}{(1.0 \text{ mole})(273 \text{ °K})} = 0.082\ \ell\text{-atm/mole-°K}.$$

We can use this numerical value for R in all calculations involving the ideal gas equation, regardless of the values of P, V, n, and T. There is one catch: the units. The units of P must be atm; the units of V must be ℓ; the units of n must be moles; the units of T must be °K. If we know any of these variables in units different from those specified, we must convert them to the appropriate units before we can use the ideal gas equation, with $R = 0.082\ \ell\text{-atm/mole-°K}$. (This is particularly troublesome with pressure, which is often expressed in terms of torr. Conversion to atm is necessary; for example,

$$\frac{380 \text{ ~~torr~~}}{760 \text{ ~~torr~~/atm}} = 0.50 \text{ atm}.$$

The value 0.50 atm, not 380 torr, must be used in the calculation.)

Example 8–9. What is the volume of a balloon containing 0.20 mole of He gas at 27 °C and 750 torr?

The temperature, pressure, and number of moles of gas are given; we need to calculate the volume of the gas. We can determine V by using the ideal gas equation,

$$PV = nRT \quad \text{or} \quad V = \frac{nRT}{P},$$

since n, R, T, and P are all known. Remember, if we are to use $R = 0.082$ ℓ-atm/mole-°K, then n must be in moles, T in °K, and P in atm. Making these conversions gives:

$$n = 0.20 \text{ mole}, \quad R = 0.082 \text{ } \ell\text{-atm/mole-°K},$$

$$T = 27 + 273 = 300 \text{ °K}, \quad P = \frac{750 \text{ torr}}{760 \text{ torr/atm}} = \frac{750}{760} \text{ atm}.$$

Substituting these values in the equation,

$$V = \frac{(0.20 \text{ mole}) (0.082 \text{ } \ell\text{-atm/mole-°K}) (300 \text{ °K})}{750/760 \text{ atm}},$$

$$V = \frac{(0.20) (0.082) (300) (760)}{750} \text{ } \ell \quad \text{or} \quad 5.0 \text{ } \ell.$$

The volume of the balloon is 5.0 ℓ.

Example 8–10. Calculate the molecular weight of 0.32 g of an ideal gas that occupies 400 ml at -29 °C and 380 torr.

We need to calculate the molecular weight of an unknown gas. The molecular weight is the number of grams of gas in one mole of the gas. Therefore, to calculate it, we need to know the number of grams of gas and the corresponding number of moles of gas present. The number of grams of gas is given: 0.32 g. We can calculate the number of moles of gas equivalent to 0.32 g from the ideal gas equation,

$$PV = nRT \quad \text{or} \quad n = \frac{PV}{RT},$$

since we know the pressure, volume, and temperature of the gas:

$$P = \frac{380 \text{ torr}}{760 \text{ torr/atm}} = 0.50 \text{ atm},$$

$$V = 400 \text{ ml} = 0.400 \text{ } \ell,$$

$$T = -29 \text{ °C} + 273° = 244 \text{ °K},$$

$$R = 0.082 \text{ } \ell\text{-atm/mole-°K}.$$

Substituting these values into the equation for n (moles):

$$n = \frac{(0.50 \text{ atm}) (0.400 \text{ } \ell)}{(0.082 \text{ } \ell\text{-atm/mole-°K}) (244 \text{ °K})} = 0.010 \text{ mole gas}.$$

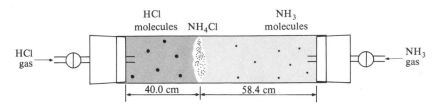

Fig. 8–16 The lighter NH_3 gas (17.0 amu) diffuses faster than the heavier HCl gas (36.5 amu). Ammonium chloride, a white solid, is formed when the NH_3 and HCl meet. Its formation indicates the distances traveled by each gas.

We can then calculate the molecular weight of the unknown gas, since we know that 0.010 mole of the gas is composed of or contains 0.32 g of the gas:

Molecular weight = number of g/mole,

Molecular weight = 0.32 g/0.010 mole
(note the units of the fraction, g/mole),

Molecular weight = 32 g/mole.

One mole of the unknown gas weighs 32 grams. What is the chemical formula of the unknown gas, assuming that it is an element?

8–8 GASES AND DIFFUSION

We know that gases spontaneously spread to fill the entire volume of their containers. We also know that foul-smelling gases such as hydrogen sulfide, H_2S (which has the odor of rotten eggs), can travel fairly rapidly from one part of a room to another. This spontaneous spreading of substances is called *diffusion*.

In gases the process of diffusion generally occurs quite readily, but some gases diffuse more quickly than others. For example, we might take a sample of H_2S gas and a sample of Cl_2 gas and simultaneously release them from their containers in one corner of a room. We could then move to the opposite corner of the room and wait to see which gas arrived first. Judging by the characteristic odors of these two gases, we would find that the H_2S arrived before the Cl_2. That is, H_2S gas diffuses more rapidly than Cl_2 gas.

▶ Actually the spreading of gases is caused by heat convection currents as well as by diffusion. We have neglected convection here, for simplicity, even though it is, in many cases, more important to gas movements than diffusion. ◀

This characteristic of variable rates of diffusion of gases was noted and studied by Thomas Graham around 1829. His experiments were, of course, better controlled and more quantitative than the experiment described above. He found that lighter

gases diffuse more rapidly than heavier ones. Using this observation, we could have predicted the results of the above experiment. The lighter H_2S gas (molecular weight = 34 amu) should diffuse faster than the heavier Cl_2 gas (molecular weight = 71 amu).

Figure 8–16 shows another example of the dependence of diffusion on mass.

The above qualitative observation is helpful, but Graham's experiments enabled him to state a more quantitative relationship, which we call *Graham's law of diffusion*: The rate of diffusion of a gas is inversely proportional to the square root of its molecular weight. We can write this law in the form of an equation,

$$r \propto \frac{1}{\sqrt{MW}},$$

where r is the rate of diffusion of the gas and MW is the molecular weight of the gas.

This quantitative statement necessarily includes the qualitative observation that we considered first: Heavier gases diffuse more slowly because of the inverse relationship between r and $\sqrt{MW}$. But it also enables us to make quantitative comparisons of rates of diffusion between two gases, using the equation

$$\frac{r_1}{r_2} = \frac{\sqrt{MW_2}}{\sqrt{MW_1}}.$$

Note the *inverse* relation between r_1 and $\sqrt{MW_1}$ and between r_2 and $\sqrt{MW_2}$. As an example, let us consider the relative rates of diffusion of the gases H_2 and O_2. The molecular weights are 2 and 32, respectively. Then

$$\frac{r_{H_2}}{r_{O_2}} = \frac{\sqrt{MW_{O_2}}}{\sqrt{MW_{H_2}}} = \frac{\sqrt{32}}{\sqrt{2}} = \sqrt{\frac{32}{2}} = \sqrt{16} = 4,$$

$$r_{H_2} = 4r_{O_2}.$$

The rate of diffusion of H_2 is 4 times as fast as the rate of diffusion of O_2. This checks qualitatively, since H_2 is the lighter gas and should therefore diffuse more rapidly.

Example 8–11. Uranium and fluorine form a gaseous compound, UF_6. It is proposed that the ^{235}U and ^{238}U isotopes be separated by gaseous diffusion of the $^{235}UF_6$ and $^{238}UF_6$ compounds. Which gas would diffuse more rapidly, and by how much?

Since $^{235}UF_6$ is the lighter gas (molecular weight of 349.0343 amu compared with 352.0412 amu for $^{238}UF_6$), it diffuses more rapidly. We can calculate the relative rates of diffusion from

$$\frac{r_{235}}{r_{238}} = \frac{\sqrt{MW_{238}}}{\sqrt{MW_{235}}} = \frac{\sqrt{352.0412}}{\sqrt{349.0343}} = \sqrt{\frac{352.0412}{349.0343}},$$

$$\frac{r_{235}}{r_{238}} = \sqrt{1.0086} = 1.0043.$$

The rate of diffusion of $^{235}UF_6$ is 1.0043 times faster than the rate of diffusion of $^{238}UF_6$.

▶ This process of separation of the ^{235}U isotope is actually the one that was used to obtain pure ^{235}U for use in atomic weapons. Huge gaseous diffusion plants costing billions of dollars are necessary to separate the uranium isotopes, since the rates of diffusion of the UF_6 compounds are so close. Until recently, this economic limitation to producing the necessary amounts of pure ^{235}U prevented the spread of atomic weapons. ◀

Example 8–12. Oxygen gas (O_2) diffuses 1.4 times faster than an unknown gas. What is the molecular weight of the unknown gas?

According to Graham's law:

$$\frac{r_{O_2}}{r_{unknown}} = \frac{\sqrt{MW_{unknown}}}{\sqrt{MW_{O_2}}}.$$

The problem states that

$$r_{O_2} = 1.4 r_{unknown} \qquad \text{or} \qquad \frac{r_{O_2}}{r_{unknown}} = 1.4.$$

Therefore

$$\frac{\sqrt{MW_{unknown}}}{\sqrt{MW_{O_2}}} = 1.4 \qquad \text{or} \qquad \sqrt{MW_{unknown}} = 1.4\sqrt{MW_{O_2}} = 1.4\sqrt{32},$$

$$MW_{unknown} = (1.4\sqrt{32})^2 = (1.4)^2(32) = 64 \text{ amu}.$$

The unknown gas has a molecular weight of 64 amu.

8–9 THE KINETIC THEORY OF GASES

We have considered several laws concerning the behaviors of gases. These laws resulted from the bright ideas of several scientists; they were tested and validated by a wealth of experimentation. But once a law or hypothesis becomes firmly established, one question always arises. Why? Why do gases diffuse, and—granted that they do—why does the lightest gas diffuse fastest? Why does a gas take the entire volume and shape of a container, while liquids and solids do not? Why do gases expand when heated and contract when compressed? Why do gases seem to mix together spontaneously and in all proportions, and yet often behave as if the other gases of the mixture were not present?

All these questions, and many more, can be asked. Try thinking of a few more; then, without looking at the following pages, try to construct a model of a gas that would explain the behaviors that we have studied and at least some of the questions that we have posed. This is no easy task, but it can be fun as well as quite instructive.

One approach to developing a model for gases might be the following. Let us consider some of the observations about gases and the laws describing their behaviors that we know to be true. Then we can postulate some things about the small particles composing the gas that seem to be necessary to permit the observed behaviors.

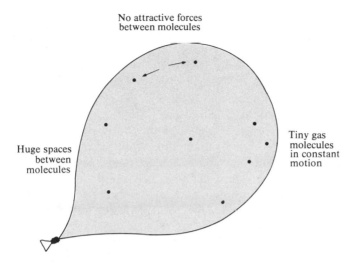

No attractive forces
between molecules

Huge spaces
between
molecules

Tiny gas
molecules
in constant
motion

Fig. 8–17 The very small gas molecules fill the entire volume of the balloon, since they are in constant motion. No attractive forces exist between the gas molecules, which are separated by large distances.

We know, for example, that gases diffuse spontaneously and rather rapidly to occupy the entire volume of their containers. This means that the particles of gas— that is, the gas molecules—must be constantly moving.

And it suggests a further postulate. Since gases diffuse rather rapidly, the pathway for diffusion must be relatively free of obstructions. How can that be? It would be true if the pathway were very large compared with the size of the gas molecules. We know, for example, that a person walking along a narrow sidewalk must often slow down and alter his path as he meets other people. But if the same number of people were walking along a sidewalk the width of a football field, their progress would be essentially unhindered and consequently faster. It seems reasonable, then, to assume that gas molecules are very small in size compared with the space in which they are moving.

The analogy with people walking along sidewalks might lead us to another postulate about gas molecules. We know that if several people are trying to stay together in a group and walk along a sidewalk, their progress is much slower than that of a single person, since their physical size causes them to encounter other walkers and groups of walkers more frequently. Hence we might hypothesize that gas molecules are moving individually and not in groups; that is, the attraction of one gas molecule for another is very low. Figure 8–17 gives a schematic representation of such gas molecules.

Now this qualitative model of a gas can be tested using other known gas properties. For example, gases are easily compressed. Is our model consistent with this fact? We have postulated that the tiny gas molecules are very far apart and do not

occur in clusters. The gas molecules, then, should be easily squeezed together. Compression simply reduces, slightly, the vast amount of empty space between molecules.

Thus our quickly and crudely derived model for gases does appear to be consistent with the compressibility of gases. Knowing this, we must subject the proposed model to many more tests. Does it explain the known gas laws? Are there any necessary changes, additions, corrections? And so on.

A theory that applies to gases and their behaviors* has been developed over a period of many years, with several outstanding scientists contributing valuable ideas and experiments. This theory, called the *kinetic theory*, actually contains the postulates or assumptions that were derived above from our analogy with people walking. The kinetic theory postulates were, of course, obtained in a different, more rigorous manner, over a much longer period of time, and have been tested and validated against many common gas behaviors.

In order to establish a simple model for use in describing, understanding, and predicting the behavior of gases, the *kinetic theory*, as it applies to gases, contains the following assumptions:

1. Gas molecules are widely separated in space; their volume is negligible when compared with the space between them.

2. There are essentially no attractive forces between gas molecules.

3. Gas molecules are in constant, rapid, random, straight-line motion, colliding with each other and with the container walls. In these collisions energy can be transferred between molecules, but there is no net decrease in the total energy of all the molecules involved in the collisions.

4. The average kinetic energy of the gas molecules is directly proportional to the Kelvin temperature. At any instant one gas molecule may have a low kinetic energy and another molecule a high kinetic energy, but at any given temperature the average kinetic energy of the total number of molecules in any gas sample is the same.

▶ The kinetic energy of any moving body is given by

$$KE = \tfrac{1}{2}mv^2$$

where m is the mass of the body and v is the velocity or speed of the body (how fast it is moving). An increase in mass increases the kinetic energy directly, i.e., doubling m also doubles KE. An increase in velocity, however, causes a larger increase in kinetic energy, since KE is proportional to the velocity squared; i.e., doubling v quadruples KE. ◀

---

*   The kinetic theory also applies to liquids and solids, as we shall see in Chapter 9. Some of the assumptions involved in the theory are different, however, for the three different states of matter.

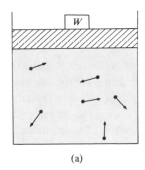
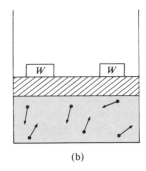

(a)   (b)

Fig. 8-18 The gas in (b) exerts a greater pressure than the gas in (a), since the molecules in (b) hit their container walls more often.

8-10 GAS LAWS AND THE KINETIC THEORY

Boyle's Law. An acceptable theory must explain known observations and facts. We must, then, be able to account for, say, Boyle's law by using the kinetic theory. Why should the pressure and volume of a gas be inversely proportional to one another?

Consider what happens to the gas pictured in Fig. 8-18. A gas in the cylinder with only one weight on the top of the piston occupies a volume of V_1. Figure 8-18a shows a few dots representing gas molecules, with arrows indicating their motion in a given direction. Now the volume available to the gas sample can be decreased to V_2 if the piston is pushed down in the cylinder, as in Fig. 8-18b. The necessary downward push is represented by an additional weight resting on the piston. It is important that none of the gas escape during the compression and that the temperature of the gas before and after compression be the same.

With these two conditions satisfied, we must determine how the pressure of the gas in the cylinder has changed. The same number of gas molecules are represented in both parts (a) and (b) of Fig. 8-18. The molecules in each case are moving with the same speed, since the temperature and hence the average kinetic energy of the molecules are unchanged. Since V_2 is less than V_1 and the gas molecules are traveling at the same speed in both cases, the molecules will make a greater number of collisions with the cylinder walls in part (b) than in part (a) [the distances between the molecules and the walls are less in (b)]. The greater number of collisions produces a greater push on the container walls in (b), and consequently, the pressure of the gas is greater in (b), while V_2 is less than V_1. Thus, the kinetic theory of gases readily accounts for the inverse relationships of gas pressure and gas volume. The kinetic theory explains Boyle's law.

▶ The pressure exerted by a gas in a container can be increased in two ways: (1) By making the total number of collisions greater, the sum of all the individual pushes is greater, even though each individual push remains unchanged. (2) By making each individual push more effective by increasing the mass of the gas particles at the same velocity or by increasing the velocity of the gas molecules of the same mass.

Increasing either mass or velocity under these conditions would involve an increase in temperature and would not, therefore, be covered by Boyle's law. ◀

Graham's Law. We have already seen that the kinetic theory accounts for the diffusion of gases by assuming that there are no attractive forces between molecules and that the molecules themselves are of negligible volume. But the theory can also be used to derive the expression for the relative diffusion rates of two different gases.

   At the same temperature, the average kinetic energy of the molecules in any gas sample is the same. We can say, therefore, that at room temperature the average kinetic energy of gas one, KE_1, is the same as the average kinetic energy of gas two, KE_2. Since

$$KE_1 = \tfrac{1}{2}m_1v_1^2 \quad \text{and} \quad KE_2 = \tfrac{1}{2}m_2v_2^2 \quad \text{and} \quad KE_1 = KE_2,$$

then

$$\tfrac{1}{2}m_1v_1^2 = \tfrac{1}{2}m_2v_2^2.$$

When we cancel the $\tfrac{1}{2}$'s and rearrange the equation, we have

$$\frac{v_1^2}{v_2^2} = \frac{m_2}{m_1} \quad \text{or} \quad \frac{v_1}{v_2} = \sqrt{\frac{m_2}{m_1}}.$$

Now if we make the logical assumption that the rate of diffusion of a gas, r, is proportional to its velocity, then

$$\frac{r_1}{r_2} = \sqrt{\frac{m_2}{m_1}},$$

which is Graham's law of diffusion for gases.

8–11 EXTENSION OF THE KINETIC THEORY

We have considered only two of the gas laws, using the kinetic theory of gases. The laws of Charles, Dalton, and Avogadro can also be explained by means of the kinetic theory. Try it. Be explicit when you indicate which assumptions of the kinetic theory apply directly.

   Actually the kinetic theory of gases has been extended considerably and is now used as a working model for liquids and solids. Modifications to the theory are necessary, of course. This is quite reasonable, since liquids and solids have characteristics and properties completely different from those of gases. But in all three states of matter, the underlying theme of molecules in constant motion—with higher energies at increased temperatures—applies. Since we know that we can condense gases to liquids and freeze liquids to solids simply by lowering the temperature of most samples, it seems reasonable to assume that intermolecular contacts and attractive forces between molecules become more important as the average kinetic energy of the molecules decreases. We shall see in Chapter 9 how the two assumptions of the kinetic theory concerning intermolecular spaces and intermolecular forces must be changed to be consistent with—and to explain—the properties of liquids and solids.

QUESTIONS

1. Define or discuss: Gas, liquid, solid, ideal gas, temperature, pressure, volume, barometer, manometer, Boyle's law, Charles's law, Dalton's law, Avogadro's law, the ideal gas law, Graham's law, STP, partial pressure, catalyst, temperature correction factor, pressure correction factor, diffusion, kinetic theory of gases.

2. Describe what would happen to the mercury level in the barometer in Fig. 8–3 if: (a) atmospheric pressure increased, (b) atmospheric pressure decreased, (c) more mercury were added to the mercury reservoir dish.

3. Describe what would happen to the mercury levels in the manometer in Fig. 8–5 if:
   a) the gas container were heated,
   b) the gas container were cooled,
   c) atmospheric pressure decreased,
   d) atmospheric pressure increased,
   e) more gas were added to the gas container,
   f) gas were removed from the container,
   g) more mercury were added to the open side of the manometer.

4. A given weight of an ideal gas occupies 2.50 ℓ at 740 torr and 300 °K. What is the volume of the gas at 755 torr and 300 °K?

5. It is found that a specific number of moles of ideal gas occupies 350 ml at 25 °C and 74 cm Hg. How many liters of gas would be present at 25 °C and 1.5 atm?

6. The volume of a given weight of an ideal gas is found to be 5.2 ℓ at 295 °K and 745 torr. What would the volume of the gas be at 350 °K and 745 torr?

7. The temperature of 2.3 g of an ideal gas occupying 1.2 ℓ is changed from 250 °K to 350 °K while the pressure remains constant. What is the final volume of the gas?

8. The volume of a certain weight of ideal gas is 225 ml at 32 °C and 745 mm Hg. The gas volume is changed to 0.500 ℓ at 32 °C. What is the final pressure of the gas?

9. At constant pressure, a 1.5-ℓ balloon (perfectly elastic) filled with an ideal gas at 300 °K is cooled to -13 °C. What is the final volume of the balloon?

10. A balloon (perfectly elastic) containing an ideal gas has a volume of 350 ml at 17 °C and 0.95 ml. The pressure remains 0.95 atm, but the volume of the balloon is changed to 400 ml by varying the temperature. What is the final temperature of the gas?

11. Using the kinetic theory, explain why: (a) Gases are miscible. (b) A gas exerts its own partial pressure when it is in a mixture of gases. (c) At the same temperature, the speed of a N_2 molecule is greater than the speed of a Br_2 molecule. (d) At the same temperature and pressure, 14 g N_2 gas and 14 g Cl_2 gas do not occupy the same volume.

12. What is the volume of 0.10 mole of an ideal gas at 300 °K and 0.30 atm?

13. Oxygen gas is collected in a 250-ml bottle by water displacement at 22 °C and 752 torr. What is the partial pressure of the oxygen gas?

14. The volume of O_2 in Question 13 is measured and found to be 200 ml. How many moles of O_2 were collected?

15. Oxygen gas is passed into an evacuated, 2.0-ℓ steel box at 27 °C until the O_2 pressure is 0.50 atm. Then 0.050 mole N_2 gas is added to the box. What is the final pressure of the gas in the box?

16. A balloon (perfectly elastic) is inflated to a volume of 450 ml with air at 739 torr and 28 °C. Then an equivalent volume of He gas is added to the balloon. What is the final gas pressure in the balloon?

17. A closed 1.2-ℓ box at 302 °K contains O_2, N_2, Ar, and CO_2 gases. The partial pressures of the gases are: P_{O_2} = 200 torr, P_{N_2} = 0.68 atm, P_{Ar} = 20 mm Hg, P_{CO_2} = 0.12 atm. What is the total gas pressure in the box?

18. A balloon (perfectly elastic) is inflated with air to a volume of 0.15 ℓ at 752 torr and 300 °K. The balloon is connected to an evacuated 500-ml flask, and all the air is transferred to the flask. What is the pressure of the air in the flask?

19. At STP an ideal gas occupies 3.36 ℓ. How many moles of gas are present?

20. Given that 0.50 mole N_2 is needed for an experiment at STP, what volume will be occupied by the N_2?

21. A 5.0-ℓ steel box contains an ideal gas at 27 °C and an atmospheric pressure of 749 torr. The box is heated to 350 °K. What is the final volume of the gas?

22. a) At STP 28 g N_2 and 32 g O_2 occupy the same volume, but 16 g O_2 and 16 g He occupy different volumes. Explain these observations, using the kinetic theory and the ideal gas law.
    b) Calculate the volumes occupied by all the gases in (a).

23. At any given temperature and pressure, equal volumes of ideal gases contain equal numbers of molecules. Explain, using the kinetic theory.

24. Calculate the ideal gas constant that has units of ℓ-torr/mole-°K.

25. Rank the following gases in order of increasing rate of diffusion: Cl_2, Ar, H_2, NH_3, NO_2, SO_3, HCl, He.

26. What is the relative diffusion rate of He to Ne?

27. In a mixture of O_2 and N_2 gases, it is found that the total gas pressure is 1.2 atm if P_{O_2} = 0.70 atm and P_{N_2} = 0.50 atm. Explain this observation, using the kinetic theory.

28. It is often found that at very high pressures or very low temperatures, the behavior of gases begins to deviate from that predicted by the laws that govern ideal gases, e.g., the ideal gas law. Using the kinetic theory, give some possible reasons for these deviations.

29. It is found that H_2 gas diffuses 5.75 times faster than an unknown gas. What is the formula weight of the unknown gas?

30. Calculate the density of 0.050 mole N_2 gas at 27 °C and 740 torr.

31. How many grams of the ideal gas CO_2 are present in a sample that occupies 820 ml at 400 °K and 0.40 atm?

32. A given weight of ideal gas at 320 °K and 740 torr has a volume of 5.0 ℓ. What will the gas volume be at STP?

33. A balloon (perfectly elastic) is inflated to 820 ml at 23 °C and 0.80 atm. What would the volume be at −3 °C and 780 mm Hg?

34. A 0.14 g sample of N_2 gas is collected by water displacement at 27 °C and 727 torr. What would be the volume of the dry N_2 gas?

35. A 0.10 ℓ sample of oxygen gas is collected over water at 22 °C and 747 torr. What volume would the dry O_2 gas occupy at STP?

36. Show how the kinetic theory accounts for the following observations: (a) Gases are easily compressible. (b) Gases exert pressure. (c) Gases diffuse and fill the entire volume of their containers. (d) Gas pressure decreases as the gas temperature is decreased, at constant volume.

37. The chemical reaction between the gases NO_2 and H_2 produces NH_3 and H_2O gases according to

$$2NO_{2(g)} + 7H_{2(g)} \rightarrow 2NH_{3(g)} + 4H_2O_{(g)}.$$

a) How many moles of NO_2 are needed to produce 7.2 g H_2O?
b) How many liters of NO_2 at STP are needed to produce 0.20 mole NH_3?
c) What volume of NH_3 can be obtained at 27 °C and 750 torr if 0.20 mole NO_2 and 14 g H_2 are available?

38. Oxygen gas is collected over water after generation from $KClO_3$ according to

$$2KClO_3 \xrightarrow{\text{heat}} 2KCl + 3O_2.$$

How many grams of $KClO_3$ are needed to produce 0.82 ℓ O_2 at 300 °K and 760 torr?

9. CHANGES OF STATE

9-1 INTRODUCTION

We can readily change steam into liquid water by allowing the steam to strike a cold surface. The beads of liquid water that condense have properties quite different from those of the gaseous water vapor that we call steam. But we also know that if the temperature is lowered below 0 °C, the liquid water freezes to give ice, or solid water. During this second transition there are also major changes in the properties of the water. What are the properties of liquids and solids? How do they differ from one another and from the properties of gases? Why is it possible to change most gases to liquids and liquids to solids simply by lowering the temperature? How can we use the kinetic theory to understand these property differences and inter-conversions of states of matter? These are just a few of the questions that we shall consider in this chapter. Write down some of the properties that you know are characteristic of liquids or solids. Ask some of your own questions, and see if the kinetic theory can give some insight into why such properties exist.

9-2 GAS–LIQUID TRANSITIONS

Liquid Properties. We have seen that the properties of gases indicate that individual gas molecules are constantly moving and are separated from one another by large distances. This makes gas volumes very dependent on pressure; gases are

242

quite compressible. It also explains why rapid diffusion is observed in gases and why they spontaneously assume the shapes and fill the entire volume of their containers.

These properties of gases can be compared with those that we generally observe for liquids:

1. Liquids are essentially incompressible.
2. Liquids maintain their own volume, but assume the shape of their containers.
3. Diffusion in liquids is much slower, in general, than diffusion in gases.
4. Liquids evaporate to give gases.

The Kinetic Theory and Liquids. Let us use the kinetic theory to consider the properties of liquids and the differences in properties between liquids and gases. Under the kinetic theory, explanations of the observed properties of liquids require assumptions different from those made for gases. Since liquids are essentially incompressible, the individual molecules composing the liquid must be close together. This is the assumption made in the kinetic theory, as it applies to liquids: Liquid molecules take up most of the space in the liquid; the free space between molecules is small. Note that this is in direct contrast to the assumption made for gases.

Liquids have their own volume, but not a characteristic shape. Individual liquid molecules, therefore, do not spontaneously move away from one another to occupy the entire container (as is the case with gases). But the molecules apparently can move around and over one another to some extent, since a liquid does spread out to take the shape of its container. The corresponding kinetic-theory assumption for liquids is: There are forces of attraction between liquid molecules, but these forces are not strong enough to fix the molecules in any one given position. This kinetic-theory assumption for liquids is also in direct contrast to that made for gases.

▶ Note that by comparing the kinetic-theory assumptions for gases and liquids, we can explain the relative densities of gases and liquids. For example, at 25 °C and 1 atm, the density of water is 0.9970 g/ml, while the density of dry air is 0.001185 g/ml. Account for this large difference, using the kinetic theory. ◀

Diffusion occurs in liquids, as it does in gases, but the rate of diffusion in liquids is in general much slower. We can understand this observation using the first two kinetic-theory assumptions for liquids as compared with gases. In liquids there are forces of attraction between individual molecules tending to hold the molecules together. Also, if a liquid molecule does break free of its neighbors, its pathway for diffusion is cluttered with many other molecules, since there is little free space in the liquid. Consequently, a liquid molecule is greatly hindered as it attempts to move through the liquid. Gaseous molecules, of course, are not subject to either of these limitations that apply to liquid molecules, and consequently diffusion in gases is much faster.

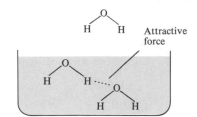

Fig. 9–1 Intermolecular forces exist between liquid molecules. For evaporation to occur, the attractive forces must be broken.

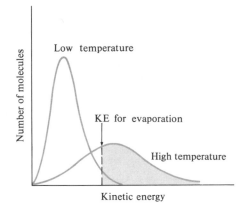

Fig. 9–2 The kinetic energy distributions of liquid molecules at two different temperatures. At higher temperatures, the number of molecules with very high kinetic energies increases, as does the average kinetic energy of the entire collection of molecules.

Evaporation. Liquids evaporate, that is, liquid molecules move spontaneously out of the liquid and become gaseous molecules. This process might, at first, appear to contradict the assumptions for liquids made by the kinetic theory. We have assumed that attractive forces, diagrammed in Fig. 9–1, exist between individual liquid molecules. Then, in order for a molecule to escape the liquid and enter the gas above the liquid, these forces of attraction between liquid molecules must be overcome. Breaking of these forces of attraction takes energy, and yet evaporation occurs spontaneously. From where do the molecules that break the inter-molecular forces in the liquid and escape as gaseous molecules get the required energy?

The kinetic theory has a built-in answer: The energy comes from the molecules themselves.

To understand that this is true, we must recall the assumption made for gases under the kinetic theory that the average kinetic energy of the molecules is directly proportional to the Kelvin temperature. This assumption also applies to liquids. But remember that at any given time any individual liquid molecule can have a high

or a low kinetic energy. Figure 9–2 shows the distribution of molecules with various kinetic energies, for a liquid at two temperatures. Note that there are a few molecules with quite high kinetic energies, energies high enough to overcome the attractive forces holding the liquid molecules together and keeping them in the liquid. These high-kinetic-energy molecules, then, can break free from the liquid and become gaseous molecules. There are more high-kinetic-energy molecules at the higher temperature.

How does the escape of these high-energy molecules affect the liquid? There are two things that are readily detected. The first is rather obvious: The amount of liquid will decrease, since liquid molecules are being lost. The second effect is less obvious, but is easily detected experimentally: There is a decrease in the temperature of the liquid. Why this second effect? Remember that temperature is a direct measure of the average kinetic energy of the molecules in the liquid. The high-kinetic-energy molecules leave the liquid. This reduces the average kinetic energy of the entire collection of molecules in the liquid, giving a corresponding decrease in the liquid's temperature.

▶ The idea of average kinetic energies being measured by the temperature, with individual energies being much different from the average, is quite similar to the grades that students receive in courses. A distribution of grades similar to those shown in Fig. 9–2 is often obtained. Then we can consider that the performance level of the students is measured by the average grade of the entire number of students. But we know that some students receive A's while other students fail. Now what happens when we select those students with A's and place them in a section by themselves, no longer counting their grades in the class average? The average grade of the class decreases; the level of performance has dropped, since the students that were performing well have been lost. ◀

Evaporation and Heat Transfer. An interesting comparison in evaporation rates can be made between evaporation from a beaker and evaporation from a Dewar flask, as diagrammed in Fig. 9–3. (We shall assume that the temperature and volume of the two containers are the same.) We can compare the two cases by following the evaporation process as if it occurred in a stepwise fashion. Again we are setting up a model to aid our thinking and understanding. The processes actually occur quite rapidly, and more randomly than our model will indicate.

▶ A *Dewar flask* is commonly known as a Thermos bottle. It consists of two concentric beakers sealed together at the top, with the space between them evacuated. Since heat energy cannot be transferred through a vacuum, heat does not flow through the sides of the Dewar, either from inside to outside or vice versa. ◀

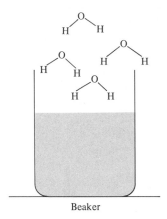

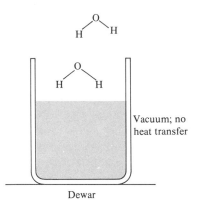

Fig. 9-3 Evaporation of water from a beaker and from a Dewar flask.

Water evaporates from a beaker at room temperature. We know that this is true, since the entire sample of water will disappear, and the beaker will become empty, in the course of a few days. At room temperature some of the water molecules at the surface of the liquid have—or gain—enough energy to break the attractive forces holding them to other water molecules. These high-kinetic-energy molecules can then evaporate. Since it is the high-energy molecules that leave the liquid, the temperature of the water falls slightly and the amount of evaporation decreases (fewer molecules are left with enough energy to escape the liquid).

Instantaneously, however, another process occurs. Heat energy is added to the water from the surroundings of the beaker, that is, from the air and the table on which the beaker sits. Heat flows into the water, since the temperature of the water is lower than the temperature of the room. And heat energy can enter the water through the walls of the beaker, the bottom of the beaker, or directly from the open liquid surface. The heat energy is absorbed by some of the water molecules,

increasing their kinetic energy and restoring the average kinetic energy of the molecules to a value which corresponds to room temperature.

Now the whole process can be repeated. Some high-kinetic-energy water molecules can evaporate, lowering the average energy of the liquid (and also the liquid level in the beaker) and the water temperature. Heat energy from the outside restores the high-kinetic-energy molecules of the water; the water temperature returns to room temperature. The liquid molecules with high kinetic energy can then evaporate, and the process continues until the liquid water is gone.

There is a fundamental difference between evaporation from a Dewar flask and the evaporation process just described for the beaker. We can envision the first step as being the same in both cases; some high-kinetic-energy water molecules evaporate from the surface of the liquid in the Dewar. This decreases the water level in the Dewar slightly, and also decreases the average kinetic energy of the molecules, causing a drop in the water temperature.

It is at this point in our model process that the beaker and Dewar are different. Heat cannot enter the Dewar flask easily. The water can absorb heat only through the open top of the Dewar.

▶ This is the reason that many Thermos-bottle tops are made small, and a good insulator is used as a plug. Then the heat flow into or out of the bottle is very limited. Hence things stay hot or cold inside the Thermos for long periods of time. ◀

Consequently, the average kinetic energy of the water molecules must remain lower than it was initially. At this lower temperature there are fewer molecules with high enough kinetic energies to escape the liquid. Loss of these molecules to the gas lowers the liquid temperature further. Soon evaporation becomes extremely slow, since, at the lower temperature, very few molecules at the water surface have enough energy to break the attractive forces of the liquid. The water level in the Dewar, which is lower than it was initially, then remains fairly constant, falling only very slowly. Also, the water temperature falls below, and remains below, room temperature. Although the water evaporates from the beaker completely within a few days, the water in the Dewar will remain for weeks.

A more dramatic example of the difference in evaporation rates in these two cases is given by liquid nitrogen. Liquid nitrogen is unimaginably cold; its temperature is about -196 °C (77 °K). When poured into a beaker, liquid nitrogen evaporates completely within a few minutes. However, an equal volume of liquid nitrogen can be stored in a Dewar for several hours.

9-3 EQUILIBRIUM VAPOR PRESSURE

Evaporation into Closed Containers. In the examples of evaporation that we have considered so far, the liquid has been allowed to evaporate freely into the air. A somewhat different situation may arise if the liquid is evaporating into a closed con-

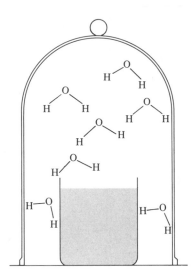

Fig. 9–4 Evaporation of water into a closed bell jar. The gaseous H_2O molecules exert a pressure against the jar walls and on the beaker and water surface; the pressure is determined by the temperature.

tainer. Such an arrangement is shown in Fig. 9–4, in which the beaker contains a liquid such as water.

What happens when the water molecules evaporate into the closed space inside the bell jar? Initially, we can imagine the process to be exactly the same as that previously described for evaporation from a beaker. The high-kinetic-energy molecules become gaseous; the liquid level drops; heat energy is absorbed by the water, restoring some molecules to high energies; these molecules then evaporate; and the process continues.

But now another consideration becomes important. The water molecules that become gaseous are free to diffuse throughout the entire volume enclosed by the bell jar. In the course of occupying this volume, the gaseous water molecules, or water-vapor molecules, collide with the bell jar, with the beaker walls, with other gas molecules and, also, with the liquid water surface. There are two things that are quite important about these collisions. First, the gaseous water molecules are exerting their own partial pressure by colliding with the container walls. Second, the water-vapor molecules are constantly hitting the liquid water surface. Quite often when this occurs, the vapor molecule sticks on the liquid surface and is attracted into the liquid, becoming a liquid water molecule again.

These processes of evaporation and condensation of water molecules occur continuously, and, after a certain length of time, the rate of evaporation becomes equal to the rate of condensation (see Fig. 9–5). When this happens, just as many water molecules are leaving the liquid as are entering the liquid from the gas. At this point we say that a *state of equilibrium* exists, a state in which two changes exactly oppose one another. At equilibrium there is no net change in the system.

In this case the two opposing changes are the evaporation and condensation of water molecules. At any one time there are as many water molecules leaving the

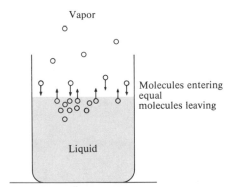

Vapor

Molecules entering
equal
molecules leaving

Liquid

Fig. 9–5 An equilibrium process. The number of molecules leaving and entering the liquid phase per unit time is equal. No net change is observed in the system.

liquid as entering it. The net number of molecules in the liquid is constant and the net number of molecules in the gas is constant. That is, there is no observable net change in the system, even though there are huge numbers of molecules that are actually moving from liquid to gas, and vice versa. For this reason, this and similar equilibrium processes are often called *states of dynamic equilibrium*.

▶ Note that, at equilibrium, the total number of water molecules in the liquid and the total number of water molecules in the gas remain unchanged. This does not say that the number of water molecules in the liquid is equal to the number of water molecules in the gas. In fact, this is rarely true. Generally, there are many more liquid molecules than gaseous ones. ◀

We have now established two things about the beaker of water evaporating into the enclosure formed by the bell jar: First, there is an equilibrium process occurring. Second, the water-vapor molecules exert a partial pressure of their own inside the bell jar. The pressure that is exerted by a vapor when it is in equilibrium with its liquid is called the *equilibrium vapor pressure* of the substance. The equilibrium vapor pressure is characteristic of a particular substance at a given temperature.

▶ The term equilibrium vapor pressure is often shortened to *vapor pressure*. Thus Table 9–2 lists the vapor pressure of water at various temperatures. The vapor pressures given are the equilibrium vapor pressures. ◀

Magnitude of Equilibrium Vapor Pressure. The value of the equilibrium vapor pressure depends on two things: (1) the nature of the liquid, (2) the temperature. Table 9–1 lists several liquids and their corresponding equilibrium vapor pressures at 25 °C. Note the very large spread of values of vapor pressures. Table 9–2 lists the equilibrium vapor pressures of water at various temperatures (see also Appendix D).

Tables 9–1 and 9–2 give experimental justification to the two generalizations above. But why should the values of vapor pressure be dependent on the type of liquid or on the temperature?

Table 9–1. Equilibrium vapor pressures of various liquids at 25 °C

Liquid	Vapor pressure, torr
H_2O	24
HNO_3	65
C_6H_6	96
CCl_4	116
CH_3OH	121
SO_3	258
HCN	736
N_2O_4	900

Table 9–2. Vapor pressure of water at various temperatures

Temperature,°C	Vapor pressure, torr
5	6.5
25	23.8
40	55.3
60	149.4
80	355.1
100	760.0

To answer these questions, let us again resort to the kinetic theory. It is perfectly reasonable that different liquids should have different vapor pressures. In each liquid there are different attractive forces between the molecules. In some liquids the intermolecular forces are strong; to break them takes considerable energy. Therefore not many molecules can escape into the vapor state, and the corresponding vapor pressure is low. On the other hand, the intermolecular forces existing in some liquids are quite weak. This gives rise to high vapor pressures, since it is rather easy for molecules to escape the liquid. Table 9–1 gives some examples. How would you rank these liquids in terms of the attractive forces between molecules?

The dependence of equilibrium vapor pressure on temperature also seems reasonable. A change in the temperature changes the average kinetic energy of the molecules. If the temperature of the liquid is increased, the average kinetic energy

of the molecules increases. This means that more liquid molecules have enough energy to overcome the intermolecular forces in the liquid and enter the vapor. With more gaseous molecules present, there are more collisions with the container walls; that is, a greater pressure is exerted. Then the vapor pressure of a liquid should increase as the temperature is increased. By similar reasoning, it follows that the vapor pressure decreases with decreasing temperature.

The Boiling Point. As the temperature of a liquid increases, the vapor pressure rises. At some characteristic temperature for each liquid, bubbles begin to form and the value of the vapor pressure becomes equal to the surrounding atmospheric pressure. The temperature at which this occurs is called the *boiling point* of the liquid.

Since vapor pressures depend on the nature of the liquid, boiling points must also be different for different liquids. The kinetic theory explanations are the same in both cases. Remember that strong intermolecular forces in the liquid make evaporation difficult, resulting in relatively low vapor pressures. Then in order for the situation to exist in which the vapor pressure is equal to atmospheric pressure, considerable energy is needed. The energy is provided by raising the temperature of the liquid to a high value. Thus, liquids with strong intermolecular forces have low vapor pressures and high boiling points.

The boiling point of a liquid also depends on the prevailing atmospheric pressure. When the pressure of the atmosphere is only 720 torr, the liquid vapor pressure need only be 720 torr for boiling to occur. Since fewer vapor molecules are necessary to produce a pressure of 720 torr than a pressure of 760 torr, the boiling point of the liquid will be lower at 720 torr. For water, for example, the boiling point is 97.7 °C at 700 torr, while it is 100 °C at 760 torr. To allow comparison of boiling points for different liquids, a *standard or normal boiling point* for liquids has been defined to be that temperature at which the vapor pressure of the liquid is equal to 760 torr (1 atm).

9-4 LIQUID–SOLID TRANSITIONS

When gases are cooled, the molecules slow down, form aggregates, and finally condense to liquids. Further cooling again slows the molecules. At some point the forces between the liquid molecules can no longer be overcome by the molecular motion (the average kinetic energy of the molecules has decreased). At this temperature the freezing or crystallization point—the liquid forms a solid, and the molecular motions are restricted to a small volume in space. The molecules of the solid are held essentially in place by the forces that exist between them.

▶ Note that there is still great molecular motion even in a solid. The difference between a solid and a liquid is the degree of restriction of molecules to a certain location. In the solid, molecules move rapidly within a localized vicinity; they are not free to move throughout the solid. In the liquid, however, molecules can move

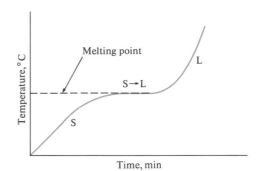

Fig. 9-6 A heating curve. The substance passes from a solid (S) through the solid–liquid transition (S → L) to a liquid (L) as heat is applied at a constant rate.

throughout the entire liquid sample, although their progress is slow when compared with that of molecules in gases. But this need not be because molecules in gases move more rapidly. Actually, at a given temperature, the average speed of molecules of the same substance existing simultaneously in the solid, liquid, and gaseous states is the same. Gas molecules can move around more easily, since there are large open spaces with no interferences. ◀

We can experimentally observe this transition between solids and liquids and determine melting points* of solids by running heating curves, as shown in Fig. 9-6. To obtain the data for a heating curve, we have to slowly heat a solid at a constant rate to a temperature above its melting point while determining the temperature of the sample at closely spaced time intervals.

If, for example, we used a lead metal (Pb) sample, we could slowly heat the metal and carefully determine its temperature as it increased from 25 °C to about 500 °C. The experimental data would appear as in Fig. 9-6. The temperature of the solid Pb would increase rather rapidly until it approached 327 °C, the melting point of Pb. Then the temperature of the sample would remain essentially constant at 327 °C for a short time (corresponding to the horizontal section of the heating curve in Fig. 9-6), even though the heating of the sample continued during this time. Only when the entire Pb sample was melted would its temperature again begin to increase above 327 °C. Further heating of the sample—now liquid lead—would increase its temperature, as shown by the heating curve.

But why does a heating curve have the characteristic shape shown in Fig. 9-6? Again we must resort to a simple model for an explanation, a model based on the kinetic theory. In the section of the curve labeled S (solid) in Fig. 9-6, the solid molecules are being heated; they are gaining kinetic energy and moving more rapidly (but their motions are still restricted to a fixed place in the solid). This increases the average kinetic energy of all the solid molecules, which corresponds to an increase in

---

*   Melting and freezing occur at the same temperature. If the transition being observed is from solid to liquid, we speak of melting and the melting point. If the transition is from liquid to solid, we refer to freezing (or crystallization) and to the freezing point.

temperature. Then during this portion of the curve we would note an increase in the sample temperature as time passed.

At some temperature, which is characteristic of the particular solid, the heat energy flowing into the solid becomes large enough to begin breaking the bonds between the molecules of the solid. At this temperature, which we call the *melting point* of the solid, the molecules begin to move much more freely, since the strong intermolecular forces that hold them in a definite location are being overcome.

But note that the heat energy is no longer being used to increase the kinetic energy of the molecules. It is being used to break the bonds between the molecules. Consequently, there is no rise in the temperature of the sample, and the heating curve levels off, as indicated by the S → L section in Fig. 9–6. So long as there are strong bonds to be broken between solid molecules, there is a constant-temperature region on the heating curve.

When all the solid–solid bonds have been broken—i.e., when the entire solid sample is melted—the heat energy flowing into the sample will again be used to increase the kinetic energy of the molecules in the sample (now liquid molecules) and the temperature will again begin to rise. This corresponds to the section labeled L (liquid) on the heating curve in Fig. 9–6. Further heating simply increases the temperature of the liquid.

But what do you think might happen when the temperature reaches the boiling point of the liquid? See if you can draw a heating curve that shows both the melting point and boiling point of a substance. Can you explain the shape of the curve and all the transitions involved?

▶ A similar kind of curve, giving essentially the same information, can be obtained if a solid is melted, then cooled, and the temperatures of the sample taken during cooling. This kind of temperature-versus-time plot is called a *cooling curve*. Try to draw a typical cooling curve. Can you give reasons for the transitions indicated, i.e., for the shape of the curve? ◀

9-5 SOLIDS

Properties of Solids. Solids have both definite volume and shape, and they are essentially incompressible. Diffusion occurs in solids but, in general, it is much slower than in liquids or gases. The shapes of solid substances are characteristic of the particular solid, since solids usually form in a definite geometric pattern or crystal structure.

▶ If the solid is in the form of a fine powder, it is difficult to observe its characteristic geometric pattern visually. Observing the powder under a microscope or with the aid of x-rays reveals its crystal pattern. ◀

We can consider the reasons for these properties of solids by using the kinetic theory. Just as we make different assumptions for gases and for liquids when we

apply the kinetic theory, we need another set of assumptions for solids. If we assume that the attractive forces between molecules in the solid are quite strong—strong enough to hold the molecules in a fixed position—then a solid substance would maintain its volume and would also have a characteristic shape.

A second assumption—that molecules in a solid are in contact with one another—can account for the incompressibility of solids. If the molecules are already touching and the empty spaces between them are very small, then it would be difficult to push the molecules closer together, i.e., to compress them.

These two kinetic-theory assumptions for solids also explain the slow diffusion observed in solids. If the molecules are in contact and are held in place by rather strong forces, it is difficult for a molecule to move through a solid. First it must break the bonds that attract it to the other surrounding molecules and hold it in position. But even if these forces were broken and the molecule were free to move, its progress would be slow, since there are so few empty spaces in the solid. The pathway for diffusion is blocked by other molecules, and, of course, these molecules are not easily moved out of the way. Hence, for a molecule to move from one point in a solid to another generally takes a much longer time than it does in liquids or gases.

▶ We know that gaseous mixing, caused by diffusion and convection currents, occurs within a room in a matter of seconds. The pleasant odors of perfume or food cooking travel rapidly throughout a room and even into other rooms of a house. Mixing in liquids is also relatively fast and occurs, for example, in a cup of coffee in a matter of a few minutes. If some cream is placed in a cup of black coffee and allowed to mix without stirring, it spreads spontaneously, but takes several minutes to mix thoroughly throughout the cup. On the other hand, diffusion of just a few millimeters in solids can often take days or even years. The rates of diffusion in solids are complicated by the fact that there are several mechanisms by which diffusion can occur. Examination of the models of several different crystal structures gives some indication of why these diffusion rates differ. The different crystal structures provide various types of empty spaces that can be involved in the diffusion process. ◀

Crystal Structures. As molecules form solids and begin to pack together, touching one another, they are held firmly by strong intermolecular forces, and form regular geometric patterns or crystal structures. The variety of crystal structures is the result of different packing arrangements. Figure 9–7 shows some of the more common crystal structures.

▶ Packing arrangements are controlled by several factors; one of great importance is the *size* of the individual ions or molecules forming the solid. In the case of NaCl, for example, the radii of the Na^+ and Cl^- ions are 0.95 Å and 1.81 Å, respectively. The Na^+/Cl^- radius ratio is $0.95/1.81 = 0.52$, and the allowed packing arrangement produces a crystal structure that is called the sodium chloride structure. Many other compounds, with similar radius ratios—such as KBr, NaBr, KCl, MgO, BaO, CaS,

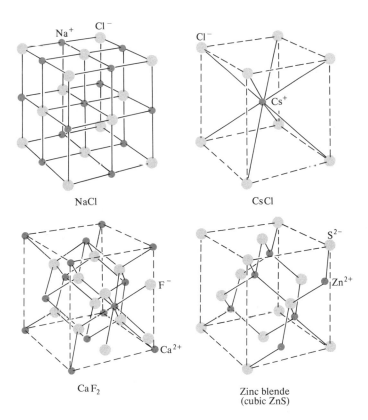

Fig. 9-7 Four common crystal structures. The sizes of all the ions have been reduced for the sake of clarity. Remember that in solids the ions are essentially touching one another.

and SrSe—also crystallize with the sodium chloride structure. On the other hand, compounds that are similar to ZnS, which has a Zn^{2+}/S^{2-} radius ratio of $0.74/1.84 = 0.40$, form solids with the zinc blende crystal structure. Other compounds such as CdS, CdTe, GaAs, AlP, InSb and PdO, which also have the zinc blende structure, have radius ratios near 0.40. Figure 9–7 illustrates the sodium chloride and zinc blende structures. For other crystal structures, different values of the radius ratio are required to make possible the characteristic geometrical arrangements. ◀

The actual crystal structure of any given compound can be experimentally determined by using the technique of x-ray diffraction. We shall not describe the details of this technique. Suffice it to say that x-rays are passed through the substance, just as they are through your body during a chest x-ray. The rays are deflected as they travel through the solid, producing characteristic patterns on a

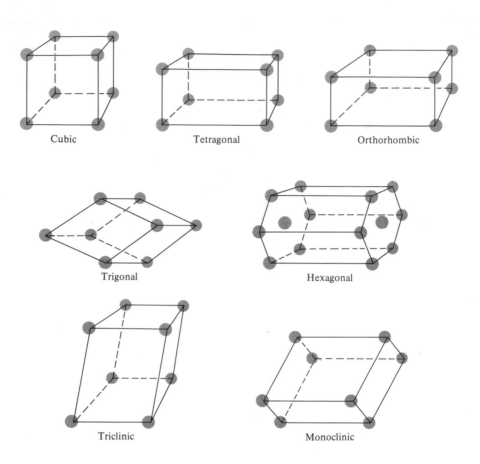

Fig. 9-8 The simplest unit cell for each of the seven crystal systems.

photograph, somewhat similar to the patterns produced in the chest x-ray. An analysis of these characteristic patterns enables a crystallographer to determine the exact crystal pattern of a solid.

Space Lattices and Unit Cells. A detailed knowledge of the crystal structure of a substance requires two things. First we must determine the regular geometrical pattern of arrangement in the solid; this is called the *space lattice*. It is simply a regular array of points, which we call *lattice points* or *lattice sites*, spread in a characteristic fashion in three dimensions. There are 14 types of space lattices (often called *Bravais lattices*) which fall in 7 different crystal systems. Figure 9-8, which shows a unit cell for each system, illustrates the relationships among these 7 systems. A *unit cell* is a simple unit of the space lattice that can be extended in

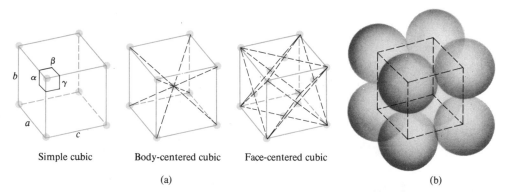

Simple cubic   Body-centered cubic   Face-centered cubic

(a)   (b)

Fig. 9–9 (a) The three unit cells of the cubic system drawn to show the symmetry. (b) A simple-cubic unit cell drawn to correspond more closely to the real solid structure.

three dimensions to reproduce the entire space lattice. Note that the differences between the various crystal systems involve the length of the unit cell edge and the angles between the unit cell edges. The three unit cell edges are all of equal length in the cubic and trigonal systems, but not in the other five systems. The angles are all 90° in the cubic, tetragonal, and orthorhombic systems, but are not all 90° in the other four systems. Table 9–3 summarizes these differences, and Fig. 9–9 shows the three unit cells of the cubic system.

Table 9–3. The seven crystal systems: their cell edges and angles

System	Unit cell edges	Unit cell angles
Cubic	$a = b = c$	$\alpha = \beta = \gamma = 90°$
Tetragonal	$a = b \neq c$	$\alpha = \beta = \gamma = 90°$
Orthorhombic	$a \neq b \neq c$	$\alpha = \beta = \gamma = 90°$
Trigonal	$a = b = c$	$\alpha = \beta = \gamma \neq 90°$
Hexagonal	$a = b \neq c$	$\alpha = \beta = 90°, \gamma = 120°$
Monoclinic	$a \neq b \neq c$	$\alpha = \beta = 90° \neq \gamma$
Triclinic	$a \neq b \neq c$	$\alpha \neq \beta \neq \gamma$

Types of Space Lattice. The second factor that must be determined about a crystal structure is the type of unit that occupies each lattice point. There can be ions occupying the lattice points, as in solid sodium chloride, in which Na^+ and Cl^- are present; or molecules at the lattice points, as in solid oxygen, in which O_2 molecules are at each lattice point; or atoms at the lattice points, as in diamond, in

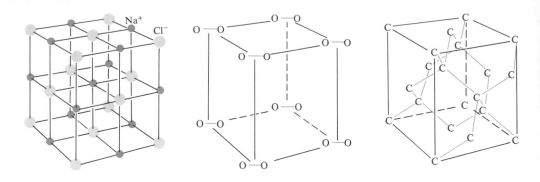

Fig. 9-10 An ionic space lattice, NACl; a molecular space lattice, solid O_2; a covalent space lattice, the diamond form of C.

which C atoms are at each site. Figure 9–10 shows diagrams of these three substances, with the appropriate units occupying the lattice points.

The properties of solids depend greatly on the type of unit that occupies the lattice sites. We need only consider the properties of the solids NaCl, O_2 (solid), and C (diamond)—to see this. Table 9–4 summarizes some of their properties.

Generalizing from these three examples, we see that solids with molecules at the lattice points, forming a *molecular space lattice*, are quite soft, have low melting and boiling points, and are good electrical and heat insulators. These properties are the result of the type of bonding that exists between molecules in this space lattice. The bonds, which are very weak, are generally called *van der Waals bonds* or *van der Waals forces*. These attractive forces are present in all matter, and are caused by the rapid and constant movement of the outer electrons present in the units attracting one another.

Table 9–4. Properties of solid NaCl, O_2, and diamond

Substance	Hardness	Melting point, °C	Boiling point, °C
NaCl	Moderate	801	1413
O_2	Soft	−218	−183
Diamond (C)	Very hard	>3550	4827

To get a better feeling for van der Waals forces, let us consider two argon atoms, as pictured in Fig. 9–11. Since the outer electrons in the argon atoms are moving so rapidly, at any given instant many of them might pile up on one side of the atom, as shown for Atom I in Fig. 9–11. This instantaneous unequal distribution of electrons makes one side of the atom have a partial negative charge (an excess of electrons) and the other side a partial positive charge; for a fraction of a microsecond the argon atom becomes a dipole.

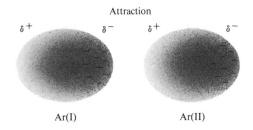

Attraction

δ^+ δ^- δ^+ δ^-

Ar(I)   Ar(II)

Fig. 9–11 The formation of van der Waals forces between two argon atoms. The instantaneous partial charges, δ^- and δ^+, of different atoms attract one another.

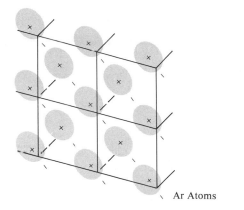

Ar Atoms

Fig. 9–12 Instantaneous cooperation among the face-centered-cubic atoms of solid argon produces the van der Waals forces of attraction.

For this extremely short time argon atom I affects argon atom II (see Fig. 9–11), and creates a similar dipole in atom II. This is the force existing between the atoms: the attraction of the partial positive charge of one atom for the partial negative charge of another. When many atoms are close together, as in solid argon, for example, they can be thought of as all acting together in forming an extended regular array of these oriented dipoles.

Figure 9–12 diagrams this, using the face-centered-cubic molecular space lattice of argon as an example. Instantaneous coupling of the argon atoms can occur across the face diagonals of the cubes, as shown, or along the cell edges.

The nature of van der Waals forces makes them very dependent on the co-operative actions of many atoms (or molecules). Therefore the atoms must be close together or the necessary interactions cannot occur. The strength of van der Waals forces decreases very rapidly as the distance between atoms increases. The co-operative actions are also greatly reduced by any form of chaotic motion of the atoms. Consequently, as the temperature of the substance increases and the atoms move faster, the van der Waals forces are destroyed. Finally, these attractive forces depend on the combined motions of many electrons. As the number of electrons on each atom (or molecule) increases, the size of the partial charge can increase, thereby increasing the attractive force between the atoms.

Van der Waals forces, then, are most effective at low temperatures, short distances, and between atoms or molecules with large collections of electrons.

Solids with ions at the lattice points form an *ionic space lattice*. These solids are hard and brittle, have high melting and boiling points, and are good insulators, since the ions are held together by the strong forces of opposite charge attraction, i.e., by ionic bonds. These attractive forces are much greater than those in molecular lattices, since the ion charges are permanent and much larger than the rapidly changing induced charges present in the molecular lattices.

Substances that have atoms at the lattice sites form *covalent network space lattices*. These lattices are held together with strong covalent bonds that extend in a three-dimensional array throughout the crystal. Using the diamond form of carbon previously shown in Fig. 9–10 as an example, note that each carbon is bonded to four other carbon atoms in a tetrahedral arrangement. This bonding scheme repeats in all three directions, so that each carbon atom occupies a site at the center of one tetrahedron and also at the corner of another tetrahedron, giving a continuous three-dimensional network of carbon atoms. Solids with a covalent network space lattice are very hard, have very high melting and boiling points, and are good insulators. Other examples are SiC, or silicon carbide (commonly known as carborundum) and SiO_2, or silicon dioxide (commonly known as quartz).

There are many solids that are similar to those with covalent network space lattices and that yet have quite different properties. Germanium (Ge) and silicon (Si), for example, have the same crystal structure as diamond, but are only moderately hard and are semiconductors; i.e., they conduct electricity moderately well, rather than being electrical insulators. Graphite, another solid form of carbon, and cadmium iodide (CdI_2) have space lattices that form in layers. In any given layer the attractive forces are covalent or ionic, but van der Waals forces hold the layers together.

These different bonding arrangements produce crystals with highly variable properties. We might consider these solids as having *intermediate network lattices*, but few generalizations can be made about their properties. For example, either atoms or ions can occupy the lattice points, with combinations of bond types being present. This yields solids that can be hard or soft, can have high or moderate melting points, and can be electrical insulators or conductors. Variations in properties are also present within a given solid. Graphite, for example, conducts electricity fairly well within layers, but poorly between layers.

Last we come to the *metallic space lattice*. Examples of this are very numerous, since all metals fall into this category. The unusual thing about metals and the metallic space lattice is the phenomenon of *conductivity*. Electrons seem to be free to move throughout the entire metal sample. This is a completely different situation from that found in the other space lattices, in which all electrons are held tightly between, or on, the atoms at the lattice sites. Even in the case of certain solids with intermediate network lattices (solids which often appear to have properties that are intermediate between those of solids that have nonmetallic and those of solids that have metallic lattices), conductivity is much more difficult to achieve than with metals. We now think of metals as substances which contain positive ions, located at appropriate lattice sites, that are embedded in a sea of electrons. These electrons are free to move throughout the entire metallic lattice.

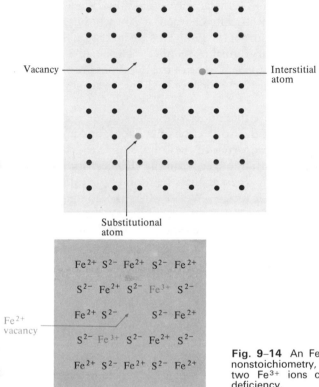

Vacancy

Interstitial atom

Substitutional atom

Fe^{2+} vacancy

Fig. 9–13 A two-dimensional lattice illustrating vacancy, interstitial, and substitutional imperfections.

$$Fe^{2+} \ S^{2-} \ Fe^{2+} \ S^{2-} \ Fe^{2+}$$
$$S^{2-} \ Fe^{2+} \ S^{2-} \ Fe^{3+} \ S^{2-}$$
$$Fe^{2+} \ S^{2-} \ \ \ \ \ \ S^{2-} \ Fe^{2+}$$
$$S^{2-} \ Fe^{3+} \ S^{2-} \ Fe^{2+} \ S^{2-}$$
$$Fe^{2+} \ S^{2-} \ Fe^{2+} \ S^{2-} \ Fe^{2+}$$

Fig. 9–14 An Fe^{2+} vacancy in iron(II) sulfide causes nonstoichiometry, $Fe_{1-x}S$. The concurrent formation of two Fe^{3+} ions compensates for the positive charge deficiency.

As an example we can consider sodium metal. At each lattice point there is a sodium ion, Na^+; for each Na^+ present there is one electron free to migrate throughout the lattice. As a result of this rather peculiar bonding scheme, metals have highly variable properties. Some are very soft, while others are very hard. Some have extremely high melting and boiling points, while others have rather low melting and boiling points. But all metals—from titanium, tungsten, and platinum to lead, gallium, and mercury—have one thing in common: *they are all conductors.*

Imperfect Lattices. When real solids form, irrespective of their particular space lattice, there are always some irregularities or imperfections that occur in the crystal. There are several ways in which this can happen. Foreign atoms can be trapped at a lattice site (or possibly between lattice sites), or atoms can be omitted from the crystal and a lattice site be left vacant (see Fig. 9–13). Actually, these cases in which imperfections occur within solids are the rule, not the exception. And the properties of the solids can be greatly affected by the presence of the imperfections.

Figure 9–14 illustrates just one of the cases in which an imperfection can change a solid. When iron (II) sulfide (FeS) forms as a solid, there are always some Fe^{2+} ions

that are left out of the lattice. That is, there are some vacant lattice sites in the FeS crystal that would ideally be occupied by Fe^{2+} ions. Now we know one thing experimentally: A sample of solid FeS is not electrically charged. (You will not receive an electrical shock if you touch it.) Since some of the Fe^{2+} ions are missing and all the S^{2-} ions are present, this imbalance of charge must be rectified. The positive charge in the lattice must be increased. This can happen if some of the Fe^{2+} ions are oxidized to Fe^{3+} ions; in fact, two Fe^{2+} ions become Fe^{3+} ions for each Fe^{2+} vacancy that is formed in the FeS solid.

The Fe^{2+} vacancy imperfection in FeS, then, does two things to the properties of the solid. First, it requires that some of the iron exist as Fe^{3+} ions, and not all as Fe^{2+} ions. This, of course, changes the properties of the solid, since Fe^{3+} behaves differently from Fe^{2+}. Second, we should no longer write the chemical formula as FeS, since there are actually more S atoms present than Fe atoms; the ratio is no longer 1 : 1, since there are some iron vacancies. We say that the compound is *nonstoichiometric*, since it does not have a simple atomic ratio. A better way of describing iron (II) sulfide would be to write the formula $Fe_{1-x}S$, where the x indicates that a small amount (often not known precisely) of Fe is missing in the solid.

9–6 WATER

We can study some of the properties of liquids and solids, and their transitions, more closely by using an example. We have selected water because of its many commonly known characteristics and properties, because of its vast significance and importance in our lives, and because of its many unusual and interesting properties. The physical properties of water, for example, are not only well known but are also used as standards for many measurements. The melting and boiling points of water are standard reference points for our temperature scales—0° and 100 °C or 32° and 212 °F —and the standard measure of volume, the liter, is defined as 1 kg of water at 4 °C.

The angular structure of the water molecule, coupled with its unshared pairs of electrons and polar oxygen–hydrogen bonds, gives water a large dipole moment (see Chapter 5). This accounts to a great extent (see Section 10–3) for its ability to dissolve large numbers of diverse substances.

▶ Water is often called the universal solvent. Although this reputation is hardly earned, you can probably think of many things that do dissolve in water. Take NaCl, sugar, tea, coffee, alcohol, and CO_2, for instance. They all dissolve in water, at least to some degree. List a few more. Can you think of some reason why all these substances—be they gases, liquids, or solids—dissolve in water? Try to devise a simple model that might explain it. Then consult Chapter 10. ◀

When some compounds, such as copper (II) sulfate, are dissolved in water and then recovered again by evaporating the water, the solids formed have water included in them and have formulas such as $CuSO_4 \cdot 5H_2O$. These chemical compounds are called *hydrates*. The water molecules in hydrates, indicated after the dot in the

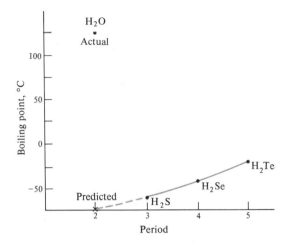

Fig. 9–15 Prediction of a −70 °C boiling point for H₂O from the data for H₂S, H₂Se, and H₂Te.

formula, are actually bonded to the parent substance. It is not adsorbed water. The water in hydrates can often be driven off (dehydration) by heating to 100 or 200 °C.

Efflorescent substances give off water of hydration spontaneously, whereas *deliquescent substances* take on water of hydration spontaneously. When a deliquescent hydrate dissolves in its own water of hydration, the substance is called *hygroscopic*. The degree of hydration of any substance depends on the properties of the substance and also on the amount of water vapor present in the surrounding atmosphere. On a very humid day, for example, a deliquescent material will pick up more water of hydration than on a day of low humidity.

Hydrogen bonding. We mentioned that water has some unusual properties. Two of these happen to be its melting point and its boiling point. But we can only realize that these two properties of water are unusual when we consider their values in relation to the values for other, similar compounds. For example, we might study the boiling points of the series of compounds formed between hydrogen and the Group VIA elements: H_2O, H_2S, H_2Se, and H_2Te. Table 9–5 gives their melting points and normal boiling points, and Fig. 9–15 plots their boiling points.

Table 9–5. Melting and boiling points of the hydrogen–Group VIA compounds

Compound	Melting point, °C	Boiling point, °C
H_2O	0.0	100.0
H_2S	−85.5	−60.7
H_2Se	−66	−41.5
H_2Te	−51	−20

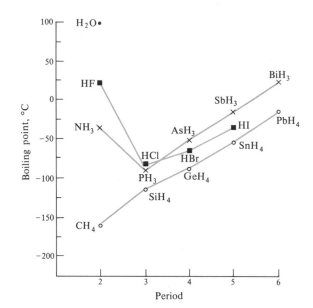

Fig. 9-16 The normal boiling points of the hydrogen compounds of Groups IVA, VA, and VIIA. As in the case for H_2O, unsystematically high boiling points are found for HF and NH_3. This is not true for CH_4.

    If we were to generalize from the three data points in Fig. 9–15, we could predict a boiling point for H_2O as shown by the extrapolated (dashed) part of the curve. But this value of about $-70\,°C$ is much too low; we know that the boiling point of water is $100\,°C$. Data such as these are quite perplexing. They immediately raise the question "Why?"

▶ You might also ask another question about the data in Fig. 9–15. Why does the boiling point of the substance increase in going down the group, from H_2S to H_2Se to H_2Te? Recalling what we said about van der Waals forces, can you explain this periodic trend in the boiling points? ◀

    Answers to such questions are, of course, generally not easy. About the only conclusion we can draw from the above data is that the intermolecular forces in liquid water are unusually high. It takes much more energy (higher temperatures) to vaporize the water molecules than would have been predicted from the data. What could be the origin of these strong forces?
    We might be able to obtain an insight into this problem if we consider the boiling points of other groups of hydrogen-containing compounds. Table 9–6 gives the melting and normal boiling points of the compounds of Groups IVA, VA, and VIIA with hydrogen, and Fig. 9–16 plots their boiling points. Note that HF and NH_3 are similar to H_2O in having higher-than-expected boiling points. But CH_4 seems to fall right in line with the other Group IVA–hydrogen compounds, i.e., its boiling point is the one that would be expected from the other data. Now we have another difference to

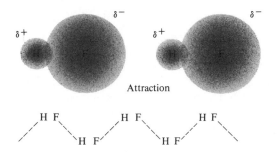

Fig. 9–17 The formation of hydrogen bonds between HF molecules.

consider. When hydrogen is bonded to the small oxygen, fluorine, or nitrogen atoms (but not carbon), the compounds have unexpectedly high boiling points.

These data and those obtained from other observations of unusual bond strengths in certain hydrogen-containing compounds, combined with the bright ideas of several chemists, have led to the concept of the *hydrogen bond*. This is a bond that exists between molecules in compounds in which hydrogen is combined with a small, highly electronegative atom. In the above cases, the small, highly electronegative elements are nitrogen, oxygen, and fluorine. In the compounds NH_3, H_2O, and HF, hydrogen bonds exist between the individual molecules; this increases the intermolecular forces and also, therefore, the boiling points. These hydrogen bonds are not present in compounds such as H_2S or HBr or PH_3 or even CH_4.

To get a better understanding of hydrogen bonds, let us consider liquid hydrogen fluoride as an example. Figure 9–17 diagrams the electron distribution in an HF molecule. The fluorine end of the molecule has a partial negative charge, since the

Table 9–6. Melting and boiling points of hydrogen-containing compounds

Compound	Melting point, °C	Boiling point, °C
CH_4	−182	−161
SiH_4	−185	−112
GeH_4	−165	−88.5
SnH_4	−150	−52
PbH_4		∼ −13
NH_3	−78	−33
PH_3	−133	−88
AsH_3	−116	−55
SbH_3	−88	−17
BiH_3		∼ +20
HF	−83	+20
HCl	−115	−85
HBr	−88	−67
HI	−51	−35

fluorine atom has a much higher electronegativity than hydrogen, and consequently attracts the bonding pair of electrons more strongly than the hydrogen atom does. This unequal sharing of electrons leaves the hydrogen end of the molecule with a partial positive charge. When many HF molecules are physically close to one another, these partial negative and partial positive charges can attract one another. This attractive force between the partial positive charge on the hydrogen end of one HF molecule and the partial negative charge on the fluorine end of a second HF molecule is a *hydrogen bond*. Four things are worth noting.

1. Hydrogen bonds exist only in compounds in which hydrogen is bonded to a small, highly electronegative atom (O, F, and N are generally the atoms involved).

2. The force existing in hydrogen bonds is opposite charge attraction.

3. Hydrogen bonds are forces (usually between molecules) that are different from the covalent-bond attractive forces within a molecule. Hydrogen bonds exist between two HF molecules; the bond between the H atom and the F atom in one HF molecule is a normal polar covalent bond. The hydrogen bond is only about one-tenth as strong as a normal covalent bond.

4. Hydrogen bonds exist in addition to van der Waals forces, which are present in all matter.

We can now explain the data shown in Figs. 9–15 and 9–16. The boiling points of H_2O, HF, and NH_3 are higher than expected because hydrogen bonding is present in these three substances. This increases the bonding forces between liquid molecules, making them more difficult to vaporize. The data also indicate that no appreciable hydrogen bonding is present in CH_4.

The "normal" trend in the boiling points of the substances in Figs. 9–15 and 9–16 can be explained by considering the other intermolecular forces present in all the liquids: the van der Waals forces. Remember that these forces increase as the sizes of the bonded atoms increase. There are more electrons present that are farther away from the positively charged nucleus in the larger atoms. These electrons can be instantaneously polarized more easily. Hence, as the molecules become larger—for example, from H_2S to H_2Se to H_2Te—the intermolecular van der Waals forces increase, and the boiling points of the compounds go up.

Ice. The concept of hydrogen bonding is also very useful in explaining several of the properties of ice, or solid water. We know that ice floats on water, and therefore that the density of ice is less than that of water. That the density of the solid form of a substance is less than the density of the liquid form is somewhat unusual. There is another unusual feature about the density of water: It has its greatest value near 4 °C (actually 3.98 °C).

▶ Recall our kinetic-theory assumptions for liquids and solids. Try to explain the density behavior of ice and water by means of the kinetic theory. What changes are needed in the theory? Do you think that the kinetic theory could have predicted this behavior for water? ◀

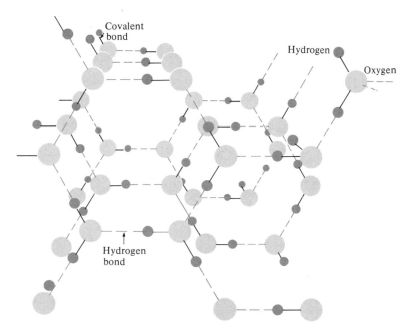

Fig. 9-18 The structure of ice. The large open spaces are caused by the hydrogen bonds that form between the essentially tetrahedral H_2O molecules.

How can hydrogen bonding explain these peculiarities in the density of water? Figure 9–18 shows the structure of ice. Note how open the structure is, i.e., how many relatively large empty spaces there are. These holes are formed because the hydrogen bonds, holding the H_2O molecules in place, are directed in certain directions (conforming to the approximate tetrahedral nature of H_2O), which necessitates the open network. In liquid water these intermolecular forces are fewer and weaker, and the individual H_2O molecules are closer together, giving the water a greater density than ice.

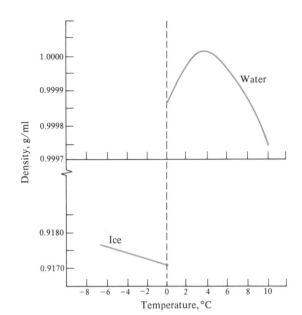

Fig. 9–19 The density of ice and water varies with temperature. Note the large change at the melting point, 0 °C, and the maximum occurring at 3.98 °C.

The explanation for the maximum in water density at 4 °C is similar. Figure 9–19 plots the density of ice and water at various temperatures. Note that ice becomes less dense as it is heated from −5 to 0 °C. This is actually the expected behavior. As heat is added to the ice, the solid H_2O molecules begin to move more rapidly, and consequently occupy a slightly larger volume. Since the amount of water remains constant and it occupies a larger volume, the density (the number of grams/cc) decreases as the temperature increases. At 0 °C the ice melts; large numbers of hydrogen bonds are broken, allowing the molecules to slide into the spaces that had been present, decreasing the volume that they occupy. This gives a sharp increase in the density.

As the liquid water is heated from 0 to 4 °C, hydrogen bonds continue to be broken and the molecules crowd more closely together, again increasing the density. But above 4 °C the heat added increases the speed of the H_2O molecules, so that they begin to occupy a larger volume, with a concurrent decrease in density, even though hydrogen bonds continue to be broken. The "normal" decrease in density with increasing temperature continues as the water is heated above 4 °C.

Phase Diagrams. The transitions between the gaseous, liquid, and solid forms of a substance can be summarized in graphical form. Figure 9–20 gives such a diagram for water; this is called a *phase diagram*. A *phase* is a physically distinct region of a mixture that exhibits its own characteristic properties.

Figure 9–21 shows an example of four separate phases. An ice cube is floating in a stoppered flask containing water and carbon tetrachloride. Note that you can see

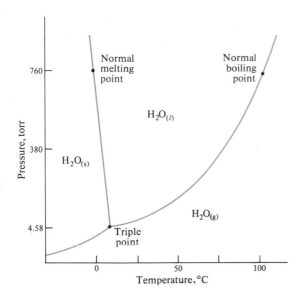

Fig. 9-20 A phase diagram for water, showing the regions of solid, liquid, and gas at various temperatures and pressures.

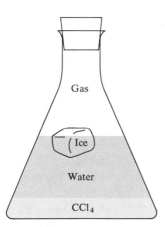

Fig. 9-21 Four phases: solid ice cube, liquid water, liquid carbon tetrachloride, and the gas above the water and ice surfaces.

boundaries between the various phases. The gas phase above the water is one separate, distinct phase (even though it is composed of several different gas molecules). The solid ice cube is a second distinct phase, and there are two liquid phases. A sharp boundary is apparent between the liquid-water phase and the liquid-carbon-tetrachloride phase.

Thus a phase diagram for water relates the gaseous, liquid, and solid phases of water. The diagram in Fig. 9–20 shows how water changes under varying conditions of pressure and temperature. We have already considered some sections of this

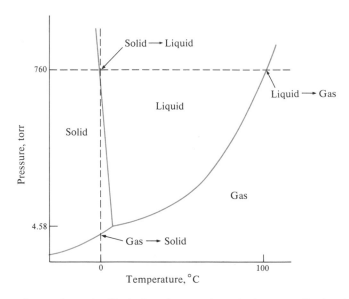

Fig. 9–22 Phase diagram for water, illustrating changes at constant pressure (horizontal dashed line) and constant temperature (vertical dashed line), and indicating the phase transitions during the changes.

diagram, without actually plotting the data. Remember that the boiling point of water changes as the prevailing pressure changes. As the pressure decreases, the boiling point decreases. This behavior is shown on the phase diagram by the line that separates the liquid and gaseous sections of the diagram. This line, then, is determined experimentally and represents a series of boiling points for water. A boiling point is determined at successively lower pressures and each point is plotted on the graph (for each point there is a different value of P and T).

We can run similar experiments on liquid water to determine its freezing point under various pressures. We might, for example, use a piston-and-cylinder arrangement, as was described for the Boyle's and Charles's law experiments in Chapter 8. A plot of a series of these experimental points would give the line in the phase diagram separating the solid and the liquid regions. Along this line, then, are the melting or freezing points of water under various pressures.

If the pressure is reduced to a low enough value (below 4.58 torr for water), liquid water cannot be obtained at any temperature. Ice transforms directly into water vapor and water vapor condenses into ice, without passing through the liquid phase. The direct transition from the solid to the gaseous state is called *sublimation*.

The three equilibrium lines—solid–liquid, liquid–gas, and solid–gas—all intersect at one point, called the *triple point*. At only this one value of temperature and pressure (0.01 °C and 4.58 torr for water) can the three phases all exist at equilibrium. Note that equilibrium is specified. In fact, all the data for a phase diagram must be obtained under equilibrium conditions (no net changes occurring in the

system), so that everyone can use the diagrams and readily duplicate the data when necessary.

Phase diagrams supply several different kinds of information. Let us consider just a few cases. Knowledge of the diagram in Fig. 9–20 enables us to predict which phase of water will be present at any given temperature and pressure. For example, at 25 °C and 380 torr, water is a liquid; at −5 °C and 580 torr, water is a solid; at 75 °C and 25 torr, water is a gas.

We can also predict what transitions will occur if the pressure on a sample remains fixed and the temperature is varied, or if the temperature is held constant and the pressure changed. The dashed lines in Fig. 9–22 indicate these two types of experiments. Under constant pressure, the change occurs horizontally across the diagram as the temperature is varied. At −25 °C and 760 torr, water is a solid. As the temperature is increased (at constant 760-torr pressure), the solid phase persists up to 0 °C. At 0 °C the solid water begins to melt, and solid and liquid water are in equilibrium. Between 0° and 100 °C, the liquid-water phase is present, and at 100 °C liquid and gaseous water are both present, at equilibrium. Above 100 °C only gaseous water is present.

▶ The process just described is the one that occurs during a heating curve. See if you can construct the curve for water, heating from −25 °C to 200 °C under constant pressure conditions. ◀

If the temperature of a water sample is held constant, say at 0 °C, and the pressure is varied, the phase changes are indicated along a vertical line on the phase diagram. Starting at a low pressure of 1 torr and 0 °C, the water sample is a gas. It remains a gas until the pressure is increased to about 3 torr (at constant temperature, 0 °C) when gaseous water condenses to form a solid–gas equilibrium mixture (no gas–liquid transition is observed). Under further increases in pressure, the water is a solid until 760 torr is reached, when the solid begins to melt and a solid–liquid equilibrium is established. At pressures higher than 760 torr, the water sample is a liquid.

This last transition illustrates an interesting aspect of water. The freezing point of water decreases as the pressure is increased. This is shown on the phase diagram by the tilt to the left of the solid–liquid equilibrium line.

Phase diagrams, such as the one described here for water, as well as one of a slightly different nature, are very useful in studying mixtures of materials and solutions. We shall describe these applications in Chapter 10.

QUESTIONS

1. Define or discuss the following terms: Gas, liquid, solid, diffusion, evaporation, kinetic energy distribution, Dewar, equilibrium vapor pressure, boiling point, melting point, normal boiling point, heating curve, space lattice, unit cell, crystal structure, molecular space lattice, ionic space lattice, covalent network space lattice, metallic space lattice, van der Waals forces, imperfect lattices, hydrate, deliquescence, hydrogen bond, phase, phase diagram, triple point.

2. Indicate in which state of matter—solid, liquid, or gas—the following substances are normally found under ordinary room temperature and pressure conditions.

a) CO_2 b) KBr   c) CCl_4 d) CaF_2 e) MgO
f) Br_2 g) HCl   h) C_6H_6 (benzene)   i) NH_3 j) Na_2CO_3

3. Using the kinetic theory, explain why liquids are essentially incompressible.

4. Experimentally it is observed that liquid germanium (Ge) increases in density as its temperature decreases. But at the freezing point, the density changes sharply, yielding solid Ge, which is less dense than liquid Ge. Explain these observations, using the kinetic theory.

5. The rate of diffusion of H atoms trapped in the CaF_2 lattice is much greater than the diffusion rate for the F^- ions. Why should this be true?

6. Why is evaporation generally called a cooling process?

7. The following substances are listed with their normal boiling points. Rank the substances from low to high in terms of their intermolecular forces.

Substance	Boiling point, °C	Substance	Boiling point, °C
a) CO	-192	e) KBr	1435
b) H_2O	100	f) CaF_2	2500
c) Ne	-246	g) SO_2	-10
d) Na	892	h) HCl	-85

8. Plot the melting points of the Group VIA–hydrogen compounds versus the period of the Group VIA element. Explain any systematic variation and any unusual feature of the data.

9. Solid krypton (Kr) has a face-centered-cubic crystal structure.

a) Draw a unit cell of solid Kr.
b) Indicate the type of unit occupying the lattice sites, the type of bonding present, and some of the expected properties.

10. Strontium oxide (SrO) crystallizes with the NaCl crystal structure. Sketch a section of the SrO space lattice, showing the units occupying the lattice sites.

11. Using the kinetic theory, explain why liquids maintain their own volume but assume the shape of their container.

12. Given the following substances and the temperature at which their vapor pressure is 100 torr, rank the substances from low to high in terms of boiling points and intermolecular forces.

Substance	Temperature, °C	Substance	Temperature, °C
a) NH_3	-68	e) MoO_3	955
b) CdI_2	640	f) KOH	1064
c) CO_2	-100	g) SiO_2	1969
d) CS_2	-5	h) SiH_4	-140

13. A partially filled, stoppered flask contains ice floating in water. Compare the average kinetic energies of H_2O molecules in the flask in the gas, liquid, and solid phases. Assume equilibrium and a temperature of 0.01 °C.

14. At $-5\,°C$, H_2O is a solid, C_2H_5OH (ethanol) is a liquid, and SO_2 is a gas.

    a) Compare the average kinetic energies of the molecules.

    b) Compare, as quantitatively as possible, the average velocities of the molecules.

15. By applying the kinetic theory, discuss diffusion rates in gases, liquids, and solids.

16. The rate of evaporation of H_2O is greater at $80\,°C$ than at $25\,°C$. Explain this observation, using the kinetic theory.

17. Assume that a perfectly insulating Dewar flask is half filled with H_2O at $25\,°C$ and is fitted with a perfectly insulating cap. Describe what would happen to the liquid and gaseous water and what conditions would be observed in the Dewar flask after a week's time, assuming no loss of H_2O from the Dewar.

18. Can a solid evaporate? If so, explain how.

19. When H_2O evaporates freely from a beaker into a room at $25\,°C$, the equilibrium vapor pressure of H_2O is not necessarily established. However, if the beaker is enclosed by a small bell jar and evaporation occurs from the same volume of H_2O at $25\,°C$, the equilibrium vapor pressure of H_2O is established. Explain.

20. On a warm summer day, the outer walls of a glass of ice water become covered with beads of water. Why does this happen?

21. The melting and boiling points of I_2 are $114°$ and $184\,°C$, respectively. Sketch a heating curve for I_2 from $25°$ to $200\,°C$.

Figure 9–23

22. Figure 9–23 is a phase diagram for CO_2. Label the axes; indicate the regions of solid, liquid, and gas; indicate the melting points; and label the triple point.

23. A beaker of boiling water is cooled to $-20\,°C$.

    a) Sketch the cooling curve that would be obtained.

    b) On the cooling curve, label each section in which the average kinetic energy of the molecules is changing and each section in which the average kinetic energy remains constant.

24. There are three Bravais lattices in the cubic system: simple cubic, face centered cubic, and body centered cubic.

    a) Draw a unit cell for these three lattices.

    b) If the simple-cubic Bravais lattice is stretched in one direction only, the simple-tetragonal Bravais lattice is obtained. Draw a unit cell for this lattice.

    c) Calculate the net number of lattice points that are contained in one unit cell of a body-centered-cubic lattice. Remember that there are other cells completely surrounding any given unit cell.

25. Solid Br_2 and I_2 have the same crystal structure, but their melting points are $-7°$ and $114\,°C$, respectively. Explain these observations.

26. Why are van der Waals forces so very dependent on temperature?

27. Iron metal crystallizes in a body-centered-cubic space lattice. Sketch a section of this lattice, indicating the units at the lattice points and the location of electrons.

28. a) Using H_2O as an example, illustrate hydrogen bonding by sketching a series of H_2O molecules identifying the hydrogen bonds and the normal H—O covalent bonds.
    b) Why would hydrogen bonds not be expected in HI?

29. a) Why does the density of ice decrease as it is heated to its melting point?
    b) Why does the density of H_2O increase between $0°$ and $4\,°C$?
    c) What change might you expect in the density of H_2S when it is melted? Why?

30. Refer to the phase diagram for H_2O in Figs. 9–20 and 9–22.

    a) What phase would be present at 1000 torr and 25 °C?
    b) Indicate the H_2O phases that are present or formed when a sample is heated from $-20\,°C$ to 100 °C at 2 torr.
    c) What would happen if the pressure on a water sample was decreased from 2 atm to 2 torr at 0 °C?

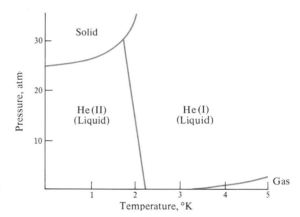

Figure 9-24

31. Consider Fig. 9–24, which is a phase diagram for He.

    a) At 1 atm and 4.5 °K, what change(s) in temperature or pressure are necessary to form liquid He(I)?
    b) What happens to liquid He(II) at 1 °K and 20 atm, if the temperature is gradually increased to 6 °K, at 20 atm?
    c) How can solid He be obtained from liquid He(II) at 1 °K?
    d) How can solid He be obtained at 1 atm pressure?

10. SOLUTIONS: TYPES AND PROPERTIES

10–1 MIXTURES

We have been studying the properties of gases, liquids, and solids in their *pure* state. For example, we have considered the equilibrium vapor pressure of water and its variation with temperature, and the phase changes of water under varying conditions of temperature and pressure. But more often than not, samples that are studied are not pure; the water has something dissolved in it. And these impurities can change the properties and behaviors of the pure substance.

We know, for example, that the freezing point of water lies below 0 °C if an antifreeze (ethylene glycol or methanol) is added to the water. This property of the water–antifreeze mixture makes it possible for automobile engines to be operated during cold weather. Mixtures such as these are very common and important in chemistry, and an understanding of their properties is based on some ideas that are already familiar, coupled with a few new concepts.

Mixtures can be grouped into two broad categories: heterogeneous and homogeneous. In *heterogeneous mixtures* the individual substances composing the mixture retain their properties. Each substance melts and boils at its own characteristic temperature; each solid has its own characteristic x-ray pattern; each has its own color; and so on. If we mix white sodium chloride with black graphite

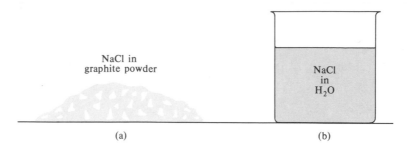

NaCl in
graphite powder

NaCl
in
H_2O

(a)   (b)

Fig. 10–1 (a) In a heterogeneous mixture of NaCl and graphite powder, the two solids can be readily separated. (b) In a homogeneous mixture—or solution—of NaCl in water, the salt particles are randomly distributed throughout the water and cannot be picked out.

powder, as shown in Fig. 10–1a, for example, we obtain a heterogeneous mixture. Looking through a magnifying glass, we could separate the small chunks of salt from the graphite powder. Heating the mixture above 801 °C would melt the sodium chloride, but not the graphite. An x-ray analysis of the mixture would indicate the presence of both sodium chloride and graphite. In other words, a heterogeneous mixture shows the properties of its individual components.

On the other hand, in *homogeneous mixtures*, the individual substances composing the mixture lose their properties, and new properties characteristic of the mixture are exhibited. The components are so thoroughly mixed together that their own properties are changed and new properties result. We can, therefore, dissolve sugar in water. It is obvious that the properties of the sugar are greatly changed. The solid sugar is no longer present. It is now mixed intimately with the water molecules, and the sugar molecules are spread throughout the water solution. If we heat the mixture, the sugar can no longer melt; the water solution will boil. But the properties of the water are also changed. Its density is greater and its boiling and melting points are different from those of pure water. In short, the homogeneous mixture has its own characteristic properties (see Fig. 10–1b).

▶ An interesting class of materials called colloids falls somewhere between heterogeneous and homogeneous mixtures. In *colloids*, small particles of a substance remain suspended in a gas or a solution indefinitely. The suspended particles form a separate phase (therefore a true solution is not present), but they are extremely small, from 1 to 100 mμ (millimicrons) in diameter (a page of this text is about 100,000 mμ thick). Larger particles give heterogeneous mixtures and smaller particles form solutions. Common examples of colloids are smoke, jellies, and cream. Colloidal particles are in continuous movement, often called *Brownian motion*, sometimes carry an electrical charge, and scatter light when the light passes through the suspension. This light-scattering property is called the *Tyndall effect*, and is commonly observed in smoke in the air. It is a result of the special size of the colloidal particles. The study of colloids is becoming increasingly important to society, since they are of prime concern in air and water pollution. ◀

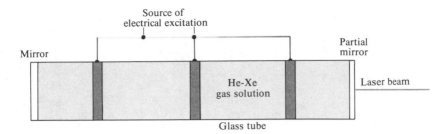

Fig. 10-2 A helium-xenon gas laser. The He–Xe gas solution fills the entire tube. The gas laser can be thought of as a "super neon light" which gives off a very narrow and highly monochromatic beam.

10-2 SOLUTIONS

A homogeneous mixture composed of two or more components is called a *solution*. The constituents composing the solution can be present in varying amounts, and the solutions may be liquid (sea water) or solid (metal alloys) or gas (air). When discussing solutions, for the sake of convenience we refer to the substance present in the larger amount as the *solvent* and to the substance present in smaller amount as the *solute*. If we add a pinch of salt to a glass of water, the water is the solvent and the salt the solute. A small amount of tin metal is a solute when it forms a homogeneous mixture with a large lead (which is the solvent) bar.

▶ This nomenclature is usually, but not always, followed. Sometimes it is more convenient to speak of a solvent that is common to a group of substances, even though it may be present in smaller amounts in a few cases. Concentrated sulfuric acid—a water solution of H_2SO_4—is in this category. We consider water to be the solvent, even though the quantity of H_2SO_4 present is greater. ◀

We shall discuss gaseous, solid, and liquid solutions separately. A large portion of chemistry is involved, in one way or another, with liquid solutions, and in particular with water solutions, which we generally call *aqueous solutions*. Although we shall cover aqueous solutions in the greatest detail, solutions of other liquids as well as gaseous and solid solutions are becoming increasingly important technologically.

Gaseous Solutions. Gases mix to form solutions in all proportions. The composition of the air is constantly changing. The amount of water vapor or carbon dioxide and even of oxygen and nitrogen continually changes, although some of the variations might be small. The operation of gas lasers depends on the preparation of certain gas mixtures, such as helium and neon or helium and xenon (see Fig. 10-2), and the lifetime of an ordinary electric light bulb strongly depends on the amount of water vapor present in the gas mixture inside the bulb.

Why do gases mix in all proportions? To answer this question we can return to the kinetic theory, as it applies to gases. This theory sets up a model for gases that

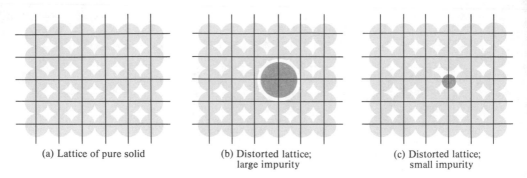

(a) Lattice of pure solid

(b) Distorted lattice;
large impurity

(c) Distorted lattice;
small impurity

Fig. 10–3 Dissolving of a particle that is much larger or smaller than the one that normally occupies a lattice site is an unfavorable process, since it leads to lattice distortion.

enables us to consider the problem in more concrete terms. The kinetic theory for gases assumes that gas particles are extremely small and do not attract one another. Therefore there are huge spaces between the gas molecules (recall, for example, Fig. 8–17). These spaces are so large that if other gas molecules were placed in the same container, their presence would not be recognized by the molecules already there. The space available for the new molecules is vast, and no intermolecular forces need be broken during the mixing process. This lack of any hindrance to mixing makes it possible for gases to mix in all proportions.

Solid Solutions. As we might expect when we consider the kinetic theory as applied to solids, the formation of solid solutions is quite limited. As with all solutions, one substance must be randomly distributed throughout another. With solid solutions, then, one solid must be homogeneously mixed with another. Actually, many common examples of solid solutions do exist. Copper dissolved in silver gives an alloy that we call sterling silver; copper in zinc is called brass. Many important solid state electronic devices, such as transistors, depend on solid solutions. Small amounts of arsenic or gallium can be dissolved in germanium, for example, to produce the electrical properties required for the operation of transistors.

On the basis of the kinetic theory, why would we expect limited solid solutions? For solids, the kinetic theory assumes strong intermolecular forces and contact between molecules. There is therefore only a small amount of free space in a solid. Most of the space is occupied by the molecules themselves. And remember that solid substances have distinct and characteristic packing arrangements, i.e., crystal structures. Thus a foreign or solute substance has many conditions to fulfill before it will dissolve in a solid solvent.

For example, Fig. 10–3 shows that the solute particle must not be too large or too small or it will disrupt the packing arrangement of the solvent by not "fitting into" its place in the lattice. The solute must also be able to satisfy the intermolecular bonding arrangements present in the solvent, or it will again disrupt the orderly packing arrangement. If either the size or the bonding conditions are not fulfilled by the solute, at least approximately, there will be little or no solid solution formed.

▶ As simple examples illustrating these two conditions, let us consider dissolving the solutes KCl, CsCl, and YCl_3 in a NaCl solvent. The KCl does form a solid solution with NaCl, but CsCl and YCl_3 do not. The Cs^+ ion is too large; it has an ionic radius of 1.69 Å versus 0.95 Å for Na^+. Consequently Cs^+ and Na^+ ions cannot pack together in the same type arrangement and CsCl does not dissolve well in NaCl. With YCl_3 the difficulty is not in the size of the ion (Y^{3+} has an ionic radius of 0.93 Å) but in the bonding characteristics. The trivalent Y^{3+} ion has too great a charge for it to replace a Na^+ ion easily. Compensation for the extra positive charge is quite difficult. Therefore YCl_3 does not dissolve in NaCl. With KCl, however, the singly charged K^+ ion has a radius of 1.33 Å and a limited solid solution can form between KCl and NaCl. ◀

Liquid Solutions. Homogeneous mixtures having a liquid as the solvent lie in the intermediate region between gas and solid solutions. A particular solvent can dissolve one solute quite well, while another solute might dissolve poorly or not at all. In any liquid solution a gas, another liquid, or a solid must be distributed randomly throughout the liquid serving as solvent.

To understand this better, we can consider dissolving solid sodium chloride in water. In some way the solid NaCl must be broken up and distributed throughout the water. The Na^+ and Cl^- ions must be spread out, surrounded by water molecules, and no longer physically close, as they are in solid NaCl.

What does the kinetic theory have to say about liquid solutions? Should their behavior be somewhere between that of gaseous and solid solutions? Judging from the assumptions made when we applied the theory to liquids, we would probably expect an intermediate behavior for liquid solutions. Remember that liquid molecules are close to one another, but that there are open spaces that make movement of molecules possible. And the intermolecular forces, although they are present, are generally not assumed to be as strong as those in solids. Thus it is possible for a solute to dissolve in a liquid solvent, since there are some spaces—although limited ones—in which the solute particle can "fit," and since the existing intermolecular forces can be overcome.

The variability with which solutes dissolve in solvents presents another interesting question. What causes a solute to dissolve well in one solvent but poorly in another solvent? To answer this question, at least partially, we must look at the process of dissolving more closely. Again we shall use a model with a series of steps postulated. As we pointed out previously, the actual processes involved in the real situation are undoubtedly more complex than those envisioned here, and certainly would not occur in a stepwise fashion, as might seem to be the case from our proposed model.

10-3 THE SOLUTION PROCESS

We can develop a model for the solution process by considering the case of solid sodium chloride dissolving in water. What forces are important in the dissolving process? What bonds need to be broken; what particles moved around? If we

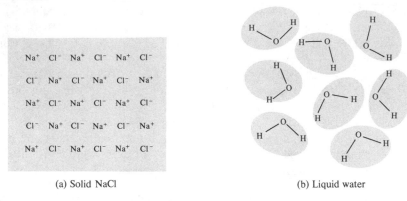

Na⁺	Cl⁻	Na⁺	Cl⁻	Na⁺	Cl⁻
Cl⁻	Na⁺	Cl⁻	Na⁺	Cl⁻	Na⁺
Na⁺	Cl⁻	Na⁺	Cl⁻	Na⁺	Cl⁻
Cl⁻	Na⁺	Cl⁻	Na⁺	Cl⁻	Na⁺
Na⁺	Cl⁻	Na⁺	Cl⁻	Na⁺	Cl⁻

(a) Solid NaCl   (b) Liquid water

Fig. 10–4 Strong ionic bonds are present in solid NaCl, while liquid H_2O molecules are held together by hydrogen bonds and van der Waals forces.

consider the structure of solid NaCl and liquid H_2O, we can propose answers to these questions. Remember that in a sodium chloride–water solution, the NaCl solute is randomly (homogeneously) mixed throughout the H_2O solvent.

Solid NaCl has an ionic crystal structure. Figure 10–4 gives a two-dimensional representation of its structure. Strong bonds exist between the Na^+ and Cl^- ions, which essentially touch one another. Then, in order for the solid NaCl to be broken up and dissolved in a solvent, these strong interionic forces must be overcome and the Na^+ and Cl^- ions separated. Since these bonds must be broken, this process takes energy.

Figure 10–4 also gives a representation of liquid H_2O. In order for solute particles to be accommodated among the liquid H_2O molecules, a spot large enough for the solute must be made. This might entail breaking some of the intermolecular bonds between the solvent molecules and their subsequent movement to make room for the solute. With NaCl the H_2O molecules must rearrange and group around the Na^+ and Cl^- ions. Since this process involves bond breaking and pushing solvent molecules out of the way, it also takes energy.

Thus there are two processes that require energy when a substance dissolves: (1) Breaking of the bonding forces in the solute, often called *solute–solute forces*. (2) Breaking of the bonding forces in the solvent, often called *solvent–solvent forces*, and moving the solvent molecules around.

A third interaction, *solvation*, provides the necessary energy for these two processes; it arises from the attraction between the solute and the solvent. If the *solvation energy*, produced by the *solute–solvent forces*, is great enough to overcome both the solute–solute and the solvent–solvent forces, then the solute will dissolve. If the solute–solvent forces are small compared with the sum of the solute–solute and solvent–solvent forces, the solute does not readily dissolve.

Figure 10–5 diagrams the situation for NaCl and H_2O. A strong solute–solvent attractive force is present, since the Na^+ and Cl^- ions on the surface of the solid NaCl can attract the polar H_2O molecules. The positively charged Na^+ ion can

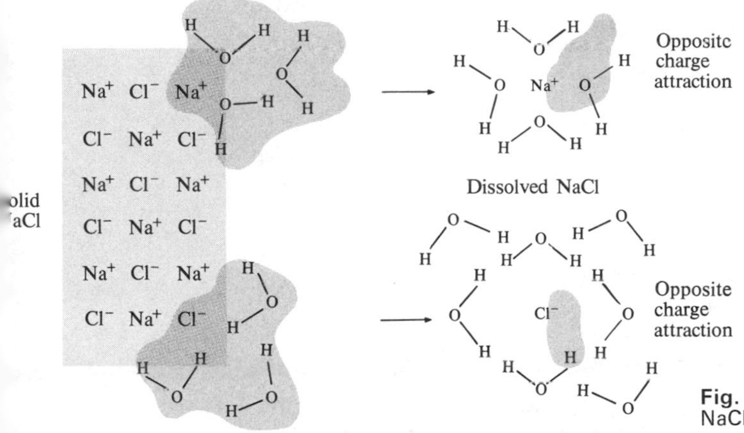

Fig. 10-5 Dissolving of solid NaCl in H₂O.

interact with the partial negative charge on the oxygen atom of a H_2O molecule, while the Cl^- ion and the hydrogens of the H_2O molecule attract one another. In fact, the Na^+ and Cl^- ions are attracted to several molecules of the H_2O. The total result of all these attractive forces is the solvation energy. For NaCl in H_2O, we know that this energy is quite large, overcoming the solute–solute and solvent–solvent forces, since NaCl dissolves easily in H_2O.

▶ The number of H_2O molecules grouped around each solute particle is generally not known. Relatively sophisticated and accurate experimental measurements are necessary to determine the number of coordinated H_2O molecules, if, indeed, exact numbers do exist. Studies are greatly complicated by the rapidly changing local environments present in the solutions. ◀

We can test this model of the solution process by applying it to another case and then trying to predict whether the solute will dissolve. What would happen if we tried to dissolve NaCl in liquid bromine, Br_2? The solute–solute forces are the same for solid NaCl as those already mentioned: strong opposite-charge attraction. But the solvent–solvent and especially the solute–solvent forces are different from those considered above for the case of NaCl dissolving in H_2O. The intermolecular forces in the Br_2 solvent are van der Waals forces, and are therefore quite weak. Even though the energy needed to overcome these weak solvent–solvent forces is small, the total energy needed to dissolve the NaCl is still large, since the ionic bonds in solid NaCl are so strong. Thus the solvation energy—the solute–solvent interaction—must also be large.

But we know that this cannot be the case, since there can be no strong attractive force built up between the nonpolar Br_2 molecules and the Na^+ and Cl^- ions on the surface of the solid NaCl. The Br_2 molecules have no full or large partial charges to attract either of the charged Na^+ or Cl^- ions. Hence essentially no attractive forces exist between solid NaCl and liquid Br_2. The solute–solvent forces are very weak,

Table 10–1. Polar solutes prefer polar solvents and nonpolar solutes prefer nonpolar solvents

Solvent	Solute solubility			
	NaCl *Ionic*	CaCl$_2$ *Ionic*	I$_2$ *Nonpolar*	CHCl$_3$ *Polar*
H$_2$O (polar)	Yes	Yes	Slight	Slight
CCl$_4$ (nonpolar)	No	No	Yes	Yes
C$_6$H$_6$ (nonpolar)	No	No	Yes	Yes

and produce a solvation energy that is too slight to overcome the very strong solute–solute forces. We would therefore predict that NaCl would not readily dissolve in liquid Br$_2$. This prediction agrees with known experimental results.

In addition to using reasoning processes such as we have just presented, chemists have developed a generalization concerning the dissolving of solutes in various types of solvents; they have done so on the basis of experience and experimentation. In general, polar solvents dissolve polar solutes and nonpolar solvents dissolve nonpolar solutes (see Tables 10–1 and 10–2). An even more general statement is "Like dissolves like." As with all generalizations, this one has many exceptions, but as a first approximation it can be useful.

10–4 SOLUBILITY

Most solutes dissolve in any given solvent only to a certain extent. In some cases very little solute dissolves and in other cases a large amount dissolves. So long as the amount of solute in the solvent can be increased, the solution is *unsaturated*. But a limit is usually reached. When the amount of solute is increased to this limit, and no more will dissolve, we have a saturated solution. A *saturated solution*, therefore, is a solution in which there is an equilibrium between dissolved and undissolved solute.

▶ In some cases—for example, sodium acetate (NaC$_2$H$_3$O$_2$)—one can obtain a solution that contains more dissolved solute than is present in a saturated solution, i.e., the solubility of the solute is exceeded. This is called a *supersaturated solution*, and is generally prepared by raising the temperature of the solution, thereby increasing the solute solubility, until all the solute dissolves, then carefully, slowly cooling the solution back to room temperature. Supersaturated solutions readily form precipitates, i.e., the excess solute easily falls out of solution, becoming a solid again, leaving a saturated solution. One can use many methods of precipitation, such as mechanical shock, scratching the inside of the container wall, or dropping a small piece of solid solute into the solution. ◀

Table 10–2. Some general solubility rules

		Soluble in H_2O		
Compound unit	Yes	No	Exceptions	Example solution reaction
1) Na^+	×		None	$NaCl_{(s)} \rightarrow Na^+_{(aq)} + Cl^-_{(aq)}$
2) K^+	×		None	$KBr_{(s)} \rightarrow K^+_{(aq)} + Br^-_{(aq)}$
3) NH_4^+	×		None	$(NH_4)_2CO_{3(s)} \rightarrow 2NH^+_{4(aq)} + CO^{2-}_{3(aq)}$
4) NO_3^-	×		None	$Ba(NO_3)_{2(s)} \rightarrow Ba^{2+}_{(aq)} + 2NO^-_{3(aq)}$
5) Cl^-	×		AgCl	$CaCl_{2(s)} \rightarrow Ca^{2+}_{(aq)} + 2Cl^-_{(aq)}$
6) SO_4^{2-}	×		$PbSO_4$, $SrSO_4$, $BaSO_4$	$Al_2(SO_4)_{3(s)} \rightarrow 2Al^{3+}_{(aq)} + 3SO^{2-}_{4(aq)}$
7) S^{2-}		×	Sulfides of Na^+, K^+, NH_4^+, Ca^{2+}, Sr^{2+}, Ba^{2+}	$CdS_{(s)} \rightarrow$ N.R.
8) OH^-		×	Hydroxides of Na^+, K^+, NH_4^+, Ca^{2+}, Sr^{2+}, Ba^{2+}	$Fe(OH)_{3(s)} \rightarrow$ N.R.

The amount of solute dissolved in a given amount of solvent to form a saturated solution is called the *solubility* of the solute in that particular solvent. When the solubility limit is reached and a saturated solution produced, the amount of dissolved solute remains constant. Since an equilibrium state is attained, as much solute is continually going into the solution as is leaving it.

Solubility Dependence. We have discussed two variables that affect solubilities: the *nature of the solute* and the *nature of the solvent*. We have seen that polar solvents tend to dissolve polar solutes, while nonpolar solvents dissolve nonpolar solutes. One other variable—temperature—has a strong effect on the solubility of many solutes. But whether an increase in temperature will increase or decrease the amount of dissolved solute is not obvious. The usual inclination—to heat a solution in order to dissolve more solute—works in a number of cases, but there are many solutions for which the opposite effect holds. Less solute is dissolved at the higher temperature.

▶ There are also situations in which temperature changes do not greatly affect solute solubilities; NaCl in H_2O is one of these cases. At 0°C 35.7 g NaCl dissolve in 100 cc of H_2O, and at 100 °C 39.1 g NaCl/100 cc dissolve. For comparison, in this same temperature range, the solubilities of NH_4NO_3 are 118 and 871 g/100 cc. ◀

This variable change of solubility with temperature is closely related to the heat evolved or absorbed during the solution process. When a substance is dissolved and heat is liberated in the process, the temperature of the solution generally rises. This solution process is *exothermic*. When heat is absorbed by dissolving a substance, an *endothermic* reaction, the temperature of the solution generally decreases. Endothermic behavior is typical of most solids dissolving in water.

Experimentally we find that increasing the temperature of the solution, in these endothermic cases, increases the solubility of the solute. We can summarize this change by writing it in equation form, using silver nitrate as an example:

$$AgNO_{3(s)} + H_2O \rightarrow solution - heat$$

or

$$Heat + AgNO_{3(s)} + H_2O \rightarrow solution.$$

The heat energy needed to break apart the $AgNO_3$ solid and form hydrated Ag^+ and NO_3^- ions—that is, ions that have H_2O molecules grouped around them—and consequently dissolve the solid, is absorbed from the water, the container, and the surrounding air. Hence the temperature of the solution drops; a beaker containing a freshly prepared silver nitrate solution feels cool when touched. Since Ag^+ and NO_3^- ions are formed when solid $AgNO_3$ dissolves, we can also write the following equation for the solution reaction:

$$AgNO_{3(s)} \xrightarrow{\;H_2O\;} Ag^+_{(aq)} + NO^-_{3(aq)}.$$

The subscripts in parentheses indicate that $AgNO_3$ is a solid and that Ag^+ and NO_3^- are *aqueous ions*; that is, they are ions dissolved in water.

▶ To be more explicit, we could write $Ag(H_2O)^+_x$ and $NO_3(H_2O)^-_x$, but since the number of H_2O molecules of hydration, x, is usually not known, the simpler notation $Ag^+_{(aq)}$ and $NO^-_{3(aq)}$ is used. Actually, people often omit the subscript (aq) and just write $Ag^+ + NO_3^-$, with the understanding that H_2O molecules are attached to each ion. ◀

Why does the addition of heat energy (an increase in the temperature of the solution) increase the solubility in the case of endothermic solution reactions? We can answer this question by using a principle, proposed by Henri LeChatelier in 1888, which concerns systems at equilibrium. Remember that an equilibrium is established when a solute has reached its solubility limit; as much solute is dissolving as is precipitating. Using $AgNO_3$ we can write this in equation form:

$$Heat + AgNO_{3(s)} \underset{\longleftarrow}{\xrightarrow{\;H_2O\;}} Ag^+_{(aq)} + NO^-_{3(aq)},$$

where the double arrows indicate an equilibrium process and H_2O written above the arrows shows that the reaction took place in the presence of H_2O.

Since solution reactions are equilibrium processes, we can apply *LeChatelier's principle*, which states that the application of any form of stress on a system at equilibrium causes a change in the system that will relieve the stress and establish a new equilibrium.

When heat is added to a beaker containing excess solid $AgNO_3$ in equilibrium with a saturated solution, a stress is applied to the equilibrium. The stress is the

addition of heat. The system can adjust to relieve the stress by using up the heat. Since the reaction is endothermic, heat can be used to break up more solid $AgNO_3$ particles, forming aqueous Ag^+ and NO_3^- ions in the process. Then the new equilibrium that is established, at the higher temperature, will have more solute dissolved and less solid $AgNO_3$ at the bottom of the beaker. The above equation describes this process in concise form. Heat, which can be considered a reactant, combines with solid $AgNO_3$ to form Ag^+ and NO_3^- ions, or dissolved $AgNO_3$. So long as solid $AgNO_3$ remains, the addition of more heat energy will continue to dissolve it, forming more Ag^+ and NO_3^- ions. Then, according to LeChatelier's principle, the solubility of a substance should increase with increasing temperature if its solution process (or reaction) is endothermic.

▶ You can, no doubt, make a prediction about reactions of exothermic solutions and their behaviors with changing temperature. Explain your prediction, using LeChatelier's principle. Note that the reverse of the reaction for dissolving $AgNO_3$—that is, precipitation of solid $AgNO_3$ from dissolved Ag^+ and NO_3^- ions— is an exothermic reaction. ◀

10-5 CONCENTRATIONS

We have been considering the amount of solute present in a particular solution, the amount of solute that dissolves to yield a saturated solution, and how the amount of dissolved solute changes as the temperature changes. By *amount*, of course, we are implying that a certain number of grams of solute are present. Actually, we can be more precise about specifying the amount of solute and also make possible comparisons of such things as solubilities of different substances if we specify the amount of a substance present *in a certain volume* of solution (see Table 10–3).

Table 10–3. Solubilities of several substances in water at 25 °C

	Solubility expressed in units of:	
Substance	*g/100 cc H_2O*	*moles/ℓ*
$BaSO_4$	0.00025	1.1×10^{-6}
NH_4Br	97	0.99
Al_2O_3	0.000095	9.3×10^{-7}
$CaCO_3$	0.0015	1.5×10^{-5}
$Cs_2C_2O_4$	283	0.80
Na_2CrO_4	87	0.54
$ZnCl_2$	432	3.2
ZrF_4	1.4	0.084

We can compare the solubilities of potassium chloride and silver chloride in water at 100 °C, for example, if we know how many grams of each substance will dissolve in a certain volume of water, say 100 cc. Experimentally, it is found that 56.7 g KCl dissolve in 100 cc H_2O at 100 °C, while 0.0021 g AgCl dissolve in 100 cc H_2O. Therefore the solubility of KCl is much greater than that of AgCl, since so much more KCl dissolves in a given volume of H_2O.

We refer to a definite amount of substance dissolved in a certain volume of solution as the *concentration* of the solution, i.e., how many grams of solute/100 cc or how many moles of solute/100 cc or how many moles of solute/liter are present. We shall define and use only two common concentration terms, molarity and normality, although there are many others.

Molarity. The molarity of a solution is the number of moles of dissolved solute in one liter of solution. Molarity has the symbol M and the units moles/ℓ. Thus, to determine the molarity of any solution, we need to determine—or know in some way—the number of moles of solute and the volume of the solution in which the solute is dissolved.

To help clarify this concept, let us consider the preparation of a six-molar solution of sodium chloride. A 6.00M NaCl solution means that 6.00 moles NaCl are present in 1.00 ℓ of solution. Then to prepare the solution, we weigh 6.00 moles of NaCl, which is 351 g NaCl, place it in a large graduated cylinder (or volumetric flask), and add enough water to give 1.00 ℓ of solution.

▶ Make certain that you can calculate this correctly. Remember that to convert moles to grams, you need the formula weight,

$$(\text{mole}) \times (\text{g/mole}) = \text{g}.$$

The thought process goes: One mole of NaCl weighs 58.5 g (23.0 + 35.5 amu = 58.5 amu). If we needed one mole of NaCl, we would take 58.5 g. But we need 6.00 moles. Therefore we need 6.00 times the weight of one mole, or

$$(6.00 \text{ moles NaCl}) (58.5 \text{ g NaCl/mole NaCl}) = 351 \text{ g NaCl.} ◀$$

Unfortunately, situations are generally not this simple. For example, how could we prepare 250 ml of a 6.4M NaCl solution? It is still necessary to have the ratio of 6.4 moles NaCl/ℓ solution, even though only 250 ml of solution are present. Then

$$\frac{x \text{ moles NaCl}}{0.250 \ \ell} = \frac{6.4 \text{ moles NaCl}}{1 \ \ell}.$$

This gives

$$x = (6.4 \text{ moles NaCl/}\ell)(0.250 \ \ell) = 1.6 \text{ moles NaCl.}$$

Therefore we can prepare 250 ml of 6.4M NaCl by dissolving 1.6 moles NaCl in enough water to give 250 ml of solution.

This type of calculation is very useful when you are working with concentrations. You can probably see that a similar procedure would be possible if you knew you had a certain number of moles of solute and you needed to calculate the volume of solution necessary to give a desired concentration. Try to make up a sample problem and solve it. Note now helpful the units can be.

Normality. Normality is a concentration term that is very similar to molarity. While molarity is the number of moles per liter of solution, *normality* is the number of gram-equivalents per liter of solution. It has the symbol N, and the units gram-equivalents/liter or g-eq/ℓ. If concentrations of solutions are expressed in terms of molarity, the number of moles of solute is specified; if concentrations of solutions are expressed in terms of normality, the number of gram-equivalents of solute is specified.

A solution with a certain normality is prepared in exactly the same way as a molar solution, except that the number of gram-equivalents of solute must be used instead of the number of moles of solute. For example, to prepare a two-normal potassium nitrate solution, 2.0N KNO_3, using only 100 ml of solution, you need to dissolve 0.20 gram-equivalent of KNO_3 in enough water to give 100 ml (0.100 ℓ) of solution:

$$\frac{0.20 \text{ g-eq } KNO_3}{0.100 \ \ell} = 2.0 \text{ g-eq } KNO_3/1.0 \ \ell = 2.0N \ KNO_3.$$

It is necessary, of course, to be able to select accurately—often by weighing—the number of gram-equivalents of solute. To do this, you generally need to know the gram-equivalent weight of the solute, which requires an understanding of the reaction involved. If you need any review, refer again to Section 7–6, in which we discussed the determination of gram-equivalent weights for oxidizing and reducing agents.

Example 10–1. How many grams of potassium hydroxide (KOH) are needed to prepare 150 ml of 0.50M KOH?

Since the concentration of the desired solution is 0.50 mole KOH/ℓ, we must calculate the number of moles of KOH that will dissolve in 150 ml (0.150 ℓ) to give a 0.50M solution. Knowing the number of moles of KOH needed and the formula weight of KOH, we can find the number of grams of KOH.

We know that the ratio of the number of moles of KOH to the volume of the solution it is dissolved in must be 0.50, since the desired concentration is 0.50 mole/ℓ = 0.50M (note the units). Then

$$\frac{x \text{ mole KOH}}{0.150 \ \ell} = 0.50 \text{ mole KOH}/\ell$$

and

$$x = (0.50 \text{ mole KOH}/\ell)(0.150 \ \ell) = 0.075 \text{ mole KOH}.$$

Therefore 0.075 mole KOH dissolved in enough H_2O to give 0.150 ℓ of solution yields 0.50M KOH.

We can now calculate the number of grams of KOH from the number of moles of KOH needed, 0.075, and the formula weight, 56 amu. One mole of KOH weighs 56 grams, but only 0.075 mole is needed. Therefore

$$(0.075 \text{ mole KOH}) (56 \text{ g KOH/mole KOH}) = 4.2 \text{ g KOH}.$$

A 0.50M KOH solution can be prepared by dissolving 4.2 g KOH in enough water to give 150 ml of solution.

Example 10–2. How many liters of 1.2M Na$_2$CO$_3$ (sodium carbonate) solution can be prepared from 0.60 mole Na$_2$CO$_3$?

The 0.60 mole Na$_2$CO$_3$ must be dissolved in enough H$_2$O so that the ratio of moles to volume is 1.2, since the desired concentration of the solution is 1.2 moles/ℓ (note the units). Then

$$\frac{0.60 \text{ mole Na}_2\text{CO}_3}{x \, \ell} = 1.2 \text{ moles Na}_2\text{CO}_3/\ell$$

or

$$x = \frac{0.60 \text{ mole Na}_2\text{CO}_3}{1.2 \text{ moles Na}_2\text{CO}_3/\ell} = 0.50 \, \ell.$$

A 1.2M Na$_2$CO$_3$ solution can be prepared by dissolving 0.60 mole Na$_2$CO$_3$ in enough water to give 0.50 ℓ of solution.

▶ How would the problem be changed if, instead of being told that 0.60 mole NaCO$_3$ was available, you were given 63.6 g Na$_2$CO$_3$ and asked to prepare the same solution? Try it. ◀

Example 10–3. Suppose that 0.37 gram-equivalent of the oxidizing agent potassium permanganate (KMnO$_4$) is dissolved in water to give 240 ml of solution. What is the normality of the KMnO$_4$ solution for a reaction in which permanganate is reduced to Mn^{2+}?

In order to calculate the normality of a solution, we need to know the number of gram-equivalents of solute and the volume of the solution. Since both the quantities are given, we can calculate the normality directly:

$$\frac{0.37 \text{ g-eq KMnO}_4}{0.250 \, \ell} = 1.5 \text{ g-eq KMnO}_4/\ell = 1.5N \text{ KMnO}_4.$$

▶ Remember that gram-equivalent weights and normalities depend on the chemical reaction involved. This is illustrated for KMnO$_4$ in Example 11–3. ◀

Example 10–4. How many gram-equivalents of iron(III) chloride are contained in 125 ml of 0.13N FeCl$_3$, for a reaction in which Fe^{3+} is reduced to Fe^{2+}?

An entire liter of 0.13N FeCl$_3$ would contain 0.13 gram-equivalent of FeCl$_3$. But 125 ml is only 0.125 ℓ. Consequently, only a fraction (0.125) of the 0.13 gram-equivalent of FeCl$_3$

could be found in 0.125 ℓ of the solution:

$$(0.125 \ell)(0.13 \text{ g-eq FeCl}_3/\ell) = 0.016 \text{ g-eq FeCl}_3.$$

Note how the units can lead directly to the proper method of calculation.
    In 125 ml of 0.13N FeCl$_3$ there is 0.016 gram-equivalent of FeCl$_3$.

Concentration Versus Amount. There is a distinct and critical difference between
the amount of a solute contained in a given solution and the concentration of the
solute in the solution. The *amount* of solute—i.e., the number of grams or moles
or gram-equivalents of the solute—in the solution depends on the total volume of
the solution, but the *concentration* does not. The concentration of a 2.0M HCl
solution is 2.0M whether we have one drop or 10 ml or 5 ℓ of the solution. But the
number of moles of HCl in each of these three cases differs greatly. There is much
less HCl in one drop of 2.0M HCl than in 5 ℓ of 2.0M HCl.

10-6 COLLIGATIVE PROPERTIES

As we have seen, the properties of a pure solvent are changed when a solute is dissolved in it. The solution formed exhibits its own distinct properties. Some changes
in behavior of solvents depend only on the number of solute particles added, i.e.,
only on the total concentration of solute particles in the solution. These are called
colligative properties. The type of solute particle being dissolved in the solution is not
important; the number of particles is critical. An equal number of atoms or molecules
or ions or conglomerates will have the same effect on the properties of the solution.
It is only necessary that the solute particles dissolve to form a true solution.
    There are two common examples of colligative properties: *freezing-point
lowering* and *boiling-point elevation.* For example, when NaCl is dissolved in water,
the freezing point of the salt-water solution formed is lower than the freezing point
of the pure-water solvent; i.e., the freezing point of the solution is below 0 °C. We
know that when salt is spread on icy sidewalks and roads, the ice melts. It melts
because the salt–ice mixture on the surface of the ice has a very low freezing point.
Hence, some of the ice melts, forming a highly concentrated salt solution. This
process continues until the entire layer of ice has melted.

▶ The ice will melt only if the temperature of the surrounding air is higher than the
freezing point of the NaCl–water solution that is forming. If the temperature is too
low or the NaCl concentration too low, the ice will not melt. Since NaCl is very
soluble in water, the freezing point of a salt solution can be lowered to below
− 20 °C. ◀

    The boiling point of a liquid also changes when a solute is dissolved in it. The
salt solution described above would have a higher boiling point than pure water.
It would boil above 100 °C. We make use of an elevated boiling point for water (as

well as a lowered freezing point) when we are running an automobile. When we add a solute such as ethylene glycol to the water that cools the motor, the solution has a boiling point above 100 °C. We can then operate the motor at higher temperatures, which increases its efficiency.

▶ The primary function of a solute such as ethylene glycol is to serve as an antifreeze, which is a direct consequence of the freezing-point-depression effect. ◀

Since aqueous solutions of solutes such as sodium chloride or ethylene glycol have lower freezing points and higher boiling points than pure water, we see that the liquid phase of the solutions is present over a greater range of temperatures than it is for pure water. This generally happens. The liquid range of a solvent is extended when a solute is dissolved in it.

Constants. When a solute dissolves in a solvent, the number of degrees that the freezing point is lowered and the boiling point raised depends on the nature of the solvent involved and the concentration of solute particles. For a given concentration of solute, the amount of freezing-point depression (or boiling-point elevation) varies from solvent to solvent. For a certain solvent, the amount of boiling-point elevation (or freezing-point depression) changes as the concentration of the solute is varied. Increasing the concentration of the solute increases the boiling point and decreases the freezing point. For a given solvent and a given solute concentration, the number of degrees that the freezing point is lowered is generally *not* equal to the number of degrees that the boiling point is raised.

It has been found experimentally that one mole of solute particles lowers the freezing point of 1000 grams of water 1.86 °C and raises the boiling point 0.512 °C. These two constants are known as the *constants of freezing-point lowering* (K_f) and *boiling-point elevation* (K_b) of water. The numerical value of these constants is a characteristic of a particular solvent, as shown by the data in Table 10–4. Remember that these constants refer to solutions in which one mole of solute is dissolved in 1000 grams of solvent.

▶ See if you can devise the experimental procedures that will yield the data necessary to determine these two constants for water. ◀

A Model. Although interactions in solutions are very complex and not yet well understood, we can get at least a qualitative idea of why a solute lowers the freezing point and raises the boiling point of a solvent. To do this we can again attempt to envision, in a simplified way, the processes that are occurring in the solution.

▶ We shall not try to make a case for the variations between solvents indicated in Table 10–4. These are difficult and poorly understood problems. What properties of

Table 10–4. Constants for several solvents for freezing-point lowering (K_f), and boiling-point elevation (K_b)

Solvent	Freezing point, °C	K_f °C/mole/1000 g H$_2$O	Boiling point, °C	K_b °C/mole/1000 g H$_2$O
Water	0	1.86	100	0.512
Carbon tetrachloride	−22.8	29.8	76.8	5.02
Benzene	5.5	5.12	80.1	2.53
Ethyl alcohol	−114.6	1.99	78.4	1.22
Naphthalene	80.2	6.80	—	—
Camphor	178.4	39.7	—	—

a solvent, both solid and liquid, do you think would be important? Note that, for a given solvent, the freezing-point depression is greater than the boiling-point elevation. Can you think of any reasons for this behavior? ◄

What happens when a liquid begins to freeze and change to a solid? The liquid molecules become essentially fixed in one region of a solid space lattice, with strong intermolecular forces existing. Remember that the formation of solid solutions is quite limited (see Section 10–2), since the lattice sites available in the solid require certain sizes and also certain bonding characteristics. Consequently, the solute particles present in the liquid solution do not, in general, dissolve as well in the solvent when it is a solid. As the liquid solvent molecules begin to freeze and form a solid, the solute particles "get in the way." The solid network of the solvent is not able to build up readily, since the solute particles present cannot easily be incorporated into the solid.

Then what must happen? The temperature of the solution must be lowered further so that the liquid molecules slow down, becoming less energetic. When the liquid solvent molecules are moving slowly enough, many of them can simultaneously form bonds necessary to establish the solid network and, in effect, trap the solute particles within the solid. In order to accomplish this, however, the temperature of the solution has to be lowered below the normal freezing point of the pure solvent, i.e., the freezing point has to be lowered by the addition of a solute.

Using this same model, we can also begin to understand the effect of the concentration of the solute on the freezing point. As the number of solute particles in the liquid solution increases, it becomes harder and harder for the solvent molecules to form a solid network, since there are more solute particles to interfere with the regular bonding and geometric characteristics of the solid. Therefore, for a solid to form, the temperature of the liquid solution must be lowered further and further. The higher the concentration of the solute, the lower the freezing point.

We can use a similar picture to consider boiling-point elevation. To make the model as simple as possible, let us consider only the evaporation of the liquid that

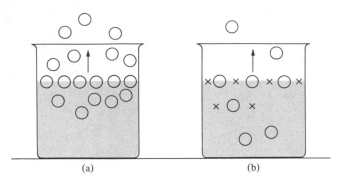

(a)   (b)

Fig. 10-6 Evaporation of water molecules from (a) pure water and (b) an aqueous solution. The molecules of the solute in (b) are indicated by crosses. More H_2O molecules escape from the surface of the pure water.

occurs at the surface of the liquid. In a pure solvent, such as water in Fig. 10-6a, the liquid molecules at the surface can escape into the gas phase if they have enough energy to overcome the intermolecular forces in the liquid. In a solution (Fig. 10-6b), the same reasoning applies, but now the number of solvent molecules present at the liquid surface has been reduced. Some solute molecules are on the surface too. Since the total number of solvent molecules at the surface is reduced, the number of molecules with sufficient energy to evaporate is also reduced. Or we can say that the number of vapor molecules formed is directly proportional to the number of liquid molecules present at the surface.

▶ Note that the solute particles in the liquid must be assumed to remain there and not evaporate as readily as the solvent molecules. ◀

What does this decrease in vapor molecules do? It lowers the vapor pressure of the solution as compared with the vapor pressure of the pure solvent. Thus at any given temperature the vapor pressure of the solution is lower than the vapor pressure of the pure solvent. In order for there to be equal vapor pressures for solution and solvent, the temperature of the solution must be higher than the temperature of the solvent. Hence, for the vapor pressure of a solution to be equal to atmospheric pressure, and boiling to occur, the solution must be at a higher temperature than the pure solvent.

The fact that boiling points of solutions are increased when concentrations of solutes are increased follows from the same model. More solute at the surface of the liquid means fewer molecules of the solvent at the surface, fewer vapor molecules produced, and a lower vapor pressure. Thus a higher temperature is needed to make the vapor pressure equal the atmospheric pressure; the boiling point is increased.

▶ The molecules of the solute on the surface of the liquid can be imagined to function as a lid over the solution. As the lid covers more and more of the surface, fewer and fewer of the molecules of the solution can escape. ◀

Some Uses. Colligative properties can be used in several ways. The following examples show that with minimal experimental procedures and sophistication— such as weighing samples and determining temperatures—we can use studies of freezing-point lowering and boiling-point elevation to calculate concentration of solutes, molecular weights of solutes, and the amount of dissociation of the solute particles.

Example 10–5. How can we determine the freezing-point-lowering constant, K_f, of water if we have 34.2 g of sugar ($C_{12}H_{22}O_{11}$) and 100 g of H_2O?

We want to determine K_f for water. What exactly is K_f? It is the number of degrees Celsius that the freezing point of 1000 grams of water is lowered when one mole of solute particles is dissolved in the water. Thus, to determine K_f, we need to add a known number of moles of solute to a known number of grams of water and determine the freezing point of the solution by, for example, running a heating or cooling curve on the solution.

Then we can dissolve 34.2 g or 0.10 mole of sugar, since

$$\text{molecular weight of } C_{12}H_{22}O_{11} = 342.30 \text{ g/mole,}$$

$$\frac{34.2 \text{ g } C_{12}H_{22}O_{11}}{342.30 \text{ g } C_{12}H_{22}O_{11}/\text{mole}} = 0.10 \text{ mole } C_{12}H_{22}O_{11}.$$

We dissolve this 34.2 g of sugar in 100 g H_2O and determine the freezing point of the solution. This procedure will give K_f directly, since 0.10 mole sugar/100 g H_2O is the same ratio as 1.0 mole sugar/1000 g H_2O. When this experiment is conducted carefully, the freezing point of the solution is found to be -1.86 °C. Then K_f for water is 1.86 °C/mole solute/1000 g H_2O.

Note that an assumption was made about the sugar molecules that are dissolved in the water. Each mole of solid sugar molecules produces one mole of dissolved sugar molecules, i.e., the molecules do not break apart or dissociate when dissolved. This happens to be a valid assumption for sugar (experimental justification is given in Section 11–2), but for many other solutes such as NaCl, NaOH, CaCl$_2$, etc., it is not true.

Example 10–6. Suppose that we dissolve 11.7 g NaCl in 200 g H_2O and find that the freezing point of the solution is -3.72 °C. What can we infer about the type of solute particles in the solution?

The question refers to the solution process when solid NaCl is dissolved in H_2O. What happens? Are NaCl solute molecules present in the solution, or does NaCl dissociate—that is, break into its constituent parts—to give Na^+ or Cl^-? There is a difference in the freezing points of the solution in these two cases. If NaCl does form ions when dissolved,

$$NaCl_{(s)} \xrightarrow{H_2O} Na^+_{(aq)} + Cl^-_{(aq)},$$

then the solution will contain two moles of solute particles (one mole of Na^+ and one mole of Cl^-) for each mole of solid NaCl dissolved.

Using the freezing point data given, we can verify the concentration of solute species actually present in the solution. Since one mole of solute lowers the freezing point of 1000 grams of water 1.86 °C, one mole of NaCl dissolved in 1000 g H_2O (or any other ratio of moles of NaCl to grams of H_2O equivalent to 1/1000) would yield a freezing point of -1.86 °C if only NaCl molecules were present. If only Na^+ and Cl^- ions were present, the freezing points would be lowered twice

as much,

$$2(-1.86 \,^\circ C) = -3.72 \,^\circ C,$$

since two moles of solute particles are present in the solution, one mole Na^+ ions and one mole Cl^- ions.

There are 11.7 g NaCl,

$$\frac{11.7 \text{ g NaCl}}{58.5 \text{ g NaCl/mole NaCl}} = 0.20 \text{ mole NaCl},$$

or 0.20 mole NaCl dissolved in 200 g H_2O, and

$$\frac{0.20 \text{ mole NaCl}}{200 \text{ g } H_2O} = \frac{1.0 \text{ mole NaCl}}{1000 \text{ g } H_2O}.$$

These two ratios are equal, indicating that the solute concentrations are the same in both cases. Therefore both solutions should have identical freezing points. The experimentally determined freezing point of the 200 g solution is $-3.72 \,^\circ C$. But if only one mole of solute is present in 1000 g H_2O, the freezing point of the solution must be $-1.86 \,^\circ C$. Thus NaCl must yield Na^+ and Cl^- ions when it dissolves, since the freezing point of the solution is twice -1.86, or $-3.72 \,^\circ C$.

Note how the use of units can simplify this calculation. The number of degrees we expect the freezing point to be lowered is given by:

$$(0.20 \text{ mole NaCl}/200 \text{ g } H_2O)(-1.86 \,^\circ C/\text{mole NaCl}/1000 \text{ g } H_2O) = \frac{(0.20)(-1.86 \,^\circ C)(1000)}{200}$$

$$= -1.86 \,^\circ C.$$

But the observed freezing point is $-3.72 \,^\circ C$:

$$\frac{-3.72 \,^\circ C}{-1.86 \,^\circ C} = 2.$$

There are 2 times as many particles present in the solution as we assumed in the first calculation. Therefore the original solution really contains 0.40 mole solute particles in 200 g H_2O. This indicates that NaCl dissociates completely into Na^+ and Cl^- ions when it dissolves.

▶ Note that, for the sake of simplicity, we have idealized the situation in Example 10-6 somewhat. Actually NaCl is not completely dissociated into Na^+ and Cl^- ions in aqueous solution. The total concentration of particles in a NaCl solution therefore is not quite 2 times the number of moles of NaCl added except in very dilute solutions. For example, the freezing point of the above solution (0.20 mole NaCl dissolved in 200 g H_2O) is about $-3.45 \,^\circ C$ rather than $-3.72 \,^\circ C$ as given, which indicates that a small amount of the NaCl is undissociated. ◀

Example 10-7. Suppose that we make an aqueous solution by dissolving 85.6 g sucrose in 500 g H_2O. We find that the freezing point of the solution is $-0.93 \,^\circ C$. From other experimental data we know that sucrose does not dissociate when it dissolves.* What is the molecular weight of the sucrose?

---

* This is an approximation. Sucrose does dissociate in aqueous solution to a very small extent.

To calculate the molecular weight, we need to know the number of grams of sucrose and the equivalent number of moles of sucrose present. Then simple division will yield the number of grams/mole. The number of grams is given as 85.6 g sucrose. We can calculate the number of moles from the freezing-point data:

$$\frac{-0.93\ °C}{-1.86\ °C/\text{mole sucrose}/1000\ g\ H_2O} = 0.50\ \text{mole sucrose}/1000\ g\ H_2O.$$

But we used only 500 g H_2O. Therefore

$$(0.50\ \text{mole sucrose}/1000\ \cancel{g\ H_2O})\ (500\ \cancel{g\ H_2O}) = 0.25\ \text{mole sucrose present.}$$

Thus the molecular weight of sucrose is

$$\frac{85.6\ g\ \text{sucrose}}{0.25\ \text{mole sucrose}} = 342\ g/\text{mole.}$$

Compare this molecular weight with that of sugar, as determined in Example 10–5. What is the chemical formula of sucrose?

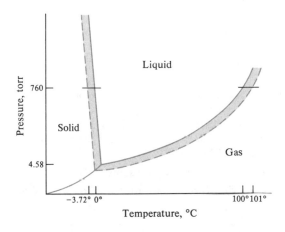

Fig. 10-7 A phase diagram for water. Solid lines indicate pure water and dashed lines a solution containing two moles of solute in 1000 grams of water. The diagram is exaggerated for the sake of clarity.

Effect on the Phase Diagram. The phase diagram for water that has already been considered is changed when solutes are added to the water. In fact, the phase diagram is a convenient way of showing the effects of freezing-point lowering and boiling-point elevation that are present in aqueous solutions. Phase diagrams for solutions are determined experimentally in the way previously described for the determination of the phase diagram of pure water.

Figure 10–7 gives a phase diagram for an aqueous solution containing two moles of solute per 1000 g H_2O (recall that $K_f = -1.86\ °C$ and $K_b = 0.512\ °C$ for H_2O). Then at 760 torr the normal freezing point of this solution is

$$2(-1.86\ °C) = -3.72\ °C$$

and the normal boiling point is

$$100\ °C + 2(0.512\ °C) = 100 + 1.024\ °C = 101\ °C.$$

At other pressures similar data yield freezing-point and boiling-point lines on the phase diagram as shown in Fig. 10–7. Note that the liquid range of the solution is greater than the liquid range of the pure water solvent.

The amount of shift in these equilibrium lines on the phase diagram is directly related to the concentration of the solute. The greater the concentration, the greater the shift in the lines and the greater the liquid range of the solution. As already mentioned, some solutes dissolve as molecules (sugar, for example), while other solutes, such as NaCl, dissociate into two or more parts which are generally charged ions, such as Na^+ and Cl^-. There are other differences between these two types of solutions. The solutions that contain ions can conduct an electric current, while those in which only molecules are present do not conduct electricity. We shall consider some of the properties of—and differences between—these types of solutions in the next chapter.

QUESTIONS

1. Define or discuss the following terms: Heterogeneous mixture, homogeneous mixture, solution, solute, solvent, colloid, solvation energy, solute–solute forces, solvent–solvent forces, solute–solvent forces, solid solutions, aqueous solutions, freezing-point lowering, colligative properties, boiling-point elevation, molarity, normality, "like dissolves like," solubility, saturated solution, endothermic, equilibrium, LeChatelier's principle, concentration, moles/ℓ, gram-equivalent, $0.012N$ NaCl, freezing-point-lowering constant, $K_b = 0.512\ °C/mole/1000\ g\ H_2O$, dissociation, $0.052M\ CaCl_2$.

2. What is the difference between a heterogeneous and a homogeneous mixture?

3. Predict in which of the two solvents, H_2O or $CHCl_3$ (chloroform), the following solutes would be most soluble:

   a) CH_4 b) CO_2
   c) $Ca(OH)_2$ d) YCl_3
   e) I_2 f) CCl_4
   g) Na_2CO_3 h) $FeCl_3$
   i) HCl j) C_6H_6

4. Chlorine gas does not dissolve well in H_2O, even though the intermolecular forces between the Cl_2 molecules are very small. Explain this observation.

5. When H_2O is saturated with Br_2 and then carbon tetrachloride is added to the solution, the CCl_4 forms a separate liquid layer which becomes yellow-brown in color when the solution is shaken. The H_2O layer simultaneously becomes lighter in color, and finally colorless. Explain.

6. Why might we expect CaF_2 and SrF_2 to form solid solutions with one another?

7. The solubility of $MgCl_2$ in H_2O is 54 g/100 cc at 20°C. What type of solution is formed when we add 6.0 moles of $MgCl_2$ to 1000 cc H_2O?

8. The solubility of solid Li_2CO_3 changes from 1.33 g/100 g H_2O at 20 °C to 1.08 g/100 g H_2O at 50 °C. Discuss the changes in the solution process with varying temperature, using LeChatelier's principle.

9. Considering the appropriate solute and solvent forces, why does Br_2 dissolve in CCl_4?

10. Predict whether the solvation energy will be high or low when the following substances are mixed with H_2O.

   a) $SO_{2(g)}$ b) $N_{2(g)}$
   c) $KF_{(s)}$ d) $YCl_{3(s)}$
   e) $SiH_{4(g)}$ f) $Cl_{2(g)}$
   g) $Ar_{(g)}$ h) $NaOH_{(s)}$
   i) $Ba(NO_3)_{2(s)}$ j) $HCl_{(g)}$

11. How can a saturated solution of $ZnCl_2$ be prepared?

12. The solubility of ammonium chlorate, NH_4ClO_3, changes from 29 to 115 g/100 cc H_2O at 0° and 75 °C, respectively. Describe what would happen if you made a $10M$ NH_4ClO_3 solution at 75 °C and cooled the solution slowly to 0°C.

13. Dissolving of KNO_3 is an endothermic process. Explain why more solid KNO_3 dissolves as the temperature of the solution is increased.

14. Given the following solution reactions:

$$KBr_{(s)} \xrightarrow{H_2O} K^+_{(aq)} + Br^-_{(aq)},$$
$$CaCl_{2(s)} \xrightarrow{H_2O} Ca^{2+}_{(aq)} + 2Cl^-_{(aq)},$$
$$Na_3PO_{4(s)} \xrightarrow{H_2O} 3Na^+_{(aq)} + PO^{3-}_{4(aq)}.$$

   a) For each mole of solid dissolved, which of the three solids given would be most effective in lowering the freezing point of H_2O? Why?
   b) If 5.0 grams of each solid were dissolved in 100 cc of H_2O, which of the three solutions would have the lowest freezing point and which the highest freezing point?

15. Consider the dissolving of carbon tetrabromide (CBr_4) (melting point = 90 °C) in chloroform ($CHCl_3$) and indicate what is meant by: solute–solute forces, solvent solvent forces, solute–solvent forces, solvation energy.

16. An entire sample of solid KCl is dissolved in H_2O, giving a cooled, unsaturated solution. Explain how solid KCl could be recovered from the solution by: (a) cooling the solution further, or (b) heating the solution.

17. The equilibrium process

$$2Na_{(s)} + Cl_{2(g)} \rightleftarrows 2NaCl_{(s)}$$

liberates heat (called the *heat of formation*). Using LeChatelier's principle, describe experimental conditions that would favor the production of $NaCl_{(s)}$.

18. Use LeChatelier's principle to predict the effect of a decrease in temperature on the following reactions:

   a) $Cu^{2+}_{(aq)} + SO^{2-}_{4(aq)} \rightleftarrows CuSO_{4(s)} + heat$
   b) $KNO_{3(s)} + heat \rightleftarrows K^+_{(aq)} + NO^-_{3(aq)}$
   c) $NaCl_{(s)} + heat \rightleftarrows Na^+_{(g)} + Cl^-_{(g)}$
   d) $Ca_{(s)} + Cl_{2(g)} \rightleftarrows CaCl_{2(s)} + heat$
   e) $Ca^{2+}_{(g)} + 2F^-_{(g)} \rightleftarrows CaF_{2(s)} + heat$

19. Calculate the freezing point of 100 g H_2O when 0.20 mole sugar is added to it.

20. When you dissolve 18.5 g KCl in 250 g H_2O, what is the freezing point of the solution?

21. Suppose that 0.30 mole $LaCl_3$ is dissolved in 600 g H_2O and the freezing point of the solution is found to be -3.72 °C. What is indicated about the nature of the solute particles?

22. Calculate the molarity of a solution containing 0.50 mole $Cu(NO_3)_2$ in 1.5 ℓ of solution.

23. Describe the experimental procedures necessary to prepare 300 ml of a $0.33M$ $BaCl_2$ solution.

24. How many moles of solid Na_2SO_4 could be recovered from 250 ml of $0.012M$ Na_2SO_4?

25. Suppose that 10.1 g KNO_3 are dissolved in enough water to give 500 ml. What is the concentration of the solution?

26. A solution contains 0.25 gram-equivalent of $KMnO_4$ per liter of solution (for reduction of MnO_4^- to Mn^{2+}). What is the normality of 12 ml of this solution?

27. For a reaction in which Cu^{2+} is reduced to Cu metal, one gram-equivalent of $CuSO_4$ weighs 80 g. How many grams of $CuSO_4$ are necessary to prepare 2.0 ℓ of $0.15N$ $CuSO_4$?

28. For a reaction in which Cu^{2+} is reduced to Cu^+, how many gram-equivalents of $Cu(NO_3)_2$ are needed to prepare 0.200 ℓ of $0.015N$ $Cu(NO_3)_2$?

29. A solution of $FeCl_3$ is prepared by dissolving 0.050 mole in enough H_2O to give 0.150 ℓ of solution. Suppose that this solution is $0.33N$ $FeCl_3$. (a) How many gram-equivalents of $FeCl_3$ are contained in every mole of $FeCl_3$? (b) What is the gram-equivalent weight of $FeCl_3$ in this case?

30. Twenty grams of an unknown solute are dissolved in 500 g H_2O. The freezing point of the solution is determined to be -7.44 °C. (a) How many moles of solute particles are present in the solution? (b) What would be the boiling point of the solution? (c) What additional information do you need to calculate the molecular weight of the solute?

31. The freezing point of 100 g of H_2O containing 5.50 g of a Group IA chloride is -2.79 °C. What is the chemical formula of the solid solute?

32. The boiling point of 100 g of H_2O containing 20.0 g of an undissociated solute is 101.0 °C. What is the molecular weight of the solute?

33. Explain why an increase in solute concentration should lower the freezing point of a solution.

34. a) Suppose that 1.83 g of $CdCl_2$ are dissolved in H_2O to give 200 ml of solution. What is the molarity of the solution?
    b) Suppose that each mole of $CdCl_2$ in (a) contains two gram-equivalents. What is the normality of the solution?

35. A $0.10N$ potassium dichromate ($K_2Cr_2O_7$) solution is being used in a particular oxidation-reduction reaction.
    a) What is the normality of 10 ml of the $K_2Cr_2O_7$ solution?
    b) How many gram-equivalents of $K_2Cr_2O_7$ are contained in 10 ml of the solution?
    c) How many gram-equivalents of $K_2Cr_2O_7$ would be needed to prepare 1.5 ℓ of the solution?
    d) There are six gram-equivalents of $K_2Cr_2O_7$ in each mole of $K_2Cr_2O_7$ in this reaction. What is the molarity of the solution?

e) Considering (d), what is the gram-equivalent weight of $K_2Cr_2O_7$?

f) How many grams of $K_2Cr_2O_7$ would you need to prepare 250 ml of $0.10N$ $K_2Cr_2O_7$?

36. Draw a labeled phase diagram for an aqueous solution containing 0.40 mole CsBr dissolved in 100 g H_2O.

11. SOLUTIONS: ELECTROLYTES

In many respects the properties of solutions depend on what happens to a substance when it dissolves. We have seen that some solutes dissociate—or break apart—in solution, while others do not. This influences, for example, the colligative property of freezing-point lowering. For every mole dissolved, a solute that dissociates yields a lower freezing point than one that does not. One mole of NaCl dissolved in 1000 grams of water produces a solution with a lower freezing point than a solution with one mole of sugar in 1000 grams of water. This effect depends only on the number of solute particles present and not on their type. Hence NaCl produces two moles of particles for each mole of NaCl dissolved, while sugar, being undissociated in solution, yields only one mole of solute particles per mole of sugar dissolved.

There are other properties of solutions, however, that depend not only on the number of particles of the solute but also on the nature of the particles. One such property is the electrical conductivity of a solution. Experimentally it is found that some solutions conduct an electric current while others do not. Observations such as these enable us to group solutions into certain categories and to determine the extent to which solutes dissociate in a given solution.

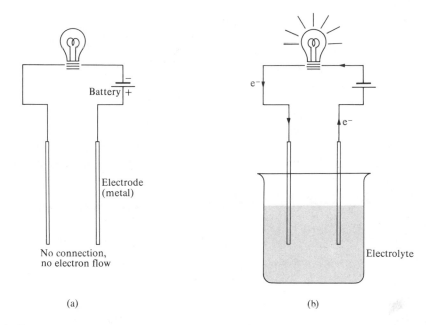

Fig. 11–1 Experimental apparatus to separate solutions into electrolytes and nonelectrolytes. (a) No current flows through the light bulb. (b) The light bulb glows, indicating a current flow.

11–1 ELECTROLYTES AND NONELECTROLYTES

Experimental Differences. A solution that conducts electricity is called an *electrolyte*; a solution that does not is a *nonelectrolyte*. The difference between these two types of solutions can be determined experimentally, using reasonably simple apparatus.

Figure 11–1 shows a possible experimental setup for doing this. Two electrically conducting wires or strips—called *electrodes*, and often made of metal—are connected to a light bulb and a battery. The battery can be envisioned as a source of electrons. When electrons flow through the wire in the light bulb, the wire heats up and begins to glow. But electrons can pass through the bulb only when they have a completed pathway both out of and back into the battery. If the metal electrodes are unconnected, as in part (a), there is no way for electrons to move from one electrode to the other. Hence there is no complete circuit, no electron flow, and no light from the bulb.

The electrodes can be dipped into a solution contained in a beaker. With some solutions the light glows; the electrical circuit is completed. In some way the solution makes possible the transfer of charge between the two electrodes. In Fig. 11–1b, the electrons enter the left electrode and flow from the right electrode. Solutions that make possible this flow of electric current are *electrolytes*.

What can we learn about the type of solute particles present in electrolytes and nonelectrolytes? One approach we might take is to test many different types of solutions, sort them into two groups—electrolytes and nonelectrolytes—and then note the types of solutes that produced electrolytes.

Table 11–1 gives a few examples of aqueous solutions. We see that pure water is a nonelectrolyte, as are solutions containing solutes that are nonpolar. But polar solutes such as HCl and ionic solutes such as NaCl produce electrolytes. From these results we might hypothesize that the nonpolar solutes dissolve in water as nonpolar, uncharged molecules that are unable to transport charge between the metal electrodes in the electrolysis cell, whereas the polar and ionic solutes dissolve to give charged particles—such as Na^+ and Cl^-—which then move through the solution and transfer charge from one electrode to the other.

Table 11–1. Aqueous solutions: electrolytes and non-electrolytes

Solution	Relative solute polarity	Solution type
H_2O, pure	—	Nonelectrolyte
NaCl	Ionic	Electrolyte
KNO_3	Ionic	Electrolyte
$CaCl_2$	Ionic	Electrolyte
HCl	Polar	Electrolyte
CO_2	Nonpolar	Nonelectrolyte
Cl_2	Nonpolar	Nonelectrolyte
CH_4	Nonpolar	Nonelectrolyte
C_6H_6	Nonpolar	Nonelectrolyte

Electrolysis. This concept of charged particles, or ions, existing in electrolytes is appealing, since charge must in some way be transferred between the metal electrodes. Another type of experiment can be conducted that will yield more information about the solute particles present in solution. We could use the experimental apparatus shown in Fig. 11–2, with both electrodes being copper wire and the electrolyte a solution of $1M$ H_2SO_4. When the copper electrodes are placed in the solution, the light bulb glows, indicating that $1M$ H_2SO_4 is an electrolyte.

But other observations can also be made. Changes take place in the solution, particularly at each electrode. A reaction is immediately evident at the copper wire connected via the light bulb to the negative side of the battery (indicated by a minus sign in Fig. 11–2); gas bubbles are rapidly given off. A simultaneous reaction occurs at the other electrode, which is connected to the plus battery terminal. However, current must pass through the solution for about 30 minutes before we can easily detect this change. After this period of electrolysis, we can make two observations. The solution is no longer clear and colorless, but has turned light blue, and the

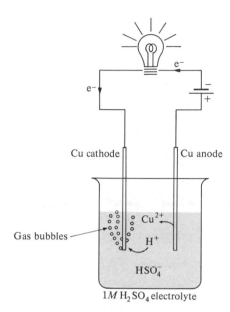

Fig. 11–2 An electrolysis cell.

copper electrode is slightly smaller. The production of chemical change such as these by the passage of electric current is called *electrolysis*.

What causes these changes at the electrodes? We can, of course, make some educated guesses, but there are further experiments that will give our postulates and arguments a firmer foundation.

▶ Without reading ahead, try out a few of your own ideas. Write down the type of gas that might be given off. Is there more than one possibility? How do you think the gas is produced at the electrode? It does not spontaneously bubble out; there is some mechanism involved. What could cause the blue color of the solution? Remember that the copper electrode is getting smaller. In considering these problems, keep in mind that electrons must necessarily be entering the electrode at which gas is produced and leaving the electrode that is getting smaller, since the solution is completing the electrical circuit and making it possible for electrons to flow both out of and into the battery. These apparent electron movements are reminiscent of the oxidation-reduction reactions that we have considered. ◀

If we knew the composition of the gas given off and the origin of the blue solution, it would be easier to decide what processes were occurring at the electrodes. We could collect the gas and then submit it to various identification tests. We might, for example, note any physical characteristics of the gas—such as color or odor—and then test other properties, such as its readiness to burn or to support combustion, its solubility, its chemical reactivity with other substances, and so on. In this case these relatively simple experiments would show that the gas given off was hydrogen.

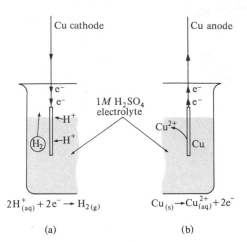

Fig. 11–3 The processes that occur at each electrode in the electrolysis cell in Fig. 11–2. (a) Hydrogen gas bubbles are given off at the cathode and (b) copper(II) ion is produced at the anode.

$$2H^+_{(aq)} + 2e^- \rightarrow H_{2(g)}$$

(a)

$$Cu_{(s)} \rightarrow Cu^{2+}_{(aq)} + 2e^-$$

(b)

▶ The tests would show that the gas is colorless and odorless, lighter than air, burns, is not soluble in water, and can form an explosive mixture with oxygen when ignited. All these properties are typical of hydrogen (H_2) gas. ◀

Now that we know that the gas is hydrogen, what possibilities are there to explain its production at the metal electrode? In some way the hydrogen must be formed from the particles present in the solution. We know that hydrogen is produced more easily from a $1M$ H_2SO_4 solution than from pure water, since water is a non-electrolyte. Then the H_2SO_4 solute must be important. In fact, if H_2SO_4 is dissociated in aqueous solution to give hydrogen ions (H^+), then we can postulate a simple mechanism for the production of H_2.

Figure 11–3a shows the H^+ ions diffusing toward the electrode into which the electrons (negative charge) are flowing. The H^+ ions pick up electrons from the electrode, and form H_2 molecules. The dissociation of H_2SO_4 and the subsequent "reaction" of H^+ ions with electrons at the copper electrode can be written as chemical equations:

$$H_2SO_{4(aq)} \rightarrow H^+_{(aq)} + HSO^-_{4(aq)}$$

and

$$2H^+_{(aq)} + 2e^- \rightarrow H_{2(g)}.$$

Similar tests are needed to determine the processes occurring at the second copper electrode (attached to the plus battery terminal). In this case electrons are leaving the electrode, and the copper wire is being consumed by the electrolysis. What can happen to the copper metal? One obvious idea is that it might dissolve in the $1M$ H_2SO_4 solution. Experimentally we can determine, however, that unless current is passing through the solution, the copper wire does not dissolve. Then we must devise a process that will dissolve the copper electrode (which will place copper

in the electrolyte) and simultaneously generate electrons which can pass out of the electrode and into the positive battery terminal. We can envision such a process, as shown in Fig. 11–3b, if we assume that copper metal (Cu) is oxidized to Cu^{2+}:

$$Cu_{(s)} \rightarrow Cu^{2+}_{(aq)} + 2e^-.$$

This equation describes a process that fulfills our criteria: The Cu electrode gets smaller and electrons are produced. The presence of Cu^{2+} ions in the solution must also agree with experimental observations. This can be tested in several ways. (Can you think of some experiments that could be used?) One easy test is the color of the solution. We can make known solutions of $CuSO_4$ and observe their colors. Any and all these experimental tests would show that Cu^{2+} is present in the solution after electrolysis (it was not present prior to electrolysis).

Then it appears that two processes are occurring simultaneously in the electrolyte, at the electrodes. The H^+ ions are being reduced to H_2 gas at the electrode that is receiving electrons from the battery,

$$2H^+_{(aq)} + 2e^- \rightarrow H_{2(g)}, \qquad \text{reduction.}$$

We call the electrode at which reduction occurs the *cathode*. The other electrode is being oxidized and furnishes electrons to the battery,

$$Cu_{(s)} \rightarrow Cu^{2+}_{(aq)} + 2e^-, \qquad \text{oxidation.}$$

We call the electrode at which oxidation occurs the *anode*.

The reactions that occur at each electrode are generally called *half-reactions*. They always contain electrons as reactants or products. The reduction half-reaction is

$$2H^+_{(aq)} + 2e^- \rightarrow H_{2(g)}.$$

The oxidation half-reaction is

$$Cu_{(s)} \rightarrow Cu^{2+}_{(aq)} + 2e^-.$$

The sum of these two half-reactions gives the total oxidation-reduction reaction that takes place. (Remember, oxidation cannot occur without a simultaneous reduction, and vice versa.)

$$
\begin{array}{ll}
2H^+_{(aq)} + 2e^- \rightarrow H_{2(g)} & \text{Reduction} \\
\underline{\quad Cu_{(s)} \rightarrow Cu^{2+}_{(aq)} + 2e^-} & \text{Oxidation} \\
2H^+_{(aq)} + Cu_{(s)} \rightarrow H_{2(g)} + Cu^{2+}_{(aq)} &
\end{array}
$$

Note that the electrons involved cancel; as many are used as are produced.

Dissociation. Because of experimental evidence such as the above, it is believed that solutes in an electrolyte are dissociated, while solutes in a nonelectrolyte remain associated as molecules. When solutes dissociate (this process is also called *ionization*), they form charged particles called *ions*. *Anions* are negatively charged, and *cations* are positively charged. Note that, in the above example, cations (H^+) are removed

from the solution at the cathode. This loss of positive charge by the solution is balanced by the reaction occurring at the anode; Cu^{2+} ions are generated. This is necessarily true in all electrolysis reactions.

The *number* of positive and negative charges in the entire solution is always equal. This charge equality can be maintained in several different ways. In some electrolysis experiments, simultaneous anion and cation loss maintains charge balance, while in other experiments anion production can counteract cation formation. The type of charge balance involved depends on the particular oxidation-reduction reaction.

11–2 STRONG AND WEAK ELECTROLYTES

Experimental Differentiation. When you test a group of solutions for electrical conductivity,. you find that some of the solutions that are electrolytes are good conductors, while other electrolytes are poor conductors (but still much better conductors than nonelectrolytes). You can make these tests using the apparatus in Figs. 11–1 and 11–2.

To differentiate between the solutions that are good electrical conductors—*strong electrolytes*—and solutions that are poor conductors—*weak electrolytes*—you need to observe the brightness of the light bulb. A bright light indicates a strong electrolyte and a dim light a weak electrolyte. A 1.0M NaCl solution, for example, is a strong electrolyte and would cause a bright glow, whereas a 1.0M NH$_4$OH solution would give a dim light, since it is a weak electrolyte. Table 11–2 lists other strong and weak electrolytes.

Table 11–2. Aqueous solutions: strong and weak electrolytes

Solutions that are strong electrolytes	Solutions that are weak electrolytes
NaCl	NH$_4$OH
CaCl$_2$	HC$_2$H$_3$O$_2$
K$_2$SO$_4$	HNO$_2$
HCl	H$_2$S
HNO$_3$	HF
H$_2$SO$_4$	
NaOH	

▶ A more exact method of determining the differences between strong and weak electrolytes would involve replacing the light bulb with a current-measuring meter, an ammeter, which could give a numerical ordering of the various electrolytes. ◀

Percent Dissociation. What could cause this difference in electrical conductivity between strong and weak electrolytes? We have seen that the conduction property of solutions is due to the presence of ions. Then a good assumption might be: The larger the number of ions in a solution, the better the conductivity of the solution. This is an easy hypothesis to test. We could prepare a series of solutions with varying concentrations of, for example, NaCl. The solution containing the highest concentration of Na^+ and Cl^- ions would be the best conductor.

The difference in conduction between strong and weak electrolytes is caused by the number of ions present in the solution. In weak electrolytes there are some—but not nearly as many—ions in solution as in strong electrolytes. Since these ions are formed by the dissociation of the solute, solutes that produce strong electrolytes must dissociate to a greater extent than solutes that give weak electrolytes.

The degree to which solutes dissociate in solution is often expressed as *percent dissociation.*

With strong electrolytes we often assume that most of the solute has broken apart into ions; that is, the solute is almost 100% dissociated. A weak electrolyte, on the other hand, has a relatively low percent dissociation. In a 1.0M NaCl solution there is close to 100% dissociation. Most of the sodium chloride exists as Na^+ and Cl^- ions; very few NaCl units, as such, exist in the solution. However, in 1.0M NH_4OH, there is only 0.4% dissociation. Most of the solute is present as NH_4OH molecules, with only a few NH_4^+ and OH^- ions present in the solution:

$$NH_4OH \xrightleftharpoons{\text{Few \%}} NH_4^+ + OH^-.$$

▶ The actual species present in solution are probably not NH_4OH molecules, but NH_3 molecules closely surrounded by (or attached to) one or more H_2O molecules. Since the true solution species is in doubt and since it provides a more consistent approach to describing reactions in which it is involved, we shall use the formula NH_4OH to represent ammonium hydroxide (instead of representations such as $NH_3 \cdot H_2O$ or $NH_3 + H_2O$). See also Section 13–14. ◀

Since, for every mole of solute added, NaCl produces many more ions than NH_4OH, solutions of NaCl are strong electrolytes, while solutions of NH_4OH are weak electrolytes.

Variations of Percent Dissociation. The degree to which a solute dissociates (or ionizes) in a solvent depends on the nature of the solute, the nature of the solvent, and the concentration of the solute.* These variations in percent dissociation are not surprising when they are considered in the light of our previous discussions on solubility. Again the various forces—solute–solute, solvent–solvent, and solute–solvent—are very important and need to be considered.

---

* Percent dissociation also depends on the temperature, but this behavior is complex and shows no general pattern.

We can illustrate the dependence of percent dissociation on the type of solute particle, in a given solvent such as water, by considering the compounds formed between hydrogen and the Group VIIA elements,

$$HX_{(aq)} \rightleftarrows H^+_{(aq)} + X^+_{(aq)}.$$

For aqueous solutions of HCl, HBr, and HI, this dissociation is essentially complete, i.e., there is about 100% dissociation:

$$HCl_{(aq)} \xrightarrow{\sim 100\%} H^+_{(aq)} + Cl^-_{(aq)},$$
$$HBr_{(aq)} \xrightarrow{\sim 100\%} H^+_{(aq)} + Br^-_{(aq)},$$
$$HI_{(aq)} \xrightarrow{\sim 100\%} H^+_{(aq)} + I^-_{(aq)}.$$

But HF is only about 1% dissociated in aqueous solution:

$$HF_{(aq)} \xrightarrow{1\%} H^+_{(aq)} + F^-_{(aq)}.$$

Consequently, in solutions of HCl, HBr, and HI there are very few neutral molecules, while in solutions of HF there are few H^+ and F^- ions and many neutral HF molecules (99% of the HF exists as molecules).

This low percent dissociation for HF, in relation to the other three compounds, is caused by the strong solute bonding forces within a single HF molecule compared with the solute–solvent interactions. We would expect a higher covalent (polar) bond energy in HF than in the other hydrogen halides, since fluorine is the most electronegative of the atoms, making HF more polar than HCl, HBr, or HI. The increased opposite charge attraction within the HF molecule strengthens the covalent bond. Consequently, HF is harder to break apart and its percent dissociation is low.

We can understand variations in the amount of dissociation in different solvents by means of similar reasoning. Consider HCl dissolved in water and in benzene (C_6H_6) as shown in Fig. 11–4, for example. Experimentally we find that an aqueous HCl solution is a strong electrolyte, while a solution of HCl in C_6H_6 is a weak electrolyte. Evidently HCl is largely dissociated in H_2O,

$$HCl_{(aq)} \xrightarrow{\sim 100\%} H^+_{(aq)} + Cl^-_{(aq)}$$

but not in C_6H_6,

$$HCl \xrightarrow{\sim 1\%} H^+ + Cl^-.$$

Considering the appropriate forces involved, we expect this result. The polar HCl molecules can interact strongly with the polar H_2O molecules and the solute–solvent forces are strong, but the interaction of HCl with nonpolar C_6H_6 molecules is quite weak and the solute–solvent forces are weak. Since it is the solute–solvent forces that are responsible for the dissociation of the solute particles, HCl does ionize in H_2O but not in C_6H_6.

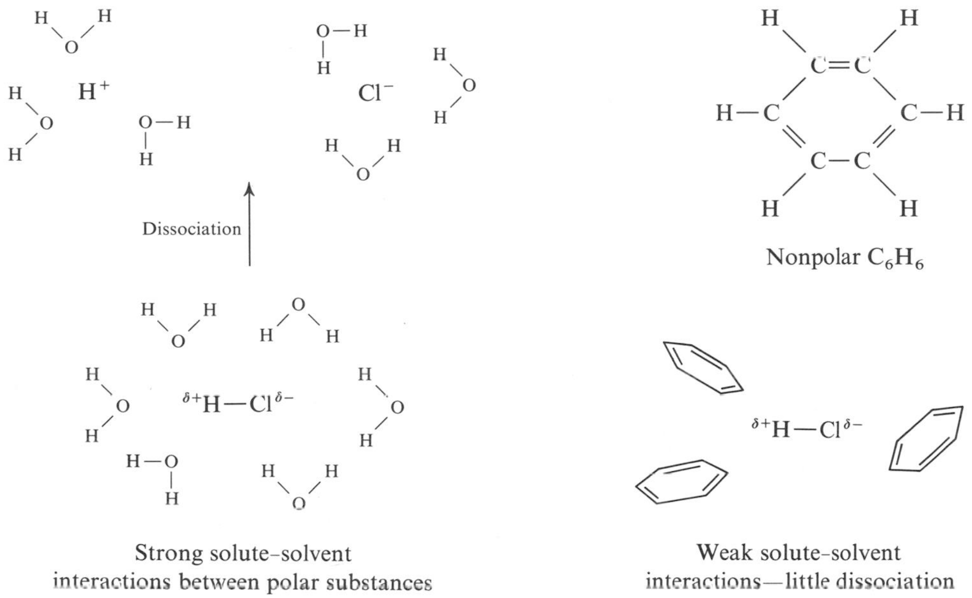

Fig. 11–4 Variations in the dissociation of HCl dissolved in H$_2$O and in benzene (C$_6$H$_6$).

The dependence of percent dissociation on concentration also seems reasonable. What effect would we expect with increasing concentration of the solute? To answer this question we can consider two limiting cases, one in which the solution is very dilute—that is, of a very low concentration—and one in which the solution is highly concentrated. For the dilute case, one molecule of solute in a large volume of solvent —say one liter—is illustrative. When this very small solute molecule dissociates into its constituent ions, there is a vast volume of solution in which they can move. The volume of this solution is so large compared with the sizes of the ions that the probability of the two ions coming in contact with one another again is extremely low. Therefore, in this infinitely dilute solution of one solute molecule per liter of solution, there is 100% dissociation. The one molecule breaks apart and remains dissociated.

The situation is much different in a highly concentrated solution. When many molecules are present and dissociating, there are always anions and cations (not necessarily from the same molecule) physically close to one another. This close proximity increases the chances of the ions associating to form a molecule. As the number of ions increases, the probability of association increases. Hence the percent dissociation of the solute molecules decreases with increasing concentration of the solution.

Then we would predict that the percent dissociation of a solute would be high at low concentrations (and 100% in infinitely dilute solutions) and would decrease as

Table 11-3. Concentration and percent dissociation

Concentration of acetic acid	Dissociation of acetic acid, %
$1M$ $HC_2H_3O_2$	0.4
$0.1M$ $HC_2H_3O_2$	1.3
$0.01M$ $HC_2H_3O_2$	4.3
$0.001M$ $HC_2H_3O_2$	15
$0.0001M$ $HC_2H_3O_2$	75

the concentration of the solution increased. Experimentally we find that this does indeed happen. As shown in Table 11-3, the percent dissociation of acetic acid increases as its concentration decreases.

Comparison of Electrolytes. Since the electrical conductivity of a solution becomes higher as the number of ions it contains increases, and since the percent dissociation of a solute depends on its concentration, any comparisons between electrolytes must be made with the solutions at the same concentration. A one-molar solution is generally used to test for and compare electrolyte behavior.

11-3 GALVANIC CELLS

In the electrolysis cells we have considered, electrical energy was used to bring about a chemical change. A battery was attached to electrodes, and a chemical reaction occurred at the electrodes and within the electrolyte.

But chemical reactions can also be used to *produce* an electric current. In fact, the batteries used in electrolysis cells operate in just this way. Two half-reactions again occur at two separate electrodes, but, instead of requiring an external source of electric energy to bring about the reactions, the half-reactions proceed spontaneously and can function as a source of electrons. Cells that use oxidation-reduction reactions to produce electric current are called *galvanic* (or *voltaic*) *cells.* All common batteries—such as dry cells used in flashlights and lead storage batteries used in automobiles—are galvanic cells, but the oxidation-reduction reactions that occur in these cells are somewhat complex and, in some cases, not completely understood. A simpler and more illustrative galvanic cell is the Daniell cell.

The Daniell Cell. This galvanic cell is made by placing a zinc metal strip or bar in a zinc sulfate ($ZnSO_4$) electrolyte and a copper metal strip or bar in a copper sulfate ($CuSO_4$) electrolyte (both $ZnSO_4$ and $CuSO_4$ are strong electrolytes and essentially 100% dissociated in aqueous solution). The two sulfate solutions are contained in the same beaker, but are separated from one another by, say, a porous partition that makes possible diffusion of ions from one side to the other but that prevents complete mixing of the solutions.

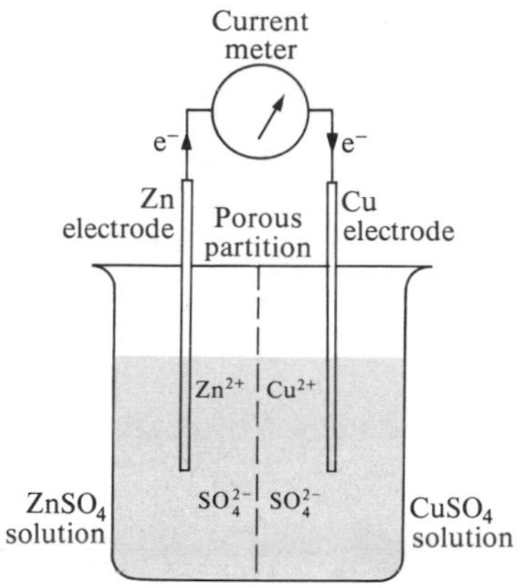

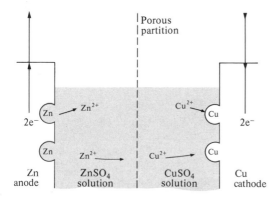

Fig. 11–5 The Daniell cell is a galvanic cell that produces an electric current spontaneously; that is, it serves
as a battery.

Fig. 11–6 The electrode reactions in the
Daniell cell. Zinc(II) ions are formed at the
anode and diffuse toward the porous partition and into the cathode compartment.
Copper(II) ions diffuse toward the copper
cathode, where they react to form more
copper metal.

▶ Actually the densities of the $ZnSO_4$ and $CuSO_4$ solutions are sufficiently different
so that the cell can be prepared without the porous partition. The $ZnSO_4$ is carefully
poured on top of the $CuSO_4$ solution. The Zn electrode is entirely within the $ZnSO_4$
solution and the Cu electrode beneath it within the $CuSO_4$ solution. ◀

Figure 11–5 is a diagram of this experimental arrangement. When the two
electrodes are connected by wires through an ammeter, the meter deflection indicates
that a current is flowing in the circuit and that electrons are emerging from the Zn
electrode and passing into the Cu electrode.

The Daniell cell, like any other galvanic cell, is a battery. The production of
electrons comes from the spontaneous oxidation-reduction reaction that takes place
when the two electrodes are connected. We can determine the two half-reactions
involved by observing the cell as it operates. After the Daniell cell has been used as
the source of electric current for some time, changes in it are readily detected. The
Zn electrode, for example, becomes small, while copper metal actually plates onto
the Cu electrode, increasing its size (Fig. 11–6).

The electrolytes also change. The number of Cu^{2+} ions in the $CuSO_4$ solution decreases in the same proportion that the Cu electrode increases in size. Since Cu^{2+} ions are being changed to Cu metal, the $CuSO_4$ electrolyte is losing positive charge. This is balanced by the diffusion of Zn^{2+} ions through the porous partition into the $CuSO_4$ solution compartment. Extra Zn^{2+} ions are available, since they are being produced at the Zn electrode. Thus two processes are happening simultaneously: The Zn electrode is being oxidized, producing Zn^{2+} ions and electrons,

$$Zn_{(s)} \rightarrow Zn^{2+}_{(aq)} + 2e^-,$$

and Cu^{2+} ions are being reduced, using electrons and producing Cu metal:

$$Cu^{2+}_{(aq)} + 2e^- \rightarrow Cu_{(s)}.$$

Since oxidation is occurring at the Zn electrode, it is the anode. The Cu electrode is the cathode, where reduction is taking place. The total Daniell cell reaction is the sum of the two half-reactions:

$$\begin{array}{c} Zn_{(s)} \rightarrow Zn^{2+}_{(aq)} + 2e^- \\ \underline{Cu^{2+}_{(aq)} + 2e^- \rightarrow Cu_{(s)}} \\ Cu^{2+}_{(aq)} + Zn_{(s)} \rightleftarrows Cu_{(s)} + Zn^{2+}_{(aq)} \end{array}.$$

This reaction occurs spontaneously; that is, Zn does reduce Cu^{2+}, and is therefore a better reducing agent than Cu.

▶ The voltage of the Daniell cell—that is, the strength or potential of the cell to produce electrons or an electric current—depends on the concentration of the Cu^{2+} and Zn^{2+} ions, and on the temperature. For one-molar $CuSO_4$ and $ZnSO_4$ solutions at 25 °C, the Daniell cell produces 1.1 volts. This is a slightly lower voltage than that of an ordinary flashlight battery. ◀

There are always some restrictions on the operation of galvanic cells. In the case of the Daniell cell, the anode must be zinc metal, the electrolyte surrounding the cathode must contain Cu^{2+}, and the two electrolyte solutions cannot be mixed. These first two requirements are absolute, since Cu^{2+} and Zn are the necessary oxidizing and reducing agents in the above chemical reaction. The third requirement —no mixing—exists only if a sustained current is needed from the cell. If Cu^{2+} ions are in the vicinity of the Zn electrode, they can pick up the electrons generated by the anode reaction and thereby plate Cu metal onto the Zn anode. This rapidly short-circuits the cell—i.e., makes it inoperative—since the Zn metal is completely covered by Cu metal and therefore can no longer enter into the reaction.

Note that a $ZnSO_4$ electrolyte is not a necessity. The solution surrounding the Zn anode must be a strong electrolyte and must not react chemically with the Zn electrode (or Zn^{2+}). Given these conditions, the anode half-reaction can occur and the cell will operate. A similar situation exists for the Cu cathode; it need not be copper. However, it does have to be an electrical conductor, cannot react with Cu

(or Cu^{2+}), and must accept copper metal plating. Given these conditions for the cathode, its half-reaction will occur.

In principle, any two oxidation-reduction half-reactions can be used as a galvanic cell.* In practice, experimental problems of some sort interfere with many of the possible choices and limit the selection of batteries to only a few that are common and practical.

11-4 OXIDATION AND REDUCTION IN SOLUTION

In the above examples of electrolysis and galvanic cells, we have been considering oxidation-reduction reactions that occur in solutions. We have seen that Zn metal and Cu^{2+} ions react spontaneously with one another to produce Zn^{2+} ions and Cu metal; and oxidation and reduction occur simultaneously. There are many other chemical reactions that involve oxidation-reduction in solutions.

▶ As we have learned from studying electrolysis cells, some oxidation-reduction reactions do not proceed spontaneously. They require the addition of energy—in this case electrical energy—to make them occur. Whether two substances will undergo a spontaneous oxidation-reduction reaction when mixed together can, of course, be determined experimentally. But it would be convenient if, in some way, we could predict that a particular reaction would or would not occur, without having to resort to an experiment each time. See if you can devise a method of comparing various reactants that will answer the question, "Will they react spontaneously?" Remember that the establishment of this kind of relative scale generally requires a standard or reference. One method that has been developed involves half-reactions and their electrode potentials, as described in Section 11-5. ◀

Sodium metal, for example, reacts very rapidly (often violently) with water, producing hydrogen gas and a solution containing sodium hydroxide (NaOH):

$$2Na_{(s)} + 2H_2O \rightarrow 2Na^+_{(aq)} + 2OH^-_{(aq)} + H_{2(g)}.$$

An aqueous solution of NaOH is a strong electrolyte. Another example illustrates the strong oxidizing ability of chlorine gas. When Cl_2 is bubbled through a solution of NaI, the element I_2 is liberated:

$$Cl_{2(g)} + 2I^-_{(aq)} \rightarrow 2Cl^-_{(aq)} + I_{2(s)}.$$

In this case Cl_2 gas is reduced (Cl_2 is the oxidizing agent) and I^- ion is oxidized (I^- is the reducing agent). Solutions of both NaI and NaCl are strong electrolytes.

Net Equations. We have written the above equations describing the reactions between Na and H_2O and between Cl_2 and a NaI solution as net equations, that is,

---

* We shall say more about selecting the oxidizing and reducing agents and the cell voltages to be expected in Section 11-5.

equations that contain only those chemical entities that are used or formed. All species that are present but are *not* changed during the reaction are left out of the equation; they are often called *spectator ions*.

We try, with net equations, to describe the situation as it exists in the solution in as realistic a fashion as possible. For example, we know from experiment that a sodium hydroxide solution is a strong electrolyte. That means that NaOH is essentially 100% dissociated in aqueous solution. When a sodium hydroxide solution is formed by some chemical reaction in water, such as sodium metal reacting with water, we write $Na^+ + OH^-$ rather than NaOH, since the two separate ions give a description that is closer to what we believe actually exists in the solution.

This is also the reason for not writing the formula units NaI and NaCl when Cl_2 reacts with a solution of NaI, producing I_2 and a NaCl solution. The molecular equation does not represent what we believe to be the conditions existing in the solution. Since solutions of NaI and NaCl are strong electrolytes, they are composed primarily of ions: $Na^+_{(aq)} + I^-_{(aq)}$ and $Na^+_{(aq)} + Cl^-_{(aq)}$. Then we can write the equation in ionic form:

$$Cl_{2(g)} + 2Na^+_{(aq)} + 2I^-_{(aq)} \rightarrow 2Na^+_{(aq)} + 2Cl^-_{(aq)} + I_{2(s)}.$$

This is the *total ionic equation* for the reaction, since it contains all the ions that exist in the solution. But note that the Na^+ ion appears as a reactant and also, unchanged, as a product. It is a *spectator ion*; it appears to sit and watch the reaction occur and not participate in it.

Just as in an equation involving numbers, we can subtract equal quantities from each side of a chemical equation. Hence we can remove $2Na^+$ spectator ions from both the reactant and product side of the equation, which yields the net equation

$$Cl_{2(g)} + 2I^-_{(aq)} \rightarrow 2Cl^-_{(aq)} + I_{2(s)}.$$

A *net equation* shows only those reactants that are used up and products that are formed during the reaction.

▶ Note that Cl_2 and I_2 are written as molecules. They do not dissociate in water solutions, but dissolve as molecules. How could we determine that this is the case experimentally? What can you say about the solubility of Cl_2 and I_2 in water? Why is their solubility so different from that of NaI and NaCl? ◀

Equations and Their Use. Oxidation-reduction equations that describe reactions occurring in solution are handled in exactly the same manner as described in Section 7-4. There is, however, one additional feature. The concentrations of the species in solution can be important.

If aluminum metal is placed in a solution containing Cu^{2+} ions, a reaction occurs, producing copper metal and Al^{3+} ions:

$$3Cu^{2+}_{(aq)} + 2Al_{(s)} \rightarrow 3Cu_{(s)} + 2Al^{3+}_{(aq)}.$$

Note that the balanced equation indicates that three moles of Cu^{2+} react with two moles of Al to produce three moles of Cu and two moles of Al^{3+}. The amount of

Al metal (the number of grams or moles of Al) that will react depends on the number of Cu^{2+} ions available. And their reacting ratio must be 2 moles Al to 3 moles Cu^{2+}. If 0.20 mole Al is added, 0.30 mole Cu^{2+} ion will react; if 0.60 mole Cu^{2+} ion is in the solution, 0.40 mole Al metal is needed for complete reaction. Remember that you can calculate the number of moles of Cu^{2+} present in a solution if you know the concentration of Cu^{2+} in moles per liter (the molarity M of Cu^{2+}) and the volume of the solution in liters.

If, for example, you know that a beaker contains 100 ml of $0.030M$ Cu^{2+} solution, you can calculate the number of grams of Al to exactly react with all the Cu^{2+}. You can find the number of moles of Cu^{2+}, since you know the concentration and the volume of the solution:

$$(\text{Concentration, moles}/\ell)(\text{Volume, } \ell) = \text{moles,}$$
$$(0.030 \text{ mole } Cu^{2+}/\ell)(0.100 \, \ell) = 3.0 \times 10^{-3} \text{ mole } Cu^{2+}.$$

From the chemical equation you know that 3.0×10^{-3} mole Cu^{2+} will react exactly with 2.0×10^{-3} mole Al metal. You can calculate the number of grams of Al equivalent to 2.0×10^{-3} mole of Al, since there are 27.0 g Al in one mole of Al:

$$(\text{mole})(\text{g/mole}) = \text{g,}$$
$$(2.0 \times 10^{-3} \text{ mole Al})(27.0 \text{ g Al/mole Al}) = 0.054 \text{ g Al.}$$

Then you must add 0.054 g Al to the beaker to react completely with all the Cu^{2+} ions in the solution.

▶ The anion in the solution is a spectator ion. It could be Cl^-, Br^-, SO_4^{2-}, etc., and must not react with Al. You could prepare the $0.030M$ Cu^{2+} solution, for example, by dissolving 0.0030 mole $CuCl_2$ in enough water to give 100 ml. What would the concentration of Cl^- be in the solution? ◀

You can determine the concentrations of solutions in the way we described in Section 10–5. In the above $0.030M$ Cu^{2+} solution, 0.0030 mole Cu^{2+} was initially present. After the reaction with 0.0020 mole Al, the concentration of Cu^{2+} is essentially zero, while the concentration of Al^{3+} has increased from zero to 0.0020 mole $Al^{3+}/0.10 \, \ell$ or $0.020M$ Al^{3+}.

We can also express these concentrations in terms of normality, the number of gram-equivalents per liter of solution. The Cu^{2+} concentration in the solution is $0.030M$. This is equivalent to $0.060N$ Cu^{2+}. We can make this conversion by determining the electron change in the reaction:

$$\begin{array}{cc} \text{Oxidation} & 3Cu^{2+} + 2Al \rightarrow 3Cu + 2Al^{3+} \\ \text{number} & +2 \qquad\qquad\qquad 0 \end{array}$$
$$\uparrow$$
$$2e^-/Cu^{2+} \text{ formula unit}$$

Since each Cu^{2+} ion can gain two electrons, each mole of Cu^{2+} contains two gram-equivalents. Therefore the normality of the Cu^{2+} solution is two times as large as the molarity.

Similar reasoning indicates that each mole of Al metal contains three gram-equivalents:

$$\text{Oxidation number} \qquad \begin{array}{c} 3Cu^{2+} + 2Al \rightarrow 3Cu + 2Al^{3+} \\ 0 \qquad\qquad\qquad +3 \\ \downarrow \\ 3e^-/\text{Al formula unit} \end{array}$$

If 0.0020 mole Al are needed for the reaction, then 0.0060 gram-equivalents of Al are taken. This is necessarily equal to the number of gram-equivalents of Cu^{2+} used, if complete reaction is to occur. In 0.10 ℓ of 0.060N Cu^{2+}, there are exactly 0.0060 gram-equivalents of Cu^{2+}:

$$(0.060 \text{ gram-equivalent } Cu^{2+}/\ell)(0.10 \,\ell) = 0.0060 \text{ g-eq } Cu^{2+}.$$

The following examples further illustrate the use of oxidation-reduction reactions in solution and the use and calculation of concentrations of solutions.

Example 11–1. Calculate the molarity of a solution of $SnCl_2$, tin(II) chloride, given that 0.019 g $SnCl_2$ is dissolved in enough water to yield 100 ml.

We need to know the number of moles of $SnCl_2$ per liter of solution. In 0.019 g $SnCl_2$ there is

$$\frac{0.019 \text{ g } SnCl_2}{190 \text{ g } SnCl_2/\text{mole } SnCl_2} = 1.0 \times 10^{-4} \text{ mole } SnCl_2.$$

Since 0.00010 mole $SnCl_2$ is dissolved in 0.100 ℓ, the molarity is:

$$\frac{0.00010 \text{ mole } SnCl_2}{0.10 \,\ell} = 0.0010 \text{ mole } SnCl_2/\ell.$$

The solution is 0.0010M $SnCl_2$.

Example 11–2. What is the normality of the $SnCl_2$ solution in Example 11–1, given that the solution is used for the reaction

$$Sn^{2+} + 2Fe^{3+} \rightarrow Sn^{4+} + 2Fe^{2+}?$$

The solution contains 0.0010 mole $SnCl_2/\ell$. We must calculate the normality or the number of gram-equivalents of $SnCl_2/\ell$. All we have to do to convert M to N is to find the number of gram-equivalents contained in each mole. We do this by determining the electron change involved in the reaction,

$$\text{Oxidation number} \qquad \begin{array}{c} Sn^{2+} \quad + \quad 2Fe^{3+} \quad \rightarrow \quad Sn^{4+} \quad + \quad 2Fe^{2+} \\ +2 \qquad\qquad\qquad\qquad +4 \\ \downarrow \\ 2e^-/Sn^{2+} \text{ formula unit} \end{array}$$

Since each mole of Sn^{2+} releases 2 moles of electrons, there are 2 gram-equivalents of Sn^{2+} in each mole of Sn^{2+}.* Therefore the normality of the $SnCl_2$ is twice as large as its molarity:

$$N\ SnCl_2 = 2(M\ SnCl_2) = 2(0.0010M\ SnCl_2) = 0.0020.$$

In this reaction a $0.0010M$ $SnCl_2$ solution is equivalent to a $0.0020N$ $SnCl_2$ solution. But what is the normality of a $0.0020M$ $FeCl_3$ solution in this reaction?

Example 11–3. A solution of potassium permanganate ($KMnO_4$) is made by dissolving 0.050 mole $KMnO_4$ in enough water to give 250 ml of solution. What is the molarity and normality of the $KMnO_4$ solution?

We can calculate the $KMnO_4$ molarity immediately:

$$\frac{0.050\ \text{mole}\ KMnO_4}{0.250\ \ell} = 0.20\ \text{mole}\ KMnO_4/\ell.$$

The solution is $0.20M$ $KMnO_4$.

But we cannot calculate the normality of the $KMnO_4$ unless we know the specific chemical reaction involved. Why is this true? Let us consider two different reactions of $KMnO_4$:

(1) $MnO_{4(aq)}^- + 4H_{(aq)}^+ + 3Ce_{(aq)}^{3+} \rightarrow MnO_{2(s)} + 3Ce_{(aq)}^{4+} + 2H_2O,$

(2) $2MnO_{4(aq)}^- + 16H_{(aq)}^+ + 10Cl_{(aq)}^- \rightarrow 2Mn_{(aq)}^{2+} + 5Cl_{2(g)} + 8H_2O.$

Both reactions are written as net ionic equations. The K^+ is a spectator ion and $KMnO_4$ is essentially 100% dissociated in solution. Note that in equation (1), MnO_4^- is converted to MnO_2, while in equation (2), MnO_4^- is converted to Mn^{2+}. To calculate the normality of the $KMnO_4$ solution, knowing that it is $0.20M$ $KMnO_4$, we need to know how many gram-equivalents of $KMnO_4$ are in one mole of $KMnO_4$. But this depends on the reaction. In equation (1),

$$
\begin{array}{ccc}
 & MnO_4^- & \rightarrow & MnO_2 \\
\text{Oxidation} & +7 & & +4 \\
\text{number} & & \uparrow \\
\end{array}
$$
$$3e^-/MnO_4^-\ \text{formula unit}$$

there are 3 gram-equivalents of MnO_4^- in one mole of MnO_4^-. Hence the normality of the $KMnO_4$ solution, with respect to this particular reaction, is

(3 gram-equivalents $KMnO_4$/~~mole~~ ~~$KMnO_4$~~) $(0.20$ ~~mole~~ ~~$KMnO_4$~~$/\ell) = 0.60N$ $KMnO_4$.

For equation (2), the situation is different:

$$
\begin{array}{ccc}
 & MnO_4^- & \rightarrow & Mn^{2+} \\
\text{Oxidation} & +7 & & +2 \\
\text{number} & & \uparrow \\
\end{array}
$$
$$5e^-/MnO_4^-\ \text{formula unit}$$

---

*   Recall the definition of a gram-equivalent: *the weight or amount of material that will accept or release one mole of electrons.* If these conversions are still unclear, refer back to Section 7–7.

In this reaction 5 gram-equivalents of MnO_4^- are present in each mole of MnO_4^-. Therefore in the chemical reaction described by equation (2), the normality of the $0.20M$ $KMnO_4$ solution is

$$(5 \text{ g-eq } KMnO_4/\text{mole } KMnO_4) \, (0.20 \text{ mole } KMnO_4/\ell) = 1.0N \, KMnO_4.$$

Thus a solution that is labeled $0.20M$ $KMnO_4$ can actually be $0.60N$ $KMnO_4$ or $1.0N$ $KMnO_4$ (and possibly other normalities), depending on the chemical reaction for which it is used. To prevent ambiguities in designating concentrations of oxidizing and reducing agents, one should label their solutions in terms of molarity and not normality.

▶ There are many cases in which the normality of a solution of a particular oxidizing or reducing agent does not change from reaction to reaction. (Can you think of a reason for this and list some possible examples?) But, as with $KMnO_4$, there are many cases in which normalities do depend on the reaction. Hence M labels are preferred. ◀

Example 11–4. According to the reaction,

$$Zn_{(s)} + Ag_{(aq)}^+ \rightarrow Zn_{(aq)}^{2+} + Ag_{(s)},$$

how many liters of $0.050M$ $AgC_2H_3O_2$ (silver acetate) are needed to exactly react with 0.020 mole of Zn?

Remember that chemical equations express reacting ratios in terms of moles.

▶ Since the net ionic equation does not contain the acetate ion, $C_2H_3O_2^-$, we have assumed that it is a spectator ion and that $AgC_2H_3O_2$ and $Zn(C_2H_3O_2)_2$ are essentially 100% dissociated in solution. ◀

If we are to find and use these specific combining ratios, the equation must be balanced. Inspection of the above equation shows that it is not balanced; the total charge on each side of the equation is not equal. Thus we must first balance the equation:

$$Zn_{(s)} + 2Ag_{(aq)}^+ \rightarrow Zn_{(aq)}^{2+} + 2Ag_{(s)}.$$

Now the equation shows that

1 mole Zn reacts with 2 moles Ag^+

and

(0.020)1 mole Zn reacts with (0.020)2 moles Ag^+

or

0.020 mole Zn reacts with 0.040 mole Ag^+.

Then 0.040 mole Ag^+ is needed to exactly react with 0.020 mole Zn.

We must therefore take enough of the $0.050M$ $AgC_2H_3O_2$ solution to provide the necessary 0.040 mole of Ag^+. Note that if one liter of the solution was used, 0.050 mole Ag^+ would be available. This is an excess of Ag^+, and therefore we know that a volume of solution less than one liter is needed:

$$0.050 \text{ mole } Ag^+/\ell \times 1.0\ell = 0.050 \text{ mole } Ag^+.$$

▶ You might recognize that just four-fifths of one liter of the $AgC_2H_3O_2$ solution is needed. These "insight" calculations are often rather easy to do when simple numbers are involved. However, when you get into more complicated decimals or fractions, as usually happens in actual circumstances, these relationships are not as easily recognized. This is one reason for developing a procedural solution, as we have done in Example 11–4. Whenever you run into complicated numbers, it helps to round them off to the nearest simple number, to make possible an easy calculation giving an approximate answer. This insight calculation will check both the answer obtained and the computational method used with the true numbers. For example, in Example 11–4, we could have started with $0.049M$ $AgC_2H_3O_2$ and 0.017 mole Zn. Then 0.034 mole Ag^+ is necessary for reaction, and the simple relationship between the number of moles needed and the number present in one liter is no longer obvious. But if we simplify the numbers to 0.03 mole and $0.05M$, we can make a quick calculation, indicating that about three-fifths of a liter, or about 600 ml of the $AgC_2H_3O_2$, will be needed. This shows the correct calculation method (0.034 ~~mole~~ ~~Ag^+~~$/0.049$ ~~mole~~ ~~Ag^+~~$/\ell$), and gives a check on the answer, $0.034/0.049 = 0.69 \ell$ = 690 ml. ◀

If we used one-half liter, then we would select only $\frac{1}{2}(0.050$ mole $Ag^+)$ or 0.025 mole Ag^+ :

$$0.050 \text{ mole } Ag^+/\ell \times 0.50 \ell = 0.025 \text{ mole } Ag^+.$$

If we used three-quarters of a liter, then we would take

$$0.050 \text{ mole } Ag^+/\ell \times 0.75 \ell = 0.038 \text{ mole } Ag^+.$$

And so on. This reasoning process shows the mathematical calculation

$$\text{Molarity} \times \text{volume in liters} = \text{moles,}$$
$$\text{moles}/\ell \times \ell = \text{moles,}$$

or

$$M \times V = \text{moles.}$$

If we know any two of these three quantities, we can calculate the other. In this case we need 0.040 mole and we have a $0.050M$ solution. Substituting these values into the equation gives

$$M \times V = \text{moles,}$$
$$0.050 \text{ mole } Ag^+/\ell \times V = 0.040 \text{ mole } Ag^+.$$

Solving for the volume of the solution yields

$$V = \frac{0.040 \text{ ~~mole~~ ~~Ag^+~~}}{0.050 \text{ ~~mole~~ ~~Ag^+~~}/\ell} = 0.80 \ell.$$

Thus we need 0.80ℓ or 800 ml of $0.050M$ $AgC_2H_3O_2$ to react with 0.020 mole Zn.

Example 11–5. How many gram-equivalents of $KMnO_4$ are needed to exactly react with 2.28 grams of $FeSO_4$, iron(II) sulfate, in solution according to the reaction

$$Fe^{2+} + H^+ + MnO_4^- \rightarrow Fe^{3+} + Mn^{2+} + H_2O?$$

Note that the equation is not balanced. Since we are working with gram-equivalents, we can use the equation in its unbalanced form. (The net equation as written also implies the species that

exist in the solution and that the potassium, K^+, and sulfate ions, SO_4^{2-}, are spectator ions.) Since

> 1 g-eq oxidizing agent reacts exactly with 1 g-eq reducing agent,

the number of gram-equivalents of MnO_4^- and Fe^{2+} (and consequently of $KMnO_4$ and $FeSO_4$) must be equal. Then if we knew the number of gram-equivalents of $FeSO_4$ in 2.28 g $FeSO_4$, we would require the same number of gram-equivalents of $KMnO_4$.

We can calculate the number of gram-equivalents of $FeSO_4$ contained in 2.28 g $FeSO_4$, considering the above reaction:

$$\text{Oxidation number} \qquad \begin{array}{ccc} Fe^{2+} & \rightarrow & Fe^{3+} \\ +2 & & +3 \\ & \downarrow & \end{array}$$

$$1e^-/Fe^{2+} \text{ formula unit}$$

Since each Fe^{2+} loses 1 electron, and therefore 1 mole Fe^{2+} loses 1 mole electrons, 1 mole Fe^{2+} is equal to 1 g-eq Fe^{2+}. Hence there is 1 g-eq $FeSO_4$ in 1 mole $FeSO_4$ and the gram-equivalent weight (the weight in grams of one gram-equivalent) is equal to the molecular weight,

$$(152 \text{ g } FeSO_4/\text{mole } FeSO_4)\,(1 \text{ mole } FeSO_4/1 \text{ g-eq } FeSO_4)$$
$$= 152 \text{ g } FeSO_4/\text{g-eq } FeSO_4 = 1 \text{ g-eq weight } FeSO_4.$$

Only 2.28 g $FeSO_4$ are available, which is only a fraction of one-gram-equivalent:

$$\frac{2.28 \text{ g } FeSO_4}{152 \text{ g } FeSO_4/\text{g-eq } FeSO_4} = 0.015 \text{ g-eq } FeSO_4.$$

Thus 2.28 g $FeSO_4 = 0.015$ g-eq $FeSO_4$.

The same number of gram-equivalents of $KMnO_4$—0.015 gram-equivalent $KMnO_4$—is required for the reaction.

Example 11–6. In Example 11–5, what would be the normality of the $KMnO_4$ solution, given that 30.0 ml were required to exactly react with the 2.28 grams of $FeSO_4$?

We can use a procedure similar to that in Example 11–4. We need to calculate the normality of the $KMnO_4$ solution; this is the number of gram-equivalents of $KMnO_4$/liter of solution. There is 0.015 g-eq $KMnO_4$ needed to react with 2.28 g $FeSO_4 = 0.015$ g-eq $FeSO_4$. The 30 ml $= 0.030 \ \ell$ of $KMnO_4$ solution provide the necessary 0.015 g-eq $KMnO_4$; that is, there is

$$\frac{0.015 \text{ g-eq } KMnO_4}{0.030 \ \ell} = 0.50 \text{ g-eq } KMnO_4/\ell.$$

The concentration of the $KMnO_4$ solution is $0.50N$ $KMnO_4$.

Note that a relationship similar to

$$M \times V = \text{moles}$$

applies to normality also:

$$N \times V = \text{gram-equivalents}.$$

In this example, the relationship used is the form

$$N = \frac{\text{gram-equivalents}}{V} = \frac{0.015 \text{ g-eq KMnO}_4}{0.030 \; \ell} = 0.50 \text{ g-eq KMnO}_4/\ell.$$

These examples indicate the use of oxidation-reduction reactions in solution and the importance of the concentrations of the various species. There are two approaches to the problems: One uses the balanced equation, the number of moles of the substances and their molarities. The second involves unbalanced equations, gram-equivalents and the normalities of the species involved.

Both these methods have their advantages, depending on the problem to be solved. Example 11–4 could also be solved by converting to normality and to gram-equivalents of Zn, which would eliminate the need for the balanced equation. However, the balancing operation is so simple in this case that it is easier to work with the equation and mole quantities. Example 11–5, on the other hand, involves a more complicated equation and also many conversions if mole quantities are used. It is therefore more easily solved using gram-equivalents. In either approach, the relationships

$$M \times V = \text{moles} \quad \text{and} \quad N \times V = \text{gram-equivalents}$$

are very important; you should master their use.

▶ Try making the necessary conversions in Examples 11–4, 11–5, and 11–6 and working the problems using a different approach, i.e., gram-equivalents for 11–4 and moles for 11–5 and 11–6. ◀

11–5 HALF-REACTIONS AND ELECTRODE POTENTIALS

Will a Reaction Occur? This is a question that is constantly present in chemistry and one that has, in general, no immediate or obvious answer. We can, of course, resort to experimental testing each time. And, in fact, if no information or experience is available for the reaction in question, experimental testing is the only answer. But for the many chemical reactions that have already been run and tested, there should be some method of cataloging that will make it possible for scientists to determine whether a particular reaction will or will not occur simply by consulting a reference book.

Earlier in this chapter we learned that copper metal and sulfuric acid do not react spontaneously to produce copper(II) ion and hydrogen gas. But we can force the reaction to occur by using the proper electrolysis conditions:

$$Cu_{(s)} + 2H^+_{(aq)} \xrightarrow{\text{Electrolysis}} Cu^{2+}_{(aq)} + H_{2(g)}.$$

On the other hand, if zinc metal is placed in sulfuric acid, hydrogen gas is liberated spontaneously according to the equation

$$Zn_{(s)} + 2H^+_{(aq)} \rightarrow Zn^{2+}_{(aq)} + H_{2(g)}.$$

A similar reaction with sodium metal proceeds so rapidly that an explosion occurs:

$$2Na_{(s)} + 2H^+_{(aq)} \longrightarrow 2Na^+_{(aq)} + H_{2(g)}.$$

These three reactions are rather simple, but they do suggest that a general "reactivity" trend might exist among the metallic elements.

▶ See if you can develop other experiments that will allow some of the elements to be ranked according to their "reacting ability" or reactivity. A qualitative scale is extremely helpful, but quantitative numbers are always more impressive. If you devise clever experiments, they will permit quantitative ranking of the reactivities. ◀

However, most chemical reactions appear to be more complex than these. For example, we would like to be able to predict whether the reaction

$$Zn_{(s)} + Cu^{2+}_{(aq)} \rightarrow Zn^{2+}_{(aq)} + Cu_{(s)}$$

will occur spontaneously or whether the reverse reaction

$$Zn^{2+}_{(aq)} + Cu_{(s)} \rightarrow Zn_{(s)} + Cu^{2+}_{(aq)}$$

will proceed more readily. Will Br_2 react with Cl^-? Will MnO_4^- react with Fe^{2+}, and if so, what will be formed? And so on.

▶ Since the Daniell cell uses the reaction between Zn and Cu^{2+} to produce an electric current, we know that this reaction occurs spontaneously, and that therefore the reverse reaction between Zn^{2+} and Cu does not proceed to any large extent. Could you predict this result from the above observations on the reactions between Cu and H^+ and between Zn and H^+? ◀

Although the answers to these questions can depend on several things and can therefore be rather complicated, several methods have been developed to predict whether a particular chemical reaction will or will not occur.

Experimental Basis. The system for predicting the possibilities of oxidation-reduction reactions in aqueous solutions is particularly useful; it is based on the principles of simple galvanic cells. In general, the procedure is as follows: A standard electrode half-reaction is chosen so that all other half-reactions can be combined with it to form a galvanic cell. If we use similar experimental conditions in every case, we can rank the half-reactions, quantitatively, in relation to the standard half-reaction by noting the electrical potential (the voltage) that is produced by the reaction, and also by determining which electrode is the anode and which the cathode.

▶ We need to measure how powerful the battery is that is produced by the reaction. We can do this by using a voltage-measuring device which measures the potential ability of the battery to produce an electric current, i.e., electrical energy in the form of a flow of electrons. ◀

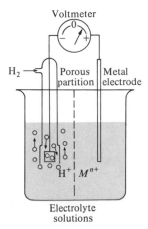

Voltmeter

H$_2$

Porous partition

Metal electrode

H$^+$ M^{n+}

Electrolyte solutions

Fig. 11–7 A galvanic cell formed between a M^{n+}/M electrode and a standard hydrogen electrode, H$^+$/H$_2$.

The procedures for setting up a galvanic cell and making the required measurements can be illustrated by means of two of the above reactions, which involve

$$Cu + 2H^+ \quad \text{and} \quad Cu^{2+} + H_2,$$
$$Zn + 2H^+ \quad \text{and} \quad Zn^{2+} + H_2.$$

From the discussion in Section 11–1 we know that the half-reactions occurring at the electrodes would be:

$$M^{2+} + 2e^- \rightleftarrows M \quad \text{and} \quad 2H^+ + 2e^- \rightleftarrows H_2,$$

where M is either Cu or Zn. Since one of the half-reactions is the same in both cases, we can choose it as the standard half-reaction and use it to compare the other two half-reactions:

$$Cu^{2+} + 2e^- \rightarrow Cu \quad \text{and} \quad Zn^{2+} + 2e^- \rightarrow Zn.$$

To ensure standard conditions in each galvanic cell, the standard half-reaction should occur at the same type of electrode in each cell. A standard hydrogen electrode usually consists of a small platinum strip over which H$_2$ gas at 1 atmosphere pressure is bubbled.

Figure 11–7 shows a simple galvanic cell that could be used for these measurements. The standard hydrogen electrode is on the left. The other metal electrode, either a Cu or a Zn metal strip, is on the right. In the cell containing the Cu electrode, the electrolyte on the left side of the porous partition is $1M$ H$^+$ ($1M$ H$_2$SO$_4$, for example) and the electrolyte on the right side of the partition is $1M$ Cu^{2+} ($1M$ CuSO$_4$, for example).

In the Zn electrolysis cell, the only difference is that $1M$ Zn^{2+} ($1M$ ZnSO$_4$, for example) is the electrolyte surrounding the Zn electrode. It is important that these electrolyte concentrations be the same in all measurements concerning these cells since, as we have seen, the percent dissociation of electrolytes—and hence the cell voltage—depends on their concentration.

With all the necessary standard experimental conditions fulfilled,* we can measure the voltage of the two cells. Experimentally we would find that the voltage produced by the Cu–H$_2$ cell is 0.34 volt, while that of the Zn–H$_2$ cell is 0.76 volt. This is one difference between the Cu^{2+}/Cu and the Zn^{2+}/Zn half-reactions.

There is also another very significant difference. In the Cu–H$_2$ cell the voltage measurement shows that the Cu metal electrode is positively charged with respect to the hydrogen electrode. In the Zn–H$_2$ cell the reverse is true; the Zn metal electrode is negatively charged when compared with the hydrogen electrode.

▶ We should expect this difference between the Cu^{2+}/Cu and Zn^{2+}/Zn half-reactions compared with the H$^+$/H$_2$ half-reaction, since we know that Zn metal reacts with H$^+$ to produce H$_2$ gas, whereas Cu metal does not. ◀

This difference in the signs of the electrodes enables us to rank, quantitatively, all three half-reactions with respect to one another. To do this we need only assign, arbitrarily, some numerical value to the H$^+$/H$_2$ half-reaction which will indicate its contribution to the total voltage of the cell. This can be any voltage that we please, since the H$^+$/H$_2$ half-reaction is a standard, with all other half-reactions referred to it.

A convenient choice for this standard value is zero volts. Then the measured voltage of any metal-H$_2$ cell is the voltage contribution of the half-reaction occurring at the metal electrode. And the relative charge on the electrodes determines the sign given to the voltage value of the metal half-reaction. For example, the measured voltage of the Cu–H$_2$ cell was 0.34 volt and the Cu electrode was positive compared with the hydrogen electrode. Since the standard half-reaction

$$2H^+ + 2e^- \rightarrow H_2$$

is assigned a zero voltage, we can assign the copper half-reaction

$$Cu^{2+} + 2e^- \rightarrow Cu$$

a voltage of $+0.34$ volt. The positive sign indicates that the Cu electrode is positively charged with respect to the H$_2$ electrode. We can treat the Zn–H$_2$ cell in the same fashion. The half-reaction

$$Zn^{2+} + 2e^- \rightarrow Zn$$

has a voltage value of -0.76 volt. Then we can rank the three half-reactions numerically in terms of their (relative) voltages:

$$
\begin{aligned}
Cu^{2+} + 2e^- &\rightarrow Cu & +0.34 \text{ volt,} \\
2H^+ + 2e^- &\rightarrow H_2 & 0.00 \text{ volt,} \\
Zn^{2+} + 2e^- &\rightarrow Zn & -0.76 \text{ volt.}
\end{aligned}
$$

---

*   These are: pressure of H$_2$ gas $= 1$ atm, $1M$ concentrations for H$^+$ and for all other species involved in the reaction, and a temperature of 25 °C. There are other more complex and sophisticated difficulties that often need to be considered. Further changes are usually required in the experimental arrangements, but no changes in fundamental principles are involved.

In the $Cu-H_2$ cell, for example, the half-reaction

$$Cu^{2+} + 2e^- \rightarrow Cu$$

occurs at the copper metal electrode and the half-reaction

$$H_2 \rightarrow 2H^+ + 2e^-$$

occurs at the hydrogen electrode. Then in solution Cu^{2+} ions migrate to the Cu electrode and pick up electrons from the electrode, thereby being reduced to give more Cu metal. At the hydrogen electrode, H^+ ions enter the solution and the electrons generated in the process are free to flow out of the platinum strip. There is thus an excess of electrons at the hydrogen electrode and an electron deficiency at the copper electrode. Hence the negatively charged hydrogen electrode and the positively charged copper electrode. In the $Zn-H_2$ cell the situation is the opposite. The

$$Zn \rightarrow Zn^{2+} + 2e^-$$

reaction occurs at the zinc electrode, producing electrons and charging the zinc electrode negatively.

Electrode Potentials. The voltages that are measured when two oxidation-reduction half-reactions are combined to give a galvanic cell are called *electrode potentials*. When these potentials are measured under the standard conditions ($1 M$ concentration for H^+ and all species involved in the reaction, 1 atm pressure for all gases, and at 25 °C), they are called *standard electrode potentials*.

▶ These are not the exact standard conditions, but are close enough for most purposes. Also there are several systems of nomenclature and another fundamentally different method of describing and using the potentials that we have called electrode potentials. All the systems produce identical answers for the sample problem, but confusion can result because of the different definition and handling of the signs of the potential values. ◀

These electrode-potential values provide a convenient method of predicting whether particular oxidation-reduction reactions will or will not occur in aqueous solution. They are all relative values, with the standard H^+/H_2 half-reaction taken as the zero reference point, i.e., assigned the value of zero volts for its standard electrode potential. Then the electrode potentials for other half-reactions fall above or below the H^+/H_2 value.

Table 11–4 gives the standard electrode potential values, E^0, at 25 °C for a number of half-reactions. Note that all the half-reactions are written as reductions. This is purely a convention to make possible rapid and easy comparisons. The half-reactions could all just as easily be written as oxidations; there would be no change in the value or the sign of the standard electrode potential.

Table 11–4. Standard electrode potentials, E^0, at 25 °C

Half-reaction	E^0, volts
$F_2 + 2e^- \rightarrow 2F^-$	2.8
$MnO_4^- + 4H^+ + 3e^- \rightarrow MnO_2 + 2H_2O$	1.69
$Ce^{4+} + e^- \rightarrow Ce^{3+}$	1.61
$MnO_4^- + 8H^+ + 5e^- \rightarrow Mn^{2+} + 4H_2O$	1.51
$Cl_2 + 2e^- \rightarrow 2Cl^-$	1.36
$Cr_2O_7^{2-} + 14H^+ + 6e^- \rightarrow 2Cr^{3+} + 7H_2O$	1.33
$MnO_2 + 4H^+ + 2e^- \rightarrow Mn^{2+} + 2H_2O$	1.23
$Br_2 + 2e^- \rightarrow 2Br^-$	1.06
$NO_3^- + 4H^+ + 3e^- \rightarrow NO + 2H_2O$	0.96
$Ag^+ + e^- \rightarrow Ag$	0.80
$Fe^{3+} + e^- \rightarrow Fe^{2+}$	0.77
$I_2 + 2e^- \rightarrow 2I^-$	0.54
$Cu^+ + e^- \rightarrow Cu$	0.52
$Cu^{2+} + 2e^- \rightarrow Cu$	0.34
$Cu^{2+} + e^- \rightarrow Cu^+$	0.15
$2H^+ + 2e^- \rightarrow H_2$	0.00
$Pb^{2+} + 2e^- \rightarrow Pb$	-0.13
$Cd^{2+} + 2e^- \rightarrow Cd$	-0.40
$Cr^{3+} + e^- \rightarrow Cr^{2+}$	-0.41
$Fe^{2+} + 2e^- \rightarrow Fe$	-0.44
$Zn^{2+} + 2e^- \rightarrow Zn$	-0.76
$Mn^{2+} + 2e^- \rightarrow Mn$	-1.18
$Al^{3+} + 3e^- \rightarrow Al$	-1.66
$Mg^{2+} + 2e^- \rightarrow Mg$	-2.37
$La^{3+} + 3e^- \rightarrow La$	-2.52
$Na^+ + e^- \rightarrow Na$	-2.71
$Ca^{2+} + 2e^- \rightarrow Ca$	-2.87
$K^+ + e^- \rightarrow K$	-2.93
$Li^+ + e^- \rightarrow Li$	-3.05

▶ As mentioned previously, there is another commonly used approach to oxidation-reduction potentials—one in which the sign of the potential is reversed when the half-reaction is reversed. Needless to say, one must be careful not to confuse these two fundamentally different definitions. ◀

Electrode Potentials. We can use the tabulated standard electrode potentials to predict whether a certain oxidation-reduction reaction will occur and also to calculate the numerical value of the standard electrode potential for the reaction—that is, the voltage of a galvanic cell—using the two half-reactions involved in the oxidation-reduction reaction.

As an example of how we can make this calculation and how we can predict a reaction, let us consider some of the above examples. The half-reactions involved in the Daniell cell and their standard electrode potential values are:

$$E^0, \text{volts}$$

$$\text{Cu}^{2+} + 2e^- \to \text{Cu} \qquad +0.34$$

$$\text{Zn} \to \text{Zn}^{2+} + 2e^- \qquad -0.76$$

Experimentally we know that this a galvanic cell and that therefore the total cell reaction

$$\text{Cu}^{2+} + \text{Zn} \to \text{Cu} + \text{Zn}^{2+}$$

proceeds spontaneously. The voltage of the cell can be measured; it is about 1.10 volts. But we can calculate this experimentally determined cell voltage from the known electrode potentials:

$$E^0_{\text{cell}} = E^0_{\text{Cu}} - E^0_{\text{Zn}} = (+0.34) - (-0.76)$$
$$= 0.34 + 0.76 - 1.10 - 1.10 \text{ volts.}$$

Then we can make the following rules for using standard electrode potentials.

1. We can calculate the standard electrode potential for the total cell reaction— that is, for the oxidation-reduction reaction in question—by subtracting the standard electrode potential for the oxidation half-reaction from the standard electrode potential value for the reduction half-reaction:

$$E^0_{\text{cell}} = E^0_{\text{red}} - E^0_{\text{ox}}.$$

2. If the standard electrode potential for the cell—i.e., the total reaction—is positive, the reaction proceeds spontaneously and the calculated voltage will be produced by a properly constructed galvanic cell. If the E^0_{cell} is negative, the reaction will not occur spontaneously and we cannot use the half-reactions for a galvanic cell.

$$E^0_{\text{cell}} + \qquad \text{Reaction yes,}$$
$$E^0_{\text{cell}} - \qquad \text{No reaction.}$$

Some examples should help to clarify the concept of electrode potentials and their use.

▶ If the E^0_{cell} turns out to be close to zero—that is, only very slightly positive or negative—this generalization is not very useful. For example, if $E^0_{\text{cell}} = 0.001$ volt, a spontaneous reaction between two reactants will occur, but the amount of product formed will be small, and consequently might be difficult to detect.

We should also mention that a positive electrode potential for a reaction indicates that the reaction will occur. But it does not tell us anything about the *speed* of the reaction. It is possible, therefore, that the electrode potential might predict a spontaneous change for a particular reaction, while our observations indicate that the reaction is not occurring. In cases such as this, the reaction proceeds so slowly (over a period of many years, for example) that changes are hard to detect. ◀

Example 11–7. Calculate the standard electrode potential of a galvanic cell, employing the half-reactions

$$Al \rightarrow Al^{3+} + 3e^-,$$
$$Ag^+ + e^- \rightarrow Ag.$$

From Table 11–4, we find that the standard electrode potential values for these two half-reactions are:

$$E^0_{Al} = -1.662 \text{ volts}, \qquad E^0_{Ag} = +0.799 \text{ volt}.$$

The standard electrode potential for the cell reaction is

$$E^0_{cell} = E^0_{red} - E^0_{ox}.$$

Since

$$Al \rightarrow Al^{3+} + 3e^-$$

is an oxidation and

$$Ag^+ + e^- \rightarrow Ag$$

is a reduction, $E^0_{ox} = -1.662$ V and $E^0_{red} = +0.799$ V. Therefore

$$E^0_{cell} = E^0_{red} - E^0_{ox} = (+0.799) - (-1.662)$$
$$= 0.799 + 1.662 = 2.461 = 2.461 \text{ volts}.$$

The standard electrode potential for the total cell reaction is 2.461 volts:

	E^0, volts
$3(Ag^+ + e^- \rightarrow Ag)$	$+0.799$
$Al \rightarrow Al^{3+} + 3e^-$	$-(-1.662)$
$3Ag^+ + Al \rightarrow 3Ag + Al^{3+}$	$+2.461$

Note that the Ag^+/Ag half-reaction has to be multiplied by 3 so that the total reaction is balanced. The electrode potentials are *not* multiplied by this number; this is an important point to remember.

Example 11–8. A solution contains Fe^{2+}. Suppose that we add a solution of Ce^{4+} to the Fe^{2+} solution. Will Fe^{3+} and Ce^{3+} be formed?

The oxidation-reduction reaction in question involves the two half-reactions:

	E^0, volts
$Ce^{4+} + e^- \rightarrow Ce^{3+}$	$+1.61$

and

	E^0, volts
$Fe^{2+} \rightarrow Fe^{3+} + e^-$	$-(+0.771)$
$Fe^{2+} + Ce^{4+} \rightarrow Fe^{3+} + Ce^{3+}$	$+0.84$

which have the standard electrode potentials shown (see Table 11–4). The standard electrode

potential for the total reaction is

$$E^0_{cell} = E^0_{red} - E^0_{ox} = F^0_{Ce} - F^0_{Fe}$$

$$= (+1.61) - (+0.771) = 1.61 - 0.771 = 0.839 = 0.84 \text{ volt.}$$

Since E^0 for the reaction is 0.84 and positive, the reaction will occur, and Ce^{4+} will oxidize Fe^{2+}, forming Fe^{3+} and Ce^{3+}.

Example 11–9. Will the reaction

$$Cr_2O_7^{2-} + 14H^+ + 6I^- \rightarrow 2Cr^{3+} + 3I_2 + 7H_2O$$

proceed spontaneously as written?

We can answer this question by calculating the standard electrode potential for the reaction. To do this we need to separate the total oxidation-reduction reaction into its two component half-reactions. The species that are involved in the oxidation-reduction process are

$$I^- \rightarrow I_2 \quad \text{and} \quad Cr_2O_7^{2-} \rightarrow Cr^{3+}.$$

That is, the oxidation number of iodine changes from -1 to 0 and that of chromium from $+6$ to $+3$. Inspection of Table 11–4 shows that the half-reactions involved are*

$$Cr_2O_7^{2-} + 14H^+ + 6e^- \rightarrow 2Cr^{3+} + 7H_2O \quad \text{and} \quad 2I^- \rightarrow I_2 + 2e^-.$$

We can obtain the total reaction and its standard electrode potential from these two half-reactions:

	E^0, volts
$Cr_2O_7^{2-} + 14H^+ + 6e^- \rightarrow 2Cr^{3+} + 7H_2O$	$+1.33$
$3(2I^- \rightarrow I_2 + 2e^-)$	$-(+0.5355)$
$Cr_2O_7^{2-} + 14H^+ + 6I^- \rightarrow 2Cr^{3+} + 3I_2 + 7H_2O$	$+0.79$

The standard electrode potential for the total reaction is

$$E^0_{cell} = E^0_{red} - E^0_{ox} = E^0_{Cr} - E^0_I$$

$$= (+1.33) - (+0.5355) = +0.7945 = 0.79 \text{ volt.}$$

Since E^0 for the reaction is positive, the reaction will proceed spontaneously.

Example 11–10. A strip of Cd metal is added to a solution containing Al^{3+} and Cu^{2+} ions. Predict what reactions, if any, will occur among these three species.

We must determine whether reactions will occur between

$$Cd + Al^{3+} \rightarrow ? \quad \text{and} \quad Cd + Cu^{2+} \rightarrow ?$$

To do this we must consider all probable reactions and calculate the standard electrode potential

---

* Although we shall not consider them here, there are methods of writing the complete half-reaction without relying on a table of data.

for each reaction. This is not difficult in this example, since the elements in question form a limited number of oxidation states. With some transition metals it's much harder to consider all possible reactions. Since Al is in Group IIIA, we may expect oxidation states of 0 and $+3$. For Cd in Group IIB, we expect oxidation states of 0 and $+2$, and for Cu in Group IB, oxidation states of 0 and $+1$, as well as the $+2$ that is given. Inspection of Table 11–4 shows that the following half-reactions are applicable to our problem:

$$
\begin{array}{ll}
& E^0\text{, volts} \\
Cu^+ + e^- \rightarrow Cu & +0.521 \\
Cu^{2+} + 2e^- \rightarrow Cu & +0.337 \\
Cu^{2+} + e^- \rightarrow Cu^+ & +0.153 \\
Cd^{2+} + 2e^- \rightarrow Cd & -0.4029 \\
Al^{3+} + 3e^- \rightarrow Al & -1.662
\end{array}
$$

Now we calculate all the reaction possibilities—that is, will Cd metal reduce Al^{3+}, Cu^{2+} and Cu^+?—by computing the standard electrode potentials for the reactions involved:

$$
\begin{array}{ll}
& E^0\text{, volts} \\
2(Al^{3+} + 3e^- \rightarrow Al) & -1.662 \\
\underline{3(Cd \rightarrow Cd^{2+} + 2e^-} & \underline{-(-0.403)} \\
2Al^{3+} + 3Cd \rightarrow 2Al + 3Cd^{2+} & -1.259
\end{array}
$$

Since E^0 is highly negative, the reaction will not occur and Al^{3+} will remain unchanged in the solution.

$$
\begin{array}{ll}
& E^0\text{, volts} \\
2(Cu^{2+} + e^- \rightarrow Cu^+) & +0.153 \\
\underline{Cd \rightarrow Cd^{2+} + 2e^-} & \underline{-(-0.403)} \\
2Cu^{2+} + Cd \rightarrow 2Cu^+ + Cd^{2+} & +0.556
\end{array}
$$

Since E^0 is positive, this reaction will occur:

$$
\begin{array}{ll}
& E^0\text{, volts} \\
Cu^{2+} + 2e^- \rightarrow Cu & +0.337 \\
\underline{Cd \rightarrow Cd^{2+} + 2e^-} & \underline{-(-0.403)} \\
Cu^{2+} + Cd \rightarrow Cu + Cd^{2+} & +0.740
\end{array}
$$

Since E^0 is positive, this reaction will also occur:

$$
\begin{array}{ll}
& E^0\text{, volts} \\
2(Cu^+ + e^- \rightarrow Cu) & +0.521 \\
\underline{Cd \rightarrow Cd^{2+} + 2e^-} & \underline{-(-0.403)} \\
2Cu^+ + Cd \rightarrow 2Cu + Cd^{2+} & +0.924
\end{array}
$$

Since E^0 is positive, this reaction will also occur.

Then we find that Cd metal will not reduce Al^{3+},

$$Cd + Al^{3+} \rightarrow \text{no reaction},$$

but it will reduce Cu^{2+} to Cu metal:

$$Cd + Cu^{2+} \rightarrow Cd^{2+} + Cu.$$

These will be the reactions observed in the solution. The reactions

$$Cd + 2Cu^{2+} \rightarrow Cd^{2+} + 2Cu^{+} \quad \text{and} \quad Cd + 2Cu^{+} \rightarrow Cd^{2+} + 2Cu$$

will probably not be observable, since any Cu^{+} formed will be reduced to Cu metal.

We can use the short list of electrode potentials in Example 11–10 to show the convenience of tabulated E^{0} values. Note that all the species on the right side of the equations are reducing agents and all species on the left side of the equations are oxidizing agents. Then good reducing agents lie at the bottom of a table of electrode potentials on the right side of the equation. Good oxidizing agents, on the other hand, lie at the top of the table on the left side of the equation. In the list in Example 11–10, Al metal is the best reducing agent and Cu^{+} the best oxidizing agent.

A table of electrode potentials also makes possible the rapid determination of whether two reactants will or will not combine. The following three half-reactions illustrate this process:

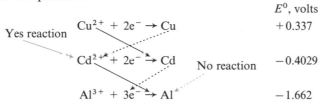

$$E^{0}, \text{volts}$$

$$Cu^{2+} + 2e^{-} \rightarrow Cu \qquad +0.337$$

$$Cd^{2+} + 2e^{-} \rightarrow Cd \qquad -0.4029$$

$$Al^{3+} + 3e^{-} \rightarrow Al \qquad -1.662$$

Reactions *do* occur between reactants connected by the diagonal lines sloping down from left to right, but do *not* occur between species connected by the diagonal lines sloping down from right to left. We have already seen, for example, that Cd metal does react with Cu^{2+} but not with Al^{3+}.

▶ Test this rule by calculating a few standard electrode potentials for oxidation-reduction reactions composed of two half-reactions from Table 11–4. Remember that electrode potential values can be given for half-reactions written as reductions, as we have done, or written as oxidations. The rules given above apply only to the half-reactions written as reductions. How would they be changed if the half-reactions were listed as oxidations? ◀

Half-reactions and their corresponding electrode potentials have many other uses, most of which we shall not consider. One further use of half-reactions that we have already encountered in working with electrode potentials is in balancing chemical equations.

▶ Standard electrode potentials are used to calculate effects of changes of concentration on reactions and are directly related to quantities such as heats of solution, equilibrium constants, and other important chemical properties helpful to practicing scientists. ◀

Half-reactions and balanced equations. Oxidation-reduction equations can be balanced by adding the two half-reactions that constitute the total reaction. We have already encountered this process in several of the examples in this chapter. The Daniell cell, for example, has the half-reactions

$$Zn \rightarrow Zn^{2+} + 2e^- \qquad \text{Oxidation}$$

and

$$Cu^{2+} + 2e^- \rightarrow Cu \qquad\qquad \text{Reduction}$$
$$\overline{Zn + Cu^{2+} \rightarrow Zn^{2+} + Cu}$$

As shown, the two half-reactions add to give the total oxidation-reduction reaction. Note that no electrons appear in the final reaction; as many electrons are released in the Zn half-reaction as are accepted in the Cu half-reaction. As we said before, this equality of electron loss and gain is necessary for balanced oxidation-reduction reactions. To ensure this equality, we must multiply the half-reactions by proper coefficients.

In the above case of the Daniell cell, the multiplying coefficients are 1, since the electron changes are already equal. This is not true for the reaction between Cr^{2+} and Cl_2, which involves the half-reactions

$$Cl_2 + 2e^- \rightarrow 2Cl^- \qquad \text{and} \qquad Cr^{2+} \rightarrow Cr^{3+} + e^-.$$

We can obtain the total balanced reaction by adding these two half-reactions after the number of electrons lost and gained are equal:

$$Cl_2 + 2e^- \rightarrow 2Cl^-$$
$$2(Cr^{2+} \rightarrow Cr^{3+} + e^-)$$

or

$$Cl_2 + 2e^- \rightarrow 2Cl^-$$
$$\underline{2Cr^{2+} \rightarrow 2Cr^{3+} + 2e^-}$$
$$Cl_2 + 2Cr^{2+} \rightarrow 2Cl^- + 2Cr^{3+}.$$

The total balanced reaction is free of electrons, since the same number are subtracted from each side of the equation.

These principles can, of course, be applied to more complicated equations.

Example 11–11. Determine the total balanced oxidation-reduction equation from the following half-reactions:

$$NO_{3(aq)}^- + 4H_{(aq)}^+ + 3e^- \rightarrow NO_{(g)} + 2H_2O_{(\ell)}$$
$$Cu_{(s)} \rightarrow Cu_{(aq)}^{2+} + 2e^-.$$

This is the reaction of Cu metal with nitric acid (HNO_3), which is a strong electrolyte in aqueous solution. We must equalize the electron loss and gain by multiplying the half-reactions by coefficients. An easy method of selecting these coefficients is to multiply each half-reaction

by the number of electrons appearing in the other half-reaction. In this case we multiply the NO_3^-/NO half-reaction by 2, which is the number of electrons involved in the Cu/Cu^{2+} half-reaction, and multiply the Cu/Cu^{2+} half-reaction by 3:

$$2(NO_3^- + 4H^+ + 3e^- \rightarrow NO + 2H_2O)$$
$$3(Cu \rightarrow Cu^{2+} + 2e^-)$$

or

$$2NO_3^- + 8H^+ + 6e^- \rightarrow 2NO + 4H_2O$$
$$\underline{3Cu \rightarrow 3Cu^{2+} + 6e^-}$$
$$2NO_3^- + 8H^+ + 3Cu + 6e^- \rightarrow 2NO + 3Cu^{2+} + 4H_2O + 6e^-.$$

Addition gives the total reaction, and the electrons cancel, since an equal number appear on each side of the equation. The balanced reaction is

$$2NO_{3(aq)}^- + 8H_{(aq)}^+ + 3Cu_{(s)} \rightarrow 2NO_{(g)} + 3Cu_{(aq)}^{2+} + 4H_2O_{(\ell)}.$$

Example 11–12. Using half-reactions, balance the equation

$$Cr_{(aq)}^{3+} + MnO_{4(aq)}^- + H_2O_{(\ell)} \rightarrow Cr_2O_{7(aq)}^{2-} + Mn_{(aq)}^{2+} + H_{(aq)}^+.$$

We can obtain the proper half-reactions from Table 11–4:

$$MnO_4 + 8H^+ + 5e \rightarrow Mn^{2+} + 4H_2O$$

and

$$2Cr^{3+} + 7H_2O \rightarrow Cr_2O_7^{2-} + 14H^+ + 6e^-.$$

By multiplying the reduction half-reaction by 6 and the oxidation half-reaction by 5, we can equalize the electron gain and loss:

$$6(MnO_4^- + 8H^+ + 5e^- \rightarrow Mn^{2+} + 4H_2O)$$
$$5(2Cr^{3+} + 7H_2O \rightarrow Cr_2O_7^{2-} + 14H^+ + 6e^-).$$

When we add these two half-reactions, we see that several species appear on both sides of the equation:

$$6MnO_4^- + 48H^+ + 30e^- \rightarrow 6Mn^{2+} + 24H_2O$$
$$\underline{10Cr^{3+} + 35H_2O \rightarrow 5Cr_2O_7^{2-} + 70H^+ + 30e^-}$$

$6MnO_4^- +$	$48H^+ +$	$10Cr^{3+} +$	$35H_2O +$	$30e^- \rightarrow$	$6Mn^{2+} +$	$5Cr_2O_7^{2-} +$	$70H^+ +$	$24H_2O +$	$30e^-$
	-48		-24	-30			-48	-24	-30
	0		11	0			22	0	0

Subtracting equal numbers of species occurring on both sides of the equation leaves the final balanced equation:

$$6MnO_{4(aq)}^- + 10Cr_{(aq)}^{3+} + 11H_2O_{(\ell)} \rightarrow 6Mn_{(aq)}^{2+} + 5Cr_2O_{7(aq)}^{2-} + 22H_{(aq)}^+.$$

The study of solutions—particularly solutions of electrolytes—is not limited to oxidation-reduction reactions. We shall consider one other very common and important group of solutions—*acids* and *bases*—in Chapter 12.

QUESTIONS

1. Define or discuss the following: Electrolyte, nonelectrolyte, weak electrolyte, strong electrolyte, electrodes, polar solutes, nonpolar solvents, electrolysis, cathode, anode, oxidation, reduction, percent dissociation, cation, anion, solute–solvent forces, galvanic cells, Daniell cell, ion diffusion in galvanic cells, spectator ions, net equations, half-reactions, electrode potential, standard hydrogen electrode.

2. For the following substances, write equations describing the dissociation reactions in aqueous solution and indicate whether the percent dissociation will be high or low:

    a) HCl, strong electrolyte   b) NH_4OH, weak electrolyte
    c) NaCl, strong electrolyte   d) $CaBr_2$, strong electrolyte
    e) H_2S, weak electrolyte

3. Predict which of the following aqueous solutions will be strong electrolytes.

    a) NaCl, high percent dissociation   b) HBr, polar solute
    c) Br_2, nonpolar solute   d) CCl_4, nonpolar solute
    e) $Zn(NO_3)_2$, ionic solute

4. A galvanic cell is made with a standard hydrogen electrode and a silver metal electrode. The anode reaction is

$$H_2 \rightarrow 2H^+ + 2e^-$$

    and the cathode reaction is

$$Ag^+ + e^- \rightarrow Ag.$$

    a) Design an appropriate experimental arrangement for the galvanic cell.
    b) What is the total cell reaction?
    c) Indicate the direction of electron flow in the electrical circuit.
    d) Indicate the direction of movement of any appropriate ions in the solution.
    e) Indicate what happens at and/or to the two electrodes.

5. The following half-reactions occur, as written, at electrodes in a galvanic cell. Tell whether the reaction occurs at the anode or at the cathode.

    a) $Cu^{2+} + 2e^- \rightarrow Cu$ b) $Zn \rightarrow Zn^{2+} + 2e^-$
    c) $2H^+ + 2e^- \rightarrow H_2$ d) $I_2 + 2e^- \rightarrow 2I^-$
    e) $Mn^{2+} + 2H_2O \rightarrow MnO_2 + 4H^+ + 2e^-$ f) $MnO_4^- + 4H^+ + 3e^- \rightarrow MnO_2 + 2H_2O$

6. Explain why the percent dissociation of a $1.0M$ HBr solution is very high, while the percent dissociation of a $1.0M$ HF solution is low.

7. Predict which of the following substances will have strong solute–solvent interactions with a water solvent.

    a) KF   b) CO_2 c) Ar   d) H_2
    e) $NaNO_3$ f) HBr   g) H_2S h) CH_4

8. The solubility of NaCl is very high in H_2O, but quite low in liquid Br_2. Why should we expect these differences in NaCl solubility?

9. A galvanic cell is constructed by placing a Cd metal electrode in a $1.0M$ Cd^{2+} solution which is separated from a hydrogen electrode and $1.0M$ H^+ ion solution by a porous partition.

When we complete the external electrical circuit, we find that the Cd metal electrode decreases in size.

a) Write probable anode and cathode half-reactions.
b) Write the total balanced cell reaction.
c) Sketch a possible cell diagram and indicate the direction of all appropriate ion movements.
d) Indicate the direction of electron flow in the external electrical circuit.
e) Calculate the voltage that would be produced by the cell.

10. Suppose that 0.55 mole $CaCl_2$ is dissolved in enough H_2O to give 100 ml of solution. What are the molarities of Ca^{2+} and Cl^-?

11. For the reaction $2Cr_{(aq)}^{2+} + Cu_{(aq)}^{2+} \rightarrow 2Cr_{(aq)}^{3+} + Cu_{(s)}$, calculate the normalities of $0.010M$ $CrCl_2$ and $0.25M$ $Cu(NO_3)_2$ solutions.

12. How many moles each of Cu^{2+} and NO_3^- ions are present in 200 ml of $0.12M$ $Cu(NO_3)_2$?

13. What volume of solution is required to prepare $0.15M$ NaOH from 12 g NaOH?

14. According to the reaction $2Br_{(aq)}^- + 2Ce_{(aq)}^{3+} \rightarrow Br_{2(\ell)} + 2Ce_{(aq)}^4$, how many moles of Br^- ion are required to exactly react with $0.050\ \ell$ of $0.020M$ $CeCl_3$?

15. For the reaction $Cu_{(aq)}^{2+} + Mg_{(s)} \rightarrow Cu_{(s)} + Mg_{(aq)}^{2+}$, answer the following questions.

a) How could $0.10N$ $Cu(NO_3)_2$ be prepared using 1.26 g $Cu(NO_3)_2$ solid?
b) How many liters of $0.10N$ $Cu(NO_3)_2$ would be required to exactly react with 0.12 gram-equivalent Mg?
c) How many gram-equivalents are contained in 200 ml of $0.20M$ $Cu(NO_3)_2$?
d) How many grams of Cu metal can be produced by reacting 4.86 g Mg and $0.300\ \ell$ of $1.00M$ $Cu(NO_3)_2$?

16. According to the reaction $Br_{(aq)}^- + Cr_2O_{7(aq)}^{2-} + H_{(aq)}^+ \rightarrow Br_{2(\ell)} + Cr_{(aq)}^{3+} + H_2O_{(\ell)}$:

a) How many moles Br^- react exactly with 0.12 mole $Cr_2O_7^{2-}$?
b) What is the normality of a $0.015M$ $Cr_2O_7^{2-}$ solution?
c) What would be the molarity and normality of a solution prepared by dissolving 10.3 g NaBr in enough H_2O to give $0.100\ \ell$?
d) How many millimeters of $0.10N$ $Cr_2O_7^{2-}$ solution are required to exactly neutralize 50 ml of $1.2M$ NaBr?
e) How many gram-equivalents of Br^- are required to exactly neutralize 50 ml of $0.012N$ $Cr_2O_7^{2-}$?
f) How many moles of Br_2 can be obtained from $0.060\ \ell$ of $1.0N$ Br^- and 300 ml of $0.10N$ $Cr_2O_7^{2-}$?

17. How many moles of $KMnO_4$ are needed to prepare $0.020M$ $KMnO_4$, given that enough water to prepare 400 ml of solution is available?

18. What is the normality of a $KMnO_4$ solution, given that 50.0 ml are required to exactly react with 125 ml of $0.20N$ $FeSO_4$?

19. How many gram-equivalents are present in $1.2\ \ell$ of $0.0055N$ $FeSO_4$?

20. A $1.20M$ CdI_2 solution is needed for an experiment. What is the maximum volume of solution that can be prepared from 109.9 g CdI_2?

21. Calculate the expected E_{cell}^0 for a galvanic cell, using the half-reactions

$$Pb^{2+} + 2e^- \rightarrow Pb \quad \text{and} \quad 2H^+ + 2e^- \rightarrow H_2.$$

22. Using data from Table 11–4, predict whether the following electrode reactions, as written, could be used for a galvanic cell:

$$Al^{3+} + 3e^- \rightarrow Al \qquad \text{cathode}$$
$$Cd \rightarrow Cd^{2+} + 2e^- \qquad \text{anode}$$

23. Predict the reactions that would be expected if a strip of Cu metal were placed in a solution containing $1.0M$ Pb^{2+} and $1.0M$ Fe^{3+}.

24. a) Write the total, balanced cell reaction for the following half-reactions:

$$NO + 2H_2O \rightarrow NO_3^- + 4H^+ + 3e^- \qquad \text{and} \qquad Ce^{4+} + e^- \rightarrow Ce^{3+}.$$

b) For the galvanic cell in part (a), calculate E^0_{cell}.

25. Will the reaction $2Cl^-_{(aq)} + Pb^{2+}_{(aq)} \rightarrow Cl_{2(g)} + Pb_{(s)}$ proceed spontaneously as written? Why?

26. Using the two half-reactions,

$$2Cr^{3+} + 7H_2O \rightarrow Cr_2O_7^{2-} + 14H^+ + 6e^-,$$
$$MnO_4^- + 4H^+ + 3e^- \rightarrow MnO_2 + 2H_2O,$$

write a balanced equation for the overall reaction.

27. For the equation $Co^{3+}_{(aq)} + Tl^+_{(aq)} \rightarrow Co^{2+}_{(aq)} + Tl^{3+}_{(aq)}$, write the two half-reactions and balance the equation.

28. Calculate the standard electrode potential for the reaction $Mg_{(s)} + I_{2(s)} \rightarrow Mg^{2+}_{(aq)} + 2I^-_{(aq)}$.

29. Select the best and poorest oxidizing agents and reducing agents from the following half-reactions:

Half-reaction	E^0, volts
$ClO_3^- + 3H^+ + 2e^- \rightarrow H_2O + HClO_2$	$+1.21$
$O_2 + 2H^+ + 2e^- \rightarrow H_2O_2$	$+0.68$
$S + 2H^+ + 2e^- \rightarrow H_2S$	$+0.14$
$Eu^{3+} + e^- \rightarrow Eu^{2+}$	-0.43

30. Using the appropriate half-reactions, balance the equation

$$H_{2(g)} + MnO_{2(s)} + H^+_{(aq)} \rightarrow Mn^{2+}_{(aq)} + H_2O.$$

31. From the following group of half-reactions, write three balanced equations for spontaneous reactions and three equations for reactions that will not proceed spontaneously:

Half-reaction	E^0, volts
$ClO_4^- + 2H^+ + 2e^- \rightarrow ClO_3^- + H_2O$	$+1.19$
$2Hg^{2+} + 2e^- \rightarrow Hg_2^{2+}$	$+0.92$
$VO^{2+} + 2H^+ + e^- \rightarrow V^{3+} + H_2O$	$+0.36$
$H_3PO_4 + 2H^+ + 2e^- \rightarrow H_3PO_3 + H_2O$	-0.28
$U^{4+} + e^- \rightarrow U^{3+}$	-0.61

32. For the half-reactions in question 31, calculate the standard electrode potential for all possible spontaneous reactions.

33. For the following oxidation-reduction reaction,

$$HNO_2 + I_2 \rightarrow IO_3^- + N_2O + H^+ + H_2O:$$

a) How many moles of HNO_2 are needed to prepare 50 ml of $0.050M$ HNO_2?
b) What is the normality of $0.0025M$ I_2?
c) How many milliliters of $0.10M$ HNO_2 react exactly with 250 ml of $0.040M$ I_2?

d) How many moles of N_2O can be obtained from 0.125 ℓ of 0.40M HNO_2 and 5.06 g I_2?

e) How many gram-equivalents of I_2 react exactly with 200 ml of 0.20M HNO_2?

f) What is the normality of a solution prepared by dissolving 5.06 g I_2 in enough H_2O to give 400 ml?

g) Write the two half-reactions; indicate which half-reaction corresponds to oxidation and which to reduction.

h) Calculate the E^0 for the reaction.

i) Does the reaction occur spontaneously? Why?

12. SOLUTIONS: ACIDS AND BASES

There are many common aqueous solutions that exhibit certain characteristic properties that make it possible to classify them as acidic or basic solutions. At first these properties were determined experimentally, with every solution being tested individually. It was found that many solutions had a distinctly sour taste, that these solutions reacted readily with many metallic elements to produce hydrogen gas, and that the solutions imparted characteristic colors to certain dye materials. For example, paper impregnated with the purple dye litmus turned pink when placed in these solutions and served as an indicator of their presence. Solutions that had these properties were called *acids*.

A second group of aqueous solutions, called *bases*, exhibited characteristic basic properties. They had a bitter taste, slippery feel, and turned litmus paper blue. And when mixed with an acid, the solutions seemed to neutralize one another; that is, the acidic and basic properties disappeared.

As we have seen before, whenever characteristic groups are noticed or certain sets of properties appear, the same questions always arise. Why? Why do the acidic and basic solution properties occur? How? How can we think about acids and bases in more concrete terms? How can we explain the behaviors and properties of acids and bases?

12-1 THE ARRHENIUS CONCEPT

There are several definitions of acids and bases. One early—and still very useful—concept is based on ideas developed by Svante Arrhenius in 1884. This definition is most useful in aqueous solutions, and is based on the dissociation of H_2O. Since water is a nonelectrolyte, we can conclude that there are very few (if any) ions present in a sample of pure water. However, if we make very precise experimental measurements of the degree of dissociation of H_2O, we find that there is a very small degree of dissociation of H_2O, according to the equation

$$H_2O \rightleftarrows H^+ + OH^-.$$

The concentrations of *hydrogen ion*, H^+, and *hydroxide ion*, OH^-, in pure water are both $1.0 \times 10^{-7} M$,

$$[H^+] = [OH^-] = 1.0 \times 10^{-7},$$

where the brackets signify concentrations expressed as *molarities* (moles per liter).

▶ Note that in pure water the H^+ and OH^- concentrations *must* be equal. Both ions are formed by the dissociation (ionization or splitting apart) of one H_2O molecule

$$\underset{H}{\overset{O}{\diagdown}} \text{H} \rightleftarrows \left(\underset{H}{\overset{O-}{\diagup}} \right)^- + H^+, \qquad H_2O \rightleftarrows OH^- + H^+.$$

The number of H^+ and OH^- ions produced are necessarily equal, and consequently, in pure H_2O, the $[H^+] = [OH^-]$. ◀

Although this percent dissociation of H_2O is very small in pure water, the H^+ and OH^- concentrations can be greatly changed by the addition of certain substances to the water. This, then, is the basis of the Arrhenius concept of acids and bases in aqueous solutions. An *acid* is any substance that increases the H^+ concentration in the solution. A *base* is any substance that increases the OH^- concentration in the solution.

Using these definitions, we can see that substances that dissociate, even to a mild degree, in H_2O to produce H^+ or OH^- will qualify as an acid or a base. Since HCl, HNO_3, and H_2SO_4 form solutions that are strong electrolytes, they have a high percent dissociation, all yielding H^+:

$$HCl \xrightarrow{H_2O} H^+ + Cl^-,$$
$$HNO_3 \xrightarrow{H_2O} H^+ + NO_3^-,$$
$$H_2SO_4 \xrightarrow{H_2O} H^+ + HSO_4^-.$$

Solutions of HCl, HNO_3, and H_2SO_4 would all be acidic, since the H^+ concentration is increased when these substances are added to H_2O. Then HCl, HNO_3, and H_2SO_4 are classified as acids. Arrhenius postulated that all acids could be written HX and could dissociate into H^+ and X^-.

On the other hand, substances such as NaOH, KOH, and $Ba(OH)_2$ are bases, since they all dissociate to increase the OH^- concentration when they are added to H_2O:

$$NaOH \xrightarrow{H_2O} Na^+ + OH^-,$$
$$KOH \xrightarrow{H_2O} K^+ + OH^-,$$
$$Ba(OH)_2 \xrightarrow{H_2O} Ba^{2+} + 2OH^-.$$

Arrhenius postulated that all bases could be written as MOH and could dissociate into M^+ and OH^-.

Common acids and bases* are often prepared by reacting appropriate metal oxides, called *anhydrides*, with water. We can obtain sulfuric acid (H_2SO_4) by dissolving sulfur trioxide gas in water:

$$SO_{3(g)} + H_2O \rightarrow H_2SO_4 \rightarrow H^+_{(aq)} + HSO^-_{4(aq)}.$$

Here SO_3 is serving as an *acidic oxide* or an *acid anhydride*. Reaction of a *basic oxide* with water can produce a base

$$Na_2O_{(s)} + H_2O \rightarrow 2Na^+_{(aq)} + 2OH^-_{(aq)}.$$

Not all acids and bases can be derived from oxides (HCl, for example).

12-2 THE BRØNSTED-LOWRY CONCEPT

An extension of the Arrhenius concept of acids and bases was proposed independently by Johannes Brønsted and Thomas Lowry in 1923. According to the Brønsted-Lowry definitions, an *acid* is a substance that is a proton donor, and a *base* is a substance that is a proton acceptor. A *proton* is an H^+ ion.

This concept of acids and bases is also quite useful when we are dealing with aqueous solutions, and it greatly increases the number of substances that can easily be classified as acids or bases.

▶ An even more general definition of acids and bases was proposed by Gilbert N. Lewis in 1923. According to the Lewis concept, an acid is a substance that is an electron-pair acceptor and a base is a substance that is an electron-pair donor. Thus reactions such as the one between boron trifluoride and ammonia can be classified as acid-base reactions:

$$
\begin{array}{ccccc}
 & \ddot{:}\!F\!\ddot{:} & H & & \ddot{:}\!F\!\ddot{:}\ \ H \\
\ddot{:}\!F\!\ddot{:}\!B & + & \ddot{:}\!N\!\ddot{:}\!H \rightarrow & \ddot{:}\!F\!\ddot{:}\!B\ \ddot{:}\ N\!\ddot{:}\!H & \blacktriangleleft \\
 & \ddot{:}\!F\!\ddot{:} & H & & \ddot{:}\!F\!\ddot{:}\ \ H \\
 & \text{acid} & \text{base}
\end{array}
$$

---

*   Nomenclature rules for acids, bases and salts are given in Appendix C.

Arrhenius acids are readily classified as Brønsted-Lowry acids. For example, perchloric acid ($HClO_4$) has a high percent dissociation,

$$HClO_4 \xrightarrow{H_2O} H^+_{(aq)} + ClO^-_{4(aq)},$$

and is an acid according to the Arrhenius definition (an increase in H^+ concentration). But $HClO_4$ is also a proton donor, and therefore also an acid according to the Brønsted-Lowry definition:

$$HClO_4 + H_2O \rightarrow H_3O^+ + ClO_4^-.$$

Note that the $HClO_4$ has donated a proton—that is, an H^+—to a H_2O molecule, forming the *hydronium ion* H_3O^+ (or $H^+ \cdot H_2O$). Since $HClO_4$ is a proton donor, it is a Brønsted-Lowry acid. Note that the only difference between the equations describing $HClO_4$ as an acid is the inclusion of H_2O as a reactant and in writing H^+ as an associated species (a hydrated ion), $H^+ \cdot H_2O$ or H_3O^+. Then what can be said about H_2O according to Brønsted-Lowry?

▶ There is independent experimental evidence that H^+ does not exist alone in water solution, but is associated with one or more molecules of water. In order to describe a solution containing H^+ in the most realistic terms, many chemists prefer the H_3O^+ description. Then the dissociation of water can be written $2H_2O \rightleftarrows H_3O^+ + OH^-$. However, the Arrhenius concept of acids is still widely used, and many chemists and chemistry textbooks often use H^+, as in $HBr \rightarrow H^+ + Br^-$. When considering aqueous solutions, we should always remember that the chemical species that actually exist in the solution are undoubtedly associated (hydrated) with one or more water molecules, even though these H_2O molecules do not specifically appear in the chemical equation. The determination of the actual number of associated H_2O molecules (if only one specific number is ever present) is a difficult experimental problem. ◀

The Brønsted-Lowry definition of an acid can be extended to include more substances than the Arrhenius definition. For example, we can represent the dissolving of ammonia gas (NH_3) in water by the equation

$$H_2O + NH_{3(g)} \rightleftarrows NH^+_{4(aq)} + OH^-_{(aq)}.$$

In this reaction H_2O has donated a proton to the NH_3 molecule and is therefore an acid. In the Arrhenius system of classification, H_2O is not considered as an acid or a base, since these definitions depend on the dissociation of H_2O.

In the two above examples of Brønsted-Lowry acids, the equations also contained Brønsted-Lowry bases. Just as in the case of acids, this definition of bases extends the Arrhenius concept. When $HClO_4$ is placed in H_2O, it is an acid,

$$HClO_4 + H_2O \rightarrow H_3O^+ + ClO_4^-.$$

The Brønsted-Lowry base in this case is H_2O, since the H_2O molecule accepts a proton. When NH_3 is dissolved in water, the H_2O acts as an acid and donates a proton to the NH_3 molecule, which acts as a base and accepts the proton:

$$H_2O + NH_3 \rightleftarrows NH_4^+ + OH^-.$$

$$\text{acid} \qquad \text{base}$$

▶ Note that, according to the Brønsted-Lowry definitions of acids and bases, H_2O can serve as either an acid or a base. ◀

If we consider the reverse of this reaction, we can also treat it as a Brønsted-Lowry acid-base reaction,

$$NH_4^+ + OH^- \rightleftarrows H_2O + NH_3,$$

$$\text{acid} \qquad \text{base}$$

since NH_4^+ donates a proton and OH^- accepts a proton. The substances NH_3 and NH_4^+ are an acid-base pair, as are H_2O and OH^-. They are called *conjugate pairs*, and are related by a proton transfer; NH_4^+ is the conjugate acid of the base NH_3 and OH^- is the conjugate base of the acid H_2O. Conjugate pairs are often designated by using subscripts in conjunction with the equation:

$$HC_2H_3O_2 + H_2O \rightleftarrows H_3O^+ + C_2H_3O_2^-.$$

$$\text{acid}_1 \qquad \text{base}_2 \qquad \text{acid}_2 \qquad \text{base}_1$$

This equation shows that the conjugate base of acetic acid ($HC_2H_3O_2$) is $C_2H_3O_2^-$ (acetate ion); H_3O^+ is the conjugate acid of the base H_2O.

12-3 STRENGTHS OF ACIDS AND BASES

Percent Dissociation. According to the Arrhenius definition, acids and bases can be classified as weak or strong according to their percent dissociation. Strong acids and bases have a high percent dissociation and are strong electrolytes. Weak acids and bases have a low percent dissociation and are weak electrolytes. We know, for example, that nitric acid (HNO_3) is a strong electrolyte with essentially 100% dissociation, while acetic acid ($HC_2H_3O_2$) is a weak electrolyte and only a few percent dissociated. Then HNO_3 is a strong acid and $HC_2H_3O_2$ is a weak acid:

$$HNO_3 \xrightarrow{\sim 100\%} H^+ + NO_3^-,$$

$$HC_2H_3O_2 \underset{\xleftarrow{\text{Few }\%}}{\longrightarrow} H^+ + C_2H_3O_2^-$$

We would describe the nitric acid solution as containing primarily H^+ and NO_3^- ions, with very few HNO_3 molecules present. In acetic acid, on the other hand, $HC_2H_3O_2$ molecules are the dominant solute species, with only a few H^+ and $C_2H_3O_2^-$ ions (hydrated) present.

Similar reasoning applies to bases. An aqueous solution of the strong base sodium hydroxide contains almost exclusively Na^+ and OH^- ions (hydrated) and

very few NaOH molecules. This is indicated by the equation

$$NaOH \xrightarrow{\sim 100\%} Na^+ + OH^-.$$

Ammonium hydroxide, however, is a weak base and has a low percent dissociation,

$$NH_4OH \xrightleftharpoons{Few \%} NH_4^+ + OH^-.$$

Consequently, the solution largely contains undissociated ammonium hydroxide and only a few NH_4^+ and OH^- ions. Table 12–1 contains a list of some common weak and strong acids and bases.

▶ Ammonium hydroxide solutions appear to contain primarily $NH_3 \cdot (H_2O)$ species rather than NH_4OH molecules. (See Sections 11–2 and 13–14.) The formula NH_4OH will be used only in those instances in which the basic dissociation is emphasized in the same way that $HC_2H_3O_2$ and HNO_3 represent the acidic nature of these acids and not the structure of the species in solution. ◀

Table 12–1. Some common acids and bases in aqueous solution

Acids	Strong or weak	Percent dissociation
HCl	Strong	High
HBr	Strong	High
HNO_3	Strong	High
H_2SO_4	Strong	High
$HClO_4$	Strong	High
$HC_2H_3O_2$	Weak	Low
HF	Weak	Low
H_2S	Weak	Low
H_2CO_3	Weak	Low
Bases		
NaOH	Strong	High
KOH	Strong	High
$Ca(OH)_2$	Strong	High
$Ba(OH)_2$	Strong	High
$Al(OH)_3$	Strong	High
NH_4OH	Weak	Low

Conjugate Pairs. The strengths of Brønsted-Lowry acids and bases are determined by comparing the ability of a particular acid or base to donate or to receive protons, respectively. If a substance readily donates a proton to H_2O, it is a strong acid.

Hydrobromic acid is in this category:

$$HBr + H_2O \rightarrow H_3O^+ + Br^-.$$

$$acid_1 \quad base_2 \quad acid_2 \quad base_1$$

The concentrations of H_3O^+ and Br^- in the solution are far greater than the HBr concentration. We can conclude, then, that HBr is a stronger acid than H_3O^+. It is also apparent that the conjugate base, Br^-, of HBr is a very weak base, since it has a very weak attraction for a proton.

Compare this HBr case to that of acetic acid,

$$HC_2H_3O_2 + H_2O \rightleftarrows H_3O^+ + C_2H_3O_2^-.$$

$$acid_1 \quad base_2 \quad acid_2 \quad base_1$$

Experimentally it is found that the $HC_2H_3O_2$ concentration in the solution is far greater than the H_3O^+ and $C_2H_3O_2^-$ concentrations. Then H_3O^+ is a better proton donor than $HC_2H_3O_2$ and, consequently, H_3O^+ is a stronger acid. Since $HC_2H_3O_2$ is a weak acid, it has a relatively strong conjugate base, $C_2H_3O_2^-$ ion, while H_2O is the weak conjugate base of the relatively strong acid H_3O^+.

In general we can say that: Strong acids have weak conjugate bases; strong bases have weak conjugate acids; weak acids have strong conjugate bases; and weak bases have strong conjugate acids.

It is also true that in reactions between Brønsted-Lowry acids and bases, the formation of the weaker acid and base is favored. In reacting HBr and H_2O, the weaker acid, H_3O^+, and weaker base, Br^-, have high concentrations. But in the reaction between $HC_2H_3O_2$ and H_2O, the concentrations of the stronger acid, H_3O^+, and base, $C_2H_3O_2^-$, are low.

Periodic Trends. We can understand, in part, the strengths of some acids and bases and the variation in strengths observed in similar groups of acids or bases by considering the size and electronegativity of the central atom that is attracting a proton. Since the sizes and electronegativities of atoms show periodic trends, the strengths of acids and bases might also be expected to exhibit periodic changes.

We know that the ionic radii of the Group VIA and Group VIIA elements increase going down the group, while the electronegativities of the elements decrease going down the group. Both these trends favor a weaker bond between the Group VIA or Group VIIA element and hydrogen, in going down a group. Thus we would predict that acid strengths should increase from H_2O to H_2S to H_2Se to H_2Te in Group VIA, with H_2Te being the strongest acid in the group. That is,

$$H_2Te \rightleftarrows H^+ + HTe^-$$

has a higher percent dissociation than any of the other acids in the group. The attraction of Te for the hydrogen atoms is the lowest of the four members in Group VIA.

Similarly in Group VIIA, the trend in acid strengths is from

$$HF \text{ to } HCl \text{ to } HBr \text{ to } HI,$$

with HI being the strongest acid (highest percent dissociation). The attraction of I for H is much lower than the attraction of F for H.

▶ In water solution, HCl, HBr, and HI are all 100% dissociated and we cannot tell the order of acidity. If we conduct experiments in a nonaqueous medium, though, the dissociation of the acids is less than 100% and we can measure the relative acid strength. ◀

For atoms of approximately the same size—for example, atoms in the same period—we can see certain trends in strengths of bases by noting only the electronegativity trend of the elements. For example, in the three bases NH_2^-, OH^-, and F^-, the electronegativity increases from N to O to F. Thus we might expect the attraction for a proton to increase in going from N to F. Consequently, we would predict that F^- is a better H^+ acceptor (base) than is NH_2^-, and that the trend in strengths of the three bases would be

$$NH_2^- \text{ to } OH^- \text{ to } F^-,$$

with F^- being the strongest base. This is exactly the *wrong* prediction. The basicity trend is the reverse, with NH_2^- being the strongest base.

This illustrates a common occurrence in science, especially when we attempt to draw generalizations from rather limited observations. We have tried to establish general acid–base behaviors, in solution, by considering properties related only to the solute. With the hydrogen compounds of the Group VIA and VIIA elements, the trends in size and electronegativity of the groups appear to follow and consequently explain the observed trend in acid strength. But when one of these variables is held approximately constant, as atom size is in the bases NH_2^-, OH^-, and F^-, it is evident that our thinking in terms of the electronegativity effects on acid or base strength is in error.* The situation is too complicated for this very simple correlation; we need to consider a more complex approach.

We can increase the complexity of our model of acid–base strengths by including not only solute–solute properties and interactions, such as arise from size and electronegativity, but also solute–solvent interactions. Apparently, as the electronegativity of an atom bonded to hydrogen increases, the shared electrons are drawn more and more closely to the electronegative atom. This makes possible greater interaction of the two charged ends of the solute molecule with the solvent (H_2O) molecules. This greater solute–solvent interaction accounts for the lower "attraction" for a proton in substances containing hydrogen bonded to a highly electronegative element. Then, in the series

$$NH_3, \quad H_2O, \quad HF,$$

---

* This is also true, of course, with the first example of the Group VIA and VIIA compounds of hydrogen. Here, apparently, the effect of the increase in atom size is large enough to establish the observed trend in the acid strength of the compounds.

we would expect the acid strength to increase from

$$NH_3 \text{ to } H_2O \text{ to } HF,$$

since F is the most electronegative element and the solute–solvent forces should be strongest with HF. This lowers the attraction between H and F in HF and it will therefore dissociate more than H_2O or NH_3. This is, of course, the experimentally observed trend with these three compounds. For the bases

$$NH_2^-, \quad OH^-, \quad F^-,$$

we would expect the base strength to increase from

$$F^- \text{ to } OH^- \text{ to } NH_2^-,$$

since F is the most electronegative atom and will have the strongest solute–solvent interactions, thereby reducing its attraction for H^+. Again, this is the observed trend.

There are two other useful generalizations of acid strengths, both concerning oxy-acids, or acids containing oxygen, such as HClO, H_2SO_4, H_3PO_4, etc. For acids with the formula and structure

$$H\text{—}O\text{—}Z,$$

the strength of the acid increases with increasing electronegativity of Z. The acid strength increases from

$$HIO \text{ to } HBrO \text{ to } HClO.$$

In compounds in which a variable number of oxygen atoms may be bonded to the central element Z, the strength of the acid increases as the number of oxygen atoms increases. In the series of oxy acids of chlorine, the acid strength increases from

$$HClO \text{ to } HClO_2 \text{ to } HClO_3 \text{ to } HClO_4.$$

It might be interesting to speculate about the reasons for these two observed trends in oxy acids.

▶ The formulas of oxy acids can be written in different ways. It is more instructive to write this series as HOCl, HOClO, HOClO$_2$, HOClO$_3$, since these formulas indicate, correctly, that the H atom is bonded to an O atom and not directly to the Cl atom. Each succeeding O is bonded directly to the central Cl atom. We have chosen to write them in the more traditional form—HClO$_4$ rather than HOClO$_3$—to avoid confusion with the formulas for bases. ◀

12–4 NEUTRALIZATION REACTIONS

Formation of Water and a Salt. When an acid and a base are mixed, they react by shifting protons. Generally, water and a salt are the products of this *neutralization reaction.*

▶ Acid–base neutralization reactions do not always produce neutral solutions, that is, solutions in which the $[H^+] = [OH^-] = 1.0 \times 10^{-7}M$. The anions of weak acids can react with H_2O, in Brønsted-Lowry fashion, to produce a slightly basic solution ($[OH^-]$ slightly greater than $1.0 \times 10^{-7}M$), and cations of weak bases can react with H_2O to produce slightly acidic solutions ($[H^+]$ slightly greater than $1.0 \times 10^{-7}M$). These reactions with H_2O are called *hydrolysis* reactions. We shall consider them in more detail in Section 13–15. ◀

A salt is composed of the cation of the base and the anion of the acid and is, except for a few cases, essentially 100% dissociated in solution. A general neutralization reaction can be written in molecular form as

$$HX + MOH \rightarrow H_2O + MX.$$

Here a proton is donated by the acid HX and accepted by the base MOH. It is not surprising that H_2O is formed in this reaction, since H^+ and OH^- ions have such a strong attraction for one another. We know that this is true, since water is very weakly dissociated,

$$H_2O \rightleftharpoons H^+ + OH^-, \quad \text{with} \quad [H^+] = [OH^-] = 1.0 \times 10^{-7}M.$$

The salt formed in this general reaction is MX. Ordinarily, salts are dissolved in the reaction solution and exist in dissociated form:

$$MX \xrightarrow{\sim 100\%} M^+ + X^-.$$

Then to recover the solid salt, $MX_{(s)}$, we usually have to evaporate the solution.

▶ In some cases a highly insoluble salt, such as $BaSO_{4(s)}$, is obtained in a neutralization reaction. Then recovery of the salt requires only pouring off the liquid solution or, better yet, filtration of the solution. Even quite insoluble salts do dissolve to a small extent. In most of these cases, that small amount of insoluble salt that does dissolve is essentially 100% dissociated. ◀

Acid–base neutralizations can occur between strong acids and bases, between weak acids and bases, or between a combination of strong and weak acids and bases. It is convenient to consider these very common reactions using the Arrhenius concept of acids and bases.

Strong Acid–Strong Base. We can describe the reaction between the strong acid HCl and the strong base NaOH by three different types of equations: the molecular equation (which indicates only the formula units of the reactants and products), the total ionic equation (which indicates all the ions or molecules that are present in the mixture, both before and after the reaction),

$$H^+ + Cl^- + Na^+ + OH^- \rightarrow H_2O + Na^+ + Cl^-,$$

and the net equation (derived from the total ionic equation by deleting all spectator ions),

$$H^+ + OH^- \rightarrow H_2O.$$

This is the net equation for most strong-acid–strong-base neutralization reactions.

▶ A different net equation for strong-acid–strong-base neutralization might occur if, for example, an insoluble precipitate were formed in the reaction. Then the solid salt would be included in the net equation, as in

Molecular equation $H_2SO_4 + Ba(OH)_2 \rightarrow 2H_2O + BaSO_{4(s)}$,

Total ionic and net equation $2H^+ + SO_4^{2-} + Ba^{2+} + 2OH^- \rightarrow 2H_2O + BaSO_{4(s)}.$ ◀

Weak Acid–Weak Base. Weak acids and bases, such as HCN and NH_4OH, are weak electrolytes. We account for this low percent dissociation in the total ionic equation by writing the reactant acid and base as undissociated species (usually as molecules),

$$HCN + NH_4OH \rightarrow H_2O + NH_{4(aq)}^+ + CN_{(aq)}^-.$$

Note that the ammonium cyanide salt formed (recovered by evaporation) is completely dissociated in solution, and is therefore written in ionic form. Inspection of this total ionic equation shows that it is also the net equation.

▶ See if you can write total ionic and net equations for the neutralization reaction that occurs when solutions of Na_2HPO_4 and NH_4OH are mixed; let us assume that you know that HPO_4^{2-} is a weak acid. ◀

Weak Acid–Strong Base. We can see from the cases considered above that the neutralization of the weak acid HNO_2 (nitrous acid) with the strong base NaOH is described by the total ionic equation

$$HNO_2 + Na_{(aq)}^+ + OH_{(aq)}^- \rightarrow H_2O + Na_{(aq)}^+ + NO_{2(aq)}^-$$

and the net equation

$$HNO_2 + OH_{(aq)}^- \rightarrow H_2O + NO_{2(aq)}^-.$$

The equations indicate that HNO_2 and H_2O have very limited dissociation, while NaOH and $NaNO_2$ are highly dissociated in solution. Note that Na^+ is a spectator ion.

Stepwise Neutralization. Some acids, such as H_2SO_4, have more than one H^+ per molecule of acid that can be dissociated or neutralized. These *polyprotic* (or polybasic) acids dissociate in steps and can be neutralized with a base in a stepwise fashion. Phosphoric acid (H_3PO_4), for example, actually produces H^+ ions in three different steps, the percent dissociation decreasing with each step:

$$H_3PO_4 \rightleftharpoons H^+ + H_2PO_4^-,$$
$$H_2PO_4^- \rightleftharpoons H^+ + HPO_4^{2-},$$
$$HPO_4^{2-} \rightleftharpoons H^+ + PO_4^{3-}.$$

There are three acids involved in these dissociations, H_3PO_4, $H_2PO_4^-$, and HPO_4^{2-}; H_3PO_4 is *triprotic* (three replaceable H^+ ions per formula unit) and $H_2PO_4^-$ is *diprotic* (two replaceable H^+ ions per formula unit). The acid strength increases from

$$HPO_4^{2-} \quad \text{to } H_2PO_4^- \quad \text{to } H_3PO_4,$$

since the H_3PO_4 has the highest percent dissociation.

If the strong base NaOH is used to neutralize a solution of H_3PO_4, three different salts can be recovered from the mixtures:

$$NaH_2PO_4 \quad \text{(monosodium dihydrogen phosphate),}$$
$$Na_2HPO_4 \quad \text{(disodium monohydrogen phosphate),}$$
$$Na_3PO_4 \quad \text{(trisodium phosphate).}$$

The molecular equations are useful in considering these stepwise reactions:

$$H_3PO_4 + NaOH \rightarrow H_2O + NaH_2PO_4,$$
$$H_3PO_4 + 2NaOH \rightarrow 2H_2O + Na_2HPO_4,$$
$$H_3PO_4 + 3NaOH \rightarrow 3H_2O + Na_3PO_4.$$

If only one mole of NaOH is added to a solution containing one mole of H_3PO_4, the salt that can be obtained on evaporation is NaH_2PO_4. But if three moles of NaOH are added to a solution containing one mole of H_3PO_4, then the obtainable salt is Na_3PO_4.

12-5 STOICHIOMETRY OF NEUTRALIZATION

The quantitative aspects, or stoichiometry, of acid–base neutralization reactions can be treated by two methods. The first involves the use of moles and a balanced equation. The second method involves acid–base gram-equivalents.

The Mole Method. The balanced equation

$$H_2SO_3 + 2NH_4OH \rightarrow 2H_2O + 2NH_{4(aq)}^+ + SO_{3(aq)}^{2-}$$

shows that two moles of NH_4OH (a weak base) are needed to exactly react with one mole of H_2SO_3 (sulfurous acid, a weak acid). The NH_4OH/H_2SO_3 reacting mole ratio is always 2 to 1. Recall that the number of moles of a solute in a given volume of a solution of a particular molar concentration is

$$M \text{ (mole/}\ell\text{)} \times V\text{ (}\ell\text{)} = \text{mole.}$$

Since acid–base neutralizations are usually carried out by mixing solutions of various concentrations, and since obtaining a balanced equation requires knowing the number of moles present (and not the concentrations), this is an important relationship.

For example, suppose that we have $0.50 \, \ell$ (or 500 ml) of a $1.0M$ H_2SO_3 solution. We know that the entire solution contains only 0.50 mole H_2SO_3,

$$M \times V = \text{mole}$$
$$(1.0 \text{ mole } H_2SO_3/\ell)(0.50\,\ell) = 0.50 \text{ mole } H_2SO_3.$$

From the above equation we see that

> 1 mole H_2SO_3 exactly reacts with 2 moles NH_4OH,
>
> (0.50)(1) mole H_2SO_3 exactly reacts with (0.50)(2) moles NH_4OH,
>
> 0.50 mole H_2SO_3 exactly reacts with 1 mole NH_4OH.

Therefore, to neutralize the entire H_2SO_3 sample, we must use 1 mole of NH_4OH. If we have available a solution that is $10M$ NH_4OH, we can calculate that only $0.10\ \ell$ (or 100 ml) of the $10M$ NH_4OH solution is required for neutralization:

$$M \times V = \text{moles}, \qquad V = \frac{\text{moles}}{M},$$

$$V = \frac{1.0\ \text{mole NH}_4\text{OH}}{10\ \text{moles NH}_4\text{OH}/\ell} = 0.10\ \ell \text{ of } 10M\ NH_4OH.$$

Then 100 ml of $10M$ NH_4OH will react completely with 500 ml of $1.0M$ H_2SO_3.

The following examples illustrate "mole-method" calculations for acid–base reactions, using slight variations of the preceding example, along with more complicated numbers.

Example 12–1. How many milliliters of a $0.050M$ HF solution will exactly react with 15 ml of $0.0020M$ $Al(OH)_3$?

We can calculate the number of moles of $Al(OH)_3$ to be neutralized, since we know the volume and concentration of the $Al(OH)_3$ solution. The balanced equation will give the required number of moles of HF. We can then determine the volume of HF, since we know the number of moles of HF and the molarity of the HF solution.

The number of moles of $Al(OH)_3$ present is

$$M \times V = \text{moles}$$

$$(0.0020\ \text{mole}/\ell)(0.015\ \ell) = 3.0 \times 10^{-5}\ \text{mole Al(OH)}_3.$$

The balanced equation gives the required reacting ratio of moles:

> $Al(OH)_3 + 3HF \rightarrow 3H_2O + Al^{3+} + 3F^-$.
>
> 1 mole $Al(OH)_3$ reacts with 3 moles HF,
>
> $(3.0 \times 10^{-5})(1)$ mole $Al(OH)_3$ reacts with $(3.0 \times 10^{-5})(3)$ mole HF,
>
> 3.0×10^{-5} mole $Al(OH)_3$ reacts with 9.0×10^{-5} mole HF.

Therefore we need 9.0×10^{-5} mole HF. Since we also know the concentration of the HF solution, this enables us to calculate the necessary volume of HF:

$$M \times V = \text{moles}, \qquad V = \frac{\text{moles}}{M},$$

$$V = \frac{9.0 \times 10^{-5}\ \text{mole HF}}{0.050\ \text{mole HF}/\ell} = \frac{9.0}{5.0} \times 10^{-3}\ \ell$$

$$= 1.8 \times 10^{-3}\ \ell = 1.8\ \text{ml}.$$

It takes 1.8 ml of $0.050M$ HF to completely neutralize 15 ml of $0.0020M$ $Al(OH)_3$.

▶ We used the molecular equation in the above example because it readily shows the reacting mole ratios between $Al(OH)_3$ and HF. We could also have used the total ionic equation. Assuming $Al(OH)_3$ to be a strong base,

$$Al^{3+} + 3OH^- + 3HF \rightarrow 3H_2O + Al^{3+} + 3F^-.$$

Note that we must use care if we consider net equations. Removing Al^{3+} from both sides of the above total ionic equation yields the net equation for the reaction,

$$3OH^- + 3HF \rightarrow 3H_2O + 3F^- \qquad \text{or, more generally,} \qquad OH^- + HF \rightarrow H_2O + F^-.$$

If we use net equations in solving quantitative stoichiometry problems, we must remember the number of OH^- (or H^+) ions that can be obtained from each molecule or formula unit of base (or acid), and take this into account. ◀

Example 12–2. What is the molarity of a $Ca(OH)_2$ solution, assuming that it takes 250 ml to completely neutralize 0.050 ℓ of 0.040M H_2SO_4?

We can calculate the $Ca(OH)_2$ molarity from the number of moles of $Ca(OH)_2$ needed and the required volume of $Ca(OH)_2$. We determine the number of moles of $Ca(OH)_2$ by the number of moles of H_2SO_4 present (volume and molarity of the H_2SO_4) and the balanced equation
The neutralization occurs according to the equation

$$H_2SO_4 + Ca(OH)_2 \rightarrow 2H_2O + Ca^{2+} + SO_4^{2-}$$

or

$$2H^+ + SO_4^{2-} + Ca^{2+} + 2OH^- \rightarrow 2H_2O + Ca^{2+} + SO_4^{2-}$$

or

$$2H^+ + 2OH^- \rightarrow 2H_2O.$$

From any of the equations we can see that

1 mole H_2SO_4 reacts with 1 mole $Ca(OH)_2$.

The number of moles of H_2SO_4 present is

$$M \times V = \text{moles}$$
$$(0.040 \text{ mole } H_2SO_4/\ell)(0.050\ \ell) = 0.0020 \text{ mole } H_2SO_4.$$

From the equation, therefore, we see that 0.0020 mole $Ca(OH)_2$ is needed for complete reaction. Since 0.0020 mole $Ca(OH)_2$ is contained in 0.250 ℓ, the molarity of the $Ca(OH)_2$ solution is

$$M \times V = \text{moles}, \qquad M = \frac{\text{moles}}{V(\ell)},$$

$$M = \frac{0.0020 \text{ mole } Ca(OH)_2}{0.250\ \ell} = 0.0080 \text{ mole } Ca(OH)_2/\ell.$$

The $Ca(OH)_2$ solution is 0.0080M.

Example 12–3. How many grams of KOH (a strong base) are needed to completely neutralize 125 ml of 0.40M H_3PO_3 (a weak acid)? Only two of the three hydrogens in H_3PO_3 are acidic.

The equation for the reaction is

$$2KOH + H_3PO_3 \rightarrow 2H_2O + 2K^+ + HPO_3^{2-}$$

or

$$2K^+ + 2OH^- + H_3PO_3 \rightarrow 2H_2O + 2K^+ + HPO_3^{2-}$$

or

$$2OH^- + H_3PO_3 \rightarrow 2H_2O + HPO_3^{2-}.$$

1 mole H_3PO_3 reacts with 2 moles KOH.

There is $(0.40 \text{ mole } H_3PO_3/\ell)(0.125 \ell) = 0.050$ mole H_3PO_3 present.

$(0.050)(1)$ mole H_3PO_3 reacts with $(0.05)(2)$ mole KOH,
0.050 mole H_3PO_3 reacts with 0.10 mole KOH.

Thus we need 0.10 mole KOH for the reaction. Since one mole of KOH weighs $39 + 16 + 1 = 56$ g, 0.10 mole KOH weighs

$$(0.15 \text{ mole KOH})(56 \text{ g KOH/mole KOH}) = 5.6 \text{ g.}$$

It takes 5.6 g KOH to completely neutralize 125 ml of $0.40M$ H_3PO_3.

Example 12–4. What is the formula weight of a triprotic weak acid, H_3Q, given that 250 ml of a solution containing 6.18 g solid H_3Q per liter of solution is completely neutralized by 0.750 ℓ of $1.00M$ NH_4OH?

The neutralization reaction is $H_3Q + 3NH_4OH \rightarrow 3H_2O + 3NH_4^+ + Q^{3-}$.
There is $(1.00 \text{ mole } NH_4OH/\ell)(0.750 \ell) = 0.750$ mole NH_4OH present.

3 moles NH_4OH react with 1 mole H_3Q,
1 mole NH_4OH reacts with $\frac{1}{3}$ mole H_3Q,
$(0.750)(1)$ mole NH_4OH reacts with $(0.750)(\frac{1}{3})$ mole H_3Q,
0.750 mole NH_4OH reacts with 0.250 mole H_3Q.

Then 0.250 mole H_3Q is present in the 250 ml of the H_3Q solution. Since one liter of the H_3Q solution contains 6.18 g H_3Q,

$$250 \text{ ml (or } \tfrac{1}{4} \ell) \text{ contains } (0.250 \ell)\left(\frac{6.18 \text{ g } H_3Q}{1.00 \ell}\right) = 1.545 \text{ g } H_3Q.$$

The weight of one mole of H_3Q is

$$\frac{1.545 \text{ g } H_3Q}{0.250 \text{ mole } H_3Q} = 61.8 \text{ g } H_3Q/\text{mole } H_3Q.$$

The formula weight of H_3Q is 61.8 g.

The Gram-Equivalent Method. Gram-equivalents are quite useful when one is working with acid–base neutralization reactions. This is particularly true because the conversions between molar and normal concentrations are more easily made with acids and bases than with oxidizing and reducing agents, although a similar process is involved.

One *gram-equivalent of an acid* is the weight of acid that will produce one mole of H^+. One *gram-equivalent of base* is the weight of base that will produce one mole of OH^-. Note that these definitions are useful in aqueous solutions, and that they require that

    1 gram-equivalent (g-eq) of acid react with 1 gram-equivalent (g-eq) of base.

This necessary consequence of the definitions of gram-equivalents is the foundation of their usefulness. Since

        1 g-eq of acid reacts with   1 g-eq of base,
    0.10 g-eq of acid reacts with 0.10 g-eq of base,
    0.29 g-eq of acid reacts with 0.29 g-eq of base,
    1.62 g-eq of acid reacts with 1.62 g-eq of base,
   11.2 g-eq of acid reacts with 11.2 g-eq of base.

Then, if we know or can determine the number of gram-equivalents of an acid that are present, we know that the same number of gram-equivalents of base are required to exactly react with the acid. This eliminates the need for the balanced equation describing the neutralization reaction.

    To use the gram-equivalent method in neutralization reactions, we usually have to calculate the number of gram-equivalents of acid or base present in a given solution. There is a relationship between gram-equivalents, volume, and normality similar to the relation between moles, volume, and molarity that we just considered:

$$N(\text{gram-equivalents}/\ell) \times V(\ell) = \text{gram-equivalents}.$$

We can interconvert the relations

$$M \times V = \text{moles} \quad \text{and} \quad N \times V = \text{g-eq}$$

by converting M to N or vice versa.

    For the case of *monoprotic* acids (such as HCl, HNO_3, etc.) and bases with only one OH^- per formula unit (such as NaOH, NH_4OH, etc.), the normality is equal to the molarity. But for cases in which there is more than one potential H^+ or OH^- per formula unit (such as H_2SO_4, H_3PO_4; $Ca(OH)_2$, $Al(OH)_3$, etc.), the normality is greater than the molarity. These conversions depend on the number of H^+ or OH^- ions that can be obtained per formula unit. From the following examples we can see that the proper conversion factors are easily obtained.

    For monoprotic acids or for bases with one OH^- ion per formula unit, only one mole of H^+ or OH^- can be produced by dissociation. We can use the strong acid HNO_3 as an example:

$$HNO_3 \xrightarrow{\sim 100\%} H^+ + NO_3^-.$$

1 mole HNO_3 produces 1 mole H^+ and   1 g-eq HNO_3 produces 1 mole H^+.

Therefore 1 mole HNO_3 = 1 g-eq HNO_3.

▶ Since one gram-equivalent is equal to one mole for monoprotic acids and for bases with one OH^- per formula unit, it is also true that the gram-equivalent weight—

that is, the weight of one gram-equivalent expressed in grams—is equal to the formula or molecular weight. This is not true, however, for acids or bases that contain more than one H^+ or OH^- per formula unit. ◄

Since molarity is moles per liter and normality is gram-equivalents per liter, and one mole is equal to one gram-equivalent, the number of moles per liter must be equal to the number of gram-equivalents per liter. For all monoprotic acids (and bases with one OH^- per formula unit), the molarity is equal to the normality:

$$HCl \rightarrow H^+ + Cl^-,$$
1 mole HCl produces 1 mole H^+,
1 g-eq HCl produces 1 mole H^+,
$$M \text{ HCl} = N \text{ HCl}.$$

If a solution is $1.25M$ HCl, it is also $1.25N$ HCl.
    This is also true for bases, such as

$$NaOH \rightarrow Na^+ + OH^-,$$
1 mole NaOH produces 1 mole OH^-,
1 g-eq NaOH produces 1 mole OH^-,
$$M \text{ NaOH} = N \text{ NaOH}.$$

A $6.2M$ NaOH solution is also $6.2N$ NaOH.
    For diprotic acids (such as H_2SO_4, H_2S, etc.) or for bases containing two OH^- ions per formula unit (such as $Ba(OH)_2$, $Zn(OH)_2$, etc.), the normality is generally two times the molarity. With H_2SO_4, for example,

$$H_2SO_4 \rightarrow 2H^+ + SO_4^{2-},$$
1 mole H_2SO_4 produces 2 moles H^+,
$\frac{1}{2}$ mole H_2SO_4 produces 1 mole H^+,
1 g-eq H_2SO_4 produces 1 mole H^+.

Therefore $\frac{1}{2}$ mole H_2SO_4 = 1 g-eq H_2SO_4.
    Thus in every one mole of a diprotic acid (or base containing two OH^- per formula unit), there are two gram-equivalents. Since one mole is equal to (or contains) two gram-equivalents for diprotic acids, the normality is two times the molarity. A solution that is $0.25M$ H_2SO_4 has a normality that is $2(0.25) = 0.50N$:

$$N \text{ H}_2\text{SO}_4 = 2(0.25M \text{ H}_2\text{SO}_4),$$
$$0.50N \text{ H}_2\text{SO}_4 = 0.25M \text{ H}_2\text{SO}_4.$$

The gram-equivalent weight of diprotic acids and of bases with two OH^- ions per formula unit is equal to $\frac{1}{2}$ the formula or molecular weight of the acid or base. For example,

$$Ca(OH)_2 \rightarrow Ca^{2+} + 2OH^-,$$
1 mole $Ca(OH)_2$ produces 2 moles OH^-,
$\frac{1}{2}$ mole $Ca(OH)_2$ produces 1 mole OH^-.

Therefore $\frac{1}{2}$ mole $Ca(OH)_2 = 1$ g-eq $Ca(OH)_2$ or 1 g-eq $Ca(OH)_2$ weighs as much as $\frac{1}{2}$ mole $Ca(OH)_2$.

Thus the gram-equivalent weight of $Ca(OH)_2$ is equal to $\frac{1}{2}$ the formula weight of $Ca(OH)_2$:

$$\text{g-eq weight } Ca(OH)_2 = \frac{\text{formula weight}}{2}$$

$$= \frac{74}{2} = 37 \text{ amu.}$$

The weight of one gram-equivalent of $Ca(OH)_2$ is 37 grams ($\frac{1}{2}$ mole of $Ca(OH)_2$).

Note the generalizing statements that usually apply:

1. The number of gram-equivalents in one mole of acid or base is equal to the subscript of the H or OH in the formula for the acid or base.

HCl	1 g-eq in 1 mole,
H_2SO_4	2 g-eq in 1 mole,
H_3PO_4	3 g-eq in 1 mole,
NaOH	1 g-eq in 1 mole,
$Ca(OH)_2$	2 g-eq in 1 mole.

▶ There are exceptions to these rules for acids or bases that contain more than one H^+ or OH^- that can be dissociated. We have seen that these dissociations often occur in steps:

$$H_2SO_4 \rightarrow H^+ + HSO_4^-, \qquad HSO_4^- \rightleftarrows H^+ + SO_4^{2-}.$$

It is possible, in some cases, to neutralize only the first H^+ that is dissociated, leaving HSO_4^- in solution. In this case H_2SO_4 is behaving as a monoprotic acid and its $N = M$, even though its chemical formula indicates that it is diprotic. ◀

2. To convert from molarity to normality, multiply the molarity by the H or OH subscript in the formula.

$$0.6M \text{ HCl} = (1)(0.6)N \text{ HCl} = 0.6N \text{ HCl,}$$
$$1.2M \text{ } H_2SO_4 = 2(1.2)N \text{ } H_2SO_4 = 2.4N \text{ } H_2SO_4,$$
$$0.03M \text{ } H_3PO_4 = 3(0.03)N \text{ } H_3PO_4 = 0.09N \text{ } H_3PO_4,$$
$$0.72M \text{ } Ca(OH)_2 = 2(0.72)N \text{ } Ca(OH)_2 = 1.44N \text{ } Ca(OH)_2.$$

▶ Note that the normality is never less than the molarity; it can only be equal to or greater than the molarity. Since $N = x \times M$, where x is usually 1, 2, or 3, the conversion of normality to molarity is $M = N/x$. A solution that is $0.74N$ $Ba(OH)_2$ is

$$0.74N \text{ } Ba(OH)_2 = \frac{0.74}{2} M \text{ } Ba(OH)_2$$

$$= 0.37M \text{ } Ba(OH)_2. \quad ◀$$

3. To obtain the gram-equivalent weight for an acid or base, divide the formula or molecular weight by the H or OH subscript in the formula.

$$\text{g-eq weight HCl} = \frac{\text{formula weight}}{1} = \frac{36.5}{1} = 36.5 \text{ g,}$$

$$\text{g-eq weight H}_2\text{SO}_4 = \frac{\text{formula weight}}{2} = \frac{98}{2} = 49 \text{ g,}$$

$$\text{g-eq weight H}_3\text{PO}_4 = \frac{\text{formula weight}}{3} = \frac{98}{3} = 33 \text{ g,}$$

$$\text{g-eq weight Al(OH)}_3 = \frac{\text{formula weight}}{3} = \frac{78}{3} = 26 \text{ g.}$$

The following examples illustrate the use of gram-equivalents, normalities, and the gram-equivalent method of problem-solving.

Example 12–5. What is the normality of a $0.26M$ H_2SO_3 solution, given that this weak acid is completely neutralized when reacted with a base?

Sulfurous acid (H_2SO_3) is a diprotic acid. When completely dissociated, one mole of H_2SO_3 will produce two moles of H^+:

$$H_2SO_3 \rightleftarrows 2H^+ + SO_3^{2-}.$$

1 mole H_2SO_3 produces 2 moles H^+,
$\frac{1}{2}$ mole H_2SO_3 produces 1 mole H^+.

Therefore

$\frac{1}{2}$ mole H_2SO_3 is equal to 1 g-eq H_2SO_3,
1 mole H_2SO_3 is equal to 2 g-eq H_2SO_3,

and

$$N\,H_2SO_3 = 2 \times M\,H_2SO_3.$$

Thus

$$0.26M\,H_2SO_3 = 2(0.26)N\,H_2SO_3 = 0.52N\,H_2SO_3.$$

Example 12–6. How many gram-equivalents of $Sr(OH)_2$ are used in a neutralization reaction that requires 0.32 mole $Sr(OH)_2$?

Since the strong base strontium hydroxide ($Sr(OH)_2$) produces two moles of OH^- ion from one mole of $Sr(OH)_2$, each mole of $Sr(OH)_2$ contains (is equivalent to) two gram-equivalents of $Sr(OH)_2$. Thus 0.32 mole $Sr(OH)_2$ is equal to $2(0.32) = 0.64$ gram-equivalent of $Sr(OH)_2$.

$$Sr(OH)_2 \rightarrow Sr^{2+} + 2OH^-,$$
1 mole $Sr(OH)_2$ produces 2 moles OH^-,
$\frac{1}{2}$ mole $Sr(OH)_2$ produces 1 mole OH^-,
1 g-eq $Sr(OH)_2$ produces 1 mole OH^-.

Therefore

$$\tfrac{1}{2} \text{ mole Sr(OH)}_2 = 1 \text{ g-eq Sr(OH)}_2$$

and

$$1 \text{ mole Sr(OH)}_2 = 2 \text{ g-eq Sr(OH)}_2,$$
$$0.32(1) \text{ mole (Sr(OH)}_2 = (0.32)2 \text{ g-eq Sr(OH)}_2,$$
$$0.32 \text{ mole Sr(OH)}_2 = 0.64 \text{ g-eq Sr(OH)}_2.$$

This reaction neutralizes 0.32 mole $Sr(OH)_2$ or 0.64 gram-equivalent $Sr(OH)_2$.

Example 12–7. How can we prepare a 0.34N H_3PO_4 solution from 100 ml of a 0.68N H_3PO_4 solution?

The 0.34N H_3PO_4 solution must contain a ratio of

$$\text{gram-equivalents } H_3PO_4/\text{liters of solution}$$

equal to 0.34. In 0.100 ℓ of 0.68N H_3PO_4 there are

$$(0.100\,\ell)(0.68 \text{ g-eq } H_3PO_4/\ell) = 0.068 \text{ g-eq } H_3PO_4.$$

We must have 0.068 g-eq H_3PO_4 in the proper volume of solution to make the ratio

$$\text{g-eq } H_3PO_4/\ell \text{ solution} = 0.34 \text{ g-eq } H_3PO_4/\ell.$$

Thus

$$\frac{0.068 \text{ g-eq } H_3PO_4}{0.34 \text{ g-eq } H_3PO_4/\ell} = 0.20 \text{ } \ell.$$

(Note how the units help to indicate the method of calculation.)

▶ If you prefer, you can set up this calculation as a proportion:

$$\frac{0.34 \text{ g-eq } H_3PO_4}{1.0 \text{ } \ell} = \frac{0.68 \text{ g-cq } H_3PO_4}{x \text{ } \ell}.$$

Then

$$x = \frac{(0.68 \text{ g-eq } H_3PO_4)(1.0 \text{ } \ell)}{(0.34 \text{ g-eq } H_3PO_4)} = 0.20 \text{ } \ell.$$

Note that the units have been included in the proportion. ◀

To obtain 0.34N H_3PO_4 from 100 ml of 0.68N H_3PO_4, we must increase the volume of the solution by adding enough water to give a total volume of 200 ml.

We can simplify this calculation by writing down the above process of "thinking through the problem" in equation form. From the discussion above, we see that the number of gram-equivalents of H_3PO_4 is the same in both the 0.68N and the 0.34N solutions; only the volumes of the solutions are different. Therefore

$$N_1 \times V_1 = \text{g-eq } H_3PO_4 = N_2 \times V_2,$$
$$N_1 \times V_1 = N_2 \times V_2.$$

Substituting the known values,

$$(0.68 \text{ g-eq } H_3PO_4/\ell)(0.100\ \ell) = (0.34 \text{ g-eq } H_3PO_4/\ell)V_2,$$

gives

$$V_2 = \frac{0.68 \text{ g-eq } H_3PO_4}{0.34 \text{ g-eq } H_3PO_4/\ell} = 0.20\ \ell.$$

The relationship between the normality and volume of two solutions is very useful in dilution problems, as shown in this example. This same relationship applies in acid–base (or oxidation-reduction*) mixing problems, as shown in the following examples.

Example 12–8. What volume of $0.12N$ $HC_2H_3O_2$ will exactly neutralize 120 ml of $0.20N$ $Ba(OH)_2$?

For exact neutralization to occur between an acid and a base, the number of gram-equivalents of acid must equal the number of gram-equivalents of base:

$$N(\text{acid}) \times V(\text{acid}) = \text{g-eq(acid)},$$
$$N(\text{base}) \times V(\text{base}) = \text{g-eq(base)},$$
$$\text{g-eq(acid)} = \text{g-eq(base)}.$$

Therefore

$$N(\text{acid}) \times V(\text{acid}) = N(\text{base}) \times V(\text{base}).$$

This is the same relationship used in Example 12–7:

$$N_1 \times V_1 = N_2 \times V_2 \qquad (V \text{ in liters}).$$

Substituting the known values,

$$(0.12 \text{ g-eq}/\ell)\ V_{HC_2H_3O_2} = (0.20 \text{ g-eq}/\ell)(0.120\ \ell),$$

$$V_{HC_2H_3O_2} = \frac{(0.20 \text{ g-eq})(0.120)}{0.12 \text{ g-eq}/\ell} = 0.20\ \ell.$$

We must use 200 ml of $0.12N$ $HC_2H_3O_2$ to neutralize 120 ml of $0.20N$ $Ba(OH)_2$.

Example 12–9. What is the normality of a H_2SO_4 solution, given that 25 ml are required to neutralize 125 ml of $0.25M$ $Al(OH)_3$?

For complete neutralization of the $Al(OH)_3$ solution, the number of gram-equivalents of H_2SO_4 must equal the number of gram-equivalents of $Al(OH)_3$. (Why is this a requirement?) Thus

$$N(H_2SO_4) \times V(H_2SO_4) = N(Al(OH)_3) \times V(Al(OH)_3).$$

But we know the molarity (M) of the $Al(OH)_3$ solution, and not its normality. We make the con-

---

*   Refer to Examples 11–4 and 11–5.

version to normality by multiplying the M of $Al(OH)_3$ by three:

$$N\ Al(OH)_3 = 3 \times M\ Al(OH)_3, \qquad 3(0.25M\ Al(OH)_3) = 0.75N\ Al(OH)_3.$$

Thus

$$N(H_2SO_4) \times (0.025\ \ell) = (0.75\ \text{g-eq}/\ell)(0.125\ \ell)$$

$$N(H_2SO_4) = \frac{(0.75\ \text{g-eq})(0.125)}{0.025\ \ell} = 3.75\ \text{g-eq}/\ell = 3.75N.$$

Complete reaction of 125 ml of $0.25M$ $Al(OH)_3$ (or $0.75N$ $Al(OH)_3$) requires 25 ml of $3.75N$ H_2SO_4. What would be the M of the H_2SO_4?

Example 12–10. What is the gram-equivalent weight of an acid, given that 1.50 grams of the acid exactly neutralize 0.750 ℓ of $0.040M$ $Zn(OH)_2$?

To calculate the gram-equivalent weight of the acid (number of grams of acid/g-eq acid), we need to determine the number of gram-equivalents of acid equal to 1.50 grams of the acid. Then we can calculate the number of grams of acid in one gram-equivalent. The number of gram-equivalents of acid is equal to the number of gram-equivalents of base ($Zn(OH)_2$) upon complete neutralization. The number of gram-equivalents of $Zn(OH)_2$ is given by

$$N(Zn(OH)_2) \times V(Zn(OH)_2) = \text{g-eq}\ Zn(OH)_2.$$

Since

$$Zn(OH)_2 \rightarrow Zn^{2+} + 2OH^-$$

and

$$1\ \text{mole}\ Zn(OH)_2\ \text{produces}\ 2\ \text{moles}\ OH^-,$$

then

$$N(Zn(OH)_2) = 2 \times M\ Zn(OH)_2, \qquad 2(0.040M\ Zn(OH)_2) = 0.080N\ Zn(OH)_2,$$

and

$$(0.080\ \text{g-eq}\ Zn(OH)_2/\ell)(0.750\ \ell) = 0.060\ \text{g-eq}\ Zn(OH)_2.$$

For neutralization, we need 0.060 g-eq acid, and 0.060 g-eq acid is equivalent to 1.50 g acid. The equivalent weight of the acid is

$$\frac{1.50\ \text{g acid}}{0.060\ \text{g-eq acid}} = 25\ \text{g acid/g-eq acid}.$$

The gram-equivalent weight of the acid is 25 grams.

▶ Again we can use a proportion for this calculation:

$$\frac{1.50\ \text{g acid}}{0.060\ \text{g-eq acid}} = \frac{x\ \text{g acid}}{1.0\ \text{g-eq acid}}$$

$$x = \frac{(1.50\ \text{g acid})(1.0\ \text{g-eq acid})}{0.060\ \text{g-eq acid}} = \frac{1.50}{0.060} = 25\ \text{g acid.} ◀$$

12–6 ACID–BASE TITRATION

We can follow the course of an acid–base neutralization reaction by using a process called *titration*. In this process a base, for example, of a known concentration, is gradually added to a selected volume of an acid of unknown concentration. We can keep track of the change in the H^+ concentration (it will decrease because of the addition of OH^-), if we wish, and determine the point of complete neutralization, the *endpoint* or *equivalence point*. At the endpoint, the number of equivalents of acid is equal to the number of equivalents of base, and, for the addition of a strong base to a strong acid, the H^+ and OH^- concentrations will both be $1.0 \times 10^{-7}M$.

▶ If either a weak acid or a weak base is involved, the H^+ and OH^- ion concentrations at the endpoint are not $1.0 \times 10^{-7}M$, since H_2O will react with the anion of a weak acid or the cation of a weak base to produce OH^- or H^+ ion concentrations, respectively, that are greater than $1.0 \times 10^{-7}M$. These are called hydrolysis reactions; see Section 13–15. ◀

We can usually detect the endpoint of a titration visually by using substances called *indicators*, such as litmus or phenolphthalein, that change color at certain H^+ ion concentrations. We can calculate the unknown acid concentration from the measured volume of base of given concentration required to neutralize the selected volume of acid. At the endpoint,

$$g\text{-eq base} = g\text{-eq acid},$$
$$N(\text{base}) \times V(\text{base}) = g\text{-eq base},$$
$$N(\text{acid}) \times V(\text{acid}) = g\text{-eq acid}.$$

Therefore

$$N(\text{base}) \times V(\text{base}) = N(\text{acid}) \times V(\text{acid}) \quad \text{and} \quad N(\text{acid}) = \frac{N(\text{base}) \times V(\text{base})}{V(\text{acid})}.$$

Since we know $N(\text{base})$, $V(\text{base})$, and $V(\text{acid})$, we can calculate $N(\text{acid})$.

To illustrate the titration procedure, let us consider what happens to the H^+ concentration as we add $0.100N$ NaOH, a strong base, to 50.0 ml of $0.100N$ HCl, a strong acid. We can obtain selected volumes of the solutions from graduated tubes, called *burets* (see Fig. 12–1).

▶ Since one mole = one gram equivalent and $N = M$ for both HCl and NaOH, the terms mole and M can be substituted for gram-equivalent and N throughout this example. This is not the case, however, when polyprotic acids or bases with more than one OH^- per formula unit are considered. ◀

The H^+ ion concentration in the $0.100N$ HCl solution, before the addition of any NaOH, is $0.100N$ H^+, since HCl is a strong acid and essentially 100% dissociated. After we have added 10.0 ml of $0.100N$ NaOH to the 50.0 ml of $0.100N$ HCl, what is the H^+ ion concentration? The addition of 10.0 ml of $0.100N$ NaOH or 1.00×10^{-3}

Fig. 12-1 A buret, a graduated cylinder fitted with a stopcock.

gram-equivalent of OH^-,

$$(0.100 \text{ g-eq NaOH}/\ell)(0.0100 \, \ell) = 1.00 \times 10^{-3} \text{ g-eq NaOH},$$

neutralizes 1.00×10^{-3} g-eq HCl (one g-eq base reacts exactly with one g-eq acid). The initial solution of 50.0 ml of 0.100N HCl contained

$$(0.100 \text{ g-eq HCl}/\ell)(0.050 \, \ell) = 5.00 \times 10^{-3} \text{ g-eq HCl}.$$

Therefore, after we have added 10.0 ml of base, there remains

$$5.00 \times 10^{-3} - 1.00 \times 10^{-3} \text{ g-eq HCl} = 4.00 \times 10^{-3} \text{ g-eq HCl}$$

in the solution, which now has a total volume of

$$50.0 \text{ ml HCl} + 10.0 \text{ ml NaOH} = 60.0 \text{ ml solution}.$$

Then the H^+ ion concentration is

$$\frac{4.00 \times 10^{-3} \text{ g-eq H}^+}{0.0600 \, \ell} = 0.0667N \text{ H}^+.$$

The H^+ ion concentration has decreased from 0.100N to 0.0667N because we added 10.0 ml of 0.100N NaOH to the 50.0 ml of 0.100N HCl.

How much will the H^+ ion concentration decrease if we add 20.0 ml of 0.100N NaOH?

$$(0.100N \text{ HCl})(0.050 \, \ell) = 5.00 \times 10^{-3} \text{ g-eq HCl present},$$
$$(0.100N \text{ NaOH})(0.020 \, \ell) = 2.00 \times 10^{-3} \text{ g-eq NaOH added},$$
$$5.00 \times 10^{-3} - 2.00 \times 10^{-3} = 3.00 \times 10^{-3} \text{ g-eq HCl remains unreacted}$$

in a total volume of $(0.0500 + 0.0200 \ \ell) = 0.0700 \ \ell$. The H^+ ion concentration is

$$\frac{3.00 \times 10^{-3} \text{ g-eq HCl}}{0.0700 \ \ell} = 0.0429N \ H^+.$$

Note that this entire calculation simplifies to:

$$(\text{Initial } N \text{ HCl})(V \text{ neutralized}) = (\text{final } N \text{ HCl})(V \text{ solution}),$$
$$(0.100N \text{ HCl})(0.0500 - 0.0200 \ \ell) = (\text{final } N \text{ HCl})(0.0500 + 0.0200 \ \ell),$$
$$\text{Final } N \text{ HCl} = \frac{(0.100N \text{ HCl})(0.0300 \ \ell)}{(0.0700 \ \ell)} = 0.0429N.$$

For neutralization of a strong acid with a strong base, we know that the H^+ ion concentration at the endpoint is $1.00 \times 10^{-7}N$. When we add 50.0 ml of $0.100N$ NaOH to 50.0 ml of $0.100N$ HCl, complete reaction occurs and the endpoint in the titration is attained ($0.100 \ N \times 0.0500 \ \ell = 5.00 \times 10^{-3}$ g-eq of both HCl and NaOH are present).

If we add NaOH beyond the endpoint, we can calculate the OH^- ion concentration, since its concentration will now be higher than the H^+ ion concentration. If we add 60.0 ml of $0.100N$ NaOH to the 50.0 ml of $0.100N$ HCl, 50.0 ml of the base will be neutralized, but 10.0 ml will remain unreacted. Therefore there will be

$$(0.100N \text{ NaOH})(0.0100 \ \ell) = 1.00 \times 10^{-3} \text{ g-eq NaOH}$$

Table 12–2. Titration of 50.0 ml of $0.100N$ HCl with $0.100N$ NaOH

Volume of 0.100N NaOH added, ml	$[H^+]$, mole/ℓ	$-\log [H^+]$
0.0	1.00×10^{-1}	1.00
10.0	6.67×10^{-2}	1.18
20.0	4.29×10^{-2}	1.37
30.0	2.50×10^{-2}	1.60
40.0	1.11×10^{-2}	1.95
49.0	1.01×10^{-3}	3.00
49.9	1.00×10^{-4}	4.00
50.0	1.00×10^{-7}	7.00
50.1	1.00×10^{-10}	10.00
51.0	1.00×10^{-11}	11.00
60.0	1.10×10^{-12}	11.96
70.0	6.00×10^{-13}	12.22
80.0	4.33×10^{-13}	12.36
90.0	3.50×10^{-13}	12.46
100.0	3.00×10^{-13}	12.52

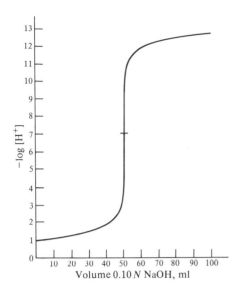

Fig. 12–2 A titration curve for the addition of 0.100N NaOH to 50.0 ml of 0.100N HCl.

in a solution whose total volume is

$$50.0 \text{ ml HCl} + 60.0 \text{ ml NaOH} = 110 \text{ ml.}$$

The OH$^-$ ion concentration is

$$\frac{1.00 \times 10^{-3} \text{ g-eq NaOH}}{0.110 \ \ell} = 0.00909N \text{ NaOH.}$$

As we shall see in Section 13–10, the corresponding H$^+$ ion concentration in the solution is

$$[\text{H}^+] = \frac{1.00 \times 10^{-14}}{[\text{OH}^-]} = \frac{1.00 \times 10^{-14}}{9.09 \times 10^{-3}} = 1.10 \times 10^{-12}M = 1.10 \times 10^{-12}N.$$

Table 12–2 gives the results of calculations similar to those above. Note that the H$^+$ concentration changes drastically between 49 and 51 ml of NaOH added, but that the change in H$^+$ concentration between 0 and 40 ml and between 60 and 100 ml of NaOH added is quite small. One other aspect of the data is also apparent. There is a huge change in the H$^+$ ion concentration, from 0.100 or 1.00 × 10^{-1}N to 3.00 × 10^{-13}N.

To facilitate plotting this data on a graph, as shown in Fig. 12–2, we usually plot $-\log[\text{H}^+]$ versus the volume of standard solution (the solution of known concentration that is being added to the solution of unknown concentration) used.* Figure 12–2 shows the dramatic change in the concentration of H$^+$ ion that occurs in the immediate vicinity of the endpoint.

---

*   As we shall see in Section 13–11, $-\log[\text{H}^+]$ is defined as the pH of a solution. Thus acid–base titration curves are generally plotted as pH vs. volume of standard solution added.

▶ If a weak acid is titrated with a strong base (or a weak base with a strong acid), the titration curve is similar, but not identical, to that in Fig. 12–2. The initial H^+ concentration is lower, since the weak acid is not completely dissociated. This raises the entire lower section of the titration curve and changes its shape slightly. Beyond the endpoint, the concentration of OH^- ion is increased slightly over the strong-acid–strong-base example in Fig. 12–2 because of hydrolysis of the anion of the weak acid. Again this raises the titration curve slightly. ◀

It should be noted that titration procedures are also applicable to oxidation-reduction reactions. In these cases the concentrations of the reactants or products, rather than of H^+ ion, must be followed (sometimes via indicators) during the course of the reactions.

QUESTIONS

1. Define or discuss the following terms: Arrhenius acid and base, Brønsted-Lowry acid and base, proton donor, hydrogen ion, anhydride, conjugate pair, neutralization, salt, strong acid, weak base, spectator ion, strong-acid–weak-base neutralization, net equation, polyprotic acid, one gram-equivalent of acid, the gram-equivalent weight of a base, titration, endpoint, indicator.

2. Indicate what is meant when one says that:
   a) HBr, H_2S, $HClO$ and HSO_4^- are acids.
   b) KOH, $Sr(OH)_2$, NH_4OH, $ZnOH^+$ are bases.

3. a) Write the molecular, total ionic, and net equations for the neutralization reaction between the weak acid $HClO$ (or $HOCl$) and the strong base KOH. Assume that all reaction products are soluble.
   b) What salt can be obtained in part (a)?

4. a) According to the Brønsted-Lowry theory, write the acid–base reaction between HI and H_2O.
   b) Indicate the acid–base conjugate pairs in part (a).

5. By writing the proper equations, show that HSO_4^- can function both as an acid and a base.

6. Write equations showing that N_2O_5 and SO_2 are the acid anhydrides of HNO_3 and H_2SO_3.

7. Calculate the molarity and normality of a NaOH solution made by dissolving 8.0 g NaOH in enough H_2O to give 0.50 ℓ.

8. Select only the strong acids from the following list.

Substance	Experimental data
HF	Weak electrolyte
HSO_4^-	Low percent dissociation
NaOH	High percent dissociation
HI	High percent dissociation
HSO_4^-	Weak electrolyte
Na_2SO_4	Strong electrolyte
$ZnOH^+$	Low percent dissociation

$HClO_4$	Strong electrolyte
KI	Strong electrolyte
$CuCl_2$	High percent dissociation
H_2S	Low percent dissociation
$La(OH)_3$	Strong electrolyte

9. a) Write the total ionic equation for the reaction between the weak acid HSO_4^- and highly dissociated KOH.

   b) Write the net equation for the reaction in part (a) and indicate all spectator ions present.

10. A $0.020M$ $Ca(OH)_2$ solution is made by dissolving 0.37 g $Ca(OH)_2$ in enough water to give 250 ml. What is the normality of the solution?

11. How many moles of $0.12M$ HCl will neutralize 0.75 ℓ of $0.25M$ TlOH?

12. Suppose that you want a $0.050M$ $KMnO_4$ solution for an experiment. If you used 3.16 g $KMnO_4$, how would you prepare the solution?

13. In the following two acid–base neutralization reactions, the conjugate pairs are given.

$$acid_1 \quad base_2 \quad acid_2 \quad base_1$$
$$HX + H_2O \rightleftarrows H_3O^+ + X^-$$
$$HQ + H_2O \rightleftarrows H_3O^+ + Q^-$$

Given that Q^- is a strong base and X^- is a weak base, rank the acids in terms of acid strength.

14. By writing appropriate equations, give the salt(s) that can be obtained when KOH and H_2SO_3 are reacted.

15. What is the H^+ concentration in the final solution when 25 ml of $0.10M$ NaOH is mixed with 75 ml of $0.020N$ H_2SO_4?

16. What is the normality of a solution prepared by dissolving 3.42 g $Ba(OH)_2$ in enough H_2O to give 200 ml of solution?

17. A neutralization reaction is carried out with 500 ml of $0.50M$ HX, a strong acid, and a strong base MOH. Given that the final solution of 750 ml has $[H^+] = 1.0 \times 10^{-7}$, what is the molarity of the MOH?

18. Write the molecular, total ionic, and net equations for the neutralization of the strong acid H_2SO_4 with the weak base NH_4OH.

19. How many gram-equivalents of H_2SO_4 are contained in 500 ml of $1.2N$ H_2SO_4?

20. Suppose that a solution is prepared by dissolving 0.20 mole $Ca(OH)_2$ in enough H_2O to give 2000 ml of solution.

   a) What is the molarity of the solution?
   b) What is the normality of the solution?
   c) What is the number of moles of $Ca(OH)_2$ present in 0.10 ℓ?
   d) What is the number of gram-equivalents of $Ca(OH)_2$ in 500 ml?
   e) What is the volume of $0.50M$ HCl needed to neutralize the entire solution?

21. What volume of $0.10N$ NaOH can be prepared from 2.0 g NaOH?

22. How many liters of $0.060M$ $HClO_4$ will exactly react with 30 ml of $0.0010M$ $Ba(OH)_2$?

23. What is the molarity of HBr, given that 50 ml are exactly neutralized by 150 ml of $0.0050N$ $Ca(OH)_2$?

24. What will be the final OH^- concentration in a solution in which 100 ml of $0.50M$ KOH, a strong base, is mixed with 0.25 ℓ of $0.020M$ $HC_2H_3O_2$, a weak acid?

25. How many gram-equivalents of $0.12N$ H_2S are needed to exactly neutralize 0.120 ℓ of $0.20N$ NaOH?

26. How many liters of $0.20M$ HCl can be prepared from 200 ml of $1.2M$ HCl?

27. What is the molecular weight of a monoprotic acid, given that 2.52 g of the acid exactly neutralize 0.200 ℓ of $0.100M$ KOH?

28. What is the equivalent weight of a base, given that 0.100 ℓ of a solution containing 5.61 g of the base is completely neutralized by 200 ml of $0.50N$ H_2SO_4?

29. How many milliliters of $0.25M$ H_2SO_4 will react exactly with 0.050 gram-equivalent of $Al(OH)_3$?

30. How many moles of the insoluble salt $BaSO_4$ are formed when 50 ml of $0.24M$ H_2SO_4 are mixed with 1.0 ℓ of $0.10M$ $Ba(OH)_2$?

31. Suppose that a solution containing 0.12 gram-equivalent of NaOH (a strong base) is mixed with 100 ml of $0.50M$ H_2SO_3 (a weak acid). How many grams of H_2O will be produced by the reaction?

32. What is the final normality of 200 ml of $1.2N$ H_2SO_4, given that the solution is mixed with 300 ml H_2O?

33. What volume of $0.020N$ $H_2C_2O_4$ can be prepared from 100 ml of $0.10M$ $H_2C_2O_4$?

34. A 250-ml solution of $0.10M$ HNO_3 (a strong acid) is titrated with 400 ml of $0.050M$ NaOH. What is the H^+ ion concentration in the final solution?

35. A 100.0 ml solution of $0.100M$ HBr (a strong acid) is titrated with $0.200M$ KOH (a strong base).
   a) Calculate the H^+ ion concentration after the addition of 0.0, 30.0, 40.0, 49.0, 49.9, and 50.0 ml of $0.200M$ KOH.
   b) Plot the data in part (a) as $-\log[H^+]$ vs. ml $0.200M$ KOH added.

36. A 50-ml sample of $0.50M$ H_2SO_3 (a weak acid) is neutralized with the strong base NaOH.
   a) Write the molecular, total ionic, and net equations describing the reaction.
   b) What salt(s) can possibly be obtained?
   c) How many moles of H_2SO_3 are present in the sample?
   d) Suppose that the H_2SO_3 is completely neutralized with NaOH. How many gram-equivalents of H_2SO_3 are neutralized?
   e) For complete neutralization of H_2SO_3, what is the normality of the $0.50M$ H_2SO_3 solution?
   f) What volume of $0.25M$ NaOH is required to completely neutralize the H_2SO_3 sample?
   g) If the H_2SO_3 sample is completely neutralized by 10 ml of a NaOH solution, what is the normality of the NaOH?
   h) If the H_2SO_3 sample is mixed with 50 ml of $1.5M$ NaOH, what is the OH^- ion concentration in the final solution?
   i) For complete neutralization of the H_2SO_3 sample, how many gram-equivalents of NaOH are required?
   j) How many moles of H_2O can be formed if the H_2SO_3 sample is mixed with 40 ml of $1.0M$ NaOH?
   k) For complete neutralization, what are the equivalent weights of H_2SO_3 and NaOH?

13. CHEMICAL EQUILIBRIUM

13-1 INTRODUCTION

We have previously encountered systems at equilibrium. Liquid water in a closed container, for example, is in equilibrium with water vapor. The amount of water in the vapor phase—producing the vapor pressure—is dependent on the temperature. Similarly, if we place some solid iodine in a closed container, iodine molecules will leave the surface of the solid iodine and enter the gas phase. If we were to measure the amount of iodine present in the gas phase shortly after placing the solid iodine in the closed container, it would be very small. With the passage of time, the vapor pressure of I_2 would increase until it reached a constant value, the equilibrium vapor pressure.

Why does the vapor pressure of I_2 remain a constant at a fixed temperature? Why doesn't it continue to increase? It might be that for some reason the solid I_2 stopped evaporating, or it might be that the tendency of the solid I_2 to evaporate is exactly balanced by the tendency of I_2 gas to condense on the surface of the solid. We can devise an experiment to distinguish between these two possibilities. If for example we introduce into the closed container some more solid I_2 containing radioactive iodine and, after a period of time, withdraw some I_2 gas, we could measure the radioactivity of the gas. We would find, in such an experiment, that the I_2 gas was radioactive and that the equilibrium between I_2 solid and I_2 gas in a

closed vessel at constant temperature is *dynamic*. The evaporation–condensation processes are taking place at the same rate and there is no *net* change:

$$I_{2(s)} \rightleftarrows I_{2(g)}.$$

For chemical reactions the situation is similar. H_2 and I_2 react very slowly at room temperature, but at higher temperatures the reaction proceeds rapidly. If we heat one mole of H_2 and one mole of I_2 in a 1-liter flask at 427 °C, HI is formed:

$$H_{2(g)} + I_{2(g)} \rightarrow 2HI_{(g)}.$$

Initially, of course, the concentration of HI is very small, but it soon begins to increase until it reaches a constant value of 0.21 mole/liter. Regardless of how much longer we wait, there is no further change in the concentration of HI at 427 °C.

If we carry out a second experiment, in which two moles of HI are heated in a 1-liter flask at 427 °C, we note that HI decomposes into H_2 and I_2:

$$2HI_{(g)} \rightarrow H_{2(g)} + I_{2(g)}.$$

When there is no further change in the concentration of the species present, we find that the HI concentration is 0.21 mole/liter. In other words, the equilibrium concentration of HI is the same for both the forward and reverse reactions. The same is true for the equilibrium concentrations of H_2 and I_2:

$$H_2 + I_2 \rightleftarrows 2HI.$$

The equilibrium concentrations are determined by the relative tendencies of the forward and reverse reactions to take place. In other chemical equilibria the situation is similar. In fact many reactions do not proceed to completion, and in most cases this is the result of an equilibrium condition. For example, acetic acid is called a weak acid because it is not extensively dissociated in water solution. This does not mean that the reaction

$$HC_2H_3O_2 + H_2O \rightarrow C_2H_3O_2^- + H_3O^+$$

does not take place, but rather that the reverse reaction

$$C_2H_3O_2^- + H_3O^+ \rightarrow HC_2H_3O_2 + H_2O$$

has a greater tendency to take place. The equilibrium nature of this reaction is usually indicated by a double arrow:

$$HC_2H_3O_2 + H_2O \rightleftarrows C_2H_3O_2^- + H_3O^+.$$

This equation does not indicate the position of the equilibrium—whether the forward reaction proceeds nearly to completion, or only to a small extent. In this chapter we shall consider ways in which the position of an equilibrium can be described and treated quantitatively.

13-2 THE EQUILIBRIUM CONSTANT

So long as we do not change the temperature, we find that the square of the equilibrium concentration of HI divided by the product of the equilibrium concentrations of H_2 and I_2 is a constant. This relationship can be expressed as

$$K = \frac{[HI]^2}{[H_2][I_2]},$$

and is experimentally verifiable. K is called the *equilibrium constant*.

▶ By convention, when we write the formula of a molecule in brackets, such as [HI], it refers to the concentration of that species in moles per liter. ◀

Another system that demonstrates a chemical equilibrium is the *dimerization* (coupling) of NO_2 to give N_2O_4:

$$2NO_{2(g)} \rightleftarrows N_2O_{4(g)}.$$

The equilibrium is such that at higher temperatures the formation of the brown NO_2 gas is favored, and at lower temperatures the colorless N_2O_4 predominates. At any fixed temperature, however, the concentrations of the two gases can be shown to fit the equilibrium expression

$$K = \frac{[N_2O_4]}{[NO_2]^2}.$$

K is the equilibrium constant for the NO_2–N_2O_4 equilibrium. It defines, for example, the concentration of N_2O_4 that can be present at equilibrium in a known amount of NO_2 (at a fixed temperature). Experiments show that the equilibrium ratio of NO_2 to N_2O_4 at 60 °C and one atm pressure is 2 : 1. Therefore, if we have two moles/liter of NO_2 to one mole/liter of N_2O_4, we can calculate K:

$$K = \frac{[N_2O_4]}{[NO_2]^2} = \frac{1}{(2)^2} = \frac{1}{4} = 0.25.$$

Knowing the equilibrium constant for this system enables us to calculate the concentration of N_2O_4 present for any concentration of NO_2 at 60 °C. For example, if we have 0.2 mole of NO_2 present in a liter, we can calculate the amount of N_2O_4 as follows:

$$K = \frac{[N_2O_4]}{[NO_2]^2},$$

$$[N_2O_4] = K \times [NO_2]^2,$$

$$[N_2O_4] = 0.25 \times (0.2)^2 = 0.25 \times 0.04 = 0.01 \text{ mole}/\ell.$$

Investigations of still other equilibria lead to a general method: To write the expression for any equilibrium constant, we write the balanced equation and set K equal to the product of the product concentrations, raised to the power of their coefficients in the equation, divided by the product of the reactant concentrations, raised to the power of their coefficients. In the general case

$$wA + xB \rightleftarrows yC + zD,$$

we write the equilibrium expression as

$$K = \frac{[C]^y [D]^z}{[A]^w [B]^x}.$$

The equilibrium between hydrogen and carbon dioxide and the products carbon monoxide and water can be expressed as

$$K = \frac{[CO][H_2O]}{[H_2][CO_2]}$$

for the equation

$$H_2 + CO_2 \rightleftarrows CO + H_2O.$$

Thus we can write the equilibrium expression for any chemical reaction merely by knowing the balanced equation for that reaction.

Example 13-1. Write the equilibrium expression for the formation of ammonia from nitrogen and hydrogen.

First we must write the balanced equation for the equilibrium:

$$3H_2 + N_2 \rightleftarrows 2NH_3.$$

The product concentration raised to the power of its coefficient is $[NH_3]^2$. The reactant concentrations similarly expressed are $[H_2]^3$ and $[N_2]$. Therefore

$$K = \frac{[NH_3]^2}{[H_2]^3 [N_2]}.$$

13-3 CALCULATIONS INVOLVING THE EQUILIBRIUM CONSTANT

The equilibrium constant can be determined by experimentation if we measure the concentrations at equilibrium at a fixed temperature. The following examples show how we can calculate the equilibrium constant, K, from experimental data.

Example 13-2. A mixture of one mole of H_2 and one mole of I_2 in a 1-liter flask is heated at 427 °C until equilibrium is attained. Measurements show the equilibrium I_2 concentration to be 0.21 mole/liter. Calculate K.

$$H_2 + I_2 \rightleftarrows 2HI.$$

Since 0.21 mole of I_2 remains at equilibrium and I_2 reacts with H_2 in a one-to-one proportion, 0.21 mole/liter must also be the equilibrium concentration of H_2.

$$[H_2] = [I_2] = 0.21 \text{ mole}/\ell.$$

From the balanced equation we can write the expression for the constant:

$$K = \frac{[HI]^2}{[H_2][I_2]}.$$

The balanced equation also tells us that every mole of I_2 that reacts must produce 2 moles of HI. Since 0.21 mole of I_2 remains at equilibrium, 0.79 mole I_2 must have reacted with 0.79 mole of H_2 to produce 2×0.79 mole HI. Thus the HI concentration must be 1.58 moles/liter, and

$$K = \frac{(1.58)^2}{(0.21)(0.21)} = 56.9.$$

▶ In this instance the equilibrium constant K is dimensionless, since all the units cancel:

$$K = \frac{(\text{mole/liter})^2}{(\text{mole/liter})(\text{mole-liter})}.$$

This is not generally true, however. Even though the units are often ignored, certain equilibrium constants have dimensions. For example the equilibrium $2NO_2 \rightleftarrows N_2O_4$ discussed above gives us:

$$K = \frac{[N_2O_4]}{[NO_2]^2} = \frac{(\text{mole/liter})}{(\text{mole/liter})^2} = \text{liter/mole.} \blacktriangleleft$$

When we know the equilibrium constant for a chemical reaction, we can use it to calculate equilibrium concentrations of the species involved, as illustrated in Example 13-3.

Example 13-3. Four moles of hydrogen iodine (HI) are heated in a 1-liter flask at 427 °C until equilibrium is established. Calculate the concentration of I_2 at equilibrium. (The same equilibrium constant must be established whether you start with HI or with H_2 and I_2.)

From the balanced equation, $H_2 + I_2 \rightleftarrows 2HI$, we know that every mole of HI that decomposes will produce 0.5 mole of I_2 and 0.5 mole of H_2. If in this instance we assume that x moles of HI have decomposed, then the equilibrium concentrations will be:

$$[HI] = (4.0 - x) \quad \text{and} \quad [I_2] = [H_2] = 0.5x$$

Thus

$$K = \frac{[HI]^2}{[H_2][I_2]} = 56.9 \quad \text{and} \quad \frac{(4.0 - x)^2}{(0.5x)(0.5x)} = 56.9.$$

Therefore

$$\frac{(4.0 - x)^2}{(0.5x)^2} = 56.9.$$

Taking the square root of both sides of the equation and solving for x, we find that

$$\frac{4.0 - x}{0.5x} = \sqrt{56.9} = 7.54,$$

$$4.0 - x = 3.77x, \quad 4.0 = 3.77x + x = 4.77x,$$

$$x = \frac{4.0}{4.77} = 0.84.$$

Thus, when a 4.0-mole sample of HI is heated at 427 °C in a 1-liter flask, 0.84 mole of HI decomposes to give 0.42 mole of I_2 and 0.42 mole of H_2 at equilibrium.

13-4 FACTORS AFFECTING EQUILIBRIA

Effect of Temperature on Equilibrium. In Chapter 10 we discussed the effect of temperature on systems at equilibrium, in light of the LeChatelier principle. The LeChatelier principle applies to chemical equilibria as well. The equilibrium constant, as we have said, is temperature dependent. The nature of the temperature dependence has to do with the relative energetics of the forward and reverse reactions. If, for the equilibrium

$$A + B \rightleftarrows C + D,$$

the forward reaction $A + B \rightarrow C + D$ is exothermic, then the reverse reaction, $C + D \rightarrow A + B$, is endothermic by the same amount. When the temperature is raised, a stress is placed on the system at equilibrium, and according to the LeChatelier principle the equilibrium shifts to remove the stress. In this instance the endothermic reaction can consume the excess heat so that the equilibrium shifts to the left, increasing the concentrations of A and B. Thus an increase in temperature favors the endothermic portion of an equilibrium more than the exothermic portion, causing a shift in the equilibrium in the endothermic direction.

For example, the formation of HI from H_2 and I_2 is exothermic:

$$H_2 + I_2 \rightarrow 2HI + 3 \text{ kcal/mole}.$$

The reverse reaction is therefore endothermic,

$$2HI \rightarrow H_2 + I_2 - 3 \text{ kcal/mole},$$

and if an equilibrium mixture of H_2, I_2, and HI is heated, the composition will change to a lower concentration of HI and larger concentrations of H_2 and I_2. Thus for the equilibrium

$$H_2 + I_2 \rightleftarrows 2HI,$$

an increase in temperature will result in a decrease in the equilibrium constant K.

It is not possible to predict the effect of a change in temperature on a system at equilibrium unless it is known which reaction is exothermic and which endothermic.

Effect of Changes in Concentration on Equilibrium. The effect of changes in concentration of one or more species in an equilibrium is easily predictable. For example, the equilibrium

$$H_2 + I_2 \rightleftarrows 2HI$$

must satisfy the equilibrium expression

$$K = \frac{[HI]^2}{[H_2][I_2]}.$$

Since the equilibrium constant K does not change (at constant temperature), an increase in the concentration of H_2 would have to result in an increase in the concentration of HI and a decrease in the concentration of I_2 in order for the equilibrium expression to be satisfied. In other words, an increase in the concentration of H_2 would cause more I_2 to react to form HI. According to the LeChatelier principle, then, the increase in the concentration of H_2 places a stress on the equilibrium. The stress is removed by a shift in the equilibrium, reducing the concentration of H_2; that is, there is a shift to the formation of more HI.

Example 13–4. A 1-liter flask at equilibrium contains 0.21 mole H_2, 0.21 mole I_2, and 1.58 moles of HI. What effect will the addition of 1 mole of H_2 have on the equilibrium?

Let us assume that the temperature is the same as in the previous example, and that

$$K = \frac{[HI]^2}{[H_2][I_2]} = 56.9.$$

When the one mole excess of H_2 is added, the equilibrium is disturbed and H_2 and I_2 are consumed as HI is formed. Let the net number of moles of H_2 reacting equal x; then at equilibrium

$$[H_2] = 1.00 + 0.21 - x = 1.21 - x, \qquad [I_2] = 0.21 - x,$$

and

$$[HI] = 1.58 + 2x.$$

Therefore

$$K = \frac{(1.58 + 2x)^2}{(1.21 - x)(0.21 - x)} = 56 \cdot 9$$

and

$$x = 0.18 \text{ mole.}^*$$

Since $x = 0.18$ mole, then the new equilibrium concentrations are

$$[H_2] = 1.21 - 0.18 = 1.03 \text{ moles}/\ell,$$
$$[I_2] = 0.21 - 0.18 = 0.03 \text{ mole}/\ell,$$
$$[HI] = 1.58 + 2(0.18) = 1.94 \text{ moles}/\ell.$$

Effect of Catalysts on Equilibrium. A catalyst has no effect on an equilibrium; it may increase the rate at which equilibrium is attained, but it can have no effect on the equilibrium concentrations, since it affects the forward and reverse reactions identically.

13-5 CHEMICAL EQUILIBRIUM IN SOLUTION

Dissociation of Water. In Chapter 12 we discussed the very weak dissociation of water. We can illustrate this dissociation as an equilibrium,

$$H_2O \rightleftarrows H^+ + OH^-,$$

and write the equilibrium expression

$$K = \frac{[H^+][OH^-]}{[H_2O]}.$$

Using very precise measurements of electrical conductivity and very pure water at 25 °C, experimenters have shown that the concentrations of H^+ and OH^- are each 10^{-7} mole/liter. Thus

$$K = \frac{[1 \times 10^{-7}][1 \times 10^{-7}]}{[H_2O]}.$$

Since the amount of water that ionizes is extremely small relative to the amount of water present, the $[H_2O]$ remains approximately constant, and we can define a new expression, the *ion product constant* K_w. K_w is the product of the concentrations of

---

*   The solution for x in this instance requires the use of the quadratic equation

$$x = \frac{-b \pm \sqrt{b^2 - 4ac}}{2a}.$$

the ions H^+ and OH^- and ignores the water concentration term. At 25 °C,

$$K_w = [H^+][OH^-] = (1 \times 10^{-7})(1 \times 10^{-7}) = 1 \times 10^{-14}.$$

The K_w must hold for any water system at 25 °C. If we increase the hydrogen ion concentration (for example, in strong acid solutions), the hydroxide ion concentration must be decreased by the same amount; if the hydroxide ion concentration increases, the hydrogen ion concentration must decrease. If we can measure one concentration, we can calculate the other as follows:

$$[H^+] = \frac{K_w}{[OH^-]}; \qquad [OH^-] = \frac{K_w}{[H^+]}.$$

Example 13–5. What is the hydroxide ion concentration of a 0.01M HCl solution?

Since HCl is a strong acid, we can assume that the dissociation into ions is complete. Therefore the hydrogen ion concentration is equal to 0.01M plus whatever concentration is contributed by the ionization of water. We know that the ionization of water is very small, however, so we can assume that all the hydrogen ions are donated by the HCl. Thus

$$[H^+] = 0.01M$$

and

$$[OH^-] = \frac{K_w}{[H^+]} = \frac{1 \times 10^{-14}}{0.01} = \frac{1 \times 10^{-14}}{1 \times 10^{-2}}.$$

Therefore

$$[OH^-] = 1 \times 10^{-12}.$$

Example 13–6. Calculate the hydrogen ion concentration of a 0.0001M NaOH solution.

As in the previous exercise, we can assume that the contribution of hydroxide ions from the ionization of water is negligible and, since NaOH is a strong base and completely dissociated in solution, all the hydroxide ions are donated by the NaOH. We can then say that

$$[OH^-] = 0.0001M = 1 \times 10^{-4} \qquad \text{and} \qquad [H^+] = \frac{K_w}{[OH^-]}.$$

Therefore

$$[H^+] = \frac{1 \times 10^{-14}}{1 \times 10^{-4}} = 1 \times 10^{-10}.$$

13–6 THE pH SCALE

From Section 13–5 we can see that the hydrogen ion concentration of an acid is greater than 1×10^{-7} and the hydrogen ion concentration of a base is less than 1×10^{-7}. In other words, the hydrogen ion concentration can be used as a quantitative measure of the acidity or basicity of water solutions. In weakly acidic or basic systems, the

Table 13–1. Relation of pH to hydrogen ion concentration

pH	$[H^+]$ in moles/liter	pH	$[H^+]$ in moles/liter
14	1×10^{-14} ($1M$ OH^-)	6	1×10^{-6}
13	1×10^{-13}	5	1×10^{-5}
12	1×10^{-12}	4	1×10^{-4}
10	1×10^{-10}	3	1×10^{-3}
9	1×10^{-9}	2	1×10^{-2}
8	1×10^{-8}	1	1×10^{-1}
7	1×10^{-7} (pure H_2O)	0	1×10^0 ($1M$ H^+)

hydrogen ion concentration is quite small and must be expressed as an exponential number. This is often cumbersome for routine work, and so for convenience let us define a new acidity scale: the *pH scale*. pH is defined as the negative logarithm of the hydrogen ion concentration:

$$pH = -\log[H^+] \quad \text{or} \quad pH = \frac{1}{\log[H^+]}.$$

(A review of logarithms can be found in Appendix A.)

Table 13–1 shows the relation of the pH scale to $[H^+]$.

Example 13–7. Calculate the pH of pure water, of $0.01M$ HCl, and of $0.0001M$ NaOH.

The hydrogen ion concentrations in these three solutions, as shown previously, are 1×10^{-7}, 1×10^{-2}, and 1×10^{-10}, respectively. For pure water,

$$pH = -\log[H^+] = -\log[10^{-7}] = -(-7) = +7.$$

Similarly, for $0.01M$ HCl,

$$pH = -\log 10^{-2} = -(-2) = 2,$$

and for $0.0001M$ NaOH,

$$pH = -\log 10^{-10} = -(-10) = 10.$$

13-7 WEAK ACIDS

Ionization Constants of Weak Acids. Weak acids and bases in water solutions can be treated quantitatively as equilibrium systems. For example, acetic acid and hydrocyanic acid are weak acids which can be represented with the equilibria

$$HC_2H_3O_2 \rightleftarrows H^+ + C_2H_3O_2^-$$

acetic acid   acetate ion

$$HCN \quad \rightleftarrows \quad H^+ + CN^-$$

hydrocyanic acid   cyanide ion

for which the equilibrium expressions can be written:

$$K = \frac{[H^+][C_2H_3O_2^-]}{[HC_2H_3O_2]}; \qquad K = \frac{[H^+][CN^-]}{[HCN]}.$$

▶ In each instance the equilibria refer to weak acids in water solution and thus measure the ability of the acid to donate a proton to a water molecule. For example,

$$HC_2H_3O_2 + H_2O \rightleftharpoons C_2H_3O_2^- + H_3O^+.$$

We have written the equilibria in the simpler form, ignoring the presence of water and using H^+ instead of H_3O^+. A stronger acid has a greater tendency to donate a proton to water, and this is reflected in a relatively larger value for the H^+ concentration. ◀

For a weak acid containing more than one acidic hydrogen, such as H_2CO_3 or H_3PO_4, more than one equilibrium expression can be written:

$$H_2CO_3 \rightleftharpoons H^+ + HCO_3^-, \qquad HCO_3^- \rightleftharpoons H^+ + CO_3^{2-},$$

$$K_1 = \frac{[H^+][HCO_3^-]}{[H_2CO_3]}, \qquad K_2 = \frac{[H^+][CO_3^{2-}]}{[HCO_3^-]}.$$

$$H_3PO_4 \rightleftharpoons H^+ + H_2PO_4^-, \qquad H_2PO_4^- \rightleftharpoons H^+ + HPO_4^{2-}, \qquad HPO_4^{2-} \rightleftharpoons H^+ + PO_4^{3-},$$

$$K_1 = \frac{[H^+][H_2PO_4^-]}{[H_3PO_4]}, \qquad K_2 = \frac{[H^+][HPO_4^{2-}]}{[H_2PO_4^-]}, \qquad K_3 = \frac{[H^+][PO_4^{3-}]}{[HPO_4^{2-}]}.$$

Example 13–8. The acetic acid in a 0.1M solution was shown to be 1.34% ionized at 25 °C. Calculate the equilibrium constant for acetic acid.

If 0.1 mole $HC_2H_3O_2$ were 100% ionized, it would yield 0.1 mole H^+ and 0.1 mole $C_2H_3O_2^-$, but since it is only 1.34% ionized, the equilibrium concentrations are

$$[H^+] = 0.1 \times 0.0134 = 0.00134 \text{ mole}/\ell,$$
$$[C_2H_3O_2^-] = 0.1 \times 0.0134 = 0.00134 \text{ mole}/\ell.$$

The un-ionized acetic acid must constitute 98.66% of the original 0.1 mole, or

$$[C_2H_3O_2^-] = 0.1 \times 0.9866 = 0.09866 \text{ mole}/\ell.$$

Therefore

$$K = \frac{[H^+][C_2H_3O_2^-]}{[HC_2H_3O_2]} = \frac{(0.00134)(0.00134)}{(0.09866)} = 1.82 \times 10^{-5}.$$

By measuring the extent of ionization of various weak acids, we can calculate the ionization equilibrium constants K (sometimes called dissociation constants, K_{Diss}). Table 13–2 lists several of these.

Table 13-2. Ionization equilibrium constants

Name	Reaction	K (25 °C)
Sulfuric acid	$H_2SO_4 \rightleftharpoons H^+ + HSO_4^-$	Large
	$HSO_4^- \rightleftharpoons H^+ + SO_4^{2-}$	1.2×10^{-2}
Sulfurous acid	$H_2SO_3 \rightleftharpoons H^+ + HSO_3^-$	1.5×10^{-2}
	$HSO_3^- \rightleftharpoons H^+ + SO_3^{2-}$	1.0×10^{-7}
Phosphoric acid	$H_3PO_4 \rightleftharpoons H^+ + H_2PO_4^-$	7.5×10^{-3}
	$H_2PO_4^- \rightleftharpoons H^+ + HPO_4^{2-}$	6.2×10^{-8}
	$HPO_4^{2-} \rightleftharpoons H^+ + PO_4^{3-}$	2.2×10^{-13}
Hydrocyanic acid	$HCN \rightleftharpoons H^+ + CN^-$	5×10^{-10}
Acetic acid	$HC_2H_3O_2 \rightleftharpoons H^+ + C_2H_3O_2^-$	1.8×10^{-5}
Formic acid	$HCHO_2 \rightleftharpoons H^+ + CHO_2^-$	1.8×10^{-4}

Calculations Involving Ionization Equilibrium Constants. Once we know the ionization equilibrium constant, we can use it to calculate equilibrium concentrations for all the species involved in the dissociation reaction.

Example 13-9. Calculate the hydrogen ion concentration of a 0.50M acetic acid solution at 25 °C.

At equilibrium the hydrogen ion concentration must equal the acetate ion concentration, since ionization of one acetic acid molecule must furnish, simultaneously, one hydrogen ion and one acetate ion:

$$HC_2H_3O_2 \rightleftharpoons C_2H_3O_2^- + H^+.$$

Let x equal the number of moles/ℓ of $HC_2H_3O_2$ ionized at equilibrium. Then

$$[H^+] = [C_2H_3O_2^-] = x.$$

The undissociated acetic acid must equal the total initial concentration minus the amount ionized, or

$$[HC_2H_3O_2] = 0.50 - x.$$

From Table 13-2 we see that $K = 1.8 \times 10^{-5}$ and

$$1.8 \times 10^{-5} = \frac{[H^+][C_2H_3O_2^-]}{[HC_2H_3O_2]}.$$

Substituting the concentration terms gives us

$$1.8 \times 10^{-5} = \frac{(x)(x)}{0.50 - x} = \frac{x^2}{0.50 - x}.$$

In order to solve this equation for the hydrogen ion concentration, x, we must use the quadratic equation (as on page 374). However, in this case we can make an approximation which simplifies the solution to the problem without significantly altering the accuracy. Since the ionization constant for acetic acid is small, the number of acetic acid molecules ionizing relative to the

number remaining undissociated is very small, that is, x is a very small number. Consequently, we say that

$$[HC_2H_3O_2] = 0.50 - x \simeq 0.50$$

and

$$1.8 \times 10^{-5} = \frac{x^2}{0.5}, \qquad 0.5(1.8 \times 10^{-5}) = x^2.$$

Therefore

$$x^2 = 0.9 \times 10^{-5} = 9 \times 10^{-6},$$

and

$$x = [H^+] = 3 \times 10^{-3}M.$$

▶ We can see why acetic acid is called a weak acid when we compare the H^+ concentration of $0.5M$ acetic acid ($[H^+] = 3 \times 10^{-3}M$) with that of $0.5M$ hydrochloric acid ($[H^+] = 0.5M$). The tendency of HCl to donate a proton to a water molecule is much greater than that of acetic acid. ◀

Example 13–10. What is the pH of a $0.2M$ HCN solution?

In order to calculate pH we must first determine the hydrogen ion concentration. We know from Table 13–2 that

$$K = \frac{[H^+][CN^-]}{[HCN]} = 5 \times 10^{-10}.$$

If we say, from HCN $\rightleftarrows H^+ + CN^-$, that the equilibrium concentrations are

$$[H^+] = [CN^-] = x \qquad \text{and} \qquad [HCN] = 0.2 - x \simeq 0.2,$$

then

$$\frac{x^2}{0.2} = 5 \times 10^{-10} \qquad \text{and} \qquad x^2 = 1 \times 10^{-10}.$$

Therefore

$$x = [H^+] = 1 \times 10^{-5}$$

and

$$pH = -\log[H^+] = -\log(1 \times 10^{-5}) = -(-5) = 5.$$

13-8 THE COMMON ION EFFECT

Changes involving ionic equilibria in solution have effects that we can predict by means of the LeChatelier principle. One of these changes—the addition of a common ion—has the effect we would expect from a change in concentration. For example, if an acetic acid solution were to be treated with acetate ion (in the form of a salt such as

sodium acetate), the equilibrium

$$HC_2H_3O_2 \rightleftarrows H^+ + C_2H_3O_2^-$$

would be shifted to the left. In other words, the addition of a common ion suppresses ionization; less acetic acid ionizes in the presence of sodium acetate.

Example 13–11. What is the hydrogen ion concentration in a solution which is $0.1M$ in acetic acid and $0.2M$ in sodium acetate?

The equilibrium concentration of acetate ion is given by the concentration of the completely dissociated sodium acetate plus the amount of acetic acid that ionizes, x. That is,

$$[C_2H_3O_2^-] = 0.2 + x.$$

The equilibrium concentration of acetic acid is equal to the initial concentration less the amount that ionizes, or

$$[HC_2H_3O_2] = 0.1 - x.$$

Since acetic acid is a weak acid, the amount that ionizes relative to the amount of acetate ion added and the amount of un-ionized acetic acid is very small. Thus we say that

$$[C_2H_3O_2^-] \simeq 0.2M \quad \text{and} \quad [HC_2H_3O_2] \simeq 0.1M.$$

Solving the equilibrium expression for $[H^+]$, we obtain

$$K = \frac{[H^+][C_2H_3O_2^-]}{[HC_2H_3O_2]}, \quad [H^+] = K \times \frac{[HC_2H_3O_2]}{[C_2H_3O_2^-]},$$

and substituting the appropriate values gives us

$$[H^+] = 1.8 \times 10^{-5} \times \frac{0.1}{0.2} = 0.9 \times 10^{-5} = 9 \times 10^{-6}.$$

In Example 13–8, we saw that the hydrogen ion concentration in a $0.1M$ $HC_2H_3O_2$ solution was $1.34 \times 10^{-3}M$. Therefore the addition of the common ion, $0.2M$ acetate, suppresses the ionization of acetic acid by a factor of more than 100.

Let us consider any weak acid, HA, for which the equilibrium can be written:

$$HA \rightleftarrows H^+ + A^-.$$

When we write the equilibrium expression and solve it for the hydrogen ion concentration, we can see that the acidity of any weak acid is determined by the ratio of the concentrations of the undissociated acid to the corresponding anion:

$$K = \frac{[H^+][A^-]}{[HA]}, \quad [H^+] = K \times \frac{[HA]}{[A^-]}.$$

Within limits, therefore, we can predictably control the pH of a solution of a weak acid by adding known amounts of a salt of that acid.

Example 13–12. How is the hydrogen ion concentration affected by the addition of one mole of sodium formate ($NaCHO_2$) to one liter of $0.50M$ formic acid?

We determine the hydrogen ion concentration of the initial solution, in the usual manner:

$$HCHO_2 \rightleftarrows H^+ + CHO_2^-,$$

$$K = \frac{[H^+][CHO_2^-]}{[HCHO_2]} = 1.8 \times 10^{-4} \qquad \text{(from Table 13–2)},$$

$$[H^+] = [CHO_2^-] = x, \qquad [HCHO_2] = 0.50 - x \simeq 0.50,$$

$$1.8 \times 10^{-4} = \frac{x^2}{0.50}, \qquad x = \sqrt{9.0 \times 10^{-5}},$$

$$x = [H^+] = 9.5 \times 10^{-3}.$$

When one mole of ionic sodium formate is added, the concentrations become

$$[CHO_2^-] = 1.0 + x \simeq 1.0M \qquad \text{and} \qquad [HCHO_2] = 0.50 - x \simeq 0.50.$$

Thus

$$[H^+] = K \times \frac{[HCHO_2]}{[CHO_2]} = 1.8 \times 10^{-4} \times \frac{0.5}{1.0} = 0.9 \times 10^{-4} = 9.0 \times 10^{-5}.$$

This corresponds to approximately a hundredfold decrease in H^+ concentration, an *increase* of 2 pH units.

13–9 BUFFERS

Systems such as those described in the previous section not only enable us to control the acidity, but they also have another important property. If a solution of a weak acid and a salt of that weak acid is treated with a strong acid or a strong base, very little change in pH occurs. Such solutions—which have the ability to consume both acids and bases—are called *buffers*. Buffered solutions are employed in systems when control of pH is important. This is particularly important in biological systems such as the blood. Small changes in the pH of the bloodstream, which could be fatal, are prevented by the presence of buffers such as the proteins and bicarbonate ion. To illustrate the action of a buffer, let us consider once more the acetic acid–sodium acetate solution.

The equilibrium is such that the addition of a strong acid will cause a shift in the equilibrium toward the undissociated acetic acid. Rather than there being a large increase in the hydrogen ion concentration (see Fig. 12–2), the hydrogen ion is consumed by reaction with acetate ion:

$$C_2H_3O_2^- + H^+ \rightarrow HC_2H_3O_2.$$

The pH does not, then, undergo significant change.

In a similar fashion, the addition of a strong base, such as sodium hydroxide, will cause the following reaction:

$$H^+ + OH^- \rightarrow H_2O.$$

The loss of hydrogen ion does not trigger a sharp increase in pH, however, because it is compensated for by a shift in the acetic acid equilibrium to the right:

$$HC_2H_3O_2 \rightarrow H^+ + C_2H_3O_2^-.$$

Weak bases combined with salts of these weak bases can also act as buffers. Ammonium hydroxide is a weak base, for example:

$$NH_4OH \rightleftarrows NH_4^+ + OH^-, \qquad K = \frac{[NH_4^+][OH^-]}{[NH_4OH]} = 1.8 \times 10^{-5}.$$

▶ It has not been demonstrated that a molecular species such as NH_4OH exists. The solution more likely contains NH_3 dissolved in water. We could then consider the equilibrium as

$$NH_3 + H_2O \rightleftarrows NH_4^+ + OH^-.$$

The equilibrium expression

$$K = \frac{[NH_4^+][OH^-]}{[NH_3][H_2O]}$$

(assuming that the water concentration is constant) could then be written as

$$K = \frac{[NH_4^+][OH^-]}{[NH_3]} = 1.8 \times 10^{-5},$$

which is equivalent to the expression used above. We should consider NH_4OH as NH_3 closely associated with one or more water molecules. ◀

A solution containing ammonium hydroxide and the strong electrolyte ammonium chloride NH_4Cl can act as a buffer, as illustrated by the following equations:

$$NH_4^+ + OH^- \rightarrow NH_4OH$$
$$NH_4OH + H^+ \rightarrow NH_4^+ + H_2O.$$

Certain species such as bicarbonate ion (HCO_3^-) and the dihydrogen phosphate ion ($H_2PO_4^-$) are *amphoteric* (capable of reacting chemically either as an acid or a base). Since they can act as weak acids or weak bases, solutions of these materials are buffers. The following equations illustrate the buffering action of bicarbonate ion:

$$HCO_3^- + OH^- \rightarrow CO_3^{2-} + H_2O$$
$$HCO_3^- + H^+ \rightarrow H_2CO_3 \rightleftarrows H_2O + CO_2$$

Standard references, such as the *Handbook of Chemistry and Physics*, contain recipes for the preparation of solutions which buffer at various pH values.

13-10 HYDROLYSIS

Water solutions of salts are not necessarily neutral. Water solutions of salts such as sodium acetate, sodium carbonate, and sodium cyanide turn litmus paper blue, but a solution of ammonium chloride turns litmus paper red. Solutions of salts such as sodium chloride, potassium sulfate, and sodium nitrate are neutral. Testing solutions of various salts in this manner has shown that salts formed from combining a weak acid and a strong base give basic solutions, salts formed from strong acids and weak bases give acidic solutions, and salts formed from strong acids and strong bases give neutral solutions.

When a salt increases the basicity of a water solution, this is reflected in an increase in the hydroxide ion concentration. Sodium acetate (the salt of the strong base NaOH and the weak acid $HC_2H_3O_2$) can cause an increase in hydroxide ion concentration by reacting with water in the following manner:

$$C_2H_3O_2^- + H_2O \rightleftarrows HC_2H_3O_2 + OH^-.$$

A reaction of this type is called a *hydrolysis reaction*. Thus, in those salts which yield basic water solutions, it is the anionic portion of the salt which reacts with the water. The generalized form of this reaction may be written as

$$A^- + H_2O \rightleftarrows HA + OH^-.$$

We can treat hydrolysis reactions of this type quantitatively by using the equilibrium expression

$$K = \frac{[OH^-][HA]}{[H_2O][A^-]}.$$

Again assuming that the amount of water reacting is negligible with respect to the total amount of water present, we can write

$$K_h = \frac{[OH^-][HA]}{[A^-]},$$

where K_h is called the *hydrolysis equilibrium constant*. We can evaluate K_h for a given reaction by measuring the equilibrium concentrations of OH^-, A^-, and HA. We can also arrive at K_h in a different manner. Since the equilibrium in question contains HA and A^- in water solution, the equilibrium ionization constant for HA must be satisfied:

$$K = \frac{[H^+][A^-]}{[HA]}.$$

Likewise, since OH^- is in water solution, the K_w for water must also be satisfied:

$$K_w = [H^+][OH^-].$$

The important point to note is that the concentration of hydrogen ions must satisfy both expressions. Solving both for $[H^+]$ gives us

$$[H^+] = \frac{K_w}{[OH^-]} \quad \text{and} \quad [H^+] = K \times \frac{[HA]}{[A^-]}.$$

Equating and rearranging, we get

$$\frac{K_w}{[OH^-]} = K \times \frac{[HA]}{[A^-]}, \quad \frac{K_w}{K} = \frac{[OH^-][HA]}{[A^-]}$$

This is identical to the expression above for K_h, and therefore

$$K_h = \frac{K_w}{K}.$$

To determine K_h for any basic salt, then, we decide which weak acid corresponds to the anion of the salt, look up the ionization constant K for the weak acid, and divide K_w by it. The important point to remember is that any time more than one related equilibrium is in operation in any one system, all the equilibrium constants must be satisfied.

In those salts which yield acidic water solutions, the situation is entirely analogous, except that it is the cationic portion of the salt which reacts with water in the hydrolysis reaction. A typical example of an acidic salt is ammonium chloride (the salt of the weak base, ammonia, and the strong acid, HCl). The ammonium ion reacts in water solution:

$$NH_4^+ + H_2O \rightleftarrows NH_4OH + H^+.$$

The equilibrium expression for the hydrolysis reaction becomes

$$K_h = \frac{[NH_4OH][H^+]}{[NH_4^+]}$$

and we see that changes in $[H^+]$ are governed by the ratio $[NH_4OH]/[NH_4^+]$.

13-11 SOLUBILITY PRODUCT CONSTANTS

There is yet another common system which is suitable for the quantitative application of equilibrium theory: the solubility of salts that are only slightly soluble. Silver chloride, for example, is a nearly insoluble salt. A saturated solution of silver chloride in contact with solid silver chloride represents an equilibrium system:

$$AgCl_{(s)} \rightleftarrows Ag^+_{(aq)} + Cl^-_{(aq)}.$$

$$\text{solid} \qquad \text{solution}$$

The usual equilibrium expression can be written:

$$K = \frac{[Ag^+][Cl^-]}{[AgCl_{(s)}]}.$$

We find, however, that the equilibrium is independent of the amount of solid silver chloride present, i.e., the concentration of solid AgCl is constant (see Section 10–4). If we add more AgCl, the concentration of ions in solution does not change. In other words, the effect of solid AgCl on the equilibrium is constant, and we can write

$$K_{sp} = [Ag^+][Cl^-],$$

in which K_{sp} is called the *solubility product constant*. We can usually write the expression merely by inspecting the formula of the salt. The solubility product constant is equal to the product of the equilibrium concentrations of each ion in the formula (in moles/liter), raised to the power of their subscripts. The following illustrations demonstrate the method:

$$CaF_2 \qquad K_{sp} = [Ca^{2+}][F^-]^2,$$
$$PbCl_2 \qquad K_{sp} = [Pb^{2+}][Cl^-]^2,$$
$$Ca_3(PO_4)_2 \qquad K_{sp} = [Ca^{2+}]^3[PO_4^{3-}]^2.$$

The following examples illustrate how we can determine solubility product constants by means of solubility data. Table 13–3 lists solubility product constants that have been determined for a few representative salts.

Table 13.3. Solubility products

CaF_2	$K_{sp} = [Ca^{2+}][F^-]^2$	$= 1.7 \times 10^{-10}$
$Mg(OH)_2$	$K_{sp} = [Mg^{2+}][OH^-]^2$	$= 1.8 \times 10^{-11}$
$AgCl$	$K_{sp} = [Ag^+][Cl^-]$	$= 2.8 \times 10^{-10}$
$AgBr$	$K_{sp} = [Ag^+][Br^-]$	$= 5.2 \times 10^{-13}$
AgI	$K_{sp} = [Ag^+][I^-]$	$= 8.5 \times 10^{-17}$
PbS	$K_{sp} = [Pb^{+2}][S^{2-}]$	$= 1 \times 10^{-29}$
$PbSO_4$	$K_{sp} = [Pb^{2+}][SO_4^{2-}]$	$= 2.0 \times 10^{-8}$
$PbCl_2$	$K_{sp} = [Pb^{2+}][Cl^-]^2$	$= 1.8 \times 10^{-4}$
ZnS	$K_{sp} = [Zn^{2+}][S^{2-}]$	$= 4.5 \times 10^{-24}$
Bi_2S_3	$K_{sp} = [Bi^{3+}]^2[S^{2-}]^3$	$= 6.8 \times 10^{-97}$
$CaCO_3$	$K_{sp} = [Ca^{2+}][CO_3^{2-}]$	$= 1 \times 10^{-8}$
CuS	$K_{sp} = [Cu^{2+}][S^{2-}]$	$= 8.7 \times 10^{-36}$
$Fe(OH)_3$	$K_{sp} = [Fe^{3+}][OH^-]^3$	$= 1 \times 10^{-36}$
HgS	$K_{sp} = [Hg^{2+}][S^{2-}]$	$= 3.5 \times 10^{-52}$

Example 13–13. It has been determined that the solubility of $PbCl_2$ in water at 20 °C is 0.99 g/100 ml. Calculate K_{sp}.

The solubility product constant for $PbCl_2$ is

$$K_{sp} = [Pb^{2+}][Cl^-]^2.$$

First we must determine the Pb^{2+} and Cl^- ion concentrations in moles/liter. If 0.99 gram $PbCl_2$ dissolves in 100 ml, then 9.9 grams must dissolve in one liter. The formula weight of $PbCl_2$ is

278 g $PbCl_2$/mole $PbCl_2$, so

$$\frac{9.9 \text{ g}}{278 \text{ g/mole}} = 0.036 \text{ mole } PbCl_2 \text{ dissolved/liter.}$$

For every mole of $PbCl_2$ dissolved, one mole of Pb^{2+} and two moles of Cl^- are formed. Therefore

$$[Pb^{2+}] = 0.036 \text{ mole/liter} \quad \text{and} \quad [Cl^-] = 2 \times 0.036 = 0.072 \text{ mole/liter.}$$

Substituting and solving for the K_{sp} of $PbCl_2$ gives us

$$K_{sp} = (0.036)(0.072)^2 = 1.8 \times 10^{-4}.$$

When we are using solubility products for purposes of determining solubility, we must remember that K_{sp} represents saturated solutions only. In other words, the solubility product can never exceed the K_{sp}, but it can be less than the K_{sp}. For example, a solution cannot contain silver ion and chloride ion so that $[Ag^+][Cl^-]$ is greater than 2.8×10^{-10}. If we take a solution of chloride ion and add silver ion to it, a precipitate will begin to form only when the solubility product is exceeded.

Example 13–14. How much silver ion can be dissolved in a $0.10M$ sodium chloride solution?

Since the solubility of silver chloride is determined by

$$K_{sp} = [Ag^+][Cl^-],$$

the maximum silver ion concentration is

$$[Ag^+] = \frac{K_{sp}}{[Cl^-]}.$$

In this instance, $[Cl^-] = 0.10M$. Therefore

$$[Ag^+] = \frac{2.8 \times 10^{-10}}{0.10} = 2.8 \times 10^{-9}M$$

is the maximum Ag^+ concentration possible in the solution. Precipitation of AgCl will begin if more Ag^+ is added.

Example 13–15. How much lead chloride can be dissolved in a $0.10M$ solution of $Pb(NO_3)_2$?

The K_{sp} for lead chloride is

$$1.8 \times 10^{-4} = [Pb^{2+}][Cl^-]^2.$$

The concentration of lead will equal the original concentration of lead nitrate plus whatever amount of lead chloride dissolves. That is,

$$[Pb^{2+}] = 0.10 + x.$$

The concentration of chloride ion will be equal to twice the amount of lead chloride that dissolves:

$$[Cl^-] = 2x.$$

Thus

$$1.8 \times 10^{-4} = (0.10 + x)(2x)^2.$$

If we assume that x is small relative to 0.10, we can write

$$1.8 \times 10^{-4} = 0.10(2x)^2 = 0.4x^2.$$

Therefore

$$x^2 = \frac{1.8 \times 10^{-4}}{0.4} \quad \text{and} \quad x = 2.1 \times 10^{-2} \text{ mole}/\ell \text{ of PbCl}_2 \text{ dissolves.}$$

In this chapter we have tried to explain and demonstrate how systems at equilibrium behave. The practicing chemist must develop a facility in the mathematical treatment of a system at equilibrium. It is important for students of chemistry to do the same. This facility is necessary for an understanding of the interrelations of materials in a system at equilibrium, and it is this understanding that is important. We can best obtain such an understanding by solving a large number of problems.

QUESTIONS

1. Write the equilibrium expression for the following reactions:

   a) $N_2O_{4(g)} \rightleftharpoons 2NO_{2(g)}$ b) $N_{2(g)} + 3H_{2(g)} \rightleftharpoons 2NH_{3(g)}$ c) $2I_{(g)} \rightleftharpoons I_{2(g)}$
   d) $2SO_{2(g)} + O_{2(g)} \rightleftharpoons 2SO_{3(g)}$
   e) $(CH_3)_3COH_{(aq)} + H^+_{(aq)} \rightleftharpoons (CH_3)_3C^+_{(aq)} + H_2O$
   f) $Zn_{(s)} + Cu^{2+}_{(aq)} \rightleftharpoons Cu_{(s)} + Zn^{2+}_{(aq)}$
   g) $HC_2H_3O_{2(aq)} \rightleftharpoons H^+_{(aq)} + C_2H_3O^-_{2(aq)}$
   h) $H_2O + C_2H_3O^-_{2(aq)} \rightleftharpoons HC_2H_3O_{2(aq)} + OH^-_{(aq)}$

2. By measuring the intensity of the brown color of NO_2, an experimenter determines that 0.30 mole of NO_2 is present in a one-liter flask in which equilibrium was established at 60 °C. Using the data in Section 13–2, calculate the amount of N_2O_4 present at equilibrium.

3. For the equilibrium described in the previous question, what are the units of the equilibrium constant K?

4. Suppose that 0.6 mole of HI is heated in a one-liter flask at 427 °C. How much will remain after equilibrium is established?

5. What would the equilibrium concentration of HI be under the conditions of question 4, given that the flask was initially charged with 2 moles of H_2 and 2 moles of I_2 and no HI?

6. The production of ammonia (NH_3) from H_2 and N_2 in the gas phase is an exothermic reaction. Use the LeChatelier principle to explain the effect on the equilibrium constant of (a) increasing the temperature, (b) adding more N_2, and (c) increasing the pressure.

7. For the reaction in question 6, explain the effects of the indicated changes on the H_2 concentration and on the NH_3 concentration.

8. Suppose that 0.2 mole of NO_2 gas is in equilibrium in a one-liter flask with 0.01 mole of N_2O_4. What will the new equilibrium concentrations be after 2 moles of NO_2 are added to the system.

9. What is the hydrogen ion concentration in the following solutions?

a) $0.001M$ NaOH

b) 20 g of NaOH in 500 ml H_2O

c) 0.02 mole of KOH in 2 liters of H_2O

10. What is the hydroxide ion concentration of the following:

a) $0.001M$ HCl

b) The solution remaining after one adds 500 ml of $2M$ HCl to 200 ml of $1M$ NaOH

c) 0.004 mole HBr in 100 ml H_2O

11. Calculate the pH of the solutions in questions 9 and 10.

12. Calculate the hydrogen ion concentrations in $0.1M$ solutions of each of the acids listed in Table 13–1.

13. Calculate the pH of each of the acid solutions in question 12.

14. What is the hydrogen ion concentration in solutions prepared from the following:

a) 10 ml of $1M$ $HC_2H_3O_2$ + 10 ml of $1M$ $NaC_2H_3O_2$

b) 10 ml of $0.5M$ HCN + 20 ml of $0.1M$ NaCN

c) 10 ml of $1M$ $HC_2H_3O_2$ + 10 ml of $0.5M$ NaOH

d) 1 ml of $1M$ NH_4Cl + 1 ml of $1M$ NH_4OH

15. Approximately how much acetic acid must be added to a $0.1M$ solution of sodium acetate to make the solution neutral?

16. Write equations which illustrate the buffering action (reaction with acid and base) of $H_2PO_4^-$ and of HPO_4^{2-}.

17. Which of these salts would form acidic solutions in water, which basic, and which neutral?

a) NaCl   b) NaCN   c) Na_3PO_4

d) $(NH_4)_2SO_4$ e) NH_4CN f) Na_2SO_4

18. Write equations which illustrate the hydrolysis reactions of the salts in the preceding question.

19. Calculate the amount of $PbCl_2$ that can dissolve in 500 ml of (a) water, (b) $0.5M$ NaCl, (c) $0.001M$ Na_2S, and (d) $1M$ Na_2SO_4.

20. Will a precipitate form if you mix saturated solutions of AgBr and AgCl?

14. CHEMICAL KINETICS: RATES OF REACTIONS

14-1 INTRODUCTION

Chapter 13 was concerned with chemical equilibria and the relationships that govern them. The position of the equilibrium, as reflected by the equilibrium constant, is determined by the relative tendency of the forward and reverse reactions to occur. For example, if the forward reaction has a greater tendency to occur than the reverse reaction, the equilibrium constant will be large. The tendency for a reaction to occur is determined by the energetics of the reaction, that is, the relative stability of the reactants and products.

Generally speaking, if a reaction is highly exothermic, it will have a great tendency to occur. For example, the gas-phase oxidation of sulfur dioxide to sulfur trioxide is exothermic:

$$2SO_{2(g)} + 2O_{2(g)} \rightleftarrows 2SO_{3(g)} + 47 \text{ kcal.}$$

Since the reverse reaction must be endothermic by the same amount, we would predict that the tendency for the forward reaction to take place would be greater than that for the reverse reaction; that is, the equilibrium should lie to the right. The equilibrium constant for this reaction is 7.1×10^{25} at 25 °C, in agreement with our prediction. Likewise the reaction of hydrogen and oxygen to yield water vapor is strongly

exothermic at room temperature,

$$H_2 + \tfrac{1}{2}O_2 \rightleftarrows H_2O + 57.8 \text{ kcal},$$

and we could expect that the equilibrium constant for this reaction would be large. We might be tempted to predict, then, that when SO_2 or H_2 is mixed with oxygen at room temperature, a rapid, exothermic reaction would take place.

When this prediction is tested experimentally, however, we find that it fails completely. Sulfur dioxide and hydrogen do not react rapidly with oxygen at room temperature. In fact, a mixture of H_2 and O_2 appears not to react at all unless it is heated or treated with a catalyst. These experimental results and many more like them force us to conclude that we cannot use equilibrium constants to predict the rate or speed of a chemical reaction. That is, a large equilibrium constant may tell us that the forward reaction has a great energetic tendency to occur, but it cannot tell us how fast the reaction will go. The study of the factors that influence the speed or rate of a chemical reaction is called *chemical kinetics*.

14–2 ACTIVATION ENERGY

We must ask ourselves why an exothermic reaction such as that between H_2 and O_2 does not take place at a measurable rate at room temperature, but *does* do so at higher temperatures. We can generalize our discussion by considering the reaction $A + B \rightarrow C + D$. We know that at room temperature A and B molecules are in constant random motion, colliding with each other often, and yet for some reason these collisions do not lead to reaction. At higher temperatures, however, we find that a significant number of the collisions between molecules do lead to reaction. Since at higher temperatures the gas molecules have greater average kinetic energy, a greater proportion of their collisions are more energetic. Figure 14–1 shows this graphically.

▶ It can be estimated that the rate of collision between molecules of A and B gas at $0\,°C$ and 1 atm is 1.1×10^{29} collisions/cm³-sec. ◀

Thus, in order for A and B to react, the collision between them must be of a certain energy; at higher temperatures the number of *effective collisions* is greater and the rate of reaction increases. In other words, in order for a reaction to take place, an energy barrier must be overcome.

In a similar fashion, we can consider a reaction of the type $X_2 \rightarrow 2X$. (An example would be the conversion of an iodine molecule into two iodine atoms.) We know that the X atoms are vibrating along the X—X bond, but only those vibrations of great enough energy lead to bond breakage. Again we see an energy barrier which must be overcome before reaction can occur.

In fact, detailed investigation has shown that nearly all reactions must overcome an energy barrier. We describe this situation by saying that the reacting species must have a certain *activation energy* before reaction can take place. Figure 14–2 diagrams

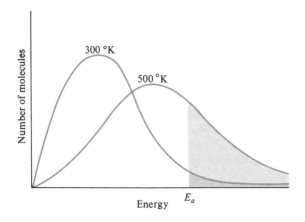

Fig. 14-1 The distribution of kinetic energies at two temperatures. The number of molecules with energy E_a or greater is proportional to the shaded area.

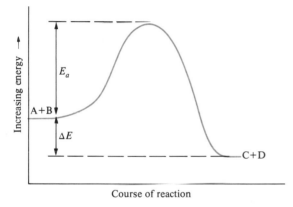

Fig. 14-2 Activation-energy diagram for the reaction A + B → C + D.

this; the energy of the reaction of A + B → C + D is shown to vary when we follow the course of the reaction from reactants to products. Since the energy of the reactants is greater than the energy of the products, the overall process is exothermic by the amount ΔE.

Before A and B can react, however, collisions between them must be energetic enough to supply the necessary *activation energy*, E_a. Relatively speaking, if the activation energy is large, the reaction will be slow; if the activation energy is small, the reaction will be fast. This concept of activation energy enables us to explain the effect of temperature on the rate of a chemical reaction: As the temperature is increased, the fraction of molecules that have the energy required for reaction increases. Thus an increase in temperature will increase the rate of a chemical reaction and a decrease in temperature will slow down a reaction.

Although there are many everyday experiences and widely used devices (such as a pressure cooker) which illustrate the effect of temperature on a reaction, the concept of activation energy gives us a unifying theory to explain the phenomena.

▶ When food is cooked, although the process is extremely complex, it is clear that chemical reactions are taking place. Many of these reactions are carried out in boiling water. The water serves as a reactant, but primarily as a medium for heat transfer. In many instances, this restricts our cooking temperature to 100 °C or below. A pressure cooker enables us to cook food faster (still in an aqueous medium) by heating the water in a closed system. The pressure is increased well above atmospheric pressure, and thus the boiling point of the water is higher. The water boils at a higher temperature and the elevated temperature increases the rates of the various reactions required to cook the food. ◀

14–3 EFFECT OF CONCENTRATION-RATE LAWS

Another factor which determines the rate of a chemical reaction is the reactant concentration. For example, the radioactive decomposition of an element is a spontaneous process and does not require that one particle collide with another. The amount of decomposition is determined only by the amount of material present and the activation energy of the process.

Suppose that we take a given volume of a solution of a sample containing carbon-14 and measure the number of beta particles emitted in one minute. If we then double the concentration of carbon-14, the number of beta particles will double. Half the concentration of $^{14}_{6}C$ yields half the number of beta particles. Several experiments of this nature soon convince us that the rate of decomposition of carbon-14 is directly proportional to the concentration of carbon-14:

$$\text{Rate} \propto [^{14}_{6}C].$$

We can change this proportion to an equation by introducing the *proportionality constant k*. The equation is called the *rate equation* or the *rate law* for the reaction

$$\text{Rate} = k\,[^{14}_{6}C]$$

and k is the *specific rate constant*. The specific rate constant k is related to the activation energy, and describes the way in which the rate of reaction depends on the concentration. If k is large, the reaction is fast; if k is small, the reaction is slow.

All reactions involving radioactive decomposition and any other reaction which can be described as

$$A \rightarrow \text{products}$$

obey a rate law: rate $= k\,[A]$. Such a reaction is called *first order*, since the rate is proportional to the concentration of the reactant raised to the *first power*.

The slightly more complicated reaction involving a collision between two reactants—such as A + B → products—is very similar. For example, we can study the

Table 14–1. Dependence of the rate of the reaction of ethyl acetate and hydroxide ion on concentration of reactants

Ethyl acetate concentration	OH⁻ concentration	Relative rate
1	1	1
1	2	2
1	3	3
2	1	2
3	1	3

reaction of ethyl acetate with sodium hydroxide

$$CH_3COOCH_2CH_3 + OH^- \rightarrow CH_3COO^- + CH_3CH_2OH$$

ethyl acetate   acetate   alcohol

by holding the concentration of hydroxide ion constant and changing the concentration of ethyl acetate.

If we double the concentration of ethyl acetate, the rate of reaction doubles; if we triple the concentration of ethyl acetate, the rate triples. In other words, the rate of the reaction is directly proportional to the concentration of ethyl acetate. Then, if we keep the concentration of ethyl acetate constant and vary the concentration of hydroxide ion, we observe the same rate–concentration relationship, that is, the rate of the reaction is also directly proportional to the concentration of hydroxide ion. In practice, this is done by measuring the rate of the reaction several times with different concentrations of reactants, as shown in Table 14–1. The rate expression obtained from such a study can be written as

$$\text{Rate} = k \, [CH_3COOCH_2CH_3][OH^-]$$

or, for the generalized reaction between A and B, in the form

$$\text{Rate} = k \, [A][B].$$

In this rate expression, the sum of the exponents of the concentration terms equals two, and the reaction is called *second order*. The specific case of a second-order reac-

tion of the type

$$A + A \rightarrow \text{product}$$

could be represented by the rate expression

$$\text{Rate} = k\,[A]^2.$$

An example would be the combination of two iodine atoms to form a diatomic iodine molecule:

$$I + I \rightarrow I_2, \qquad \text{Rate} = k\,[I]^2.$$

14-4 EFFECT OF A CATALYST ON REACTION RATE

We find out by experiment that we can often increase the rate of a reaction by adding certain materials that are not consumed during the course of the reaction. For example: (a) Sulfur dioxide and oxygen react at room temperature to form sulfur trioxide if the gases are brought into contact in the presence of finely divided platinum. (b) Ethyl alcohol can be dehydrated to ethylene in the presence of sulfuric acid. And (c) 2-pentene reacts with hydrogen to form pentane when the mixture is in contact with powdered nickel:*

$$2SO_2 + O_2 \xrightarrow{\text{Pt}} 2SO_3,$$

$$\underset{\substack{\text{ethyl} \\ \text{alcohol}}}{CH_3CH_2OH} \xrightarrow{\text{H}_2\text{SO}_4} \underset{\text{ethylene}}{CH_2{=}CH_2} + H_2O,$$

$$H_2 + \underset{\text{2-pentene}}{CH_3CH_2CH{=}CHCH_3} \xrightarrow{\text{Ni}} \underset{\text{pentane}}{CH_3CH_2CH_2CH_2CH_3}.$$

In these reactions the platinum, sulfuric acid, and nickel act as *catalysts*. None of the reactions would proceed to a measurable degree at room temperature without the aid of the catalysts. Although the specific mode of action of the catalyst must be determined separately for each individual reaction, in general, the result of catalytic action is a lowering of the activation energy (recall Fig. 14–2). When there is a lower activation energy, more molecules are capable of reaction and the rate of reaction is consequently increased.

▶ The catalyst actually enables the reaction to proceed in a different manner, by a different pathway that has a lower activation energy. ◀

---

* Ethyl alcohol, ethylene, 2-pentene, and pentane are organic compounds. We shall discuss them in detail in Chapter 16.

Thus the three main factors which affect the rates of chemical reactions are all associated with activation energy. The temperature and the concentration of reactants determine the number of molecules that have the activation energy necessary for reaction. Anything we do to increase this number—by raising the temperature or increasing the concentration—increases the rate of reaction. We can accomplish the same thing by changing the reaction pathway and lowering the activation energy by means of catalysts, thereby increasing the number of molecules that have enough energy for reaction.

14-5 KINETICS AND REACTION MECHANISMS

So far, when we have talked about chemical reactions, we have been concerned with the reactants and the products, but not with the way in which reactants become products. The detailed step-by-step process that takes place during a chemical reaction is called the *reaction mechanism*. Obviously, to understand fully what is going on in a chemical reaction, we must know the reaction mechanism.

In many cases we can deduce a likely reaction mechanism by a careful consideration of kinetic data. For example, the conversion of methyl bromide to methyl alcohol by the action of hydroxide ion might proceed by at least two probable pathways:

$$CH_3Br + OH^- \rightarrow CH_3OH + Br^-$$

methyl   methyl

bromide   alcohol

Mechanism A

$$OH^- + CH_3\!-\!Br \rightarrow HO\cdots CH_3 \cdots Br^- \rightarrow CH_3\!-\!OH + Br^-$$

Mechanism B

$$(1) \qquad CH_3\!-\!Br \rightarrow CH_3^+ + Br^-$$

methyl

carbonium

ion

$$(2)\ CH_3^+ + OH^- \rightarrow CH_3\!-\!OH$$

Mechanism A illustrates the situation in which the hydroxide ion approaches the methyl bromide molecule and forms the carbon-oxygen bond at the same time as the carbon-bromine bond is being broken. Bond formation and bond breakage occur simultaneously, and the process is a one-step reaction.

Mechanism B, however, is a two-step process. In the first step, the carbon-bromine bond is broken to give bromide ion and a species called the methyl carbonium ion. In a subsequent second step, this methyl carbonium ion reacts with hydroxide ion to form the product methyl alcohol. Here bond breakage and bond formation do not occur simultaneously, but in distinct, separate processes.

Is it possible for us to distinguish between these two mechanisms? Would we expect these processes to differ in terms of their chemical kinetics? Mechanism A is a one-step process, and so we would expect the rate of reaction to be directly proportional to the concentration of methyl bromide and to the concentration of hydroxide ion. In other words, the reaction would be first order in each of these species, and should therefore obey the second-order rate law,

$$\text{rate} = k\,[CH_3Br][OH^-].$$

Mechanism B, however, presents a more complicated problem. Since it is a two-step mechanism, either one of the steps could control the rate of reaction. For instance, if step 1 were very fast relative to step 2, then the slow step 2 would represent the bottleneck in the conversion to methyl alcohol and would control the rate of the reaction. Similarly, if step 1 were slow relative to step 2, it would control the overall rate. The slow step in a multistep mechanism is called the *rate-determining step*.

Assume for a moment that step 1 is the rate-determining step in mechanism B. The rate, then, must be independent of hydroxide ion concentration (since hydroxide ions do not take part in the slow step) and must obey the first-order rate law, rate $= k\,[CH_3Br]$. On the other hand, if step 2 is the slow step, the rate law should be: rate $= k\,[CH_3^+][OH^-]$. We cannot measure the concentration of methyl carbonium ion, but we can say that it must be proportional to the methyl bromide from which it is formed. The rate law then becomes: rate $= k\,[CH_3Br][OH^-]$, indistinguishable from that of mechanism A.

This does not mean, however, that we are stalemated. We should still be able to distinguish between the two mechanisms, basing our analysis on chemical intuition. In mechanism B, the first step involves breaking a covalent bond, a process which in many other cases is known to be relatively slow. The second step involves reaction of two oppositely charged ions to form a covalent bond. This reaction is very similar to the extremely fast neutralization reaction $H^+ + OH^- \rightarrow H_2O$. It is likely, then, that in mechanism B, step 1 is the slow, rate-determining step.

Thus mechanism A should obey the second-order rate law, rate $= k\,[CH_3Br]$ $[OH^-]$ and mechanism B requires first-order kinetics, rate $= k\,[CH_3Br]$. Experimentally we find that, when we keep the temperature constant, the rate of the reaction is directly proportional to the concentration of methyl bromide as well as to the concentration of hydroxide. In other words, mechanism B cannot be correct. This does not necessarily mean that mechanism A is the correct one, but only that it is consistent with the kinetic data.

Once mechanistic information is available, we must use caution when we apply it to other analogous systems. For example, the conversion of tertiary-butyl bromide, $(CH_3)_3CBr$, to tertiary-butyl alcohol, $(CH_3)_3COH$, looks on the surface to be a reaction identical to that of methyl bromide with hydroxide ion. In fact, kinetic studies show that the reaction

$$(CH_3)_3CBr + OH^- \rightarrow (CH_3)_3COH + Br^-$$

is first order, rate $= k \left[(CH_3)_3CBr\right]$, and a two-step mechanism, similar to mechanism B above, is postulated for this reaction:

(1) $(CH_3)_3CBr \xrightarrow{\text{Slow}} (CH_3)_3C^+ + Br^-$

              tert-butyl   tert-butyl
              bromide   cation

(2) $(CH_3)_3C^+ + OH^- \xrightarrow{\text{Fast}} (CH_3)_3COH$

                               tert-butyl
                               alcohol

Sometimes we encounter rate laws that are of nonintegral (fractional) order. The reaction of hydrogen with bromine is a case in point:

$$H_2 + Br_2 \rightarrow 2HBr$$

We might expect that HBr is formed directly from a sufficiently energetic collision of H_2 and Br_2. We would predict that such a one-step mechanism would show kinetics that are first order in both H_2 and Br_2. That is, rate $= k\left[H_2\right]\left[Br_2\right]$. Experimentally we find that the rate law is: rate $= k\left[H_2\right]\left[Br_2\right]^{\frac{1}{2}}$. The reaction is indeed first order in H_2, but only one-half order in Br_2. This means that the dependence of the rate of reaction on the concentration of bromine is not as we predicted, and therefore our proposed mechanism must be incorrect.

The rate law would fit a mechanism in which the slow step was

$$H_2 + Br\cdot \rightarrow HBr + H\cdot,$$

since a bromine atom is essentially half a bromine molecule, and the rate law, rate $= k\left[H_2\right]\left[Br\cdot\right]$, would be kinetically indistinguishable from rate $= k\left[H_2\right]\left[Br_2\right]^{\frac{1}{2}}$. In effect, the experimentally determined rate law forces us to write a mechanism in which bromine atoms are first formed and then react with a hydrogen molecule in the rate-determining step. Such a mechanism is

(1) $Br_2 \longrightarrow 2Br\cdot$
(2) $H_2 + Br\cdot \xrightarrow{\text{Slow}} HBr + H\cdot,$
(3) $H\cdot + Br_2 \longrightarrow HBr + Br\cdot,$
(4) Steps 2 and 3 repeat themselves: 2, 3, 2, 3, 2, 3, . . . etc.

Note that each step in this mechanism gives a product which has an unpaired electron (called a *free radical*) and that step 3 produces the bromine atom for step 2. In theory, then, once the initially formed bromine atoms react in step 2, no other bromine molecule need break into atoms. That is, only one bromine molecule need participate in step 1 in order to convert all the H_2 and Br_2 to HBr. A reaction which proceeds by a mechanism of this type is called a *free-radical chain reaction*.

In practice, although one bromine atom can cause the formation of several thousand HBr molecules, its effect is limited by side reactions which do not produce free radicals, and hence stop the chain. Such reactions, which are called *termination*

steps, include

$$Br\cdot\ +\ Br\cdot\ \rightarrow Br_2,$$
$$H\cdot\ +\ H\cdot\ \rightarrow H_2,$$
$$H\cdot\ +\ Br\cdot\ \rightarrow HBr.$$

▶ The rate law for the H_2 + Br_2 reaction is more complicated than we have shown; that is,

$$Rate = \frac{k\,[H_2][Br_2]^{1/2}}{1 + k'\,([HBr]/[Br_2])}.$$

The simpler $1\frac{1}{2}$ order rate law that we have used holds only at the beginning of the reaction and in the presence of a large excess of Br_2. Similarly, the reaction

$$H_2 + I_2 \rightleftarrows 2HI$$

was once considered to be a classical example of a second-order reaction involving a one-step collision of the two species involved. Recent studies show that this second-order reaction is more complicated than originally thought and probably proceeds by a mechanism similar to the H_2 + Br_2 reaction. This illustrates two points:

1. We cannot predict the rate law. We must actually determine it experimentally.
2. The rate law does not prove that a proposed mechanism of a reaction is correct, but only that it is consistent with the data. We must be prepared to change our minds. ◀

14–6 CATALYSTS AND REACTION MECHANISMS

As mentioned previously, a catalyst can be used to speed up a chemical reaction. For example, H_2 and O_2 do not react at a measurable rate at room temperature, but the reaction does proceed if the gases are mixed and allowed to come into contact in the presence of finely divided platinum. When platinum is present, the reaction takes place at a much faster rate than it does in the absence of the platinum catalyst. Therefore more molecules must have the kinetic energy necessary to overcome the activation-energy barrier. Since we have not changed the temperature at all and the distribution of kinetic energies (recall Fig. 14–1) must therefore be the same under both conditions, the effect of the catalyst must be to *lower* the activation energy.

With the activation energy lowered, more molecular collisions produce the energy required for reaction than in the uncatalyzed reaction. The question we must ask is: How does a catalyst go about lowering the activation energy of a reaction? Since this activation energy depends on the specific reaction in question, we must investigate the details of the part played by a catalyst in each individual type of reaction.

One thing does seem clear: A catalyst speeds up a reaction by altering the reacting pathway. In other words, a catalyst changes the mechanism of a reaction so that the slow step in the reaction has a lower activation energy than the slow step in the uncatalyzed process.

For example, consider the reaction which involves the dehydration of ethyl alcohol to give ethylene:

$$CH_3CH_2OH \xrightarrow[400-500\ °C]{} CH_2=CH_2 + H_2O$$

ethyl   ethylene
alcohol

In the uncatalyzed reaction, the slow step in the reaction probably involves the cleavage of the carbon-oxygen bond to form the ethyl radical and the hydroxyl radical:

$$CH_3CH_2\!\!-\!\!OH \rightarrow CH_3CH_2\cdot + \cdot OH$$

ethyl   hydroxyl
radical   radical

The activation energy for this step is quite high, and temperatures of 400 to 500 °C are required for the reaction. The dehydration, however, can be carried out at much lower temperatures (~ 200 °C) if we employ an acid catalyst:

$$CH_3CH_2OH \xrightarrow{H^+} CH_2=CH_2 + H_2O$$

The mechanism for this acid-catalyzed reaction must, therefore, be different from the uncatalyzed case. It is thought to include a very fast reaction between the acid and the alcohol, followed by a slow loss of water and a subsequent fast step to give the product ethylene and regenerate the catalyst:

(1) $CH_3CH_2OH + H^+ \xrightarrow{Fast} CH_3CH_2\overset{+}{O} \overset{\diagup H}{\diagdown H}$

protonated
alcohol

(2) $CH_3CH_2\!\!-\!\!\overset{+}{O}\overset{\diagup H}{\diagdown H} \xrightarrow{Slow} CH_3CH_2^+ + H\!-\!O\!-\!H$

ethyl
carbonium
ion

(3) $CH_3CH_2^+ \xrightarrow{Fast} CH_2=CH_2 + H^+$

In this instance, as well as in the uncatalyzed reaction, the rate-determining step involves the cleavage of the carbon-oxygen bond. The difference is that the cleavage here leads to different products (ethyl carbonium ion and water) and has a lower activation energy. Apparently it is easier to cleave the carbon-oxygen bond in the protonated alcohol than in the unprotonated alcohol. The function of the acid catalyst in this reaction is to provide the protonated alcohol.

14-7 CONCLUSION

We could go to great lengths in describing how kinetic data can be used to support or disprove mechanisms of reactions. One of the most complicated and challenging current problems is the study of the details of the function of enzymes, the catalysts of living systems. The kinetics and mechanisms of the photochemical reactions that occur in polluted air and water are also of great import.

The examples in this chapter, however, should suffice to show how a study of kinetics can supply some very detailed information about chemical reactions.

QUESTIONS

1. Sketch an activation-energy diagram similar to that in Fig. 14–2 for an endothermic reaction.

2. Explain the effects of increasing the temperature, increasing the concentration, and adding a catalyst on the rate of a chemical reaction.

3. The reaction $NO_2 + CO \rightarrow CO_2 + NO$ is a one-step process; what is the rate law? What is the order of the reaction, and what effect will doubling the NO_2 concentration have on the rate?

4. For the reaction $2NO_2 + F_2 \rightarrow 2NO_2F$, the experimentally determined rate law is: rate $= k\,[NO_2][F_2]$. Write a reasonable mechanism for this reaction.

5. For the reaction $CH_3CH_2Br + KOH \rightarrow CH_2{=}CH_2 + H_2O + KBr$, three mechanisms can be proposed:

   A. 1) $CH_3CH_2Br \rightarrow CH_3CH_2^+ + Br^-$
      2) $CH_3CH_2^+ + OH^- \rightarrow CH_2{=}CH_2 + H_2O$

   B. $OH^- + CH_3CH_2Br \rightarrow H_2O + CH_2{=}CH_2 + Br^-$

   C. 1) $OH^- + CH_3CH_2Br \rightarrow H_2O + {}^-CH_2CH_2Br$
      2) $^-CH_2CH_2Br \rightarrow CH_2{=}CH_2 + Br^-$

   For mechanisms A and C, assume that either one of the two steps can be the slow step. What conclusions can you draw, given that the rate law is found to be: rate $= k\,[CH_3CH_2Br]$?

6. For the reaction in question 5, the experimental rate law was found to be: rate $= k\,[CH_3CH_2Br][OH^-]$. What conclusions can you now draw about the mechanism of the reaction?

7. It is known that processes which involve cleavage of a carbon-deuterium bond (C—D) are slower than those in which a C—H bond is broken. A mechanism for the reaction of benzene (C_6H_6) with nitronium ion (NO_2^+) has been proposed:

Benzene

2)

The rate law, rate $= k\,[C_6H_6][NO_2^+]$, is consistent with either step 1 or step 2 being rate determining. What conclusions can you draw from the fact that the reaction of deuterated benzene (C_6D_6) with NO_2^+ proceeds at the same rate as with ordinary benzene?

15. METALS AND NONMETALS

15-1 INTRODUCTION

We have previously discussed the properties of several elements. In Chapter 4 we considered the periodic properties of the Group IA and IIA metals, and in Chapter 7 we talked about the very important nonmetals hydrogen and oxygen, along with the noble gas elements. In this chapter we shall discuss representative chemical properties, reactions, and compounds of metallic and nonmetallic elements. We shall illustrate the properties of typical metallic-type elements using the Group IVA metals, the transition metal chromium (Cr), and various complexes of metal ions as examples. Nonmetallic properties are represented by the Group IVA elements carbon and silicon (Si) and the Group VIIA halogens. We shall briefly cover intermediate, or semimetal, behaviors, using as examples the Group IVA elements silicon and germanium (Ge).

We have tried throughout this chapter to discuss the chemistry of the elements in terms of the concepts covered in previous chapters, such as bonding theories, molecular geometry, solution properties, equilibrium relationships, kinetics, and so on. A review of the various concepts, as they are encountered, is often very helpful.

15-2 TYPICAL PROPERTIES

We indicated in Chapter 4 that most of the known elements (more than three-quarters of them) are metals, and that these elements are found at the left and center portions of the periodic table, with the nonmetallic elements occupying positions on

the extreme right of the table. Elements do, of course, vary widely with respect to chemical and physical properties, but there are several properties which are typical of almost all the metallic-type elements. These properties serve to separate the elements into two classes, metals and nonmetals, and into a third class, often called *semimetals*.

▶ The elements B, Si, Ge, As, Sb, Te, Po, and At fall into the category of semimetals (sometimes called *metalloids*) and separate the metals from the nonmetals in the periodic table in a diagonal fashion. When we are discussing particular properties, it is sometimes convenient to include P, Bi, and Se as semimetals. ◀

Metals are elements with high electrical and thermal conductivity, characteristic luster, and the ability to be rolled or pounded into sheets or drawn into wires. They tend to donate electrons and to form positive ions during chemical reactions; their oxides are basic. Nonmetals, on the other hand, are good electrical and thermal insulators (very poor conductors), have low luster, or are gases at room temperature and pressure. They tend to accept electrons during reaction, forming negative ions; their oxides are acidic.

The Group IVA elements C, Si, Ge, Sn, and Pb discussed in this chapter illustrate many of the properties of nonmetals, semimetals, and metals, as well as a typical group trend, generally shared by Group A elements, of increasing metallic properties in going from the top to the bottom of the group.

Table 15–1. Some properties of the Group IVA elements

Element	Atomic number	Electron configuration	Atomic radius, Å	First ionization potential, eV	Electro-negativity	Electrical conductance, (microhms)$^{-1}$
C	6	$(He)2s^2 2p^2$	0.914	11.3	2.5	0.0007
Si	14	$(Ne)3s^2 3p^2$	1.32	8.1	1.8	0.10
Ge	32	$(Ar)3d^{10}4s^2 4p^2$	1.37	8.1	1.8	0.022
Sn	50	$(Kr)4d^{10}5s^2 5p^2$	1.62	7.3	1.8	0.088
Pb	82	$(Xe)4f^{14}5d^{10}6s^2 6p^2$	1.75	7.4	1.8	0.046

Table 15–1 presents several atomic properties of the Group IVA elements. Carbon shows typical nonmetallic properties, such as high ionization potential, high electronegativity, and low electrical conductivity. Tin and lead, which are metals, have correspondingly low ionization potentials, low electronegativities, and high electrical conductivities. The semimetals silicon and germanium exhibit intermediate properties.

The chemical properties and compounds of the Group IVA elements also illustrate the differences between metals and nonmetals. These properties are described in the following sections, along with the properties of the Group VIIA nonmetals and the transition metal chromium.

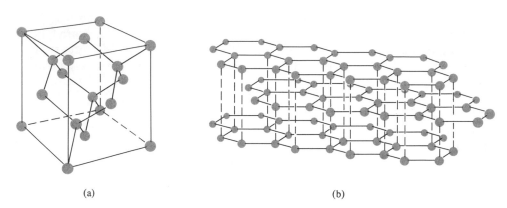

(a)   (b)

Fig. 15-1   Crystal structures of two carbon allotropes: (a) diamond and (b) graphite.

NONMETALS

15-3 CARBON AND SILICON

Element Properties: Carbon. Elemental carbon occurs in two different forms, diamond and graphite, often called *allotropes*, which means that they are different forms of an element in the same physical state. One major difference between diamond and graphite is in their crystal structures. Diamond has a three-dimensional network structure, in which each carbon atom is tetrahedrally bonded to four other carbon atoms, as shown in Fig. 15-1(a). This strongly interlocked structure causes diamond to be colorless and transparent, with high refractivity, and gives it its characteristic properties of extreme hardness, very high melting and boiling points (about 3500 and 4200 °C, respectively), very low electrical conductivity, and chemical inertness.

Graphite has a much different crystal structure, diagrammed in Fig. 15-1(b); each carbon atom is chemically bonded to only three other carbon atoms. This produces a layered structure composed of hexagonal rings of carbon atoms, with individual layers of atoms held together by van der Waals forces. These weak interlayer forces explain graphite's relative softness, its slippery feel, and its tendency to break into sheets.

Graphite has other properties, much different from those of diamond, that we can explain by using the ideas of chemical bonding covered earlier. Unlike diamond, graphite is a black solid, with a slight metallic luster and also a reasonably high electrical conductivity. Experiments conducted by means of x-ray diffraction reveal different C—C bond distances in diamond (1.5 Å) compared with graphite (1.42 Å within the layers and 3.40 Å between layers). The shorter C—C bond length within a layer in graphite suggests some multiple bond character between these carbon atoms.

Figure 15-2(a) shows structures for a graphite layer using the required number of double bonds to fulfill the octet rule (see Section 6-2). Note that one single electron-dot

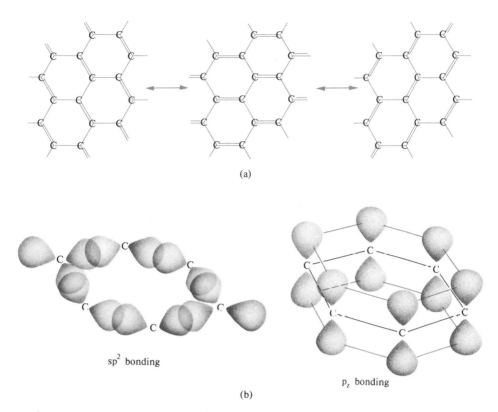

(a)

sp² bonding

(b)

p_z bonding

Fig. 15–2 Bond formation in graphite: (a) Three resonance structures. (b) sp² overlap forming σ-bonds and p_z overlap forming π-bonds produce a hybrid bonding structure.

or Lewis structure cannot be written for graphite; three identical structures are required. We have called these *resonance structures* (see Section 6–4). The actual electronic structure is envisioned as a mixture of these three separate structures, producing a resonance hybrid in which each C—C bond is a $1\frac{1}{3}$ bond. This multiple character of the bonds within layers in graphite accounts for graphite's shorter C—C bond length compared with the bond length of diamond. The large C—C distance between the layers in graphite agrees with the proposed van der Waals attractive forces, since these forces are quite weak when compared with the covalent bonding attractions present within the layers of carbon atoms.

We can explain the dark color, luster, and the electrical conductivity of graphite more readily using the concept of chemical bonding that involves hybrid atomic orbitals (see Section 6–4). As shown in Fig. 15–2(b), the sp² hybrid orbitals of carbon are directed in just the proper directions (in a plane with 120° bond angles, the shape of an equilateral triangle) to be used by each carbon atom within a layer for bonding to sp² hybrid orbitals of three other carbon atoms. This bonding with sp² hybrid orbitals forms three sigma (σ) bonds—bonds directed along the internuclear axis—

and uses three of the four available valence electrons for each carbon atom
(C $1s^2 2s^2 2p^2$ electron configuration).

The fourth bonding electron can be envisioned as occupying the $2p_z$ orbital on
each carbon atom; it is not involved in σ-bond formation. But these $2p_z$ orbitals of
the individual carbon atoms can interact with one another, forming a π-orbital—
electron density above and below the internuclear axis—which extends throughout
the entire layer (a multicenter π-orbital). Since each $2p_z$ orbital contains only one
electron (and can have a maximum of two electrons) and the π-orbital extends over
the entire layer of carbon atoms, electrons within this π-system are not localized between any two atoms; they are delocalized and free to move throughout the entire
layer. It is these free electrons that give graphite properties similar to those of some
metals: a dark, opaque, somewhat shiny solid with fair electrical conductivity. Note
that the electrical conductivity of graphite is fairly high within the layer, but is very
low between layers. This experimental observation is in agreement with the proposed
bonding model.

Element Properties: Silicon. The element silicon (Si) is the second most abundant
element in the earth's crust. Combined with oxygen, it accounts for sand and quartz,
and in combination with oxygen and various metals, it accounts for practically all
glass, rocks, and soil. Silicon crystallizes in the diamond structure and, consequently,
is quite hard, brittle, and has high melting and boiling points (about 1420 and
2400 °C, respectively). Unlike diamond, however, elemental silicon is a black,
opaque, somewhat lustrous solid, with fair electrical conductivity. These metallic
properties, which we shall discuss further in Section 15–8, often cause us to categorize silicon as a semimetal. On the other hand, silicon's chemical reactions and
compounds are essentially nonmetallic in nature, as we shall see.

Compounds of Carbon. Carbon is a very unusual element, in that it readily forms
compounds in which carbon atoms are bonded to one another in the form of chains or
rings. This property, called *catenation*, accounts for the huge number of carbon compounds that are considered in the study of *organic chemistry*. We shall cover this
branch of chemistry in Chapters 16 and 17. In this chapter we shall use only a few
typical examples of carbon compounds and reactions to illustrate its nonmetallic
character.

Compounds with Oxygen. Compounds of carbon with oxygen are very common and
quite important. We have discussed the formation of carbon monoxide and carbon
dioxide previously, and have indicated their importance in air pollution (Section 7–5).
Figure 15–3 shows the geometries of these two molecules.

The CO molecule contains a triple bond, one σ-bond formed by overlap of the sp
hybrid orbitals of the carbon and oxygen atoms and two π-bonds from the $2p_y$ and $2p_z$
orbitals. Carbon dioxide has one σ-bond between the C and each O atom and two
π-bonding systems involving all three atoms, yielding C—O bonds that are slightly
stronger than normal C—O double bonds. Note that the octet rule is obeyed for both

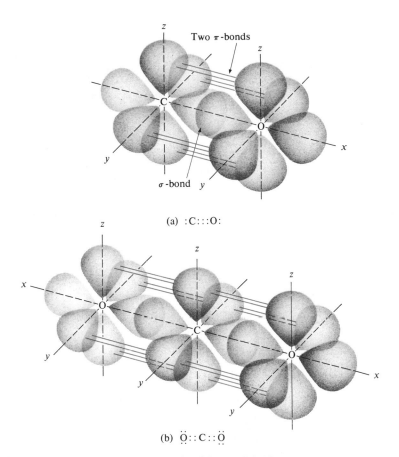

(a) :C:::O:

(b) $\ddot{O}::C::\ddot{O}$

Fig. 15–3 Formation of σ- and π-bonds in: (a) CO, and (b) CO_2.

molecules, without the use of resonance.

When we dissolve CO_2 in water, we obtain a weakly acidic solution (CO_2 is an acidic oxide). This can be explained by the two simultaneous equilibria

$$CO_2 + H_2O \rightleftarrows H_2CO_3 \rightleftarrows H^+ + HCO_3^-.$$

Experimentally it is found that more than 99% of the undissociated acid exists as hydrated CO_2 molecules, that is, $CO_2 + H_2O$. Apparently carbonic acid (H_2CO_3) dissociates to H^+ and HCO_3^- as rapidly as it is formed.

The dissociation of the bicarbonate ion (HCO_3^-),

$$HCO_3^- \rightleftarrows H^+ + CO_3^{2-},$$

is quite low, $K = 4.8 \times 10^{-11}$, indicating that HCO_3^- is a very weak acid.* Cor-

---

*   Acid or base ionization equilibrium constants are designated by K (see Chapter 13).

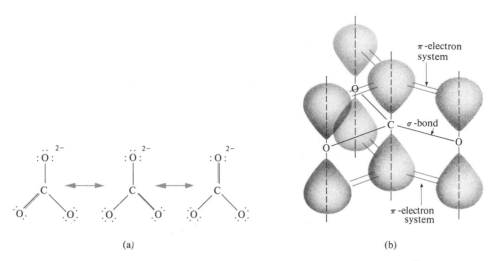

Fig. 15-4 Bonding in the carbonate ion: (a) Three resonance structures for CO_3^{2-}. (b) σ- and π-bond formation in CO_3^{2-}.

respondingly we know that the carbonate ion, CO_3^{2-}, hydrolyzes quite readily. The hydrolysis constant (K_h) for the reaction

$$CO_3^{2-} + H_2O \rightleftarrows HCO_3^- + OH^-$$

is the ion product of water (K_w) divided by the ionization constant for the weak acid involved (K for HCO_3^-) which gives

$$K_h = \frac{K_w}{K} = \frac{1 \times 10^{-14}}{4.8 \times 10^{-11}} = 2.1 \times 10^{-4}.$$

Thus solutions of CO_3^{2-} ion are mildly basic because of the hydrolysis of CO_3^{2-}.

Two salts that can be obtained from solutions of CO_2 in water are very important commercially (and are produced by the Solvay process). Sodium bicarbonate ($NaHCO_3$) is used extensively as baking soda, and sodium carbonate (Na_2CO_3) is used in the glass and soap industries. Another compound, calcium carbonate ($CaCO_3$), is the important industrial material called limestone.

The carbonate ion requires either the application of resonance theory or hybrid orbitals to account for the observed bonding properties. Figure 15-4(a) shows three resonance structures for CO_3^{2-}. The resonance hybrid, then, would contain three identical C—O bonds whose strength would be equal to about $1\frac{1}{3}$ bond. Figure 15-4(b) shows that a π-electron system formed by overlap of the carbon and oxygen $2p_z$ orbitals extends over all four atoms above and below the ion plane. The σ-bonds are formed using sp^2 hybrid orbitals of the carbon atom.

Compounds with Nitrogen. A common group of compounds in which carbon is bonded to nitrogen is called the *cyanides*. One method of preparation involves the

carbonates, as illustrated for sodium cyanide by the equation

$$Na_2CO_{3(s)} + 4C_{(s)} + N_{2(g)} \rightleftarrows 2NaCN_{(s)} + 3CO_{(g)}.$$

The gas hydrogen cyanide (HCN) is very poisonous (as are the metal cyanides, such as AgCN, $Hg(CN)_2$, etc.), and when dissolved in water is a weak acid governed by the dissociation constant $K = 5 \times 10^{-10}$,

$$HCN_{(aq)} \rightleftarrows H^+_{(aq)} + CN^-_{(aq)}.$$

The cyanide ion CN^- has a structure similar to that of CO, shown in Fig. 15-3(a), and forms many very stable complexes with metal ions, such as $[Ag(CN)_2]^-$, $[Ni(CN)_4]^{2-}$, $[Fe(CN)_6]^{3-}$, and $[Fe(CN)_6]^{4-}$. These complexes have structures— linear, square planar, and octahedral (see Fig. 15-12)—that are typical of many complex ions (see Section 15-7). The stability of the cyanide complexes is illustrated by the dissociation of $[Ag(CN)_2]^-$,

$$[Ag(CN)_2]^- \rightleftarrows Ag^+ + 2CN^-,$$

for which the ionization constant is $K = 1.8 \times 10^{-19}$.

Compounds with Halogens. Carbon forms a series of compounds with the Group VIIA halogens which can be considered as substituted hydrocarbons (see Chapter 16). For example, the hydrogen atoms in the hydrocarbon methane (CH_4) can be replaced by halogen atoms, forming compounds such as $CHCl_3$ (chloroform or trichloromethane) and CCl_4 (carbon tetrachloride or tetrachloromethane).

An amazing series of compounds called *fluorocarbons* are formed when the substituting halogen is fluorine. Tetrafluoromethane (CF_4) is typical of these materials because of its extreme chemical inertness. It does not, for example, burn in air, as does methane, nor does it react with strong acids such as boiling nitric (HNO_3) or concentrated sulfuric (H_2SO_4), nor does it react with strong oxidizing agents such as potassium permanganate ($KMnO_4$), or with reducing agents such as hydrogen or carbon at high temperatures.

The commercially important plastic known as Teflon, used extensively for coating cooking utensils, is a high-polymer fluorocarbon. It is formed by the polymerization of tetrafluoroethylene (C_2F_4), which is similar to the formation of the polymer polyethylene from ethylene, C_2H_4 (see Section 16-4). We can imagine the polymerization process, forming polytetrafluoroethylene or Teflon, to proceed via the unstable intermediate shown here in brackets.

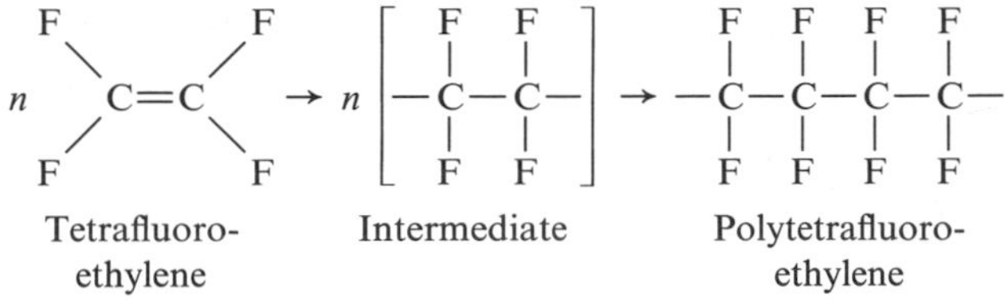

Tetrafluoro-   Intermediate   Polytetrafluoroethylene   ethylene

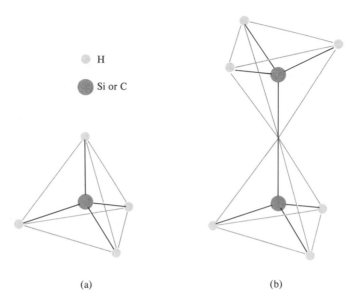

Fig. 15-5   (a) Tetrahedral SiH_4 or CH_4, and (b) Si_2H_6 or C_2H_6, tetrahedra joined by a Si—Si or C—C bond.

Another series of halogen-substituted hydrocarbons is also very important commercially. The compound difluorodichloromethane (CCl_2F_2), which is known as Freon, is a stable, odorless, nontoxic gas that is used as the refrigerant in most current refrigerators (replacing the toxic gas sulfur dioxide, SO_2, used previously). Other members of this series of compounds are used in the laboratory as cryogenic liquids of various (low) temperatures.

Compounds of Silicon.   Most of the important silicon compounds contain oxygen, but several compounds with hydrogen and with the halides resemble the analogous compounds of carbon.

Compounds with Hydrogen.   The hydrosilicons or *silanes* have the general formula Si_nH_{2n+2}, as compared to the hydrocarbons or *alkanes* (see Chapter 16) with the general formula C_nH_{2n+2}. The silanes also resemble the alkanes structurally, as shown in Fig. 15-5. There are many more alkanes than silanes, however, since the number of carbon atoms that can bond together (n in the general formula) is very large, while the range of known silanes includes only those with $n = 1$ to $n = 6$ (SiH_4, Si_2H_6, Si_3H_9, Si_4H_{10}, Si_5H_{12}, Si_6H_{14}). The silanes are also much less stable than the alkanes; they spontaneously burn in air,

$$2Si_2H_{6(g)} + 7O_{2(g)} \rightarrow 4SiO_{2(s)} + 6H_2O_{(g)},$$

to form the very stable compound silicon dioxide (SiO_2).

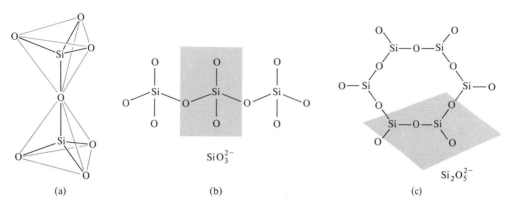

Fig. 15–6 Silicate structures: (a) SiO_4^{4-} tetrahedra sharing one oxygen produce $Si_2O_7^{6-}$ units. (b) Chain of SiO_3^{2-} units. (c) Three shared oxygens produce a sheetlike structure, with the fourth oxygen either above or below the plane of the other three.

Compounds with Halogens. The silicon tetrahalides can be prepared by direct-combination reactions between the elements

$$Si_{(s)} + 2X_{2(g)} \rightarrow SiX_4.$$

At room temperature and pressure, SiF_4 is a gas, $SiCl_4$ and $SiBr_4$ are liquids, and SiI_4 is a solid. Although these silicon halides have the same chemical formulas and molecular geometries as the corresponding carbon compounds, their chemical properties are quite different. Unlike the carbon compounds, the silicon tetrahalides hydrolyze readily, yielding SiO_2:

$$SiX_4 + 2H_2O \rightarrow SiO_{2(s)} + 4H^+ + 4X^-.$$

With SiF_4 a reaction with the HF product produces fluorosilicic acid, a strong acid:

$$SiF_{4(g)} + 2HF_{(aq)} \rightarrow 2H^+ + SiF_6^{2-}.$$

The hexafluorosilicate(IV) ion, SiF_6^{2-}, and the silicon tetrafluoride molecule have octahedral and tetrahedral geometries, respectively.

Compounds with Oxygen. The compounds of silicon with oxygen are based on a tetrahedral SiO_4^{4-} unit. Many compounds called *silicates* are formed when the SiO_4^{4-} units share different numbers of their oxygen atoms, as shown in Fig. 15–6. Discrete SiO_4^{4-} anions are known to exist in compounds such as zirconium silicate, $ZrSiO_4$ (the mineral zircon). One oxygen atom (a bridging oxygen) can be shared by two SiO_4^{4-} units to form $Si_2O_7^{6-}$ (Fig. 15–6a). Sharing of two oxygen atoms by the SiO_4^{4-} tetrahedra yields a polymeric chainlike structure of SiO_3^{2-} anion units (Fig. 15–6b). The mineral spodumene, $LiAl(SiO_3)_2$, is an example of this general series of compounds, called *pyroxenes*.

▶ Silicones have structures based on the SiO_3^{2-} repeating units, but with the two unshared (negatively charged) silicate oxygens (see Fig. 15–6b) replaced by $-CH_3$ (methyl) groups. ◀

Sharing of three oxygen atoms by the SiO_4^{4-} units yields two-dimensional sheets, as in mica, with a $Si_2O_5^{2-}$ repeating unit (Fig. 15–6c). When all four oxygen atoms are shared by the SiO_4^{4-} tetrahedra, a three-dimensional network is formed with the formula SiO_2 (silica). The crystal structure is based on the diamond structure shown in Fig. 15–1a. Crystalline SiO_2 is called quartz, and has a high melting point. Vitreous—i.e., noncrystalline—SiO_2, which has a lower softening point than quartz, is made into various types of laboratory glassware.

SiO_2 is a very stable compound and is quite unreactive with common chemical reagents. For example, it is unattacked by acids, with the exception of hydrofluoric acid (aqueous HF). It reacts with this acid to form SiF_4 gas and complex ions such as SiF_6^{2-}:

$$SiO_{2(s)} + 4HF_{(aq)} \rightarrow SiF_{4(g)} + 2H_2O_{(\ell)},$$
$$SiO_{2(s)} + 6HF_{(aq)} \rightarrow SiF_{6(aq)}^{2-} + 2H_{(aq)}^+ + 2H_2O_{(\ell)}.$$

It is evident that quartz or glass are unsatisfactory containers for HF solutions.

15–4 GROUP VIIA: THE HALOGENS

Group Properties. Table 15–2 gives some atomic properties of the halogens, and Table 15–3 lists characteristic properties of the diatomic species X_2. We have seen previously (Section 6–2) how the ns^2np^5 outer electron configuration can account for the fact that Group VIIA elements ordinarily occur as diatomic molecules:

$$:\overset{..}{Br}\cdot + \cdot\overset{..}{Br}: \rightarrow :\overset{..}{Br}:\overset{..}{Br}:$$

The electron configurations also explain the increase in atomic radius in going from top to bottom (F to I) in the group. (Recall Section 6–4, in which we said that outer electrons are located in energy levels with higher quantum numbers, which increases the size of the atoms, even though the nuclear charge increases as you go down the group.)

▶ Other group trends are evident in Table 15–2. The ionization potentials, for example, decrease from F to I, i.e., down the group. We can understand this trend by considering the increase in atomic radius together with the increasing number of electrons and the corresponding increase in nuclear charge. Even though the positive charge on the nucleus is considerably larger in I ($+53$) as compared to F ($+9$), the electrons in the outer 5p orbitals of I are not attracted by an appreciably larger positive charge than the electrons in the outer 2p orbitals of F. The increased nuclear charge in I is screened (i.e., effectively neutralized) from the outer 5p electrons by the other intervening electrons: those in the 3s, 3p, 3d, 4s, 4p, 4d, and 5s sublevels. Then

Table 15–2. Atomic properties of the halogens

Element	Atomic number	Electron configuration	Covalent radius, Å	Ionization potential, eV	Electron affinity, eV	Electro-negativity	Electrode potential, E^0, V*
F	9	$1s^2 2s^2 2p^5$	0.72	17.42	3.45	4.0	2.87
Cl	17	$(Ne)3s^2 3p^5$	0.99	13.01	3.61	3.0	1.36
Br	35	$(Ar)3d^{10} 4s^2 4p^5$	1.14	11.84	3.36	2.8	1.09
I	53	$(Kr)4d^{10} 5s^2 5p^5$	1.33	10.45	3.06	2.5	0.54

*E^0 values are given for the half-reaction $\frac{1}{2}X_2 + e^- \rightarrow X^-$.

Table 15–3. Properties of diatomic halogens

Molecule	State at room temp. and pressure	Color	Melting point, °C	Boiling point, °C
F_2	Gas	Pale yellow	-219.6	-188.2
Cl_2	Gas	Yellow-green	-101.0	-34.7
Br_2	Liquid	Red-brown	-7.2	58
I_2	Solid	Violet-black	113.7	183

the electrons in the 5p orbitals in I are attracted by about the same positive nuclear charge as the 2p electrons in F. Since the 5p electrons are located much farther from the attractive positive charge, they are more easily removed from the atom, and I has a lower ionization potential than F. ◀

A group trend is evident for the melting and boiling points (Table 15–3) of the diatomic molecules. We can understand an increase in melting point from F_2 to I_2 by considering the attractive forces that exist between the molecules. Since nonpolar covalent bonds are present in both F_2 and I_2, the forces that bind one F_2 molecule to another F_2 molecule or one I_2 molecule to another I_2 molecule are van der Waals forces (Section 9–5). Van der Waals forces depend on instantaneous polarization of electron charge distributions; these forces increase with more readily polarized electron groups. Since I_2 is a much larger molecule than F_2, with many more electrons present, the entire distribution of electrons in I_2 can be polarized more easily than that in F_2, creating a larger instantaneous dipole in the case of I_2. Consequently, the attractive force between I_2 molecules is greater than the attractive force between F_2 molecules, which results in a higher melting (or boiling) point for I_2.

Electrode Potentials. The standard electrode potentials (E^0) given in Table 15–2 indicate that the diatomic halogens are very good oxidizing agents. In fact, F_2 is one of the best oxidizing agents known. Again a group trend is evident. Can you think of any explanations for the trend? (See Section 11–5.)

We can use the E^0 values for the halogens to predict reactions that might occur between them. For example, we would predict that Br_2 would be formed if Cl_2 gas were bubbled through an aqueous solution containing Br^- ion. The appropriate half-reactions and standard electrode potentials are (see Section 11–5):

$$\tfrac{1}{2}Cl_2 + e^- \to Cl^-, \qquad E^0 = 1.36 \text{ V},$$
$$\tfrac{1}{2}Br_2 + e^- \to Br^-, \qquad E^0 = 1.09 \text{ V}.$$

The total reaction would be

$$\tfrac{1}{2}Cl_2 + Br^- \to \tfrac{1}{2}Br_2 + Cl^-,$$

which has a standard electrode potential of

$$E^0 = E^0_{red} - E^0_{oxid} = 1.36 - 1.09 = 0.27 \text{ V}.$$

Since E^0 for the reaction is positive, the reaction will occur as written (again recall Section 11–5).

A similar approach shows that F_2 will react with Cl^-, Br^-, and I^- to produce Cl_2, Br_2, and I_2, respectively; that Br_2 will react with I^- to give I_2, and that I_2 will not oxidize Br^-, Cl^-, or F^- to the diatomic elements.

We can often follow the course of these reactions by adding carbon tetrachloride to the test tube or beaker containing the X^- ion in aqueous solution. For example, if we add Cl_2 to a solution of NaI, I_2 is produced, according to the equation

$$Cl_2 + 2I^- \to I_2 + 2Cl^-.$$

We can easily detect the presence of I_2 by mixing the CCl_4 and H_2O solutions, say, by shaking the test tube or by stirring vigorously. When I_2 is dissolved in CCl_4, it imparts a deep red-violet-to-purple color to the CCl_4 liquid. This is a very sensitive test for I_2. Since I_2 is very soluble in CCl_4 and only slightly soluble in H_2O (Why is this true? See Section 10–3), mixing of the two liquids makes it possible for any I_2 produced by the above reaction to dissolve in the CCl_4 and, consequently, to be detected.

Preparation of the Elements. The above reactions between Cl_2 and Br^- or I^- are used to prepare the free elements Br_2 and I_2. Sea water and oil-well brines contain relatively high concentrations of Br^- and I^-. When Cl_2 gas is passed through these solutions, Br_2 and I_2 are liberated (according to the above equations).

Production of F_2 and Cl_2, on the other hand, is usually accomplished by an electrolytic technique. One reason for this preparation method is the very high electrode potential values for the F_2-F^- and Cl_2-Cl^- half-reactions; an appropriate oxidizing agent is not available, even though the concentrations of F^- and Cl^- ions in sea water and oil-well brines might be relatively high. Elemental F_2 is usually pre-

pared by the electrolysis of molten, anhydrous potassium hydrogen fluoride, KHF_2:

$$2KHF_{2(\ell)} \xrightarrow[\text{heat}]{\text{Electrolysis}} F_{2(g)} + H_{2(g)} + 2KF_{(\ell)}.$$

The principal method of obtaining Cl_2 is by the electrolysis of aqueous sodium chloride solutions:

$$2Na^+_{(aq)} + 2Cl^-_{(aq)} + H_2O_{(\ell)} \xrightarrow{\text{Electrolysis}} Cl_{2(g)} + H_{2(g)} + 2Na^+_{(aq)} + 2OH^-_{(aq)}.$$

Note that sodium hydroxide and hydrogen are also products. In addition, smaller quantities of Cl_2 are obtained from the electrolysis of molten NaCl:

$$2NaCl_{(\ell)} \xrightarrow[\text{heat}]{\text{Electrolysis}} 2Na_{(\ell)} + Cl_{2(g)}.$$

Note that sodium metal is also a product. Can you explain why Na metal is not produced when an aqueous NaCl solution is electrolyzed (see Section 11 5)?

Compounds of the Halogens

Metal Halides: Bonding. The type of bonding in metal halides depends on the metallic element involved. Metals in Groups IA or IIA form well-known ionic halides. These compounds—such as NaCl, MgF_2, $CaCl_2$, etc.—exhibit properties that are typical of solid ionic lattices: They have relatively high melting and boiling points, they are brittle, they are good electrical insulators, etc. (see Section 9–5).

The crystal structures of many metal halides are quite common among other types of chemical compounds as well. Hence the halide structure often serves as a model by which we may classify other compounds. For example, KF, MgO, and MnS all crystallize with the sodium chloride structure (see Section 9–5). In addition, $SrCl_2$, CeO_2, and CdF_2 have the CaF_2 or fluorite crystal structure, and many compounds also have structures that are similar to (and classified as) the CsCl, or the $CaCl_2$, or the CdI_2 crystal structure.

Metal Halides: Solubility. We have seen that, in general, chlorides, bromides, and iodides are quite soluble in water (see Section 10–3). Fluorides, on the other hand, are often quite insoluble. We can consider the solubility question in more quantitative terms by using the solubility product constant, K_{sp}. A low value of K_{sp} indicates very low solubility, while a high value of K_{sp} indicates high solubility. For example, the K_{sp} for MgF_2 is 8×10^{-8}. We can calculate the number of moles of MgF_2 that will dissolve in one liter of H_2O. Let this concentration be x moles/ℓ. Then, at equilibrium, x moles Mg^{2+}/ℓ and $2x$ moles F^-/ℓ will be present in solution, since

$$MgF_{2(s)} \rightleftarrows Mg^{2+}_{(aq)} + 2F^-_{(aq)}.$$

We know that

$$[Mg^{2+}][F^-]^2 = K_{sp} = 8 \times 10^{-8}$$

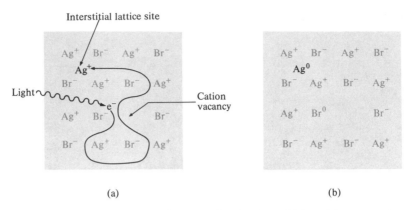

(a)   (b)

Fig. 15–7 (a) Light absorbed by AgBr solid containing Ag$^+$ interstitials and Ag$^+$ vacancies. (b) Formation of an initial Ag metal deposit after the process in (a) occurs.

must be fulfilled at equilibrium. Therefore,

$$(x)(2x)^2 = 8 \times 10^{-8}, \qquad 4x^3 = 8 \times 10^{-8},$$
$$x = \sqrt[3]{20 \times 10^{-9}} = 2.7 \times 10^{-3} = 3 \times 10^{-3} M \; Mg^{2+}.$$

Then 0.003 mole MgF_2 (about 0.2 g MgF_2) will dissolve in one liter of H_2O. We see that MgF_2 is relatively insoluble in water.

Uses of Metal Halides in Photography. Metal halides of somewhat less ionic character than the Group IA and IIA halides are formed by silver. The silver halides AgCl and AgBr are very important in photography. In fact, the usual photographic process depends entirely on a photochemical reaction that occurs within a AgBr or AgBr–AgI emulsion (a suspension of silver halide microcrystals in, for example, gelatin). The details of the changes that take place within the AgBr emulsion when it is exposed to light are not yet understood, but the general features of the process are known.

When visible light strikes the AgBr, an electron is displaced from a Br$^-$ ion, forming a neutral Br:

$$Br^- + light \rightarrow Br + e^-.$$

The electron produced by absorption of this light energy moves through the AgBr until it becomes trapped at an interstitial lattice site, i.e., a position in the lattice between the normal lattice sites (see Fig. 15–7a and Section 9–5). The interstitial silver ion Ag_I^+ (the subscript I stands for *interstitial*) is converted to an uncharged silver metal atom, located at an interstitial position,

$$Ag_I^+ + e^- \rightarrow Ag_I.$$

Since the normal defects in AgBr solid lattices are Ag_I^+ interstitials, and since the Ag_I^+ can move (by diffusion; see Sections 8–8 and 9–5) relatively easily throughout the AgBr solid, a small group or conglomerate of neutral Ag atoms can form. This group

of neutral Ag atoms creates a very small (invisible) amount of silver metal, forming a "latent image," as diagrammed in Fig. 15–7b. This tiny amount of photochemically produced Ag metal then serves as a nucleus for the formation of larger quantities of Ag during the developing stage of the photographic process.

Uses of AgI. Currently silver iodide is also of great interest because of its use in weather modification. Rain or snow can seemingly be caused to occur (at selected times and in desired places) by spraying small particles of AgI (and possibly other solid materials) into some types of clouds (the process is called cloud-seeding). At first it was thought that the small AgI particles acted as condensation nuclei for the water vapor in the clouds primarily because of a similarity in crystal structure between AgI and ice. Although AgI appears to be the most successful cloud-seeding substance found to date, this original hypothesis of the role played by the similarity of crystal structures is now being questioned and investigated extensively.

▶ Studies also indicate that seeding of hurricanes with AgI diminishes their intensity to a significant extent. These experiments will be continued with selected tropical storms.

It is interesting to contemplate the possible effects that an extensive program of AgI cloud-seeding, designed to partially control weather patterns, will have. Much of the AgI returns to the earth, principally via rain and snow. What changes, if any, will these increased silver and iodine concentrations have on our water supplies, fields of wheat and corn, and so on? What effects, if any, will the AgI particles remaining in the atmosphere have? Will they cause a serious air pollution problem? Answers to these questions are, of course, not known. In fact, the actual relevance of the questions over the next several years is not even obvious. But they are questions that must be asked, studied, and answered—hopefully before a problem comes into being. ◀

Uses of PbI_2. Another metal halide, lead(II) iodide (PbI_2), may also be significant in modifying weather patterns. The presence of lead and iodine (as well as bromine) in automobile exhaust fumes appears to have increased the atmospheric concentrations of these elements greatly over the last several years. It is possible that minute particles of PbI_2 (or PbBrI)—particles that are present in the atmosphere because of the combustion of leaded gasoline—nucleate ice crystals and cause rain or snow that would not have fallen at all, or at least would not have fallen within a particular locality so often.

Laboratory studies of the ice-nucleation properties of PbI_2, similar to those conducted with AgI, indicate that PbI_2 does serve as an ice-nucleating agent, although it does not seem to be as efficient as AgI.

Acids Formed by the Halogens: *Hydrohalic acids.* The most common acids formed by the Group VIIA halogens are aqueous solutions of the hydrogen halides: HF, HCl, HBr, and HI. The monomolecular gaseous species HCl, HBr, and HI dissolve in water to yield strong acids (see Section 12–3). That is, they are essentially 100%

ionized or dissociated in water solution, as indicated by their high ionization equilibrium constants ($K > 1$; see Section 13–8).

On the other hand, HF forms an associated gaseous species (several molecules bonded together) and dissolves in water to yield a weak acid. Association of gaseous (and liquid) HF molecules is attributed to hydrogen bonding forces (see Section 9–6, and recall the unusually high melting and boiling points of HF). That aqueous HF is a weak acid is reflected in its rather small ionization constant, $K = 5.7 \times 10^{-4}$.

One common method of preparing hydrogen fluoride and hydrogen chloride gases uses the reaction between concentrated sulfuric acid (H_2SO_4) and CaF_2 or NaCl solids. For example,

$$NaCl_{(s)} + H_2SO_{4(\ell)} \rightarrow NaHSO_{4(s)} + HCl_{(g)}.$$

Hydrochloric acid is obtained by dissolving the HCl gas in water. HBr and HI are prepared from their sodium salts and phosphoric acid, H_3PO_4 (which is a less powerful oxidizing agent than H_2SO_4). For example,

$$NaI_{(s)} + H_3PO_{4(\ell)} \rightarrow HI_{(g)} + NaH_2PO_{4(s)}.$$

Table 15–4. Oxychloride acids

Acid formula	Acid name	Anion name	Oxidation state of Cl	Ionization constant, K
HClO	Hypochlorous	Hypochlorite	+1	3.2×10^{-8}
HClO$_2$	Chlorous	Chlorite	+3	1.1×10^{-2}
HClO$_3$	Chloric	Chlorate	+5	Strong acid
HClO$_4$	Perchloric	Perchlorate	+7	Strong acid

Oxyacids. The halogens also form several oxyacids, as illustrated by the oxychloride acids in Table 15–4. Note that HClO and $HClO_2$ are weak acids, while $HClO_3$ and $HClO_4$ are strong acids, and that the chemical formulas, as we have written them, are somewhat misleading when the actual bonding in the acids is considered.

Figure 15–8 shows the electron-dot (Lewis) structures of the oxyacids of chlorine, along with the geometries of the corresponding anions. The hydrogen atoms in the oxyacids are bonded to an oxygen atom rather than to the chlorine atom, as implied by the chemical formulas in Table 15–4.

Hypochlorous acid is a weak acid (as are all hypohalous acids), and is prepared in low concentrations by reacting Cl_2 with water:

$$Cl_{2(g)} + H_2O \rightarrow H^+_{(aq)} + Cl^-_{(aq)} + HClO_{(aq)}.$$

The equilibrium concentrations of hydrogen ion and hypochlorite ion in HClO solutions are low:

$$HClO \rightleftarrows H^+ + ClO^-.$$

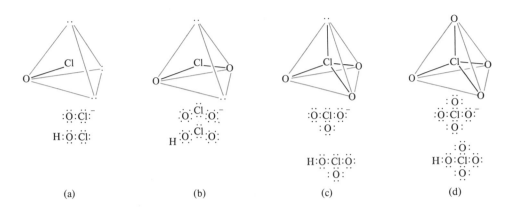

(a)   (b)   (c)   (d)

Fig. 15-8 Oxychloride anions and acids are based on tetrahedral geometries: (a) ClO^- and $HClO$; (b) ClO_2^- and $HClO_2$; (c) ClO_3^- and $HClO_3$; (d) ClO_4^- and $HClO_4$.

We can calculate the H^+ and ClO^- concentrations in a $0.010M$ solution of $HClO$, since we know that $K = 3.2 \times 10^{-8}$ (see Section 13-9). If we assume that x moles/ℓ of $HClO$ ionize, then at equilibrium

$$\frac{[H^+][ClO^-]}{[HClO]} = K = 3.2 \times 10^{-8},$$

$$\frac{(x)(x)}{(0.010 - x)} = 3.2 \times 10^{-8}.$$

Neglecting x with respect to 0.010, we have

$$x^2 = 3.2 \times 10^{-10} \quad \text{and} \quad x = \sqrt{3.2 \times 10^{-10}} = 1.8 \times 10^{-5}.$$

Therefore

$$[H^+] = 1.8 \times 10^{-5}.$$

Since we know that $pH = -\log[H^+]$, then

$$pH = -\log 1.8 \times 10^{-5} = 5.0 - 0.255 = 4.7.$$

In a $0.010M$ solution of $HClO$, the $[H^+] = 1.8 \times 10^{-5}M$ and the $pH = 4.7$.

Salts of Oxyacids: *Hypochlorites.* The hypochlorite ion is used commercially as a bleach; for example, in aqueous solutions such as $NaClO$ (Chlorox). The hypochlorite ion can be prepared by electrolyzing cold $NaCl$ solutions and mixing the electrode reaction products:

$$2Cl^- \rightarrow Cl_2 + 2e^- \qquad\qquad \text{Anode half-reaction}$$
$$2e^- + 2H_2O \rightarrow 2OH^- + H_2 \qquad \text{Cathode half-reaction}$$
$$\overline{2Cl^- + 2H_2O \rightarrow Cl_2 + 2OH^- + H_2} \qquad \text{Cell reaction}$$

The reaction that takes place when the products are mixed,

$$Cl_2 + 2OH^- \rightarrow ClO^- + Cl^- + H_2O,$$

produces the desired ClO^- ion in solution.

It has been found that, in aqueous solution, the hypochlorite ion exchanges oxygen atoms with water molecules. This reaction can be observed, and its process followed, by labeling the oxygen initially present in the ClO^- ion. In the case of oxygen, this label can be the ^{18}O isotope. Then the following reaction takes place:

$$Cl^{18}O^- + H_2O \rightarrow ClO^- + H_2^{18}O.$$

Chlorates and Perchlorates. The standard electrode potentials in Table 15–5 show that HClO is not the only oxychloride that is a strong oxidizing agent; all the half-reactions have rather high positive E^0 values. The chlorates and perchlorates are all strong oxidants, and perchloric acid ($HClO_4$) is one of the strongest of the common acids. Care is required, however, when one is working with $HClO_4$ and with chlorate and perchlorate salts, since these compounds sometimes explode.

Table 15–5. Standard electrode potentials for oxychloride half-reactions in acid solution

Half-reaction	E^0, V
$HClO_2 + 2H^+ + 2e^- \rightarrow HClO + H_2O$	$+1.64$
$HClO + H^+ + e^- \rightarrow \frac{1}{2}Cl_2 + H_2O$	$+1.63$
$ClO_2 + H^+ + e^- \rightarrow HClO_2$	$+1.28$
$ClO_3^- + 3H^+ + 2e^- \rightarrow HClO_2 + H_2O$	$+1.21$
$ClO_4^- + 2H^+ + 2e^- \rightarrow ClO_3^- + H_2O$	$+1.19$

Actually a standard laboratory preparation of O_2 gas takes advantage of the tendency of potassium chlorate to decompose. At high temperatures and in the presence of the catalyst manganese(IV) oxide, the following reaction occurs:

$$2KClO_{3(s)} \xrightarrow[MnO_2]{Heat} 2KCl_{(s)} + 3O_{2(g)}.$$

At slightly lower temperatures and without the presence of MnO_2, the decomposition yields perchlorates and chlorides:

$$4KClO_{3(s)} \xrightarrow{Heat} 3KClO_{4(s)} + KCl_{(s)}.$$

Although $KClO_4$ has less tendency to decompose than $KClO_3$, both compounds are used extensively in flares, rocket-propellant mixtures, and fireworks.

As is true for the hypochlorite ion, the chlorate ion in aqueous solution also exchanges oxygen atoms with water molecules. We can gain an insight into the mechanism of this oxygen exchange by studying the kinetics of the reaction, i.e., by

experimentally determining the rate law for the reaction. The rate of the reaction

$$Cl^{18}O_3^- + H_2O \rightarrow Cl^{18}O_2O^- + H_2^{18}O$$

is given by

$$\text{Rate} = k\,[Cl^{18}O_3^-][H^+]^2,$$

where k is the *rate constant* for the reaction (see Section 14–3) and $[H_2O]$ is assumed to be constant. This third-order rate law indicates that the reaction is catalyzed by H^+, and is consistent with the following reaction mechanism:

$$2H^+ + Cl^{18}O_3 \rightleftarrows H_2Cl^{18}O_3^+ \qquad \text{fast}$$
$$H_2Cl^{18}O_3^+ + H_2O \rightleftarrows H_2Cl^{18}O_2O^+ + H_2^{18}O \qquad \text{slow}$$
$$H_2Cl^{18}O_2O^+ \rightleftarrows 2H^+ + Cl^{18}O_2O^- \qquad \text{fast}$$

METALS

15 5 TIN AND LEAD: GROUP IVA METALS

Properties. Both tin (Sn) and lead (Pb) show typical metallic properties. They can be rolled and shaped readily, and are good electrical conductors. Tin is commonly used as a protective coating for iron in "tin" cans, and lead is used for electrodes and terminals in batteries, and as sinkers on fishing lures. Tin is very shiny, even after long exposure to the atmosphere, and relatively hard, while lead is rather soft and loses its metallic shine if it is in contact with air for only a short time.

Preparation. Tin metal is generally electroplated from aqueous solutions containing Sn^{2+} ion. An iron metal sheet, serving as the cathode, can be plated and used to make "tin" cans.

Lead is generally prepared from its primary mineral source, galena, which is principally PbS. The sulfide is roasted in oxygen gas and the resulting oxide, PbO, is reduced at high temperature with carbon:

$$2PbS_{(s)} + 3O_{2(g)} \xrightarrow{\text{Heat}} 2PbO_{(s)} + 2SO_{2(g)},$$
$$2PbO_{(s)} + C_{(s)} \xrightarrow{\text{Heat}} 2Pb_{(\ell)} + CO_{2(g)}.$$

The resulting Pb metal can be purified further by a process called zone refining (see Section 15–8). Because of this purification technique, many metals—Sn and Pb among them—are now available in highly pure form, 99.9999% Pb, for example.

Oxidation States. The Group IVA metals Sn and Pb have ns^2np^2 outer electron configurations. Hence the $+2$ and $+4$ oxidation states are known for both elements, but the $+4$ state of Pb is not very stable. Only a few compounds of $+4$ lead are important, while Pb^{2+}, Sn^{2+}, and Sn^{4+} compounds are common. In aqueous solution, the $+4$

oxidation state readily converts to Sn^{2+} and Pb^{2+}, emphasizing the greater stability of the $+2$ oxidation state for both elements.

The trend in Group IVA of decreasing stability of the $+4$ oxidation state going down the group (compare Si and Pb, for example) is attributed to the increased stability of the ns^2 electron pair for Sn and Pb; that is, the increased stability at higher atomic numbers. Combining the fact of increasing nuclear charge as you go down the group with the probability distribution plots (see Section 5–8) for s electrons, can you give a qualitative explanation of the increased stability of the $6s^2$ electron pair of Pb compared to the $4s^2$ electron pair of Ge?

These differences in the stability of oxidation states of Sn and Pb are illustrated by the standard electrode potentials given in Table 15–6 and by the compounds formed with oxygen, sulfur, and the halogens.

Table 15–6. Standard electrode potentials for Sn and Pb half-reactions

Half-reaction	E^0, V
$PbO_2 + 4H^+ + 2e^- \rightarrow Pb^{2+} + 2H_2O$	$+1.46$
$Sn^{4+} + 2e^- \rightarrow Sn^{2+}$	$+0.15$
$Pb^{2+} + 2e^- \rightarrow Pb$	-0.13
$Sn^{2+} + 2e^- \rightarrow Sn$	-0.14

Compounds with Oxygen. Oxides of Sn and Pb can be prepared by direct-combination reactions between the elements:

$$Sn_{(s)} + O_{2(g)} \xrightarrow{\text{Heat}} SnO_{2(s)},$$
$$2Pb_{(\ell)} + O_{2(g)} \xrightarrow{\text{Heat}} 2PbO_{(s)}.$$

Note that these reactions yield tin(IV) oxide and lead(II) oxide. Tin(II) oxide (and also PbO) can be prepared by heating the hydroxide:

$$Sn^{2+}_{(aq)} + 2OH^-_{(aq)} \longrightarrow Sn(OH)_{2(s)},$$
$$Sn(OH)_{2(s)} \xrightarrow{\text{Heat}} SnO_{(s)} + H_2O.$$

Lead(IV) oxide can be formed by the oxidation of PbO by hypochlorite ion in alkaline solution:

$$PbO_{(s)} + OH^-_{(aq)} + H_2O \rightarrow [Pb(OH)_3]^-_{(aq)},$$
$$[Pb(OH)_3]^-_{(aq)} + ClO^-_{(aq)} \rightarrow PbO_{2(s)} + Cl^-_{(aq)} + OH^-_{(aq)} + H_2O.$$

As indicated by the E^0 value in Table 15–6, PbO_2 is a strong oxidizing agent.

The monoxides SnO and PbO or the hydroxides $Sn(OH)_2$ and $Pb(OH)_2$ are amphoteric, and therefore dissolve in either acid or base. For example,

$$Pb(OH)_{2(s)} + 2H^+_{(aq)} \rightarrow Pb^{2+}_{(aq)} + 2H_2O,$$
$$Pb(OH)_{2(s)} + OH^-_{(aq)} \rightarrow [Pb(OH)_3]^-_{(aq)}.$$

Compounds with Sulfur. Reactions of Sn and Pb with S are similar to those considered for oxygen:

$$Sn + S \xrightarrow{\text{Heat}} SnS_2, \qquad Pb + S \xrightarrow{\text{Heat}} PbS.$$

Tin(II) sulfide (and PbS) may be precipitated from aqueous solution with H_2S:

$$Sn^{2+}_{(aq)} + H_2S_{(aq)} \rightarrow SnS_{(s)} + 2H^+_{(aq)}.$$

Lead(IV) sulfide (PbS_2) is not known. The disulfide of tin, SnS_2, is soluble in high concentrations of either H^+ or S^{2-}; that is, it is *thioamphoteric*:

$$SnS_{2(s)} + H^+_{(aq)} \rightarrow Sn^{2+}_{(aq)} + H_2S_{(aq)},$$
$$SnS_{2(s)} + S^{2-}_{(aq)} \rightarrow SnS^{2-}_{3(aq)}.$$

Compounds with Halides: *Preparation.* The tetrahalides of Sn and Pb are known (except for $PbBr_4$ and PbI_4), but they are relatively volatile and unstable substances with bonding that is primarily covalent. Lead(IV) chloride, for example, decomposes at about 100 °C to lead(II) chloride and chlorine:

$$PbCl_{4(\ell)} \xrightarrow{\text{Heat}} PbCl_{2(s)} + Cl_{2(g)}.$$

Since the dihalides of Sn and Pb have considerably more ionic character than the tetrahalides, they have higher melting and boiling points, are much less volatile, and are saltlike solids at room temperature and pressure. Tin(II) halides can be prepared by the reaction of Sn with the appropriate hydrohalic acid. For example,

$$Sn_{(s)} + 2H^+_{(aq)} + 2Cl^-_{(aq)} \rightarrow SnCl_{2(s)} + H_{2(g)}.$$

The lead dihalides are obtained by direct combinations of the elements or by precipitation from aqueous solution, since they are relatively insoluble in water. For example,

$$Pb_{(\ell)} + Cl_{2(g)} \xrightarrow{\text{Heat}} PbCl_{2(s)},$$
$$Pb^{2+}_{(aq)} + 2I^-_{(aq)} \longrightarrow PbI_{2(s)}.$$

Geometries. We can understand the geometries of gaseous molecules of SnX_4, PbX_4, SnX_2, and PbX_2 by using the approach to chemical bonding described in Section 6–5.

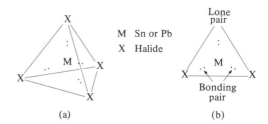

Fig. 15-9 (a) Tetrahedral geometry of Sn(IV) or Pb(IV) halides. (b) Angular geometry of Sn(II) or Pb(II) halides.

As shown in Fig. 15-9, the tetrahalides form tetrahedral molecules and the dihalides are V-shaped or angular molecules. According to this bonding theory, electron pairs are located around a central atom in an arrangement that minimizes the repulsions between the negatively charged electrons. A tetrahedral arrangement with bond angles of about 109° is the preferred geometry for four electron pairs, as shown in Fig. 15-9a, while the three groups of electron pairs—two bonding pairs and one lone pair—present in the dihalides have a preferred triangular arrangement about a central atom, as illustrated in Fig. 15-9b. The bond angle is probably near 120° (the ideal trigonal angle) and possibly slightly less than 120° (because of the stronger repulsions between the lone electron pair and the bonding electron pairs compared to the repulsion between the bonding pairs). These bonding descriptions are in good agreement with the experimental data on the Sn and Pb di- and tetrahalides.

▶ Compare SnX_2 and PbX_2 with the halides of Group II elements. The compounds $CaCl_2$, $SrBr_2$, ZnI_2, $CdCl_2$, etc., all form linear molecules, since no lone electron pair is present (see Section 6-5). ◀

15-6 CHROMIUM: A TRANSITION METAL

Properties. We shall use chromium to illustrate properties that are typical of *transition metals*. These metals are elements filling an inner energy level and generally located in Groups IIIB to IIB in the periodic table. Transition metals are good conductors of heat and electricity (Group IB elements are outstanding in this respect) and, in general, are hard, with high melting and boiling points (Group IIB elements are exceptions here). Possibly their most characteristic properties, however, involve their chemical compounds. Their compounds are usually colored and often paramagnetic (because of unpaired electrons), and large numbers of compounds are involved, since many transition metals exhibit multiple, positive oxidation states.

Table 15-7 lists properties of chromium. Note the large number of oxidation states formed, the high melting and boiling points, the relatively high electrical conductivity, and the rather unusual outer electron configuration, $3d^5 4s^1$ rather than $3d^4 4s^2$. We attribute the $3d^5 4s^1$ electron structure to the increased stability of the two half-filled subshells; that is, 5 of 10 possible d electrons and 1 of 2 possible s electrons.

Chromium metal is used extensively as a protective and decorative coating for iron and other metals in numerous appliances and in automobiles (chrome plating). The protective aspect of Cr plating is not understood completely since Cr is a relatively active metal (see the electrode potential values in Table 15-10). Apparently a

Table 15-7. Properties of chromium, $_{24}$Cr

Electron configuration	$(Ar)3d^54s^1$
Oxidation states*	$+2, +3, (+4), (+5), +6$
Electronegativity	1.6
Ionization potentials, eV†	
First	6.76
Second	16.49
Third	30.95
Fourth	50
Fifth	73
Sixth	91
Seventh	161
Eighth	185
Atomic radius, Å	1.30
Cr^{3+} radius, Å	0.69
Cr^{6+} radius, Å	0.52
Melting point, °C	1890
Boiling point, °C	2480(?)
Electrical conductance, microhms^{-1}	0.078
Density, g/ml	7.19
Crystal structure	Body-centered-cubic

*   Values in parentheses indicate unstable oxidation states.

†   Values obtained by vacuum ultraviolet spectroscopy.

very thin and very passive protective layer of an oxide, possibly Cr_2O_3, is formed on an exposed Cr metal surface. This invisible layer prevents further attack on the metal.

Preparation. Chromium plating is done by electrolyzing aqueous chromic acid (see below) solutions containing H_2SO_4. To prepare these electroplating solutions and to produce Cr metal in its pure form, manufacturers must use a metallurgical process to separate Cr from the other elements that occur in Cr-containing ores. Chromium is generally produced from the mineral chromite ($FeCr_2O_4$). This separation and purification process, which is typical of metals, is given schematically in Fig. 15-10.

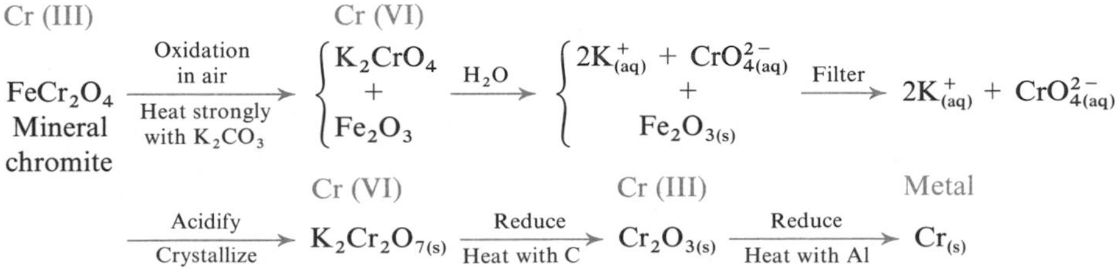

Fig. 15-10 A schematic flow chart for the preparation of chromium metal from its principal mineral $FeCr_2O_4$.

Note that Cr(III) (in $FeCr_2O_4$) is first oxidized to Cr(VI), which precipitates as a chromate (CrO_4^{2-}) from basic solution. When acidified, the chromate is converted to the dichromate ($Cr_2O_7^{2-}$; see below) which can be reduced to Cr(III) with carbon. Then the powerful reducing agent aluminum metal is used to obtain Cr metal from Cr(III).

Reactions and Compounds. The important reactions and compounds of chromium involve only the +2, +3, and +6 oxidation states. Table 15–8 presents representative compounds of these three oxidation states, and Table 15–9 lists products of Cr with common reagents.

Table 15–8. Representative species of the +2, +3, and +6 oxidation states of chromium

Cr(II)	Cr(III)	Cr(VI)
CrO, black	Cr_2O_3, green	CrO_3, red
CrS, black	$CrCl_3$, violet	$CrO_2Cl_{2(\ell)}$, dark red
$CrCl_2$, white	$Cr(OH)_3$, green-gray	$CrO_{4(aq)}^{2-}$, yellow
$Cr(OH)_2$, yellow-brown	$[Cr(OH)_4]_{(aq)}^-$, deep green	$Cr_2O_{7(aq)}^{2-}$, red-orange
$CrSO_4$, blue	$Cr_2(SO_4)_3$, blue-violet	

Table 15–9. Products of reactions of Cr with common reactants

Reactant	Product	Product color
O_2	Cr_2O_3	Green
Br_2	$CrBr_3$	Green
S	CrS	Black
$HCl_{(aq)}$	$Cr_{(aq)}^{2+} + H_2$	Blue
$H_2O_{(steam)}$	$Cr_2O_3' + H_2$	Green
$NaOH_{(aq)}$	$[Cr(OH)_6]_{(aq)}^{3-} + H_2$	Violet

Table 15–10. Standard electrode potentials for some chromium half-reactions

Half-reaction	E^0, V
$Cr_2O_7^{2-} + 14H^+ + 6e^- \rightarrow 2Cr^{3+} + 7H_2O$	+1.33
$Cr_2O_7^{2-} + 14H^+ + 12e^- \rightarrow 2Cr + 7H_2O$	+0.30
$Cr^{3+} + e^- \rightarrow Cr^{2+}$	−0.41
$Cr^{3+} + 3e^- \rightarrow Cr$	−0.74
$Cr^{2+} + 2e^- \rightarrow Cr$	−0.91

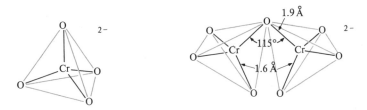

Fig. 15-11   Diagrams of the tetrahedral CrO_4^{2-} ion and the $Cr_2O_7^{2-}$ ion in $(NH_4)_2Cr_2O_7$.

The Oxidation State Cr(VI).   The $+6$ oxidation state involves, primarily, the chromate and dichromate ions, CrO_4^{2-} and $Cr_2O_7^{2-}$. These two ions may be interconverted by changing the H^+ ion concentration in solution. Addition of H^+ to a CrO_4^{2-} ion solution forms dichromate ion according to the reaction

$$2H^+ + 2CrO_4^{2-} \rightarrow Cr_2O_7^{2-} + H_2O.$$

The CrO_4^{2-} ion has tetrahedral geometry, while the $Cr_2O_7^{2-}$ ion is composed of two tetrahedra sharing one oxygen atom, as shown in Fig. 15–11 for $(NH_4)_2Cr_2O_7$.

The dichromate ion is also a very good—and quite common—oxidizing agent in acid solution, as indicated by the standard electrode potentials given in Table 15–10. For example, by comparing the data given in Tables 15–5 and 15–10, we can predict that $Cr_2O_7^{2-}$ will be a better oxidizing agent than ClO_3^-, and that if we mix $Cr_2O_7^{2-}$ and ClO_3^- in an acid solution, Cr^{3+} and ClO_4^- will be formed:*

$$Cr_2O_7^{2-} + 14H^+ + 6e^- \rightarrow 2Cr^{3+} + 7H_2O$$
$$\underline{3(ClO_3^- + H_2O \rightarrow ClO_4^- + 2H^+ + 2e^-)}$$
$$Cr_2O_7^{2-} + 3ClO_3^- + 8H^+ \rightarrow 2Cr^{3+} + 3ClO_4^- + 4H_2O$$

Oxides.   The oxides of chromium illustrate a common trend of changing from basic to acidic nature as the oxidation state of the metal increases. Chromium(II) oxide (CrO) is quite basic,

$$CrO_{(s)} + 2H^+_{(aq)} \rightarrow Cr^{2+}_{(aq)} + H_2O,$$

while chromium(III) oxide, Cr_2O_3, is amphoteric,

$$Cr_2O_{3(s)} + 6H^+_{(aq)} \rightarrow 2Cr^{3+}_{(aq)} + 3H_2O,$$
$$Cr_2O_{3(s)} + 2OH^-_{(aq)} + 3H_2O \rightarrow 2Cr(OH)_4^-{}_{(aq)},$$

and chromium(VI) oxide (CrO_3) is quite acidic,

$$CrO_3 + 2OH^- \rightarrow CrO_4^{2-} + H_2O.$$

---

*   You should be able to balance this equation, and most others, with little difficulty (see Sections 7–7 and 11–5 for review), and also solve quantitative problems based on balanced equations, as in question 36 at the end of this chapter.

The Oxidation States Cr(II) and Cr(III). Tables 15–8 and 15–9 list compounds of Cr(II) and Cr(III), and Table 15–10 indicates that Cr(II) is a good reducing agent. In aqueous solution, Cr(II) is readily oxidized to Cr(III) by atmospheric oxygen. However, the Cr(II) oxidation state can be stabilized by forming a complex such as $[Cr(NH_3)_6]^{2+}$. Complex compounds, formed from complex ions, account for a large number of Cr(II) and Cr(III) compounds. We shall use these compounds to introduce the study of the vast number of complex ions and complex compounds.

15-7 COMPLEX COMPOUNDS

A complex ion or complex compound is composed of a *central metal cation* bonded to several (often four or six) anions or molecules, or a mixture of anions and molecules. The groups attached to the central metal are called *ligands*. In the two complex ions $[Cr(H_2O)_6]^{3+}$ and $[Cr(OH)_6]^{3-}$, the central metal ion is Cr^{3+} and the ligands are H_2O molecules in the first case and OH^- (hydroxide) ions in the second. With few exceptions, ligands have at least one free electron pair that can be used in bonding to the central metal. For example, hydroxide ion has three available electron pairs, cyanide ion two, H_2O two, and ammonia one:

$$:\!\ddot{O}\!:\!H^- \qquad :\!C\!:\!:\!:\!N\!:^- \qquad H\!:\!\ddot{O}\!: \qquad H\!:\!\ddot{N}\!:\!H$$
$$\qquad\qquad\qquad\qquad\qquad\quad H \qquad\quad H$$

The amounts of ionic and covalent bonding present in complex ions, however, varies greatly from ion to ion.

The number of ligand atoms directly bonded to the central cation, or the number of coordination positions around the central metal, is called the *coordination number* of the central metal ion. In the complex ions $[Cr(NH_3)_6]^{3+}$, $[Zn(NH_3)_4]^{2+}$, and $[Ag(NH_3)_2]^+$, the coordination numbers for Cr^{3+}, Zn^{2+}, and Ag^+ are 6, 4, and 2, respectively.

The ammonia molecule is an example of a *unidentate* ligand, since it bonds to the central cation via only one atom. Some ligands can bond to the central metal through two atoms located at different places within the ligand molecule or ion. These are *bidentate ligands*.* For example, oxalate ion, $C_2O_4^{2-}$, and ethylenediamine, $NH_2CH_2CH_2NH_2$, are bidentate ligands:

---

*   *Multidentate ligands*, which can coordinate at 2, 3, 4, 5, or 6 positions, are also known.

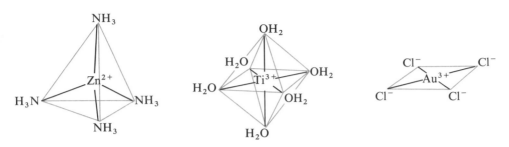

Fig. 15-12 Typical geometries of complex ions: tetrahedral $[Zn(NH_3)_4]^{2+}$, octahedral $[Ti(H_2O)_6]^{3+}$, and square planar $[AuCl_4]^-$.

▶ Complex ions and compounds have many practical uses. They are, for example, used in metallurgical refining processes, in various chemical analysis procedures, and in water-softening. Recently, complex compounds of platinum, such as $[Pt(NH_3)_2Cl_4]$ and $[Pt(NH_3)_2Cl_2]$, have been found to inhibit both leukemia and sarcoma malignancies in mice. In addition, the study of complexes has advanced our understanding of chemical bonding considerably. ◀

Structure. Most complex ions of the transition metals have coordination numbers of four or six, with ligand groups distributed in three typical geometric arrangements: tetrahedral, square planar, and octahedral. Of these three, the tetrahedral and octahedral symmetries are the most common and important. Figure 15–12 gives several examples.

For Cr^{2+} and Cr^{3+} central metal ions, complexes are of octahedral geometry, with a coordination number of six. Examples are $[Cr(NH_3)_6]^{2+}$ and $[Cr(CN)_6]^{4-}$ for Cr(II) and $[CrCl_6]^{3-}$ and $[Cr(H_2O)_4Cl_2]^+$ for Cr(III). Note, in this last example, that two different ligands are present. This is a very common occurrence; as many as six different ligand groups can be coordinated to a central metal cation with coordination number six. You can see why such a vast number of complex ions and complex compounds can be prepared.

In both tetrahedral and square planar complex ions, the coordination number is four. Tetrahedral complexes are common for Zn^{2+} and Cd^{2+} ions (see Fig. 15–12), as in $[ZnCl_4]^{2-}$ and $[Cd(CN)_4]^{2-}$, while square planar complex ions are common for Ni^{2+} and Pt^{2+}, for example $[Ni(CN)_4]^{2-}$ and $[Pt(NH_3)_2Cl_2]$ (see Fig. 15–12).

Isomers. Complex compounds exhibit many different types of *isomers* (that is, compounds with the same chemical formula but with different arrangements of atoms, and therefore different properties). One of the most important of the general classification of isomers is *stereoisomerism*. Stereoisomers are compounds with identical ligands arranged in different geometric patterns. One type of stereoisomerism is

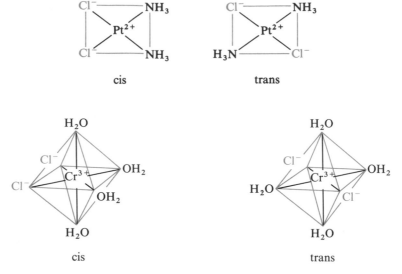

Fig. 15–13 *Cis* and *trans* isomers of the square planar complex compound [Pt(NH₃)₂Cl₂] and the octahedral complex ion [Cr(H₂O)₄Cl₂]⁺.

called *geometric isomerism* (or *cis-trans isomerism*), which is illustrated in Fig. 15–13 for the square planar $[Pt(NH_3)_2Cl_2]$ complex compound and for the octahedral complex ion $[Cr(H_2O)_4Cl_2]^+$.

▶ Note that tetrahedral complexes cannot form *cis-trans* isomers because of the geometric relationships of the ligands. Convince yourself of this by trying to draw (or build with models) geometric isomers of the tetrahedral $[Zn(NH_3)_2Cl_2]$ compound. ◀

A second type of stereoisomerism is *optical isomerism*. Optical isomers are non-superimposable mirror images of one another. You can think of a pair of gloves as an analogy for optical isomers. The gloves are mirror images and are not super-imposable. Such pairs of isomers are called *enantiomorphs* or *mirror images*.

Figure 15–14 shows examples of octahedral optical isomers. Most of the well-characterized optical isomers of complex ions and compounds involve octahedral symmetry and bidentate ligands.

Enantiomorphs have identical properties, except for their interaction with polarized light. One isomer rotates the plane of the polarized light to the right (*dextro*rotatory) while the other optical isomer rotates it an equal amount to the left (*levo*rotatory). Optical isomers are very important in organic chemistry. They are covered in detail in Sections 17–4 and 17–5.

Reactions: *Complex Stability versus Reaction Rate.* Many of the transition metal complexes undergo rapid reactions in which their ligand groups are replaced by other

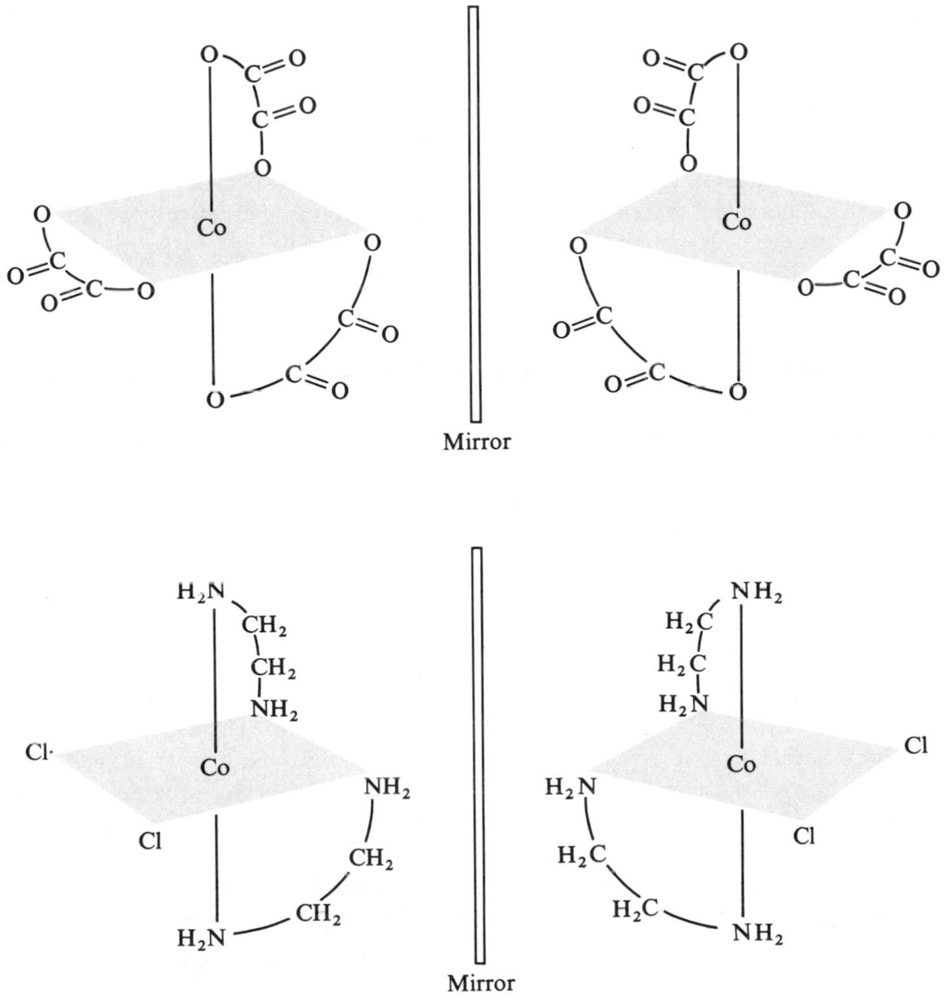

Fig. 15–14 $[Co(C_2O_4)_3]^{3-}$ and $[Co(NH_2CH_2CH_2NH_2)_2Cl_2]^+$, optical isomers of the octahedral complexes. The oxalate ion, $C_2O_4^{2-}$, and ethylenediamine, $NH_2CH_2CH_2NH_2$, are bidentate ligands.

ligands. These are called *labile* complexes. On the other hand, some central metal ions (primarily Cr^{3+} and Co^{3+}) form complex ions that undergo substitution reactions slowly. These are *nonlabile* or *inert* complexes. For example, if the very stable square planar complex ion $[Ni(CN)_4]^{2-}$ is mixed with an aqueous solution containing labeled CN^- ions, exchange of the CN^- ions between the complex and the solution occurs so rapidly that the speed of the reaction cannot be measured by current techniques. The complex ion $[Co(NH_3)_6]^{3+}$, however, persists for days in an acid solution, with very slow replacement of the NH_3 ligands with H_2O ligands. The

complex $[Co(NH_3)_6]^{3+}$ is nonlabile, even though the equilibrium constant for the reaction,

$$[Co(NH_3)_6^*]^{3+} + 6H^+ + 6H_2O \rightarrow [Co(H_2O)_6]^{3+} + 6NH_4^+,$$

is very large: $K \approx 10^{25}$.

These observations vividly illustrate the difference between the energetics and the kinetics of chemical reactions. The $[Ni(CN)_4]^{2-}$ ion is energetically very stable, as indicated by the very small equilibrium constant $K \approx 10^{-22}$ for the dissociation reaction:

$$[Ni(CN)_4]_{(aq)}^{2-} \rightleftarrows Ni_{(aq)}^{2+} + 4CN_{(aq)}^-.$$

And yet the CN^- ion replacement in the complex occurs so rapidly that it is un-measurable. The kinetic lability is very high, even though the energetic stability is high. The reverse situation is true for $[Co(NH_3)_6]^{3+}$ in acid solution. Kinetic lability (replacement of NH_3 with H_2O ligands) is low even though the reaction is very favor-able energetically ($K \approx 10^{25}$ for the above reaction).

This is generally true for chemical reactions: There is no necessary correspon-dence between the energetic favorability of a reaction and the speed with which the reaction occurs.

Complexes and Oxidation State Stability. The presence or formation of complex ions can also change the stability of a particular oxidation state of an element or change the stability of a central metal ion toward oxidation or reduction. The Co(III) oxidation state, for example, liberates oxygen gas from aqueous solutions, since $[Co(H_2O)_6]^{3+}$ is such a strong oxidizing agent. Co(III) can be obtained in an aqueous solution, however, if NH_3 is present so that the complex ion $[Co(NH_3)_6]^{3+}$ is formed. This highly stable complex does not oxidize water.

We can illustrate increasing stability toward reduction by complex formation with complex ions of Zn(II). The standard electrode potentials given for the appro-priate half-reactions indicate increasing stability from $Zn_{(aq)}^{2+}$ to $[Zn(NH_3)_4]^{2+}$ to $[Zn(CN)_4]^{2-}$:

$$Zn_{(aq)}^{2+} + 2e^- \rightarrow Zn_{(s)}, \qquad E^0 = -0.76 \text{ V},$$
$$[Zn(NH_3)_4]^{2+} + 2e^- \rightarrow Zn_{(s)} + 4NH_3, \qquad E^0 = -1.04 \text{ V},$$
$$[Zn(CN)_4]^{2-} + 2e^- \rightarrow Zn_{(s)} + 4CN^-, \qquad E^0 = -1.26 \text{ V}.$$

Note that this increased stability of the Zn(II) oxidation state parallels the in-crease in stability of the complex ion formed. The dissociation of the complex ions is governed by an equilibrium constant,

$$[Zn(NH_3)_4]^{2+} \rightleftarrows Zn^{2+} + 4NH_3, \qquad K \approx 10^{-10},$$
$$[Zn(CN)_4]^{2-} \rightleftarrows Zn^{2+} + 4CN^-, \qquad K \approx 10^{-18},$$

which reflects the stability of the complex. Other values of ionization constants for complex ions are given in Table 15–11.

Table 15–11. Equilibrium ionization constants for some complex ions

Complex ion dissociation	K
$[Cu(NH_3)_4]^{2+} \rightleftarrows Cu^{2+} + 4NH_3$	1.0×10^{-12}
$[Co(NH_3)_6]^{2+} \rightleftarrows Co^{2+} + 6NH_3$	4.0×10^{-5}
$[Co(NH_3)_6]^{3+} \rightleftarrows Co^{3+} + 6NH_3$	6.3×10^{-36}
$[Ag(NH_3)_2]^{+} \rightleftarrows Ag^{+} + 2NH_3$	6×10^{-8}
$[Ag(CN)_2]^{-} \rightleftarrows Ag^{+} + 2CN^{-}$	1.8×10^{-19}
$[Hg(CN)_4]^{2-} \rightleftarrows Hg^{2+} + 4CN^{-}$	4×10^{-42}

These equilibrium constants can be used to calculate the equilibrium concentrations involved in solutions containing the complex ion. For example, using the value for K from Table 15–11, we can calculate that the Ag^+ ion concentration in a $0.10M$ $[Ag(CN)_2]^-$ solution is $1.7 \times 10^{-7}M$. Since this Ag^+ concentration is so small, the formation of complex ions such as $[Ag(CN)_2]^-$ are useful in dissolving relatively insoluble precipitates such as AgCl ($K_{sp} = 1.7 \times 10^{-10}$).

Table 15–12. Names of some complex compounds*

Formula	Name
$[Ag(NH_3)_2]^{+}$	Diamminesilver(I) ion
$[Cr(H_2O)_6]^{3+}$	Hexaaquochromium(III) ion
$[PtCl_6]^{2-}$	Hexachloroplatinate(IV) ion
$[Fe(CN)_6]^{4-}$	Hexacyanoferrate(II) ion
$[Co(NH_3)_3(NO_2)_3]$	Trinitrotriamminecobalt(III)
$K[Ni(H_2O)_3Cl_3]$	Potassium trichlorotriaquonickelate(II)
$[Ni(H_2O)_5Cl]Cl$	Chloropentaaquonickel(II) chloride
$[Cr(NH_3)_6][Cr(CN)_6]$	Hexaamminechromium(III) hexacyanochromate(III)

*   Often common or traditional names are used. For example, $[Fe(CN)_6]^{4-}$ is called ferro-cyanide ion and $[Fe(CN)_6]^{3-}$ is called ferricyanide ion.

Nomenclature. Since there are so many complex ions and complex compounds, a system for naming them has been developed. We shall give only a few of the rules here. They will enable you to name the simple and more common complexes. Table 15–12 gives some examples.

1. In a complex compound, the cation is named first.

2. Within a complex ion, the constituents are named in this order: anions, neutral molecules, central metal ion.

3. Anion ligands are given -o endings: Cl^-, *chloro*; CN^-, *cyano*; NO_2^-, *nitro*; $C_2O_4^{2-}$, *oxalato*.

4. Neutral ligands have unchanged or special names: H_2O, *aquo*; NH_3, *ammine*; CO, *carbonyl*.

5. The number of ligands of a particular type is indicated by a prefix: *di-*, *tri-*, *tetra-*, etc. (for 2, 3, 4, etc.).

6. The central metal ion oxidation state is given as a Roman numeral in parentheses after the name of the complex ion.

7. If the complex ion is an anion (negatively charged), the ending *-ate* is given to the central cation.

Bonding in Complexes.   There are several approaches to chemical bonding that are currently being applied to the bonding in complex ions and complex compounds. These theories are undergoing constant change and revision, as are all chemical bonding theories. We shall cover only two approaches here: the valence-bond theory and the crystal-field theory. Both accounts are very brief and serve only as introductions to the theories.

Valence-Bond Theory.   The valence-bond theory describes the bonding in complexes in terms of the donation of electron pairs by the ligands to hybrid orbitals of the central metal ion. This approach is an extension of many of the ideas presented in Chapter 6, electron-dot structures, atomic orbitals, hybrid atomic orbitals, and even the octet rule. This theory was used by chemists prior to the crystal-field theory described below. It has the advantage of being easily visualized and used, but it also has several deficiencies that the crystal-field theory overcomes.

Table 15–13. Hybrid atomic orbitals: their formation and symmetry

Atomic orbitals combined	Hybrid orbital	Symmetry of hybrid orbital
$s + p_x$	sp	Linear
$s + p_x + p_y$	sp^2	Trigonal
$s + p_x + p_y + p_z$	sp^3	Tetrahedral
$d_{x^2-y^2} + s + p_x + p_y$	dsp^2	Square planar
$d_{z^2} + d_{x^2-y^2} + s + p_x + p_y + p_z$	d^2sp^3	Octahedral

Bond formation, according to the valence-bond theory, is the sharing, or overlap, of empty atomic orbitals of the central metal with filled orbitals of the ligands. The geometric configuration, or symmetry, of the resulting complex is governed by the number and type of available central metal orbitals. When we recall the geometry of particular hybrid orbitals (see Section 6–5) given in Table 15–13, we see that the very common tetrahedral and octahedral complex ions require that sp^3 and d^2sp^3 hybrid orbitals, respectively, be made available by the central metal. Ligands that donate electrons to these hybrid orbitals will be located in the correct geometric positions around the central ion, i.e., the complex would have the geometry that is experimentally observed.

As an example of this bonding approach, let us consider the complex ion $[Cr(NH_3)_6]^{3+}$. (Can you name it?) We know (from experiments) that the ion is octahedral and that the Cr^{3+} ion has a coordination number of six. To form an octahedral complex, the six NH_3 ligand molecules donate six electron pairs (one pair per NH_3 group) to the empty d^2sp^3 hybrid orbitals of the Cr^{3+} ion. Note that six electron pairs, 12 total electrons, completely fill the d^2sp^3 orbitals. We can diagram this bonding scheme for $[Cr(NH_3)_6]^{3+}$ in a very convenient fashion:

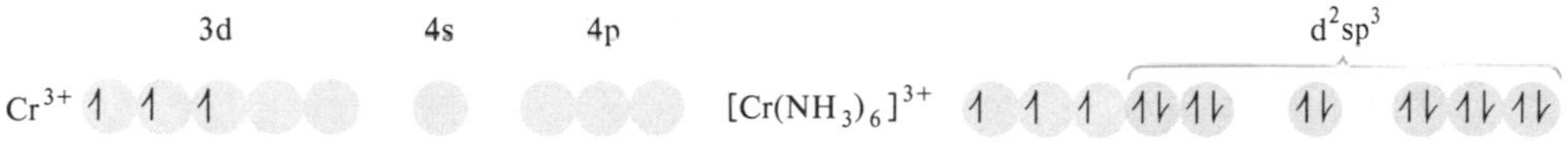

The Cr^{3+} ion has a $3d^3$ outer electron configuration. These three electrons enter three of the five available 3d orbitals of the Cr^{3+} central metal ion. The other two 3d orbitals, along with the empty 4s and the three empty 4p orbitals, can each accept an electron pair from one of the NH_3 ligands. We can think of this process as the formation of a completely filled set of d^2sp^3 hybrid orbitals, which gives the required octahedral orientation to the NH_3 groups around the Cr^{3+} central metal ion.

We can extend this idea to the other transition metals. The electron configuration for Mn^{3+} is $3d^4$. What symmetry would we expect for the $[Mn(CN)_6]^{3-}$ complex ion?

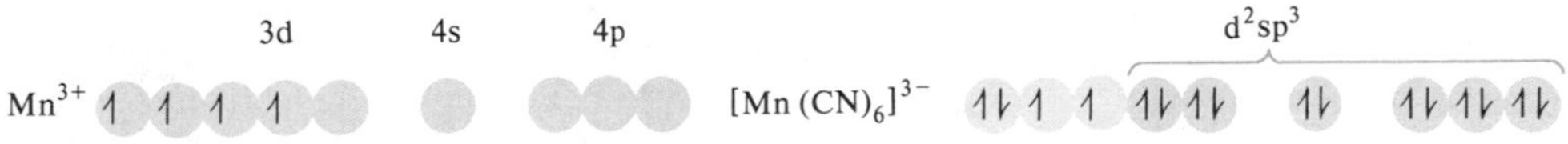

Using the above diagrammatic approach, we would again predict octahedral geometry for the complex ion; that is, d^2sp^3 hybrid formation for the central Mn^{3+} ion. In this case, one of the unpaired 3d electrons in Mn^{3+} must be paired to make possible d^2sp^3 hybrid bonding in the complex.

Then we have another criterion for the theory: The number of unpaired electrons predicted by the theory must correspond to the number observed experimentally. In $[Cr(NH_3)_6]^{3+}$ and $[Mn(CN)_6]^{3-}$, the valence-bond theory predicts three and two unpaired electrons, respectively; these predictions agree with experimental results.

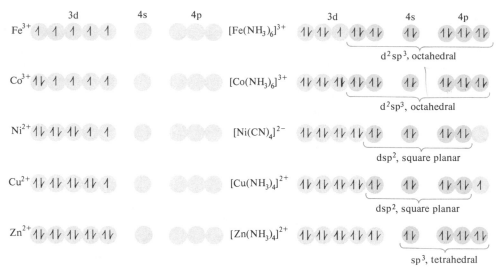

Fig. 15–15 Valence-bond diagrams for octahedral (d^2sp^3), square planar (dsp^2), and tetrahedral (sp^3) complex ions. Note the number of unpaired electrons in each case.

Figure 15–15 gives several other valence bond diagrams for octahedral, tetrahedral (sp^3), and square planar (dsp^2) complex ions. In all these cases, the bonding description agrees with the experimentally observed symmetry and number of unpaired electrons.

Unfortunately this rather attractive simple bonding picture for complexes has defects. For example, from Fig. 15–15 we would expect all Fe^{3+} complexes with octahedral symmetry to have one unpaired electron. In fact, the octahedral complex $[FeF_6]^{3-}$ has five unpaired electrons, the same number as the free Fe^{3+} ion.

We also see from Fig. 15–15 that $[Cu(NH_3)_4]^{2+}$ has square planar geometry (dsp^2 hybridization) with one unpaired electron. But this unpaired electron must be promoted from its original place in a 3d orbital into a higher-energy 4p orbital. Since this promotion requires energy, why does $[Cu(NH_3)_4]^{2+}$ form the dsp^2 square planar configuration rather than the tetrahedral sp^3 geometry?

There are many similar problems with the valence-bond approach to bonding in complexes. Some of these difficulties have been handled by making alterations and additions to the theory. But there is one glaring weakness in the theory: It cannot, even in a qualitative sense, explain the optical properties that are exhibited by complex ions and compounds. We cannot explain the various colors of almost all transition metal complexes on the basis of the valence-bond theory. This severe drawback is completely overcome by the crystal-field theory.

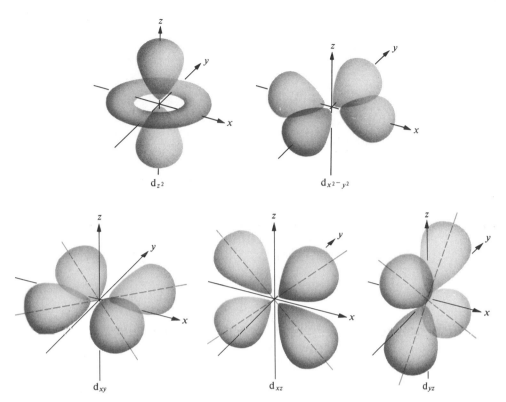

Fig. 15-16   Shapes of the five d orbitals.

Crystal-Field Theory. The crystal-field theory, as applied to chemical bonding in transition metal complexes, is based on the effects that the negative electron distributions in the ligands have on the d orbitals of the central metal ion. We have learned that the atoms of elements, or their positive or negative ions, have a set of energy levels—1s, 2s, 2p, 3s, etc.—of increasing energy, and that these sublevels are composed of a certain number of orbitals, one for each s, three for each p, five for each d, seven for each f, and so on (see Section 5–8). Then if we focus our attention on the d orbitals of the transition-metal elements, there are five separate orbitals to consider. In a free, gaseous atom or ion (a gaseous Ti^{3+} ion in a vacuum, for example), all five d orbitals have the same energy. We say that the orbitals are *degenerate.*

    We have also seen that the atomic orbitals have a different shape, i.e., their distributions of electrons about the nucleus have different geometries (see Section 5–8). All s orbitals are spherical, while p orbitals are somewhat dumbbell shaped. The three different p orbitals are identical in shape, but are pointed in different directions in space, the p_x along the x-axis, the p_y along the y-axis, and the p_z along the z-axis.

    A similar, but slightly more complex, situation exists for the d orbitals; their geometries are given in Fig. 15–16. Note that three of the d orbitals are identical in

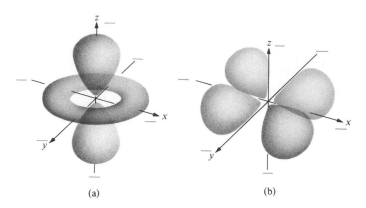

(a)   (b)

Fig. 15-17   (a) Octahedrally positioned point charges and their relation to the d_z orbital. (b) d_{xy} orbital interaction with an electrical field that has an octahedral point charge.

shape—the d_{xy}, d_{xz}, and d_{yz}—but have different orientations in space. The other two d orbitals—the d_{z^2} and $d_{x^2-y^2}$—have different shapes.

Now we have two things to consider about the d orbitals: (1) In a free ion, they all have identical energies, and (2) they all have either different orientations in space or different shapes. In a vacuum an ion has no local surroundings, and the d orbitals all have the same energy. When an ion occurs in a solution or in a solid, however, it has close neighbors. It is the effect of these surrounding groups on the energies of the d orbitals—which have different shapes and therefore different interactions with the surroundings—that is the basis of crystal-field theory.

We shall consider briefly only octahedral geometries, i.e., transition metal ions in octahedral complexes such as $[PtCl_6]^{2-}$. Similar arguments can be made for tetrahedral, square planar, or any other symmetry, but the results are less easily visualized. Crystal-field theory assumes that ligands act like point negative charges, i.e., that the ligands have no physical extent and that the negative charge associated with them is located at one point in space. Then for $[PtCl_6]^{2-}$, each Cl^- ligand is assumed to be a point charge; these six point charges are located in an octahedral arrangement around the Pt^{4+} ion.

Now these negative point charges interact with the d orbitals of the Pt^{4+} ion, and since the d orbital symmetries are different, the interactions are different. Figure 15-17a shows the location of charges in relation to the d_{z^2} orbital and Fig. 15-17b the point charge location with respect to the d_{xy} orbital. The charges lie along the axes, and can therefore interact strongly with any electron distribution located in the d_{z^2} orbital. Since this is a like-charge interaction (both negative charges), there is a repulsive force and an increase in energy.

The d_{xy} maximum electron density, on the other hand, is located between the axes, and, consequently, between the negative point charges (the ligands). Thus the interaction between the d_{xy} orbital and the surrounding octahedrally located charges is not as great as it is for the d_{z^2} orbital.

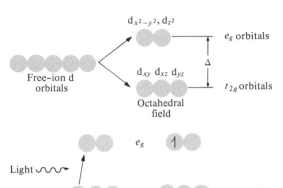

Fig. 15-18 Crystal-field splitting of the d orbitals in octahedral symmetry.

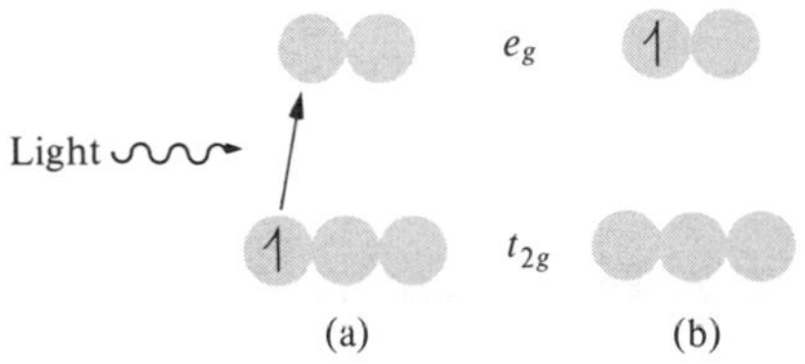

Fig. 15-19 Light energy is absorbed by the $[Ti(H_2O)_6]^{3+}$ ion in its ground state in (a), producing an excited electronic state in (b).

Actually the d orbitals are split into two groups by an octahedral crystal field, i.e., octahedrally located point charges, since the interactions of the d_{xy}, d_{xz}, and d_{yz} orbitals with the point charges are identical and the interactions of the d_{z^2} and $d_{x^2-y^2}$ orbitals are also identical—but different from those of the d_{xy}, d_{xz}, and d_{yz}.* The d_{xy}, d_{xz}, and d_{yz} orbitals are called the t_{2g} *orbitals*. They lie lower in energy than the d_{z^2} and $d_{x^2-y^2}$, which compose a second set called the e_g *orbitals*.

This energy splitting of the d orbitals by an octahedral crystal field is diagrammed in Fig. 15-18. The e_g orbitals are separated from the t_{2g} orbitals by an energy difference generally referred to as the *crystal-field splitting energy*. It is often given the symbol Δ.

This bonding approach, applied to complexes, can make the same types of predictions and offer explanations for the same kinds of behaviors of complexes as was true for the valence bond theory, and in a more quantitative fashion. But the crystal-field theory has one more major advantage: It can explain the optical properties of transition metal complex ions.

Figure 15-19 diagrams the electron transition that can account for the optical absorption spectrum, and consequently the color, of $[Ti(H_2O)_6]^{3+}$. When visible light energy is absorbed by the $[Ti(H_2O)_6]^{3+}$ complex ion, the single 3d electron is excited from the t_{2g} orbitals into the e_g orbitals. The amount of energy absorbed corresponds to the energy separation between the e_g and the t_{2g} levels, the crystal field splitting energy Δ. In the case of H_2O ligands for the Ti^{3+} ion, the energy Δ has a value which causes the $[Ti(H_2O)_6]^{3+}$ solution to appear light violet in color. For different ligands, however, different values of Δ are obtained. Hence different absorption spectra and therefore different complex ion colors are expected—and observed— for various ligands with the same central metal ion.

---

*   That the d_{z^2} and $d_{x^2-y^2}$ interactions with octahedrally placed negative charges are identical is not completely obvious pictorially, but it can be proved mathematically.

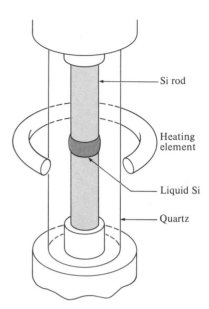

Si rod

Heating
element

Liquid Si

Quartz

Fig. 15–20 Zone refining process for Si, showing
the "floating zone" technique in a vacuum or an
inert atmosphere. No crucible is used, in order to
reduce contamination; surface tension sustains the
liquid zone. The radio frequency heater is moved
along the sample, which is sometimes rotated.

SEMIMETALS

15–8 SILICON AND GERMANIUM

Element Properties.   Both Si and Ge are hard, brittle elements with high melting and
boiling points; both crystallize with the diamond structure. Their most characteristic property is their electrical conductivity; both are semiconductors. Their electrical
conductivity is higher than that of insulators—diamond, NaCl, etc.—but lower than
that of metals—Sn, Pb, Cu, etc. This electrical property enables Si and Ge to be used
in many solid-state electrical devices, such as transistors, and has led to their great
industrial importance. In order to take advantage of this unusual property, however,
manufacturers must have Si and Ge available in very pure form.

Preparation.   The elements can be prepared by reducing SiO_2 or GeO_2 with carbon
at high temperatures:

$$GeO_{2(s)} + C_{(s)} \xrightarrow{\text{Heat}} Ge_{(s)} + CO_{2(g)}.$$

Highly pure Ge or Si can be obtained by decomposition of the volatile tetrahalides, in
a complex vapor preparation procedure summarized by

$$GeCl_{4(g)} \xrightarrow{\text{Heat}} Ge_{(s)} + 2Cl_{2(g)}.$$

Purification.   Ultra-high-purity Si or Ge is often obtained by a process known as
zone refining, diagrammed in Fig. 15–20. This procedure involves moving a small

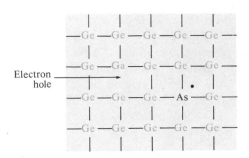

Fig. 15-21 Germanium containing Group IIIA (Ga) and Group VA (As) impurities. An excess of electrons is associated with the As impurity and a deficiency of electrons with the Ga impurity.

liquid zone very slowly through a solid Si or Ge bar (or rod). (This liquid zone is obtained by spot-heating the bar with a heating element.) Impurities that are present dissolve preferentially in the liquid zone and are thereby swept from one end of the sample to the other. When the liquid zone reaches the end of the bar, the heating element is rapidly (and automatically) moved back to its first position at the opposite end of the bar. This process is repeated many times until the impurity content in the sample has been reduced to a very low level.

▶ In fact, Si and Ge are among the purest substances known. They are prepared with impurity concentrations of less than one part per billion. This is about 10^7 times purer than ordinary reagent-grade chemicals, which have impurity levels around one part per thousand. ◀

Variation of Electrical Properties. Actually most of the Si and Ge used in solid-state technology is slightly impure, but the impurities are selectively added to the Si and Ge in well-known concentrations. It is the presence of these special impurities that causes the desired electrical properties of the Si and Ge host materials. Figure 15–21 shows a schematic representation of Group IIIA and Group VA impurity additions to Ge.

n-Type Ge. Solid Ge has the diamond crystal structure (see Fig. 15–1a), in which each Ge atom is bonded to four other Ge atoms and all four available valence electrons are used in bonding. These bonding electrons are held tightly between the bonded Ge atoms and are not free to move throughout the solid material. When a Group VA element such as arsenic (As) replaces one of the Ge atoms in the crystal lattice, however, the As atom uses only four of its five available electrons in bonding to the surrounding four Ge atoms. The remaining electron of the substituted As atom is not bound very tightly to the As and is free to roam about the Ge solid, with very little activation energy. This free electron is shown schematically in Fig. 15–21. When electrons are the mobile charge carriers in semiconductors, we call them *n-type* (*n* for negative) semiconductors.

p-Type Ge. Similarly, substitution of a Group IIIA element, such as Ga, for a Ge atom provides an electron deficiency, since the Ga atom has only three available bonding electrons. Then one Ga—Ge bond lacks an electron. This *electron hole* serves as an attraction for any other electrons in the lattice, and therefore appears to be positively charged. Note that as electrons move into the hole from other bonds (see Fig. 15–21), the hole migrates through the material in a direction opposite to that of the moving electrons. Semiconductors with hole charge carriers are called *p-type* (*p* for positive) semiconductors.

Chemical Compounds. The compounds of Si illustrate its nonmetallic nature, as discussed in Section 15–3. The compounds of Ge, however, show primarily metallic-type properties. They are similar to the compounds of Sn, but the $+4$ oxidation state is somewhat more stable. Table 15–14 presents typical reactions of Ge.

Table 15–14. Reactions of germanium

$$Ge + 2X_2 \rightarrow 2GeX_4$$
(any
halogen)
$$Ge + O_2 \rightarrow GeO_2$$
$$Ge + 2S \rightarrow GeS_2$$
$$3Ge + 4H^+ + 4NO_3^- \rightarrow 3GeO_2 + 4NO + 2H_2O$$
$$Ge + 2OH^- + 4H_2O \rightarrow [Ge(OH)_6]^{2-} + 2H_2$$

QUESTIONS

1. Predict which elements in Group VA would be most metallic and which elements would show the most nonmetallic properties.

2. Explain why graphite has different values of electrical conductivity along different directions in a graphite crystal.

3. Ozone (O_3) is an allotrope of oxygen gas and has an angular geometry. Write an electron-dot (Lewis) structure for O_3, and any resonance structures if necessary.

4. Explain how van der Waals forces are important in graphite and why they are not important in diamond.

5. The molecule ethylene (CH_2CH_2) is planar and the H—C—H bond angles are 120°. Describe the bonding present in the molecule by indicating any hybrid bond formation, the number of σ- and π-bonds formed, and the bond order for the C—C bond.

6. Distinguish between σ- and π-bonds; give an example of each.

7. Calculate the H^+ ion concentration in a 0.10M NaHCO$_3$ solution.

8. Calculate the H^+ concentration in a 0.10M Na$_2$CO$_3$ solution.

9. Devise a method of preparation for Mg(CN)$_2$ solid.

10. A beaker contains 1.0ℓ of solution with 0.010 mole AgCl solid on the bottom of the beaker in equilibrium with the solution. Describe as quantitatively as possible what will happen if you dissolve 0.10 mole NaCN in the solution. Assume that the volume does not change when you add the NaCN, and that the NaCN is 100% dissociated in solution.

11. Write an equation for the reaction that should occur when Si_3H_9 is heated in air.

12. Sketch the SiF_4 molecule and the SiF_6^{2-} ion, showing clearly their geometric arrangements.

13. Show how the sharing of three oxygen atoms by SiO_4^{4-} units yields a repeating unit of $Si_2O_5^{2-}$.

14. Predict a trend in electronegativity in Group VIIA; give reasons to support your prediction.

15. Explain why Cl_2 is a gas and I_2 a solid under ordinary room temperature and pressure conditions.

16. Suppose you have a test tube containing a $0.10M$ NaBr solution and some CCl_4. Describe in words and with any appropriate equation(s) what would happen if Cl_2 gas were to be bubbled through the solution, followed by brisk mixing of the aqueous and CCl_4 solutions.

17. Devise a preparation method for I_2 from sea water containing I^- ion.

18. Sketch the crystal structures of MgO and CeO_2.

19. Calculate the F^- ion concentration in a saturated aqueous solution of CaF_2, $K_{sp} = 1.7 \times 10^{-10}$.

20. Calculate the Mn^{2+} ion concentration in a $0.10M$ NaOH solution, given that K_{sp} for $Mn(OH)_2$ is 2×10^{-13}.

21. Using MnS (NaCl crystal structure) as an example, describe what is meant by interstitial, cation vacancy, anion vacancy, and Fe^{2+} substitutional impurity.

22. (a) Write the net equation for the neutralization reaction between the strong base $Ca(OH)_2$ and the weak acid HClO. (b) At the equivalence point in the reaction in (a), will the pH of the solution be above 7.0, below it, or exactly 7.0? Why?

23. What is the H^+ ion concentration and the pH in 100 ml of a $0.10M$ HF solution that also contains 0.010 mole NaF? Assume that NaF is completely dissolved and dissociated.

24. Using the rate law for oxygen exchange between ClO_3^- and H_2O given in Section 15–4, explain: (a) why the rate law indicates that the reaction is catalyzed by H^+; (b) why the proposed mechanism is consistent with the rate law.

25. Sketch an electrolysis cell that might be usable in plating tin on thin iron sheets.

26. In Group IIIA, the $+3$ oxidation state is generally the most stable for all the members except thallium; for example, Tl^+ is more stable than Tl^{3+}. What explanation can you give for the increased stability of the Tl(I) oxidation state?

27. Suppose that you have an aqueous solution containing Sn^{4+} ion. What would happen if powdered Pb metal were mixed with the solution? Justify your answer.

28. Give preparation schemes for $SnBr_{2(s)}$ and for $PbBr_{2(s)}$.

29. a) Predict and sketch the geometry of $GeCl_{4(g)}$.
    b) Predict and sketch the geometry of $GaCl_{3(g)}$.
    c) Predict and sketch the geometry of $ZnCl_{2(g)}$.
    d) Predict and sketch the geometry of $SnCl_{2(g)}$.

30. (a) What oxidation states would you predict for vanadium? Why? (b) Give the electron configurations for each oxidation state in (a).

31. Which ion should give a colorless aqueous solution: Sc^{3+}, V^{3+}, Mn^{3+}, Fe^{3+}, Co^{3+}? Explain your answer.

32. Consult Table 15–7 and answer the following:

    a) Why is there a relatively constant increase in the ionization potentials for chromium from the third to the sixth IP, and then a large increase between the sixth and seventh?

    b) Why do the radii of the Cr atom, Cr^{3+} ion, and Cr^{6+} ion decrease, in that order?

    c) Sketch a unit cell for Cr.

33. Suppose that you have a beaker containing an aqueous solution of Cr^{3+} and Cl^-. Describe what would happen, indicating solution colors, precipitate formation and color, precipitate dissolving, etc., if a solution of NaOH were added drop by drop, with mixing, until the resulting solution was quite basic. Write any pertinent equations.

34. If $Cr_2O_7^{2-}$ is mixed with Br^- in the presence of H^+, a chemical reaction produces Br_2, Cr^{3+}, and H_2O. Write a balanced equation for the reaction.

35. a) Show quantitatively that in an acid solution MnO_4^- will oxidize Cr^{3+} to $Cr_2O_7^{2-}$ and form $MnO_{2(s)}$.

    b) Write the balanced equation for the reaction in (a).

36. In an acid solution, chromium metal reacts with Pb^{2+} to form Cr^{3+} and Pb metal. How many grams of Pb can be produced if 0.40 g Cr are placed in 100 ml of a $0.10M$ Pb^{2+} solution?

37. What is the concentration of Ag^+ ion in a $0.10M$ $[Ag(NH_3)_2]^+$ solution ($K = 6 \times 10^{-8}$)?

38. Suppose that CN^- ion is added to a $0.10M$ Hg^{2+} solution until the CN^- ion concentration is $0.15M$ CN^-. What will be the final equilibrium concentration of Hg^{2+}? (K for $[Hg(CN)_4]^{2-}$ is 4×10^{-42}).

39. Give and sketch one example each of complex ions with octahedral, tetrahedral, and square planar geometry.

40. Sketch any geometric and/or optical isomers for:

    a) $[AuBr_2Cl_2]^-$, square planar

    b) $[Co(H_2O)_4(NH_3)_2]^{3+}$, octahedral

    c) $[Zn(H_2O)_2(NH_3)_2]^{2+}$, tetrahedral

    d) $[Co(C_2O_4)_2Cl_2]^{3-}$, octahedral

41. Name the complex ions or compounds:

    a) $K_2[PtBr_2Cl_2]$

    b) $[Co(H_2O)_4(NO_2)_2]NO_3$

    c) $[Cr(C_2O_4)Cl_4]^{3-}$

    d) $[Cu(NH_3)_4]^{2+}$

    e) (a) through (d) in question 40.

42. Using valence bond theory, predict the hybridization involved, the geometry, and the number of unpaired electrons for the complex ions:

    a) $[Cr(H_2O)_6]^{3+}$ b) $[Mn(CN)_6]^{4-}$ c) $[Fe(CN)_6]^{4-}$ d) $[Cu(NH_3)_4]^{2+}$ e) $[Cd(NH_3)_4]^{2+}$

43. a) Using valence-bond theory, explain why all six-coordinated Fe^{3+} complexes should have only one unpaired electron.

    b) The complex ion $[FeF_6]^{3-}$ is known to have five unpaired electrons. Can you make an alteration in the valence-bond theory that will permit five unpaired electrons and still maintain octahedral symmetry? Note that sp^3d^2 hybridization is equivalent to d^2sp^3 hybridization, as far as symmetry is concerned.

44. Show, using a sketch, what is meant by an octahedral crystal field.

45. a) Why does an octahedral crystal field split the five d orbitals into two sets, one containing two orbitals and another three orbitals?

    b) Define the crystal-field-splitting energy Δ in octahedral symmetry.

46. Using the crystal-field theory, explain why $[V(H_2O)_6]^{4+}$ might be expected to form a colored solution.

47. By using the procedure described for octahedral crystal fields, see if you can determine how the d orbitals will be split in tetrahedral and square planar crystal fields.

48. a) Describe a zone refining process.
    b) Why do impurities usually dissolve preferentially in the liquid zone during zone refining?

49. Describe the difference between *n*-type and *p*-type semiconductors.

50. Predict the type of semiconductivity that would be observed for gallium arsenide (GaAs) that contains substitutional zinc impurities.

16. ORGANIC CHEMISTRY

16-1 INTRODUCTION

The growth of the chemistry of carbon compounds, *organic chemistry*, in the last century has been remarkable. Although no accurate count is available, approximately a million different organic compounds are known, and the number increases daily. Many of these compounds have been isolated from natural sources, such as trees, flowers, animals, oil deposits, etc. More compounds have been prepared in the laboratory, utilizing chemical reactions that have been developed in the 150-year lifetime of organic chemistry.

▶ In the early years of organic chemistry, science was not the isolated discipline it is today. The workers of the early 1800's were natural philosophers and their work was always conducted in the light of the prevailing philosophical and theological thought. The obvious association of organic compounds with living systems led to a philosophical position which stated that organic compounds could be synthesized only in the live plant or animal or from other compounds obtained from the living source. Thus when, in 1828, Wohler announced the synthesis of urea (an organic compound present in urine) from nonliving (inorganic) sources, it caused quite a stir. In a few years the synthesis of urea was corroborated, and new organic compounds were prepared from inorganic sources. The doctrine was destroyed and the separation between the chemist and the philosopher–theologian widened. ◀

The sections that follow will be concerned with the properties of carbon which make possible this array of compounds and some of the important reactions which make possible the transformation of known organic compounds into new compounds.

16–2 STRUCTURE AND BONDING

Knowledge of structure and bonding—that is, the way carbon is bonded to other atoms in a molecule—is prerequisite to understanding organic chemistry. If you think about the electronic configuration of carbon, $1s^2 2s^2 2p_x 2p_y$, and the fact that carbon forms covalent bonds by sharing electrons, you might predict that carbon would form two bonds with, for example, two chlorine atoms. These bonds would result from a sharing of the two unshared 2p electrons in carbon with the unshared 3p electron in each of the chlorine atoms (Cl has the structure $1s^2 2s^2 2p^6 3s^2 3p_x^2 3p_y^2 3p_z$). In fact, dichlorocarbene (CCl_2) is an extremely unstable and reactive molecule which cannot be isolated at ordinary temperatures. Carbon tetrachloride (CCl_4), however, is a well known, stable organic compound. Methylene (CH_2) is also an unstable species, but methane (CH_4) is quite stable. Investigation of the known organic compounds shows that, except in very rare instances, carbon forms compounds by sharing electrons with other atoms to form four bonds.

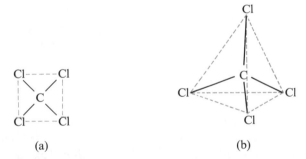

(a)   (b)

Fig. 16–1 Two geometrical arrangements of CCl_4 having zero dipole moments. (a) The square planar arrangement has the carbon in the center and four equal C—Cl bonds directed toward the corners of a square. (b) The bonds are directed toward the corners of a tetrahedron.

Another piece of experimental evidence gives us more information about the structure of carbon compounds: the zero dipole moment of carbon tetrachloride (CCl_4). Carbon and chlorine differ on the electronegativity scale by 0.5, and thus any given carbon-chlorine bond should have a measurable dipole moment. In order to explain the lack of a dipole moment for CCl_4, we have to postulate a geometrical structure in which the individual C-Cl dipoles are directed so as to cancel each other.

Two common geometrical structures require a zero dipole moment for CCl_4: a square-planar structure (part (a) of Fig. 16–1) and a tetrahedral structure (part (b) of Fig. 16–1). Van't Hoff and LeBel in 1874 simultaneously proposed the tetrahedral

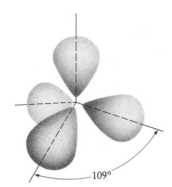

Fig. 16–2 An sp³ hybrid molecular orbital with the bonding lobes pointing toward the corners of a tetrahedron.

structure for carbon compounds to explain the phenomenon of optical activity, which we shall discuss in Chapter 17. This proposal was later substantiated by studies of electron diffraction.

▶ A rectangular-planar structure would also satisfy the requirements, but the arguments put forth for the square-planar structure hold for the rectangular as well. ◀

Since carbon cannot form four bonds utilizing only the unfilled p-orbitals, the simplest explanation requires that the s-orbital in carbon become available for bonding. Therefore, as in Chapter 6, we write the electronic configuration of carbon as $1s^2 2s 2p_x 2p_y 2p_z$. There are now four orbitals (an s- and three p-orbitals), each containing one electron which can be shared to form a bond. Calculations have shown that bonding involving these four separate orbitals is improbable from the standpoint of energy, but that a *hybrid* orbital obtained by *mixing* these four orbitals is favorable from the standpoint of energy.

This *hybrid orbital* contains four lobes capable of forming bonds. Each lobe can be considered to have one-fourth s-character and three-fourths p-character. Such an orbital is called an sp³ hybrid orbital, and has the four bonding lobes directed toward the corners of a tetrahedron, which adequately explains the bonding and geometry found in compounds such as CCl_4 (see Fig. 16–2). Thus carbon is said to form four bonds using sp³ hybrid orbitals directed toward the corners of a tetrahedron with bond angles of 109.5°.

THE HYDROCARBONS

Basing our remarks on the structural theory discussed above, we shall now consider several representative types of organic compounds. We shall classify these compounds according to similarity in structure and bonding. It should not be surprising that this also represents a classification according to chemical and physical properties, since these properties are determined by the structure and bonding involved.

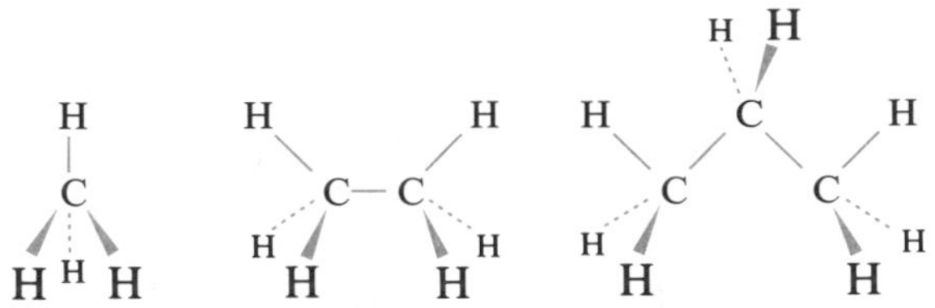

Methane   Ethane   Propane

Three-dimensional formulas

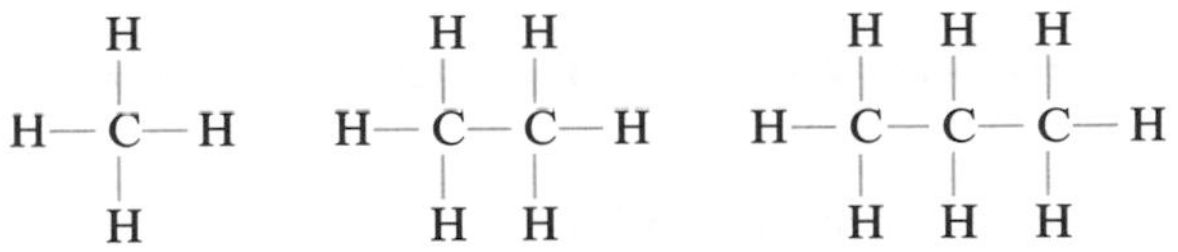

Two-dimensional projection formulas

$$CH_4 \qquad\qquad CH_3CH_3 \qquad\qquad CH_3CH_2CH_3$$

Shorthand formulas

Fig. 16-3 Three ways of illustrating the structures of the first three alkanes.

16-3 ALKANES

Traditionally the first class of organic compounds studied is a group of compounds referred to as *hydrocarbons*, containing only hydrogen and carbon. There are several subclasses of hydrocarbons, separated according to the bonding arrangement of the carbon and hydrogen atoms. The simplest class of hydrocarbons, called the *alkanes*, have carbon atoms bonded with sp^3 hybrid orbitals to hydrogen atoms or other carbon atoms. Each carbon atom forms four single bonds with either hydrogen or a second carbon atom. This ability of carbon to covalently bond to itself is the reason for the large number of organic compounds.

When we are writing the structural formulas of organic compounds, it is inconvenient to use diagrams which emphasize the three-dimensional tetrahedral nature of the carbon atoms. For this reason, we ordinarily use two-dimensional projection formulas or a shorthand notation, as in Fig. 16-3, and we shall employ them here when possible.

It is hard for the beginning student of organic chemistry to acquire a facility in writing and visualizing structural formulas.* When you are studying organic

---

*   The difficulty is a result in large part of using two-dimensional projection formulas to illustrate three-dimensional species.

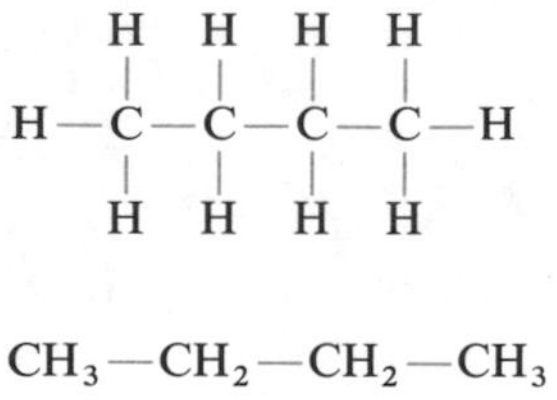

$$H \begin{matrix} H & H & H & H \\ | & | & | & | \\ -C & -C & -C & -C \\ | & | & | & | \\ H & H & H & H \end{matrix} H$$

$$CH_3 - CH_2 - CH_2 - CH_3$$

n-butane

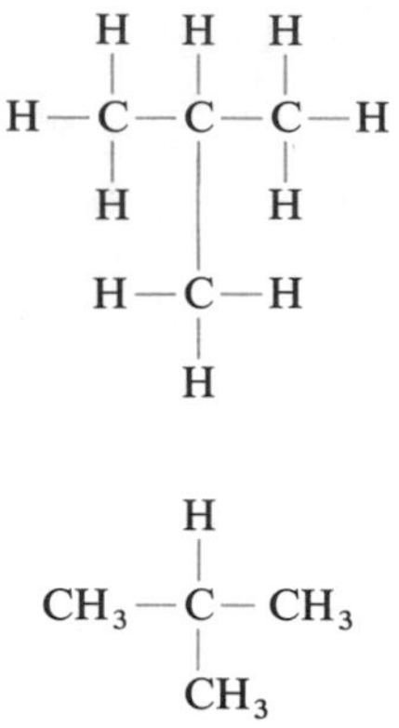

$$H \begin{matrix} H & H & H \\ | & | & | \\ -C & -C & -C \\ | & | & | \\ H & & H \\ & | & \\ & H-C-H \\ & | & \\ & H & \end{matrix} H$$

$$CH_3 - \overset{\overset{\displaystyle H}{|}}{\underset{\underset{\displaystyle CH_3}{|}}{C}} - CH_3$$

Isobutane
(2-methylpropane)

Fig. 16–4 Two different structures for butane (C_4H_{10}).

$$CH_3 - CH_2 - CH_2 - CH_2 - CH_3$$

n-pentane

$$CH_3 - \overset{\overset{\displaystyle CH_3}{|}}{\underset{\underset{\displaystyle H}{|}}{C}} - CH_2 - CH_3$$

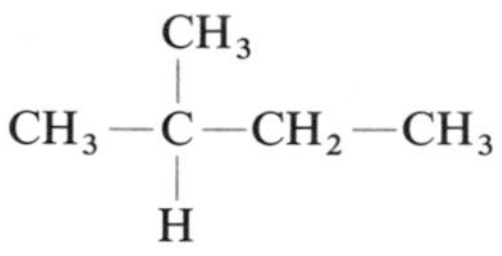

Isopentane
(2-methylbutane)

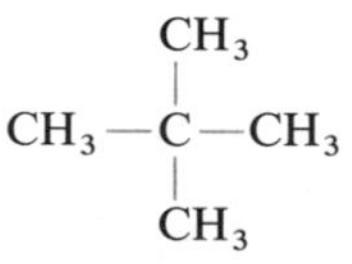

$$CH_3 - \overset{\overset{\displaystyle CH_3}{|}}{\underset{\underset{\displaystyle CH_3}{|}}{C}} - CH_3$$

Neopentane
(2,2-dimethylpropane)

Fig. 16–5 Three isomers of pentane (C_5H_{12}).

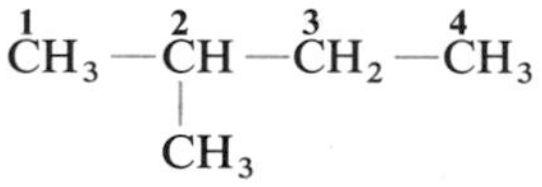

$$\overset{1}{C}H_3 - \overset{2}{C}H - \overset{3}{C}H_2 - \overset{4}{C}H_3$$
$$\quad\quad |$$
$$\quad\quad CH_3$$

2-methylbutane

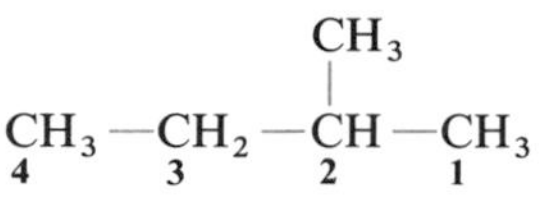

$$\underset{4}{CH_3} - \underset{3}{CH_2} - \overset{\overset{\displaystyle CH_3}{|}}{\underset{2}{CH}} - \underset{1}{CH_3}$$

2-methylbutane

$$\overset{4}{C}H_3 \quad CH_3$$
$$| \quad\quad |$$
$$\underset{3}{CH_2} - \overset{2}{CH}$$
$$\quad\quad |$$
$$\quad\quad \underset{1}{CH_3}$$

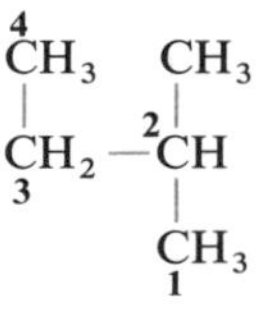

2-methylbutane

Fig. 16–6 Three ways of writing the structure of 2-methyl-butane.

chemistry, you should consume much scrap-paper in writing and rewriting structural formulas. You have to learn, for example, that it is the order of attachment of the atoms that distinguishes a given compound rather than the manner in which a projection formula may be written on paper.

For example, you may illustrate the structure of propane in several different ways and still represent only one compound, determined by the number and order of attachment of the atoms.

$$
\begin{array}{ccc}
& \text{H} & \\
& | & \\
& \text{H--C--H} & \\
\text{H} & | & \\
| & & \\
\text{H--C--C--H} & \quad\quad & \\
| \;\; | & & \\
\text{H} \; \text{H} & &
\end{array}
$$

Propane

Note that, in each representation, both end carbon atoms are bonded to three hydrogen atoms and a second carbon atom, while the central carbon atom is bonded to two hydrogen atoms and two carbon atoms.

The alkanes can be represented by a general formula, C_nH_{2n+2}. Each member of this family differs from the preceding member by a CH_2 group. Such a series of compounds is referred to as a *homologous series*. The first three members—methane, ethane, and propane—are shown in Fig. 16–3.

Isomerism

The next-higher compound in the series, butane (C_4H_{10}), introduces a new concept. It is possible to write two (and only two) different structural formulas for C_4H_{10}, as shown in Fig. 16–4. Note that a different order of bonding is represented, and not just a different way of writing the same molecule. In fact, two butanes exist, *n*-butane and isobutane (2-methylpropane). The *n*-butane (*n* = normal) is the *straight-chain compound* (boiling point 0 °C), while isobutane is a name for the *branched compound* (boiling point − 12 °C). The phenomenon of having two or more compounds with the same empirical formula is called *isomerism* (meaning composed of the same parts) and *n*-butane and isobutane are referred to as *isomers*.

As we continue with higher-alkane homologs, the number of possible isomers increases. For example, Fig. 16–5 shows the three pentane (C_5H_{12}) isomers. Again we must be careful, however. As shown in Fig. 16–6, we can write 2-methylbutane* in other ways, but if we consider the order of attachment of the atoms in the molecule, we can see that only one compound is represented.

---

* The nomenclature of simple organic compounds is explained in Appendix C.

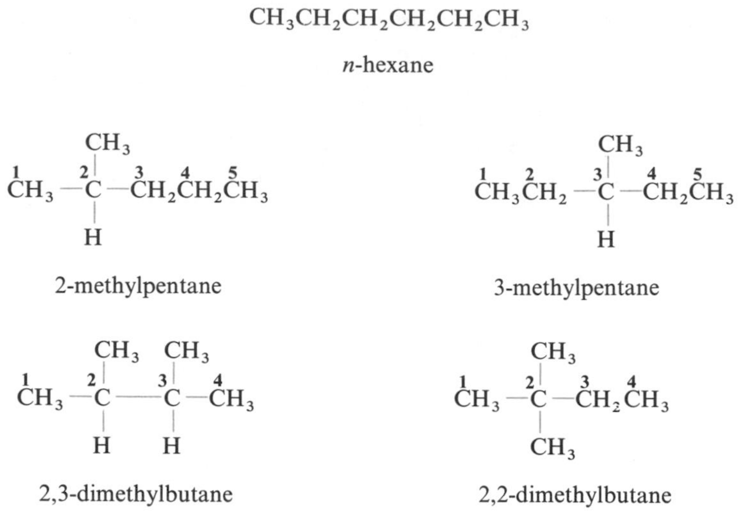

$$CH_3CH_2CH_2CH_2CH_2CH_3$$

n-hexane

$$\overset{1}{CH_3}-\overset{2}{\underset{H}{\overset{CH_3}{C}}}-\overset{3}{CH_2}\overset{4}{CH_2}\overset{5}{CH_3}$$

2-methylpentane

$$\overset{1}{CH_3}\overset{2}{CH_2}-\overset{3}{\underset{H}{\overset{CH_3}{C}}}-\overset{4}{CH_2}\overset{5}{CH_3}$$

3-methylpentane

$$\overset{1}{CH_3}-\overset{2}{\underset{H}{\overset{CH_3}{C}}}-\overset{3}{\underset{H}{\overset{CH_3}{C}}}-\overset{4}{CH_3}$$

2,3-dimethylbutane

$$\overset{1}{CH_3}-\overset{2}{\underset{CH_3}{\overset{CH_3}{C}}}-\overset{3}{CH_2}\overset{4}{CH_3}$$

2,2-dimethylbutane

Fig. 16–7 Five isomers of hexane (C_6H_{14}).

The hexanes (C_6H_{14}) include five isomers (Fig. 16–7), and as the number of carbon atoms increases (heptanes, octanes, nonanes, etc.), the number of possible isomers grows rapidly, as shown in Table 16–1.

Table 16–1. Increase in the number of possible alkane isomers as the number of carbon atoms increases

Number of C atoms	Number of isomers
7	9
8	18
9	35
10	75
15	4347
20	366,319
30	4 billion
40	6×10^{13}

Cycloalkanes

Since the carbon atoms in alkanes can form long chains, it is not surprising that they can also form ring compounds called *cycloalkanes*. A few representative structures of cycloalkanes are shown in Fig. 16–8.

A ring such as cyclopropane cannot achieve the normal 109.5° tetrahedral bond angle usually formed by carbon. Thus the ring is said to be *strained*. These strained

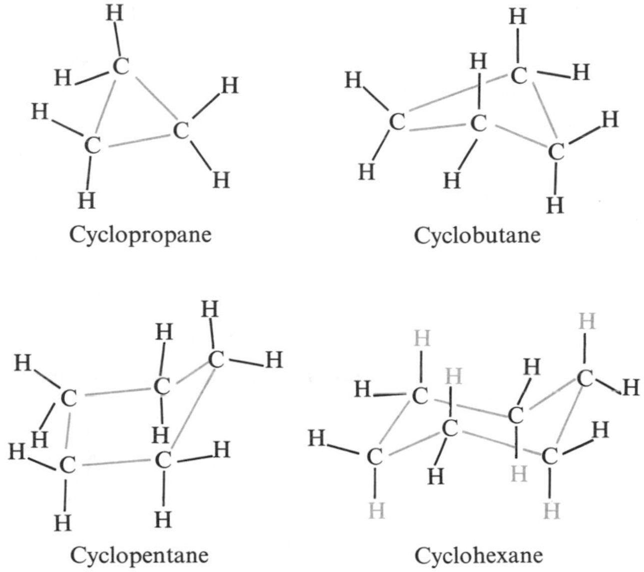

Fig. 16-8 Three-dimensional representations of cycloalkanes.

compounds are more reactive (less stable) than the corresponding open-chain alkanes. The strain decreases as the size of the ring increases until, at cyclohexane, the tetra-hedral bond angle is again possible.

▶ The C—C—C bond angles in cyclohexane are actually somewhat larger than the ideal tetrahedral angle. This distortion is a result of the repulsive interactions of the various hydrogens and their bonding electrons. The colored hydrogens in Fig. 16–8, called *axial* hydrogens are closer together than the black hydrogens, called *equatorial*. If we replace a hydrogen with a larger group the interaction increases. The larger the group the greater the tendency for it to occupy an equatorial position. ◀

Reactions of Alkanes

The alkanes are the least reactive of the classes of organic compounds, although some of the reactions they undergo are of great importance. Methane, for example, is the main constituent of natural gas and is an important fuel:

$$CH_4 + 2O_2 \rightarrow CO_2 + 2H_2O + 212.8 \text{ kcal/mole.}$$

The reaction of an alkane such as methane with chlorine in the presence of light is an industrially important reaction which illustrates a general concept. We can write the reaction out in terms of its *mechanism*, that is, the detailed step-by-step

processes involved in the conversion of reactants to products:*

(1) $Cl_2 \rightarrow 2Cl\cdot$
(2) $CH_4 + Cl\cdot \rightarrow \cdot CH_3 + HCl$
(3) $\cdot CH_3 + Cl_2 \rightarrow CH_3Cl + Cl\cdot$
                                    Chloro-

                                    methane
(4)   Steps 2 and 3 are repeated: 2, 3, 2, 3, 2, 3 etc.

     Similarly

$$CH_3Cl + Cl\cdot \rightarrow \cdot CH_2Cl + HCl$$
$$\cdot CH_2Cl + Cl_2 \rightarrow CH_2Cl_2 + Cl\cdot$$
                                    Methylene

                                    chloride

    For example, the reaction of methane with chlorine, initiated by heat or light, is comprised of several steps; each step generates the reactant for a subsequent step. Ultimately the methylene chloride (CH_2Cl_2) will be converted to chloroform ($CHCl_3$) and finally to carbon tetrachloride (CCl_4). Try working out these last steps in the reaction. A reaction of this type is called a *free-radical chain reaction*, since one of the products of each step is a reactant for another step. (The term "free radical" refers to a chemical species containing an unpaired electron.)

    Important sources of hydrocarbons are natural gas, petroleum, and coal. By the process of catalytic cracking, the petroleum industry converts the high-molecular-weight hydrocarbons in oil to smaller fragments, which are used as gasoline and as raw materials to make plastics and other chemicals.

16–4 ALKENES

Structure

The alkenes are a homologous series of compounds with the general formula C_nH_{2n}, the simplest example being ethylene (C_2H_4). The alkenes are said to be *unsaturated*, because they have two fewer hydrogen atoms than the alkanes (general formula C_nH_{2n+2}). The alkenes must therefore exhibit something other than the sp^3 hybrid bonding found in alkanes. Experiments have shown that ethylene is a *planar* molecule, containing a double bond between the two carbon atoms. It has also been shown that ethylene reacts with hydrogen in the presence of a catalyst to form ethane, (under conditions which do not cause cleavage of the alkane carbon–carbon single bond).

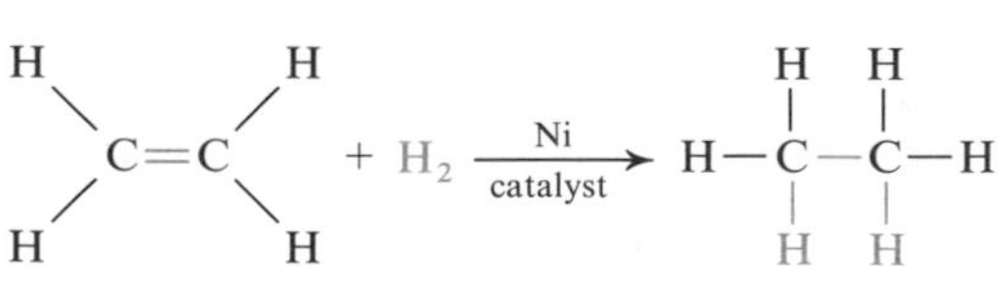

Ethylene   Ethane

---

*   See Chapter 14 for more details on mechanisms of reactions.

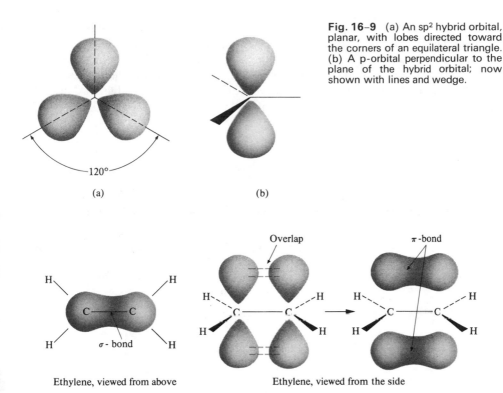

Fig. 16–9 (a) An sp² hybrid orbital, planar, with lobes directed toward the corners of an equilateral triangle. (b) A p-orbital perpendicular to the plane of the hybrid orbital; now shown with lines and wedge.

Fig. 16–10 Diagrammatic representation of the bonding in ethylene.

These data suggest that one of the carbon–carbon bonds in ethylene is similar to the carbon–carbon bond in alkanes such as ethane, but the second carbon–carbon bond is more reactive and of a different type.

Modification of the theory which we used to explain tetrahedral bonding in the alkanes can accommodate the above experimental data. Consider the electronic configuration of carbon, in which one of the 2s electrons has been elevated to the vacant $2p_z$ orbital $(1s^2 2s 2p_x 2p_y 2p_z)$. Hybridization (mixing) of the 2s orbital with two of the 2p orbitals affords a new hybrid orbital designated sp^2.

Calculations have shown that this new sp^2 hybrid orbital, shown in Fig. 16–9, is planar, with each of its *three lobes* directed toward the corners of an equilateral triangle (trigonal). The remaining 2p orbital, which is not involved in hybridization, is oriented perpendicularly to the plane of the sp^2 orbital.

We can explain the structure of ethylene if we say that each carbon atom exhibits sp^2 hybridization. Figure 16–10 shows two of the three sp^2 bonding lobes of each carbon atom bonding with hydrogen atoms and the third lobe forming one-half of a carbon–carbon bond. The p-orbitals are said to overlap to form a cloud of electrons above and below the plane of the atoms. The electron cloud above and below the plane of the molecule, resulting from overlap of the p-orbitals, is referred to as a

π-*bond*. The second carbon–carbon bond formed between two sp^2 lobes is called a σ-*bond*. The σ-bond is similar to the carbon–carbon bonds found in alkanes, and is relatively unreactive. The electrons in the π-bond are more loosely held and more accessible. Thus the π-bond is more reactive. This bonding theory therefore succeeds in explaining both the geometry and the reactivity of alkenes. Table 16–2 lists the names and structures of some representative alkenes.

Table 16–2. Some representative alkenes

Name	Structure	Melting point, °C	Boiling point, °C
Ethylene	$CH_2\!\!=\!\!CH_2$	−169	−102
Propylene	$CH_3CH\!\!=\!\!CH_2$	−185	−48
1-butene	$CH_3CH_2CH\!\!=\!\!CH_2$		−6.5
1-pentene	$CH_3CH_2CH_2CH\!\!=\!\!CH_2$		30
1-hexene	$CH_3CH_2CH_2CH_2CH\!\!=\!\!CH_2$	−138	63.5
Isobutylene	$\overset{\displaystyle CH_3}{\overset{\displaystyle \vert}{CH_3\!\!-\!\!C\!\!=\!\!CH_2}}$	−141	−7
3-methyl-1-butene	$CH_2\!\!=\!\!CHCHCH_3$	−135	25
2, 3-dimethyl-2-butene	$(CH_3)_2\!\!-\!\!C\!\!=\!\!C(CH_3)_2$	−74	73
Cyclopentene	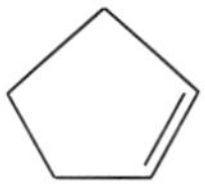	−93	46
Cyclohexene	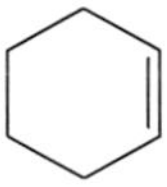	−104	83

In addition to the type of structural isomerism mentioned in the section on alkenes, there are two features of alkenes which lead to further isomerism. The first, involving the position of the double bond, is illustrated with two butene isomers.

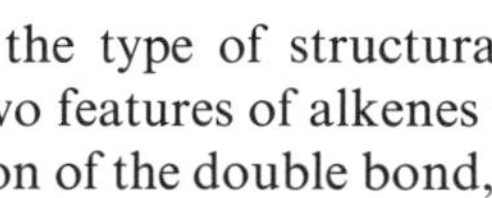

1-butene

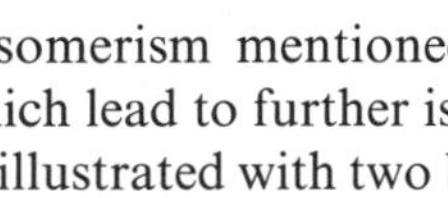

2-butene

The second type of isomerism involves the geometry of the substituents (the attached atoms or groups of atoms) about the double bond. We shall discuss it in Chapter 17.

Source and Preparation

The lower-molecular-weight alkenes such as ethylene, propylene, butene, etc., are by-products of the production of gasoline, and are of great commercial importance. Ethylene, for example, is used in the synthesis of ethyl alcohol by a process which is competitive with the fermentation method. It is also used in large quantities to produce *polyethylene*, long-chain molecules (polymers) formed by joining several hundred ethylene units. Polyethylene is employed in many ways, ranging from kitchenware and plastic sandwich bags to coatings for the bottoms of skis.

▶ The properties of polyethylene can obviously vary considerably. One property which is constant, however is its low chemical reactivity. This is, of course, a desirable feature. We would not be satisfied, for example, if the running surface of our skis reacted with water. So long as the item is in use, we appreciate the chemical inertness. But when it has served its purpose, how do we dispose of it? We now have the situation in which people dispose of their garbage or leaves or grass clippings, which can be degraded in the ground, by wrapping them in plastic bags—which cannot be degraded. In cleaning our yards we are actually contributing to pollution. ◀

Alkenes of more complex structure are often not commercially available, but must be prepared (synthesized) in the laboratory. This may be done by many different methods. Two of the most common, involving the dehydration of alcohols and the dehydrohalogenation (loss of the elements of a hydrogen halide such as HBr) of alkylhalides, are these:

$$\begin{array}{c} \text{H} \\ | \quad | \\ -\text{C}-\text{C}- \\ | \quad | \\ \quad \text{OH} \end{array} + \xrightarrow[\text{Dehydration}]{\text{H}_2\text{SO}_4} \quad \diagbox \text{C}=\text{C} \diagbox + \text{H}_2\text{O}$$

An alcohol

$$\begin{array}{c} \text{H} \\ | \quad | \\ -\text{C}-\text{C}- \\ | \quad | \\ \quad \text{Br} \end{array} + \text{KOH} \xrightarrow[\text{Dehydrohalogenation}]{} \quad \diagbox \text{C}=\text{C} \diagbox + \text{H}_2\text{O} + \text{KBr}$$

An alkyl-
halide

Reactions of Alkenes

The reactions of alkenes are, in general, reactions which result in conversion of the sp^2 hybridized carbon atoms to sp^3 hybridization. This is accomplished by cleavage

of the π-bond with formation of two σ-bonds. Such reactions are called *addition reactions*, since they result in the addition of the reagent to the alkene. Here are three examples:

$$
\underset{\text{Propene}}{
\begin{array}{c}
\text{H} \\ \diagdown \\
\end{array}
C=C
\begin{array}{c}
\text{H} \\ \diagup \\ \text{CH}_3
\end{array}
}
\;+\; \text{HBr} \;\rightarrow\;
\underset{\text{2-bromopropane}}{
\text{H}-\underset{\underset{\text{H}}{|}}{\overset{\overset{\text{H}}{|}}{\text{C}}}-\underset{\underset{\text{Br}}{|}}{\overset{\overset{\text{H}}{|}}{\text{C}}}-\text{CH}_3
}
$$

$$
\underset{\text{1-butene}}{
\begin{array}{c}
\text{H} \\ \diagdown \\
\end{array}
C=C
\begin{array}{c}
\text{H} \\ \diagup \\ \text{CH}_2\text{CH}_3
\end{array}
}
\;+\; \text{Br}_2 \;\rightarrow\;
\underset{\text{1,2-dibromobutane}}{
\text{H}-\underset{\underset{\text{Br}}{|}}{\overset{\overset{\text{H}}{|}}{\text{C}}}-\underset{\underset{\text{Br}}{|}}{\overset{\overset{\text{H}}{|}}{\text{C}}}-\text{CH}_2\text{CH}_3
}
$$

$$
3\;
\begin{array}{c}
\text{H} \\ \diagdown \\
\end{array}
C=C
\begin{array}{c}
\text{H} \\ \diagup \\ \text{H}
\end{array}
\;+\; 2\text{KMnO}_4 + 4\text{H}_2\text{O} \;\rightarrow\; 3\,
\underset{\underset{\text{Ethylene glycol}}{}}{
\text{H}-\underset{\underset{\text{OH}}{|}}{\overset{\overset{\text{H}}{|}}{\text{C}}}-\underset{\underset{\text{OH}}{|}}{\overset{\overset{\text{H}}{|}}{\text{C}}}-\text{H}
}
+ 2\text{MnO}_2 + 2\text{KOH}
$$

The addition reactions with Br_2 and $KMnO_4$ are also of interest as qualitative tests for alkene unsaturation—that is, to detect the presence of an alkene double bond. Bromine is colored, but the dibromo addition product is usually colorless. Similarly, the magenta MnO_4^- is converted to MnO_2, a dark-brown precipitate. Thus the reactions are accompanied by visual changes which indicate the presence of alkenes. Since bromine can react with alkanes in the presence of light and permanganate oxidizes alcohols at elevated temperatures, care must be taken in carrying out these tests for unsaturation.

16–5 ALKYNES

Structure

The alkynes are a class of hydrocarbons which also exhibit unsaturation according to the bromine and permanganate tests. The simplest member of this series of compounds is acetylene (C_2H_2). Quantitative hydrogenation experiments show that one mole of acetylene reacts in the presence of a catalyst with two moles of hydrogen, producing ethane:

$$
\underset{\text{Acetylene}}{C_2H_2} \;+\; 2H_2 \;\xrightarrow{\text{PtO}_2}\; \underset{\text{Ethane}}{CH_3CH_3}
$$

This indicates that acetylene has a *triple bond*, described as including one σ-bond

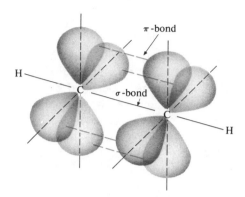

π-bond

H

σ-bond

H

Fig. 16–11 A diagrammatic representation of the bonding in acetylene. Two π-bonds are formed by p-orbital overlap and the σ-bond is formed by sp hybrid orbital overlap.

and two π-bonds. As illustrated in Fig. 16–11, the carbon atoms in acetylene are said to exhibit sp hybridization; the two unhybridized p-orbitals on each atom are mutually perpendicular, and overlap with the parallel p-orbitals on the second carbon atom to form a cylinder of π-electrons around the C-C σ-bond.

Acetylene, the most important of the alkynes, is prepared commercially by the hydrolysis of calcium carbide (CaC_2), by the dehydrogenation of ethylene, or by partial combustion of methane:

$$CaO + 3C \xrightarrow{2000\,°C} CaC_2 + CO$$
$$CaC_2 + 2H_2O \longrightarrow H-C{\equiv}C-H + Ca(OH)_2$$

Reactions of Alkynes

The reactions of alkynes are analogous to those of alkenes, except that there are two π-bonds which can react:

$$CH_3-C{\equiv}C-H + Br_2 \rightarrow$$

Propyne

$$\underset{\text{1,2-dibromopropene}}{\overset{CH_3}{\underset{Br}{>}}C{=}C\overset{Br}{\underset{H}{<}}}$$

$$\underset{Br}{\overset{CH_3}{>}}C{=}C\underset{H}{\overset{Br}{<}} + Br_2 \rightarrow CH_3-\overset{Br}{\underset{Br}{C}}-\overset{Br}{\underset{Br}{C}}-H$$

1,1,2,2-tetrabromo-
propane

16–6 AROMATIC HYDROCARBONS

Structure

In the history of organic chemistry there have been a few outstanding problems, the solutions of which have dramatically advanced theoretical knowledge in the field. One of these was the structural problem of carbon compounds elucidated by van't Hoff and LeBel and discussed at the beginning of this chapter. Another has come to be called the "benzene problem."

Benzene, one of a class of compounds called *aromatic compounds*, has the formula C_6H_6, which suggests extensive unsaturation. Treatment with permanganate gives no reaction, nor does simple treatment with bromine. But in the presence of a catalyst such as Fe, a *substitution* reaction (bromine replacing hydrogen) rather than an addition reaction takes place.

$$C_6H_6 + Br_2 \xrightarrow{\text{Fe}} C_6H_5Br + HBr$$

Benzene   Bromo-
                           benzene

Catalytic hydrogenation of benzene, forming cyclohexane, proceeds with difficulty, indicating that benzene is more stable than ordinary unsaturated alkenes and alkynes.

$$C_6H_6 + 3H_2 \xrightarrow{\text{Catalyst}}$$

Benzene

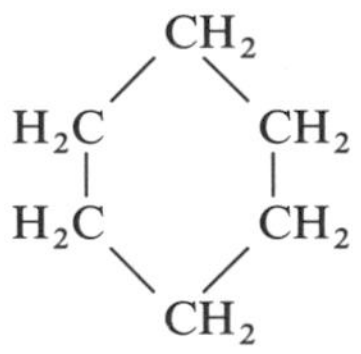

Cyclohexane

Studies in x-ray diffraction, as well as spectroscopic studies, show that benzene is flat and hexagonal, with C—C—C and C—C—H bond angles of 120° and with a hydrogen attached to each carbon atom.

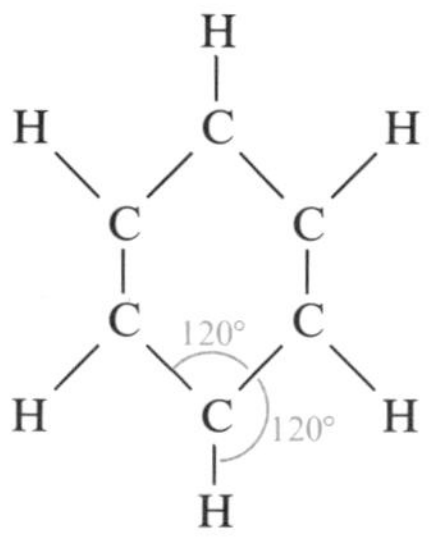

The planarity and bond angles suggest that the carbon atoms are sp² hybridized. One of the early structures suggested by a nineteenth-century German chemist named August Kekulé included alternating double and single bonds.

H
|
H   C   H
 \\   / \\   /
  C   C
  ‖   ‖   or   [hexagon structure]
  C   C
 /   \\   / \\
H   C   H
      |
      H

This structure, which was proposed before the x-ray studies were made, is not satisfactory, however. Carbon–carbon double bonds are now known to be shorter than
carbon–carbon single bonds. Thus the original Kekulé structure would be distorted
and inconsistent with the x-ray data.

We cannot write one conventional structure to explain the properties and
structure of benzene. In other words, benzene cannot be represented by structure
A or structure B, but can be considered a *resonance hybrid* of the two.

A   B

The double-headed arrow is a symbol for resonance hybridization, and does not
mean that A and B are in equilibrium; it means that benzene is neither A nor B,
but some intermediate hybrid of the two.

We can visualize this concept by considering the alternative orbital representation
of the molecule in Fig. 16–12. The carbon atoms are sp² hybridized and coplanar
in a ring. Perpendicular to each carbon atom is a p-orbital which can overlap with
two adjacent p-orbitals. The result is a cloud of π-electrons above and below the

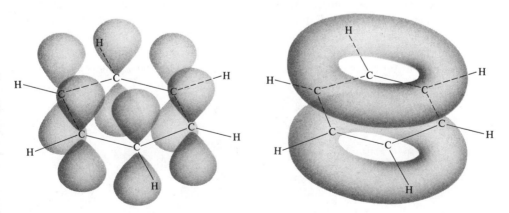

Fig. 16–12 Orbital diagram of benzene.

plane of the ring. The π-electrons are *delocalized* and no longer reside between any two carbon atoms. Thus benzene does not have double bonds of the alkene type. It has, instead, carbon–carbon bonds that are midway between single and double bonds, that is, 1.5 bonds.

Benzene is sometimes written as a hexagon with a circle inside to emphasize the delocalization of the π-electrons, but more often the original Kekulé structure is written, and the delocalization is assumed.

Resonance Energy

The fact that the π-electrons in benzene are delocalized (rather than existing as they do in alkene double bonds) indicates that the delocalized structure must be more stable than the hypothetical alternating double-single bond structure. Evaluation of this difference in stability would give a measure of the effectiveness of delocalization of electrons in stabilizing a molecule.

Several attempts, both experimental and theoretical, have been made to evaluate this *delocalization energy* or *resonance energy*. One method involved a study of the heats of hydrogenation of cyclohexene and benzene.

$$\text{Cyclohexene} + \text{H}_2 \rightarrow \text{Cyclohexane} + 28.6 \text{ kcal/mole}$$

If benzene had the structure indicated by the original Kekulé formula, one might expect three times as much energy (85.8 kcal/mole) to be evolved during hydrogenation of benzene to cyclohexane, since three double bonds are involved. The experimental value for the heat of hydrogenation of benzene is 49.8 kcal/mole. Thus benzene, due to the delocalization of π-electrons, is said to be stabilized by 36.0 kcal/mole.

Delocalization is also seen in other hydrocarbons, some of which are shown in Fig. 16–13; these compounds have chemical properties similar to those of benzene.

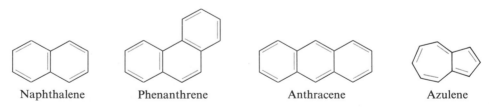

Naphthalene   Phenanthrene   Anthracene   Azulene

Fig. 16–13 Some representative aromatic hydrocarbons. Only one resonance structure is shown; π-bonds are shown in color.

FUNCTIONAL GROUPS

There are many important organic compounds in which carbon forms covalent bonds with atoms of substances other than carbon and hydrogen. These atoms—usually F, Cl, Br, I, O, S, N, and P—often differ substantially from carbon, in electronegativity and in chemical reactivity. Organic compounds including such atoms or groups of atoms have chemical and physical properties determined by the particular atom(s) and the nature of the bonding.

Thus a certain part of an organic molecule may in large part determine the properties of that compound. This portion of the molecule is termed a *functional group*. Organic molecules that have the same functional group have similar properties, and hence may be classified according to the functional group.

16–7 ALCOHOLS AND PHENOLS

Compounds containing the hydroxyl functional group —OH covalently bonded to a carbon atom are called *alcohols*. When the carbon atom bearing the hydroxyl group is part of an aromatic ring, the compound is termed a *phenol*.

$$H-\overset{\overset{\displaystyle H}{|}}{\underset{\underset{\displaystyle H}{|}}{C}}-OH$$

Methanol
(methyl alcohol)

OH

Phenol

The properties of alcohol are unlike those of inorganic hydroxides, due to the strong covalent bond between carbon and oxygen. Although alcohols do not have a great tendency to ionize, they are polar molecules, and in many respects are similar to water.

Ethanol (CH_3CH_2OH) has a molecular weight of 46 and propane $(CH_3CH_2CH_3)$ a molecular weight of 44, yet ethanol has a boiling point of 78.5 °C, while propane is a gas at room temperature and pressure (boiling point -42.2 °C). We can explain this large difference in boiling points in the same way as we explained the high boiling point of water in Section 9–6; that is, it has to do with hydrogen bonding.

$$O-H-----O-H-----O-H-----O-H \qquad R = CH_3CH_2$$
$$\;\;|\qquad\qquad\quad|\qquad\qquad\quad|\qquad\qquad\quad|$$
$$\;\;R\qquad\qquad\quad R\qquad\qquad\quad R\qquad\qquad\quad R$$

The extensive association between molecules in the liquid phase, due to hydrogen bonding, requires more energy to break up than non-hydrogen-bonded materials of similar molecular weight, and the boiling point is correspondingly higher.

Alcohols ($K \simeq 10^{-18}$) are slightly less acidic than water ($K = 10^{-14}$) (see Chapter 13) but phenols ($K \simeq 10^{-10}$) are more acidic than water and act as weak acids. Phenols form salts when reacted with bases such as NaOH or Na_2CO_3, but not with weaker bases such as $NaHCO_3$.

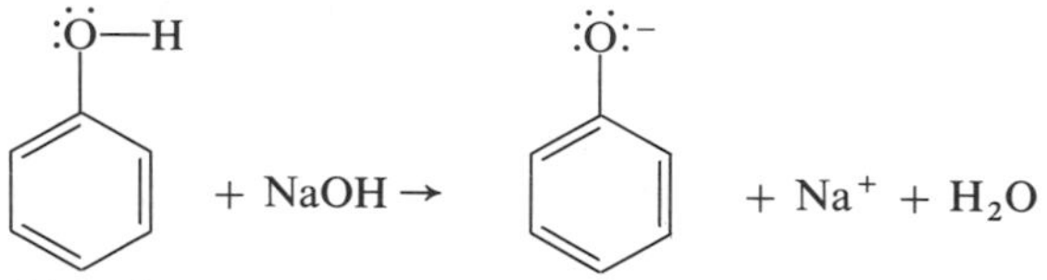

$$:\overset{..}{O}\!-\!H \qquad \qquad :\overset{..}{O}\!:^-$$

Phenol   + NaOH →   Sodium phenoxide (a salt)   + $Na^+ + H_2O$

One explanation offered for the increased acidity of phenols over alcohols is the increased stability of the anion formed. The alkoxide ion formed when an alcohol ionizes holds the negative charge directly on the oxygen atom, readily available to accept a proton, and the equilibrium is far to the left.

$$R\!-\!O\!: \rightleftarrows R\!-\!\overset{..}{O}\!:^- + H^+ \qquad R = \text{alkyl}$$
$$|\qquad\qquad\qquad\qquad\qquad\qquad \text{group}$$
$$H \quad \text{Alkoxide ion}$$

In the phenoxide ion, however, the pairs of electrons on the oxygen atom are able to interact with the π-electrons in the aromatic ring.

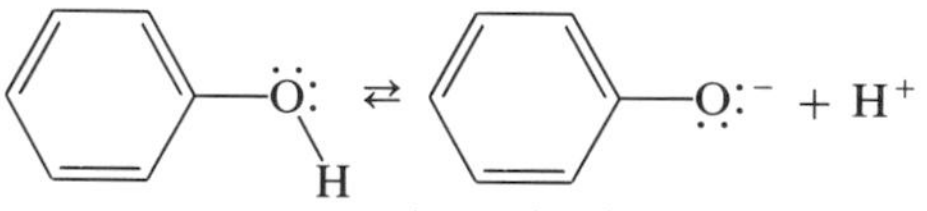

$$\rightleftarrows \qquad \qquad -O\!:^- + H^+$$

Phenoxide ion

Such interactions cause the negative charge to be *delocalized* and thus to be less available for bonding with a proton than if the charge were restricted to the oxygen atom. In other words, the phenoxide ion is considered to be a *resonance hybrid* of several structures shown in Fig. 16–14, in which the negative charge is distributed about the aromatic ring.

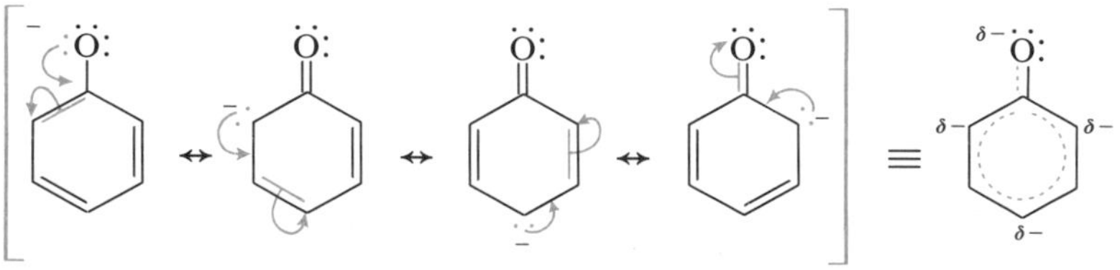

Fig. 16–14 The phenoxide ion may be considered as a resonance hybrid of several structures, effectively delocalizing the negative charge.

The resultant hybrid structure, as in the case of benzene, is more stable than any of the contributing structures, and this stabilization of the phenoxide ion by delocalization tends to favor its formation, thus increasing the acidity of phenols as compared with alcohols.

Preparation of Alcohols

Methods of preparing alcohols are numerous. The lower-molecular-weight alcohols such as ethyl alcohol, isopropyl alcohol, and tertiary butyl alcohol are prepared from the hydration of the corresponding alkenes available from the cracking of petroleum.

$$CH_3CH_2OH \qquad CH_3-\overset{\overset{\displaystyle H}{|}}{\underset{\underset{\displaystyle OH}{|}}{C}}-CH_3 \qquad CH_3-\overset{\overset{\displaystyle CH_3}{|}}{\underset{\underset{\displaystyle OH}{|}}{C}}-CH_3$$

Ethyl alcohol   Isopropyl alcohol t-butyl alcohol

The reaction can be illustrated by the case of ethyl alcohol:

$$CH_2 = CH_2 + H_2O \underset{}{\overset{300°;\ H_3PO_4}{\rightleftharpoons}} CH_3CH_2OH$$
Ethylene   Ethyl alcohol

The product is obtained as a dilute water solution which can be distilled to give a constant boiling mixture of water and alcohol containing 95.6% alcohol by weight. For special purposes the remaining water can be removed chemically or by distillation with benzene to form *absolute alcohol*. Ethyl alcohol, produced in this fashion and rendered unfit for human consumption (denatured) by the addition of other chemicals, is quite inexpensive and is used in large amounts by the chemical industry as a solvent and as a starting material for synthesis of other materials.

Ethyl alcohol is also called *grain alcohol*, and has been produced for centuries as a beverage. In most cases the starting materials are starches from corn, wheat, potatoes, etc., which are converted in the presence of natural catalysts, called *enzymes*, to glucose.

$$(C_{12}H_{22}O_{11})_x + xH_2O \xrightarrow{\text{Enzymes}} 2xC_6H_{12}O_6$$
Starch   Glucose

The glucose is then converted, with the aid of enzymes present in yeast, to ethyl alcohol and carbon dioxide by a complex series of reactions.

$$C_6H_{12}O_6 \xrightarrow[\substack{\text{Several} \\ \text{steps}}]{\text{Enzymes}} 2CH_3CH_2OH + 2CO_2$$
Glucose   Ethanol

Methanol (CH_3OH), the smallest member of the homologous series of alcohols, is called *wood alcohol* because it can be prepared by vigorously heating wood in the absence of air. This process is too expensive for commercial purposes, and a catalytic process utilizing carbon monoxide and hydrogen is employed in the commercial synthesis of methanol.

$$CO + H_2O \xrightarrow[\text{ZnO-CrO}_3]{400\,°C;\ 300\ atm} \underset{\text{Methanol}}{CH_3OH}$$

Methanol is used primarily as a solvent and in the synthesis of other materials. It was once commonly used as an automobile antifreeze, but its low boiling point (65 °C) and the higher operating temperatures of the cooling systems in modern automobiles have reduced the practicality of methanol for this purpose. Ethylene glycol

$$\begin{array}{cc} CH_2 - CH_2 \\ | \quad\quad | \\ OH \quad OH \end{array}$$

with a boiling point of 197 °C, is now commonly used as an antifreeze. Methanol is much more poisonous than ethanol. Many deaths and cases of blindness have resulted when people have used methanol as an alcoholic beverage. Its poisonous effects are also apparent when a person inhales methanol fumes extensively or absorbs methanol through the skin.

Reactions of Alcohols

Since alcohols are very weak acids, very strong bases or very reactive chemicals are required to form alkoxide ions. Three representative reactions are:

$$CH_3CH_2OH + NaNH_2 \rightarrow CH_3CH_2O^- \ Na^+ + NH_3$$
$$CH_3CH_2OH + NaH \rightarrow CH_3CH_2O^- \ Na^+ + H_2$$
$$CH_3CH_2OH + Na \rightarrow CH_3CH_2O^- \ Na^+ + \tfrac{1}{2}H_2$$

Most alcohols burn in the presence of air, with the formation of carbon dioxide and water. It is the alcohol which burns when brandy is flamed, for example. When there is milder chemical oxidation, the reaction which occurs is determined by the structure of the alcohol.

Alcohols which have one carbon atom (or none, in the case of methanol) attached to the carbon atom holding the hydroxyl group are called *primary* alcohols. A

secondary alcohol is one in which there are two carbon atoms attached to the carbinol carbon atom, and a *tertiary* alcohol has three such carbon atoms.

$$
R-CH_2OH \qquad
\begin{array}{c} R \quad H \quad R \\ \diagdown \;\; | \;\; \diagup \\ C \\ | \\ OH \end{array}
\qquad
\begin{array}{c} R \quad R \quad R \\ \diagdown \;\; | \;\; \diagup \\ C \\ | \\ OH \end{array}
$$

Primary   Secondary   Tertiary

Primary alcohols, when treated with oxidizing agents such as CrO_3, are converted to aldehydes.

$$
R-CH_2OH \xrightarrow{\text{Oxidation}} R-\overset{\overset{\displaystyle O}{\|}}{C}-H
$$

Primary   An aldehyde
alcohol

$$
R-\overset{\overset{\displaystyle O}{\|}}{C}-H \xrightarrow{\text{Oxidation}} R-\overset{\overset{\displaystyle O}{\|}}{C}-OH
$$

Aldehyde   A carboxylic
                                acid

Since the aldehyde is usually susceptible to further oxidation to the corresponding carboxylic acid, this is not a good method for the preparation of aldehydes, unless the aldehyde can be removed from the reaction mixture as it is formed.

$$
R-\overset{\overset{\displaystyle H}{|}}{\underset{\underset{\displaystyle OH}{|}}{C}}-R \xrightarrow{\text{Oxidation}} R-\overset{}{\underset{\underset{\displaystyle O}{\|}}{C}}-R
$$

Secondary   A ketone
alcohol

Oxidation of secondary alcohols is like that of primary alcohols, except that the product of the first step, a ketone, is stable to further mild oxidation.

Since no hydrogen atoms are attached to the carbinol carbon atom in tertiary alcohols, and since a carbon–carbon single bond is not readily broken under oxidative conditions, tertiary alcohols are stable to mild oxidation and give no reaction.

A solution of zinc chloride in concentrated hydrochloric acid may be used to qualitatively differentiate primary, secondary, and tertiary alcohols of low molecular weight. In this test, the *Lucas test*, the alcohol is dissolved in the reagent ($ZnCl_2$–HCl). Tertiary alcohols react very rapidly at room temperature to give an insoluble layer of alkyl chloride. Secondary alcohols require several minutes before

the insoluble layer appears, and primary alcohols form alkyl chlorides only on heating.

$$\begin{matrix} R \\ | \\ R-C-OH \\ | \\ R \end{matrix} + HCl \xrightarrow[-H_2O]{ZnCl_2} \begin{matrix} R \\ | \\ R-C-Cl \\ | \\ R \end{matrix} \quad \text{Rapid formation}$$

$$\begin{matrix} R \\ | \\ R-C-OH \\ | \\ H \end{matrix} + HCl \xrightarrow[-H_2O]{ZnCl_2} \begin{matrix} R \\ | \\ R-C-Cl \\ | \\ H \end{matrix} \quad \text{Several minutes}$$

$$\begin{matrix} H \\ | \\ R-C-OH \\ | \\ H \end{matrix} + HCl \xrightarrow{ZnCl_2} \text{No reaction at room temperature}$$

When alcohols are treated with concentrated sulfuric acid, water is split out and an alkyl, hydrogen sulfate, is formed.

$$CH_3CH_2OH + H_2SO_4 \rightarrow CH_3CH_2OSO_3H + H_2O$$
$$\text{Ethyl hydrogen}$$
$$\text{sulfate}$$

The alkyl hydrogen sulfate is unstable at elevated temperatures, eliminating a molecule of sulfuric acid and forming the corresponding alkene:

$$CH_3CH_2-OSO_3H \rightarrow CH_2{=}CH_2 + H_2SO_4.$$

Thus the overall reaction is sometimes written as a dehydration reaction catalyzed by sulfuric acid. This is the reverse of the reaction mentioned above for the preparation of alcohols:

$$CH_3CH_2-OH \xrightarrow[150\,°C]{H_2SO_4} CH_2{=}CH_2 + H_2O.$$

16-8 ETHERS

At lower temperatures the alkyl hydrogen sulfate, obtained from alcohols and sulfuric acid, may react with excess alcohol, resulting in the formation of an ether.

$$CH_3CH_2-O-SO_3H + CH_3CH_2OH \xrightarrow{130\,°C} CH_3CH_2-O-CH_2CH_3 + H_2SO_4$$
$$\text{Diethyl ether}$$

The overall reaction may be considered an acid-catalyzed condensation reaction, joining two molecules of alcohol with the expulsion of a molecule of water. It is usually written as a one-step process:

$$2CH_3\!-\!OH \xrightarrow{\text{H}_2\text{SO}_4} CH_3\!-\!O\!-\!CH_3 + H_2O.$$
$$\text{Dimethyl ether}$$

The structure of ethers, $R\!-\!O\!-\!R'$, differs from that of alcohols, $R\!-\!OH$, significantly. The lack of a hydroxyl group destroys the possibility of hydrogen bonding and reduces the polarity of the molecule. For example, ethyl alcohol (C_2H_6O) has a boiling point of 78.5 °C at one atm pressure, but dimethyl ether (C_2H_6O) is a gas at room temperature. Low-molecular-weight alcohols are soluble in polar solvents such as water, but ethers are insoluble in water. The reduced polarity is accompanied by low chemical reactivity, and thus ethers, particularly diethyl ether, are very good solvents for organic reactions.

The use of sulfuric acid to prepare ethers is limited to ethers which are symmetrical, that is, the groups attached to the oxygen must be the same. For example, if a mixture of methyl alcohol and ethyl alcohol is treated with sulfuric acid, a mixture of ethers results:

$$CH_3OH + CH_3CH_2OH \xrightarrow[130\,°C]{\text{H}_2\text{SO}_4} \begin{array}{c} CH_3\!-\!O\!-\!CH_3 \\ + \\ CH_3CH_2\!-\!O\!-\!CH_3 \\ + \\ CH_3CH_2\!-\!O\!-\!CH_2CH_3 \end{array}$$

In order to prepare unsymmetrical ethers efficiently, one must use other reactions. Here is one such reaction, involving an alkyl iodide and an alkoxide, for preparing methyl ethyl ether:

$$CH_3I + CH_3CH_2ONa \rightarrow CH_3\!-\!O\!-\!CH_2CH_3 + NaI$$

| Methyl iodide | Sodium ethoxide | Methyl ethyl ether |

In addition to their importance as organic solvents, certain ethers have important medical uses. Diethyl ether, commonly called "ether," was first used as a general anesthetic at Massachusetts General Hospital in 1846. In spite of its flammability and certain undesirable side effects that follow its use, diethyl ether remains the most important general anesthetic in use today.

Divinyl ether, $CH_2\!\!=\!\!CH\!-\!O\!-\!CH\!\!=\!\!CH_2$, is more rapid in its anesthetic action than diethyl ether and does not have some of the unpleasant side effects. However, the action of divinyl ether is of short duration and prolonged use of it may cause liver damage. Divinyl ether has been effectively employed as an anesthetic for children.

16-9 ALDEHYDES AND KETONES

The characteristic functional group of aldehydes and ketones is the carbonyl group,

The carbon atom is sp^2 hybridized, and so one of the carbon–oxygen bonds is a
σ-bond and the second is a π-bond. Since oxygen is more electronegative than carbon,
the electrons in the double bond are attracted toward oxygen, creating a dipole:

Aldehydes have the general formula

$$
\begin{array}{cc}
\overset{\displaystyle O}{\underset{\displaystyle \parallel}{}} & \overset{\displaystyle O}{\underset{\displaystyle \parallel}{}} \\
R-C-H \quad \text{and ketones} & R-C-R.
\end{array}
$$

It should therefore not be surprising that the reactions of aldehydes and ketones
are similar. Often the only difference is in the rate of the reaction. The ketone
reactions are often slower due to the large size of the R group compared to hydrogen.
This increased bulk can often block the approach of a reagent to the carbonyl group,
slowing the rate of the reaction.

   Aldehydes and ketones can be prepared from the oxidation of primary and
secondary alcohols, respectively. As mentioned previously, care must be taken in
the oxidative preparation of aldehydes, since the aldehydes are readily oxidized to
the corresponding carboxylic acids. One method which has been used successfully
for low-molecular-weight aldehydes is to carry out the reaction at a temperature
above the boiling point of the aldehyde, so that the aldehyde may be distilled from
the reaction mixture as soon as it is formed, preventing further oxidation.

   The fact that aldehydes are so readily oxidized means that they are good reducing
agents. This is the basis for qualitative tests for aldehydes. *Benedict's solution* (containing copper sulfate, sodium citrate, and sodium carbonate) is one reagent used in
testing for aldehydes. Benedict's solution may be considered a basic solution of
CuO to which a complexing agent (citrate ion) has been added to prevent the precipitation of copper(II) hydroxide, $Cu(OH)_2$. The aldehyde reduces the copper(II)
ion to copper(I), which forms a yellow to red precipitate of Cu_2O, indicating a positive
test for the aldehyde,

$$
\underset{\text{Formaldehyde}}{H-\overset{\displaystyle O}{\overset{\displaystyle \parallel}{C}}-H} + 2CuO \rightarrow \underset{\text{Formic acid}}{H-\overset{\displaystyle O}{\overset{\displaystyle \parallel}{C}}-OH} + Cu_2O_{(s)}.
$$

The Benedict's test is not foolproof; the results in certain instances can be misleading and must be interpreted with care.

A second mild oxidizing reagent used in testing for aldehydes is *Tollen's reagent*. Tollen's reagent is prepared by adding a dilute solution of ammonium hydroxide to a silver nitrate solution. Initially a brown precipitate of silver oxide is formed, which dissolves on further addition of ammonium hydroxide, forming the diammine silver complex ion.

$$2Ag^+ + 2NO_3^- + 2NH_3 + H_2O \rightarrow Ag_2O_{(s)} + 2NH_4^+ + 2NO_3^-$$
$$Ag_2O_{(s)} + 4NH_3 + H_2O \rightarrow 2Ag(NH_3)_2^+ + 2OH^-$$

The reagent can be considered as a basic solution of silver oxide, Ag_2O. In the presence of an aldehyde, Tollen's reagent is reduced to free silver, and if the reaction is carried out in a clean test tube, a silver mirror forms on the bottom of the tube:

$$\underset{\text{Formaldehyde}}{H-\overset{\overset{\displaystyle O}{\|}}{C}-H} + Ag_2O \rightarrow \underset{\text{Formic acid}}{H-\overset{\overset{\displaystyle O}{\|}}{C}-OH} + 2Ag_{(s)}$$

Ketones are stable to mild oxidation. Conversion to a corresponding carboxylic acid requires cleavage of a carbon–carbon σ-bond, and this occurs only under vigorous oxidative conditions.

Formaldehyde,

$$H-\overset{\overset{\displaystyle O}{\|}}{C}-H,$$

is the simplest aldehyde and can be prepared by the oxidation of methanol under conditions which preclude further oxidation to formic acid.

$$CH_3OH \xrightarrow{\text{Oxidation}} H-\overset{\overset{\displaystyle O}{\|}}{C}-H$$

Formaldehyde is soluble in water. A 40% aqueous solution, called *formalin*, is used as a disinfectant, a preservative of biological specimens, and as starting material for the production of certain plastics.

The next higher homolog is acetaldehyde,

$$CH_3\overset{\overset{\displaystyle O}{\|}}{C}-H,$$

which can be prepared by oxidizing ethyl alcohol, and, in the form of a trimer $(CH_3CHO)_3$, has been used as a sedative and a hypnotic (sleep-producer).

Chloral,

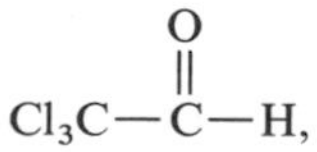

$$\underset{\text{}}{Cl_3C-\overset{\displaystyle O}{\overset{\displaystyle \|}{C}}-H,}$$

is a liquid aldehyde which reacts with water to form a crystalline hydrate called chloral hydrate. Chloral hydrate, a poisonous hypnotic sometimes called knockout drops, is formed by the addition of water to the π-electrons of the carbonyl double bond. The positive hydrogen of water becomes attached to the negative oxygen of the carbonyl group, while the negative oxygen of water is attached to the positive carbonyl carbon atom:

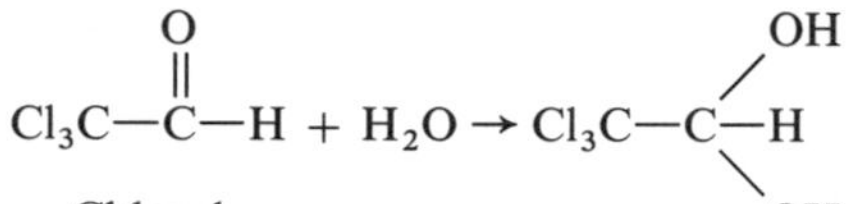

$$Cl_3C-\overset{\displaystyle O}{\overset{\displaystyle \|}{C}}-H + H_2O \rightarrow Cl_3C-\overset{\displaystyle OH}{\underset{\displaystyle OH}{C}}-H$$

Chloral   Chloral hydrate

Although chloral is unusual in that it forms a stable hydrate, many aldehydes exist in water solution in equilibrium with the hydrate. In most cases, however, trying to isolate the hydrate affords only the free aldehyde:

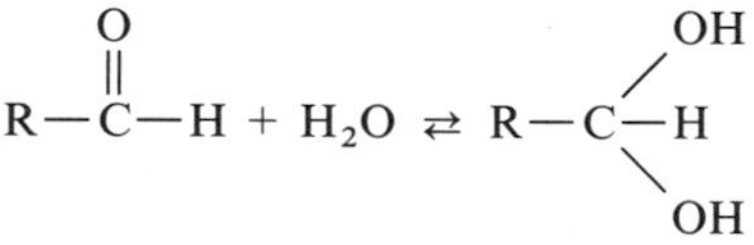

$$R-\overset{\displaystyle O}{\overset{\displaystyle \|}{C}}-H + H_2O \rightleftarrows R-\overset{\displaystyle OH}{\underset{\displaystyle OH}{C}}-H$$

A reaction similar to hydrate formation occurs between aldehydes and alcohols. The *hemiacetal* formed from this reaction is also generally unstable and cannot be isolated.

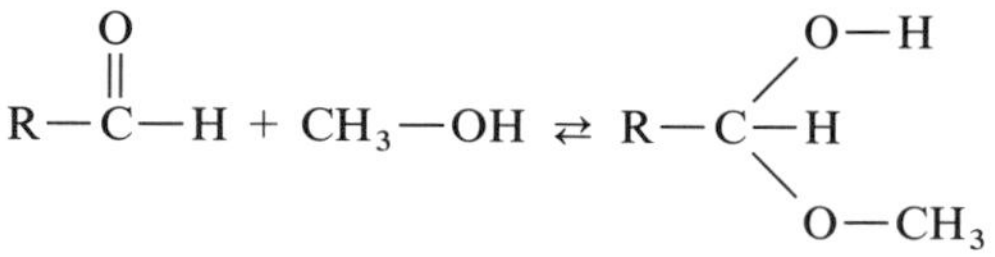

$$R-\overset{\displaystyle O}{\overset{\displaystyle \|}{C}}-H + CH_3-OH \rightleftarrows R-\overset{\displaystyle O-H}{\underset{\displaystyle O-CH_3}{C}}-H$$

Hemiacetal

However, if the reaction is carried out under *anhydrous* conditions with an acid catalyst, the hemiacetal may be converted to the *acetal*, via the intermediate methyl-oxonium ion.

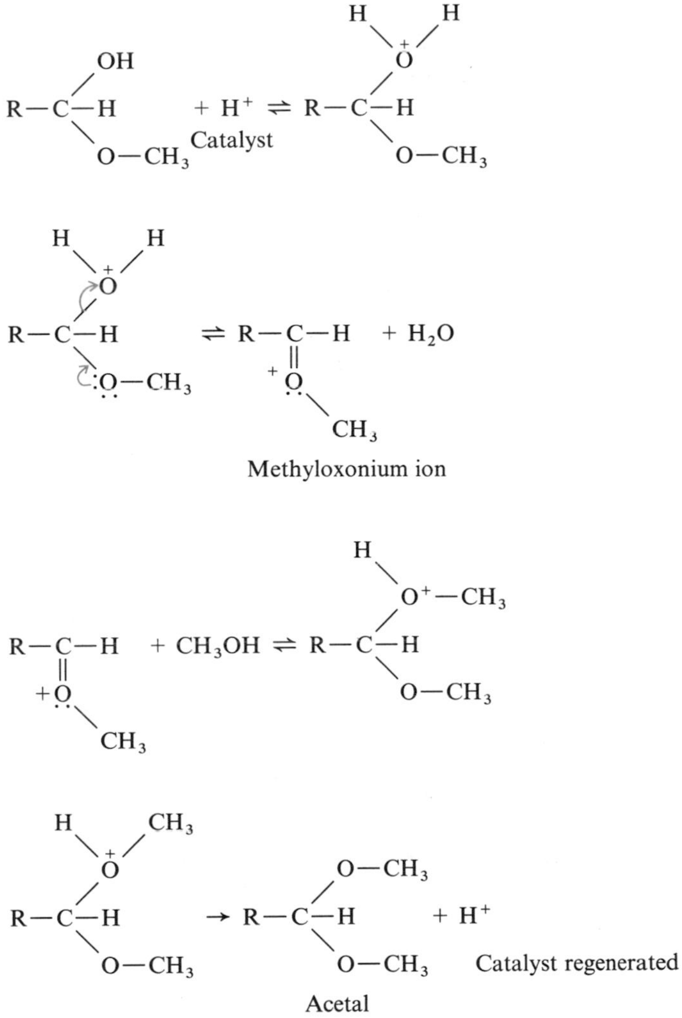

$$
\underset{\substack{| \\ \text{O}-\text{CH}_3}}{\overset{\substack{\text{OH} \\ |}}{\text{R}-\text{C}-\text{H}}} \; + \; \text{H}^+ \; \underset{\text{Catalyst}}{\rightleftharpoons} \; \underset{\substack{| \\ \text{O}-\text{CH}_3}}{\overset{\substack{\text{H} \quad \text{H} \\ \overset{+}{\text{O}} \\ |}}{\text{R}-\text{C}-\text{H}}}
$$

$$
\underset{\substack{\overset{..}{\text{C}}\!:\!\text{O}-\text{CH}_3}}{\overset{\substack{\text{H} \quad \text{H} \\ \overset{+}{\text{O}} \\ |}}{\text{R}-\text{C}-\text{H}}} \; \rightleftharpoons \; \underset{\substack{+ \overset{..}{\text{O}} \\ \diagdown \\ \text{CH}_3}}{\overset{\substack{\| \\ }}{\text{R}-\text{C}-\text{H}}} \; + \; \text{H}_2\text{O}
$$

Methyloxonium ion

$$
\underset{\substack{+\overset{..}{\text{O}} \\ \diagdown \\ \text{CH}_3}}{\overset{\|}{\text{R}-\text{C}-\text{H}}} \; + \; \text{CH}_3\text{OH} \; \rightleftharpoons \; \underset{\substack{| \\ \text{O}-\text{CH}_3}}{\overset{\substack{\text{H} \\ \diagdown \\ \text{O}^+-\text{CH}_3 \\ |}}{\text{R}-\text{C}-\text{H}}}
$$

$$
\underset{\substack{| \\ \text{O}-\text{CH}_3}}{\overset{\substack{\text{H} \quad \text{CH}_3 \\ \overset{+}{\text{O}} \\ |}}{\text{R}-\text{C}-\text{H}}} \; \rightarrow \; \underset{\substack{| \\ \text{O}-\text{CH}_3}}{\overset{\substack{\text{O}-\text{CH}_3 \\ | }}{\text{R}-\text{C}-\text{H}}} \; + \; \text{H}^+ \quad \text{Catalyst regenerated}
$$

Acetal

The overall reaction which does not indicate the *mechanism* (stepwise process) of the reaction is written as follows:

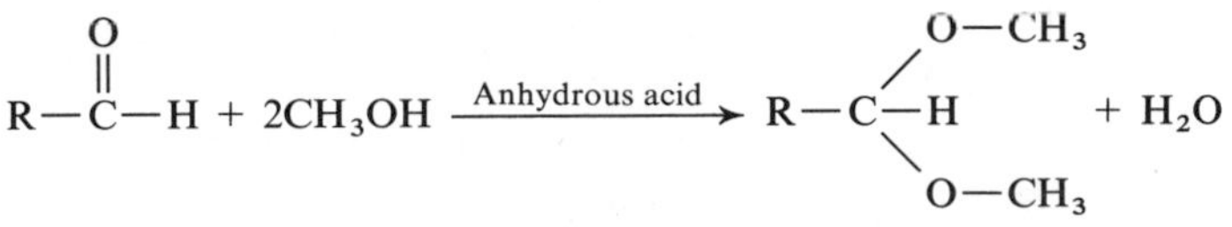

$$
\underset{}{\overset{\substack{\text{O} \\ \|}}{\text{R}-\text{C}-\text{H}}} + 2\text{CH}_3\text{OH} \; \xrightarrow{\text{Anhydrous acid}} \; \underset{\substack{| \\ \text{O}-\text{CH}_3}}{\overset{\substack{\text{O}-\text{CH}_3 \\ |}}{\text{R}-\text{C}-\text{H}}} \; + \; \text{H}_2\text{O}
$$

Acetone, the simplest ketone, is an important industrial solvent. It can be prepared by the oxidation of isopropyl alcohol:

$$CH_3-\overset{\overset{\displaystyle H}{|}}{\underset{\underset{\displaystyle OH}{|}}{C}}-CH_3 \xrightarrow{\text{Oxidation}} CH_3-\overset{\overset{\displaystyle O}{\|}}{C}-CH_3$$

Isopropyl alcohol   Acetone

Ketones can be converted to ketals by the action of anhydrous acid on a solution of the ketone in water-free alcohol:

$$R-\overset{\overset{\displaystyle O}{\|}}{C}-R + CH_3OH \underset{}{\overset{H^+}{\rightleftharpoons}} R-\overset{\overset{\displaystyle OH}{\diagup}}{\underset{\underset{\displaystyle O-CH_3}{\diagdown}}{C}}-R$$

Hemiketal

$$R-\overset{\overset{\displaystyle O-H}{\diagup}}{\underset{\underset{\displaystyle O-CH_3}{\diagdown}}{C}}-R \quad + CH_3OH \xrightarrow{H^+} R-\overset{\overset{\displaystyle O-CH_3}{\diagup}}{\underset{\underset{\displaystyle O-CH_3}{\diagdown}}{C}}-R$$

Ketal

Both ketals and acetals can be converted back to the corresponding carbonyl compounds by the action of *aqueous acid*.

$$R-\overset{\overset{\displaystyle O-CH_3}{\diagup}}{\underset{\underset{\displaystyle O-CH_3}{\diagdown}}{C}}-R \quad + H_2O \xrightarrow{H^+} R-\overset{\overset{\displaystyle O}{\|}}{C}-R + 2CH_3OH$$

Since acetals and ketals are stable to most reagents other than aqueous acid, they are used as protective agents. The carbonyl group can be protected during the course of a chemical reaction by first forming the ketal or acetal and then regenerating the carbonyl group with aqueous acid after the desired reaction has been completed.

Although hemiacetals and hemiketals are usually unstable, they are readily formed by sugars and, as we shall see in Chapter 18, hemiacetal and acetal formation play a large part in the structures of carbohydrates.

Carbonyl compounds are among the most important intermediates in synthetic organic chemistry; the reactions they undergo are too numerous to mention here. For further information, we refer you to the sources listed at the end of this chapter.

16–10 CARBOXYLIC ACIDS

The carboxylic acid functional group is

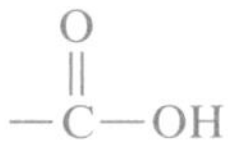

$$\begin{matrix} O \\ \parallel \\ -C-OH \end{matrix}$$

often written as $-CO_2H$ or $-COOH$, and a general formula may be written as

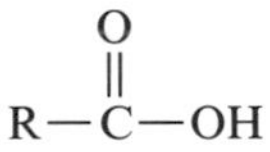

$$\begin{matrix} O \\ \parallel \\ R-C-OH \end{matrix}$$

When R— is an alkyl group such as CH_3— or CH_3CH_2—, the acid is called a *fatty acid*, since some of the higher homologs may be obtained from chemical degradation of naturally occurring fats and oils.

Carboxylic acids are generally weaker acids than mineral acids but much stronger acids than alcohols or phenols. We can explain the acidity of carboxylic acids by means of resonance theory. When we write the equilibrium for the dissociation of a carboxylic acid, we can see that any factors which cause the equilibrium to proceed in the direction of dissociation will increase the acidity of the acid.

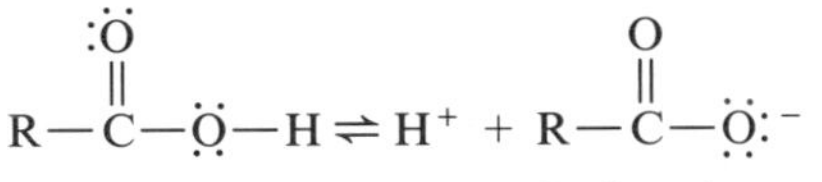

$$R-\overset{\overset{\displaystyle :\ddot{O}}{\parallel}}{C}-\ddot{O}-H \rightleftharpoons H^+ + R-\overset{\overset{\displaystyle O}{\parallel}}{C}-\ddot{O}:^-$$

Carboxylic   Carboxylate
acid   ion

The carboxylate ion (the conjugate base of the carboxylic acid) has the π-electrons of its carbon–oxygen double bond located in a position which makes possible overlap with one of the unshared pairs of electrons on the negatively charged oxygen atom. This overlap leads to delocalization of the pair of electrons, and therefore the carboxylate anion is a resonance hybrid, as shown in Fig. 16–15, in which the negative charge is dispersed between the two oxygen atoms. This hybrid structure is more stable than either structure which has the negative charge located on one oxygen atom. Since the carboxylate ion is stabilized, it has a greater tendency to form and the corresponding acid is stronger.

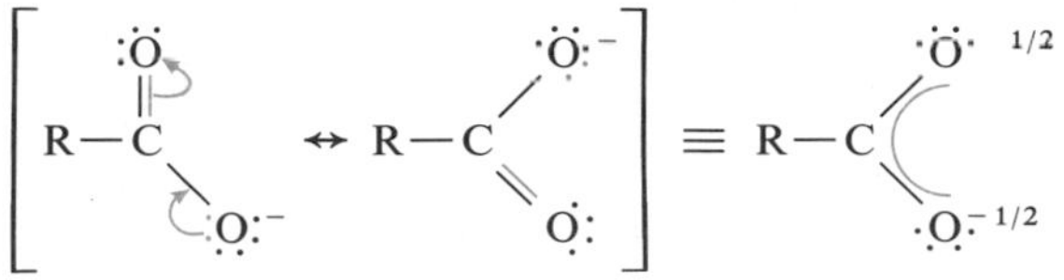

Fig. 16–15 Resonance stabilization of the carboxylate anion.

Other factors affect the relative acidity of carboxylic acids. For example, the carbon–oxygen double bond in the carboxyl group is polarized in the same way as the carbonyl group in aldehydes and ketones:

$$-\overset{\overset{\displaystyle \delta-:\ddot{O}}{\|}}{\underset{\delta+}{C}}\rightleftharpoons O \leftharpoonup H$$

This polarization, which places a partial positive charge on the carboxyl carbon atom, causes the electrons in the carbon–oxygen single bond to be attracted toward the carbon. Such attraction weakens the oxygen–hydrogen bond and increases the tendency of hydrogen to leave as a proton, that is, it increases the acidity.

Thus anything which increases the positive character of the carboxyl carbon atom increases the acidity. For example, acetic acid ($HC_2H_3O_2$ or CH_3COOH) is a weak carboxylic acid (acid ionization constant, $K = 1.8 \times 10^{-5}$). If one of the methyl hydrogens is replaced by a more electronegative chlorine atom to give chloroacetic acid, the acidity increases ($K = 1.4 \times 10^{-3}$). This is due to a shift of the electrons in the Cl—C, C—C, C—O bonds toward the electronegative chlorine atom.

$$\underset{\underset{\displaystyle H}{|}}{\overset{\overset{\displaystyle H \quad O}{| \quad \|}}{Cl \rightleftharpoons C \rightleftharpoons C \rightleftharpoons O}} \qquad \text{Arrows indicate shift of electron density.}$$

Chloroacetic acid

If all three of the methyl hydrogens in acetic acid are replaced with chlorine atoms to give trichloracetic acid (Cl_3CCOOH), the acidity is further increased ($K = 3 \times 10^{-1}$). Trifluoroacetic acid (F_3CCOOH) has an acid strength which rivals that of the strong mineral acids, being completely dissociated in water solution, presumably due to the very high electron-attracting capability of fluorine.

As mentioned previously, carboxylic acids may be prepared via the aldehydes by the oxidation of primary alcohols.

$$CH_3OH \xrightarrow[H_2SO_4]{CrO_3} \overset{\overset{\displaystyle O}{\|}}{H-C-OH} \qquad\qquad CH_3CH_2OH \xrightarrow[H_2SO_4]{CrO_3} \overset{\overset{\displaystyle O}{\|}}{CH_3C-OH}$$

Formic acid   Acetic acid

$$CH_3CH_2CH_2OH \xrightarrow[H_2SO_4]{CrO_3} \overset{\overset{\displaystyle O}{\|}}{CH_3CH_2C-OH}$$

Propionic acid

Benzyl alcohol   Benzoic acid

Reactions of Carboxylic Acids

The carboxylic acids behave in a normal acidic manner in neutralization reactions; they react with bases to form salts and water:

$$CH_3\overset{\displaystyle O}{\overset{\|}{C}}-OH + KOH \rightarrow CH_3\overset{\displaystyle O}{\overset{\|}{C}}-OK + H_2O$$

Potassium
acetate

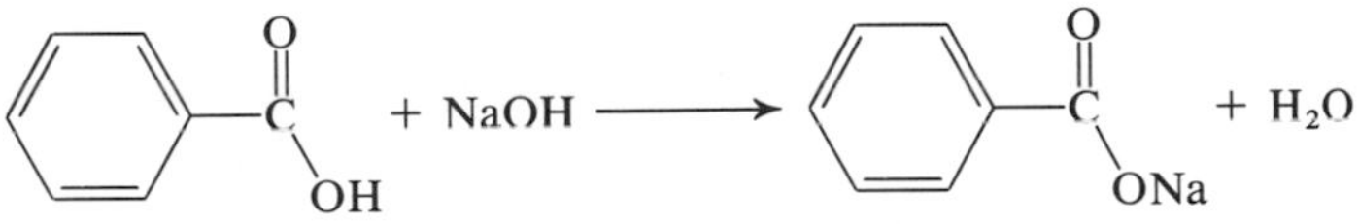

$+ NaOH \longrightarrow$ ... $+ H_2O$

Sodium
benzoate

In the presence of concentrated, strong acid, carboxylic acids react with alcohols to form *esters* and water, a process called *esterification*:

$$RCOOH + R-OH \underset{}{\overset{H^+}{\rightleftharpoons}} R\overset{\displaystyle O}{\overset{\|}{C}}-OR + H_2O$$

An ester

In this manner ethyl acetate, an important solvent for lacquers, and one which is sold as a nail-polish remover, can be synthesized from acetic acid and ethyl alcohol:

$$CH_3\overset{\displaystyle O}{\overset{\|}{C}}-OH + CH_3CH_2OH \xrightarrow[-H_2O]{H^+} CH_3\overset{\displaystyle O}{\overset{\|}{C}}-O-CH_2CH_3$$

Ethyl acetate

Methyl salicylate, also called oil of wintergreen, is a natural constituent of many plants used in perfumery, as a flavoring for candies, and as a liniment. It is prepared commercially by the esterification of salicylic acid with methanol.

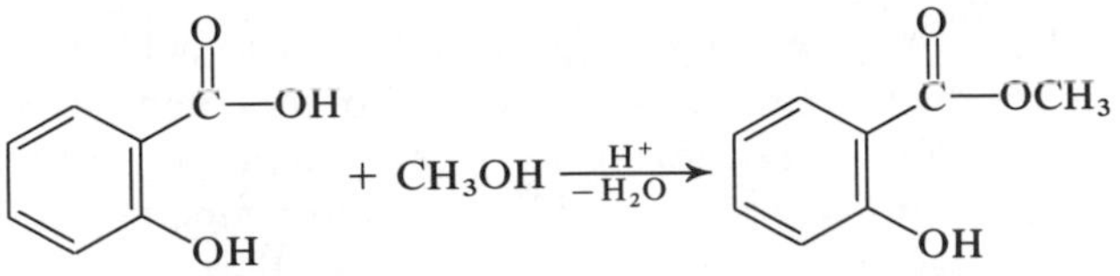

$+ CH_3OH \xrightarrow[-H_2O]{H^+}$

Salicylic acid   Methyl salicylate

The lower-molecular-weight esters are generally volatile and have a pleasant odor. The odors of many fruits and flowers, for example, are due to their constituent volatile esters. Natural fats and oils, which we shall discuss in Chapter 18, are esters of high-molecular-weight fatty acids.

Esterification is an equilibrium reaction. However, by using an excess of acid or alcohol or by using an acid which is also a dehydrating agent (concentrated H_2SO_4 effectively removes the water as it is formed), one can shift the equilibrium toward the ester. When the conditions are changed to afford an excess of water (for example, in dilute acid), the reaction can be reversed and esters can be hydrolyzed to the component acid and alcohol:

$$CH_3\overset{\overset{\displaystyle O}{\|}}{C}-OCH_2CH_3 + H_2O \xrightarrow{H^+} CH_3\overset{\overset{\displaystyle O}{\|}}{C}-OH + CH_3CH_2OH$$

When carboxylic acids are heated with a strong dehydrating agent such as P_4O_{10} or H_2SO_4, two molecules of the acid are condensed to form an acid anhydride and water:

$$R-\overset{\overset{\displaystyle O}{\|}}{C}-OH + H-O-\overset{\overset{\displaystyle O}{\|}}{C}-R \xrightarrow[\text{Heat}]{H_2SO_4} R-\overset{\overset{\displaystyle O}{\|}}{C}-O-\overset{\overset{\displaystyle O}{\|}}{C}-R + H_2O$$
$$\text{Anhydride}$$

Anhydrides, particularly those of low molecular weight, are very reactive. Acetic anhydride, formed by the condensation of two molecules of acetic acid, reacts violently with water and can produce painful skin burns.

$$CH_3\overset{\overset{\displaystyle O}{\|}}{C}-OH + HO-\overset{\overset{\displaystyle O}{\|}}{C}CH_3 \xrightarrow[-H_2O]{P_4O_{10}} CH_3-\overset{\overset{\displaystyle O}{\|}}{C}-O-\overset{\overset{\displaystyle O}{\|}}{C}-CH_3$$
$$\text{Acetic anhydride}$$

$$CH_3-\overset{\overset{\displaystyle O}{\|}}{C}-O-\overset{\overset{\displaystyle O}{\|}}{C}-CH_3 + H_2O \longrightarrow 2CH_3\overset{\overset{\displaystyle O}{\|}}{C}-OH + \text{heat}$$

Vapors of acetic anhydride are irritating to the mucous membranes and the eyes, and care must be taken in handling this material.

The increased reactivity of anhydrides with respect to acids makes anhydrides valuable reagents for the preparation of esters and other compounds. Acetic anhydride reacts with alcohols and phenols to form esters spontaneously. No catalyst is required and the reaction takes place rapidly at room temperature.

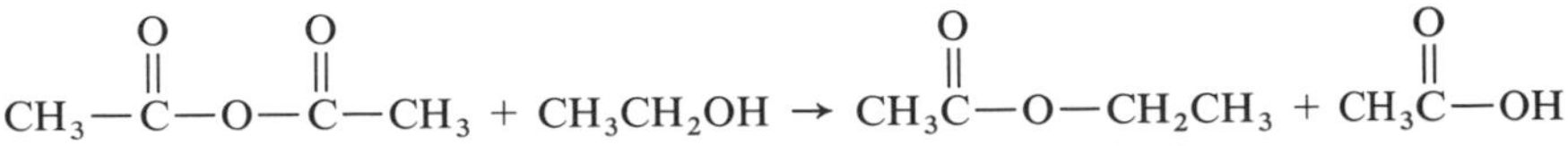

$$CH_3-\overset{\overset{\displaystyle O}{\|}}{C}-O-\overset{\overset{\displaystyle O}{\|}}{C}-CH_3 + CH_3CH_2OH \rightarrow CH_3\overset{\overset{\displaystyle O}{\|}}{C}-O-CH_2CH_3 + CH_3\overset{\overset{\displaystyle O}{\|}}{C}-OH$$

Acetic anhydride   Ethyl acetate

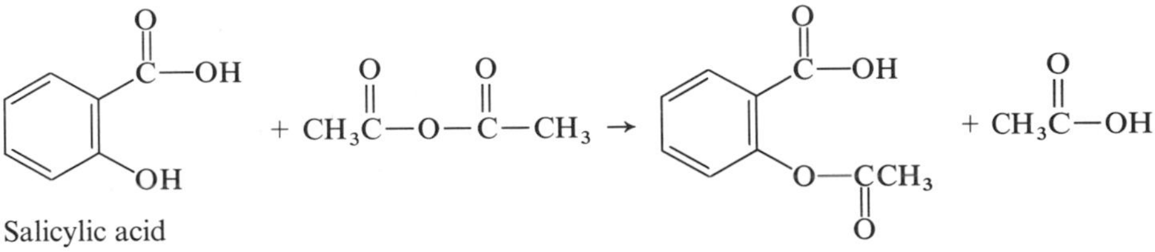

Salicylic acid

Acetyl salicylic
acid (aspirin)

16-11 AMINES

Amines may be considered derivatives of ammonia (NH_3), in which one or more of the hydrogen atoms has been replaced by an alkyl or aromatic group. If two hydrogens remain attached to the nitrogen atom, the amine is called *primary*; if one hydrogen is attached to the nitrogen, it is *secondary*, and no hydrogens, *tertiary*:

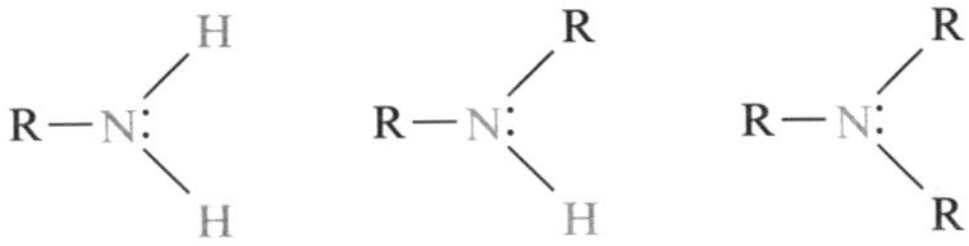

Primary   Secondary   Tertiary
amine   amine   amine

Amines are weak bases, since the unshared pair of electrons on the nitrogen atom is available for bonding with a proton.

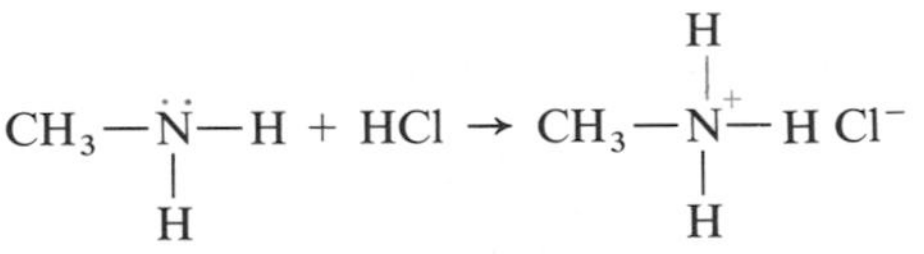

$$CH_3-\overset{\overset{\displaystyle \cdot\cdot}{N}}{\underset{H}{|}}-H + HCl \rightarrow CH_3-\overset{\overset{\displaystyle H}{|}}{\underset{H}{\overset{+}{N}}}-H \; Cl^-$$

Methyl amine   Methyl ammonium
                      chloride

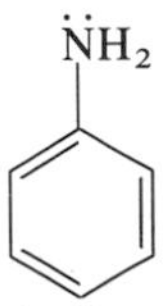

Aniline

In general, the greater the substitution on the nitrogen, the greater the strength of the base. Thus the order of basicity may be listed as $NH_3 < RNH_2 < R_2NH < R_3N$. If the nitrogen atom is attached to an aromatic ring, as in aniline, the strength of the base is reduced. Aromatic amines are weaker bases than ammonia.

The explanation for this reduced base strength of the aromatic amines is similar to the explanation previously offered for the stabilization of the phenoxide ion (recall Section 16–7). The unshared pair of electrons on the nitrogen can overlap with the π-electron orbitals in the aromatic ring and undergo the resonance delocalization illustrated in Fig. 16–16. The unshared electrons are no longer localized on the nitrogen atom and are thus less available for bond formation with a proton; the basicity is reduced.

We can prepare amines by causing an alkyl iodide to react with ammonia. The ammonia displaces iodide ion from the alkyl iodide, forming an alkyl ammonium iodide salt, which can be converted to the free primary amine with sodium hydroxide:

$$CH_3I + NH_3 \rightarrow CH_3\overset{+}{N}H_3I^-$$

Methyl   Methyl
iodide   ammonium
     iodide

$$CH_3\overset{+}{N}H_3I^- + NaOH \rightarrow CH_3NH_2 + H_2O + NaI$$

Secondary and tertiary amines can be prepared by similar reaction sequences,* as follows:

$$CH_3NH_2 + CH_3I \rightarrow CH_3-\underset{\overset{+}{}}{\overset{\overset{\displaystyle CH_3}{|}}{N}}H_2I^-$$

Dimethyl-
ammonium
iodide

$$CH_3\underset{\overset{+}{}}{N}H_2I^- + NaOH \rightarrow CH_3-\overset{\overset{\displaystyle CH_3}{|}}{N}\diagdown_{H} + H_2O + NaI$$

Dimethylamine

$$(CH_3)_2NH + CH_3I \rightarrow (CH_3)_3\overset{+}{N}HI^-$$

Dimethyl-   Trimethyl-
amine   ammonium
      iodide

$$(CH_3)_3\overset{+}{N}HI^- + NaOH \rightarrow (CH_3)_3N + H_2O + NaI$$

Trimethyl-
amine

---

*   The reaction is not easy to control, however, and the reaction of ammonia with methyl iodide may lead to a mixture of primary, secondary, and tertiary amines.

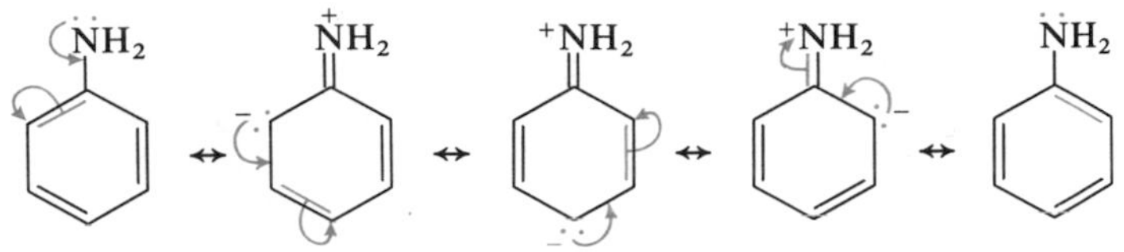

Fig. 16–16 Resonance delocalization of the electron pair on the nitrogen atom in aniline.

In addition to the reaction of amines with acids to form salts, one of the most important reactions of primary and secondary amines is with carboxylic acids or anhydrides to form *amides*. Dimethyl formamide, for example

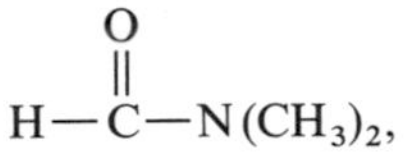

$$H-\overset{\overset{\displaystyle O}{\|}}{C}-N(CH_3)_2,$$

is an important solvent.

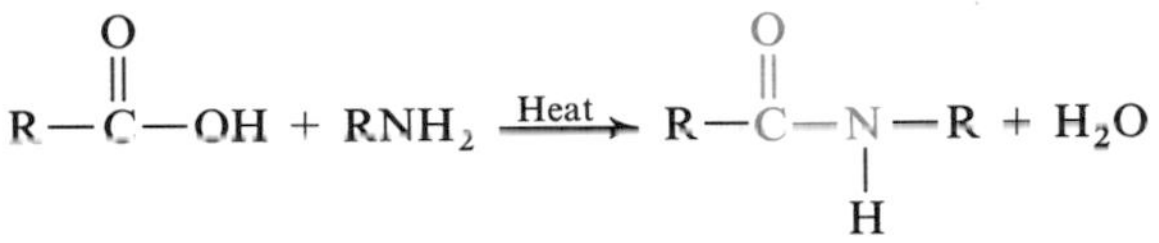

$$R-\overset{\overset{\displaystyle O}{\|}}{C}-OH + RNH_2 \xrightarrow{\text{Heat}} R-\overset{\overset{\displaystyle O}{\|}}{C}-\underset{\underset{\displaystyle H}{|}}{N}-R + H_2O$$

An amide

$$CH_3-\overset{\overset{\displaystyle O}{\|}}{C}-O-\overset{\overset{\displaystyle O}{\|}}{C}-CH_3 + NH_3 \longrightarrow CH_3\overset{\overset{\displaystyle O}{\|}}{C}-NH_2 + CH_3\overset{\overset{\displaystyle O}{\|}}{C}-OH$$

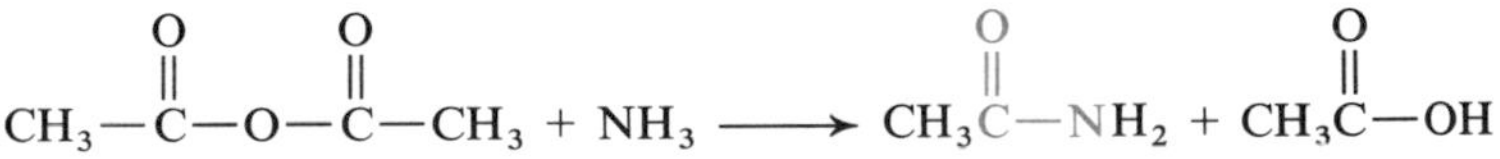

Acetamide

Nylon, a long-chain polymeric amide, can be prepared from adipic acid and hexamethylene diamine:

$$x\text{HO}\overset{\overset{\displaystyle O}{\|}}{C}-(CH_2)_4-\overset{\overset{\displaystyle O}{\|}}{C}-OH + xH_2N-CH_2(CH_2)_4-CH_2NH_2$$

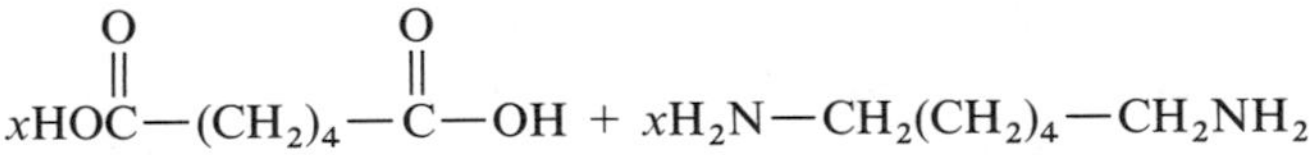

Adipic acid   Hexamethylene diamine

$$H-O-\overset{\overset{\displaystyle O}{\|}}{C}-(CH_2)_4-\overset{\overset{\displaystyle O}{\|}}{C}\left(-\underset{\underset{\displaystyle H}{|}}{N}-(CH_2)_6-\underset{\underset{\displaystyle H}{|}}{N}-\overset{\overset{\displaystyle O}{\|}}{C}-(CH_2)_4-\overset{\overset{\displaystyle O}{\|}}{C}\right)_{x-1}-\underset{\underset{\displaystyle H}{|}}{N}-(CH_2)_6-NH_2$$

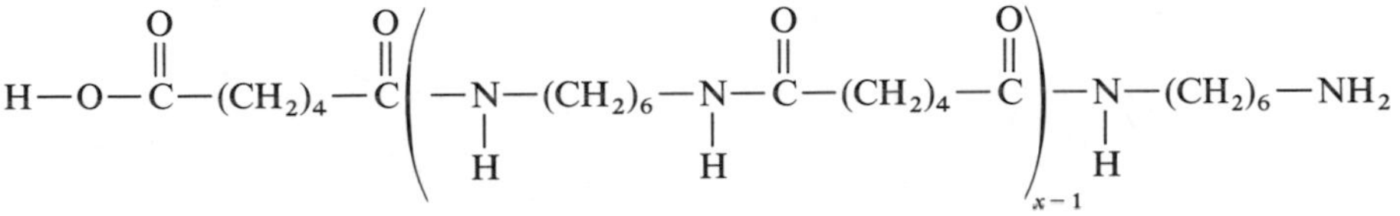

Nylon (*x* is a large number)

Amide formation is also an important biological reaction. Amino acids are joined by amide links to form proteins. We shall discuss this reaction in Chapter 18.

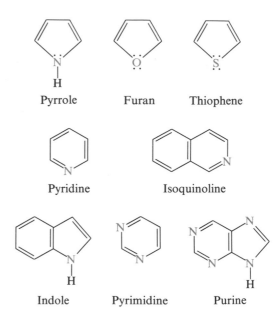

Fig. 16-17 Some representative heterocyclic compounds.

16–12 HETEROCYCLIC COMPOUNDS

Cyclic compounds which contain one or more atoms other than carbon in the ring are called *heterocyclic*. Figure 16–17 shows a number of representative heterocyclic compounds. Many heterocyclic compounds containing nitrogen are naturally occurring materials called *alkaloids*. Certain synthetic heterocyclic materials are used as medicinal agents. Pyrrole and pyrrole derivatives are found in coal tar, along with pyridine and pyridine derivatives. As we shall see later, pyrrole forms a portion of the heme molecule, which is responsible for the fact that hemoglobin transports oxygen. The chlorophyll molecule also contains pyrrole rings.

The pyridine ring system is found in nicotinic acid (one of the B vitamins), as well as in the isomeric tobacco alkaloids nicotine and anabasine:

Nicotine   Anabasine

The isoquinoline ring system is found in many naturally occurring alkaloids, including morphine, codeine, and heroin.

Indole is a product of intestinal putrefaction (but is also used in perfumes). Indole acetic acid is an important plant hormone. Psilocybin, the active constituent of the hallucinogenic Mexican ceremonial mushroom, and psychedelic amide derivatives of lysergic acid (obtained from the rye fungus, ergot) also contain the indole group.

Indole acetic acid   Psilocybin   Lysergic acid

Derivatives of purine and pyrimidine, found in DNA and RNA, are responsible for the transmission of genetic information (see Chapter 18 for further discussion).

16–13 SUMMARY

This introductory section has of necessity been very limited. The only types of compounds and reactions we have mentioned are those which are prerequisite to an understanding of the subsequent chapters. However, do not think that the subject is limited. A comprehensive treatment of organic chemistry would take a lifetime, and a reasonably complete survey would require a full-year course. For a more complete treatment of the subject, we urge you to refer to either of the first two references at the end of this chapter.

SUGGESTIONS FOR FURTHER READING

Roberts, S. D., and M. C. Caserio, *Basic Principles of Organic Chemistry*, New York: W. A. Benjamin, 1965

Morrison, R. T., and R. N. Boyd, *Organic Chemistry*, second edition, Boston: Allyn and Bacon, 1966

Conrow, K., and R. N. McDonald, *Deductive Organic Chemistry: A Short Course*, Reading, Mass.: Addison-Wesley, 1966

Cram, D. J., and G. S. Hammond, *Organic Chemistry*, second edition, New York: McGraw-Hill, 1962

House, H. O., *Modern Synthetic Reactions*, New York: W. A. Benjamin, 1965

Gutsche, C. David, *The Chemistry of Carbonyl Compounds*, Englewood Cliffs, N.J.: Prentice-Hall, 1957

QUESTIONS

1. Sketch diagrams illustrating chloroform ($CHCl_3$) in square-planar, rectangular-planar, and tetrahedral shapes. Do the same for CH_2Cl_2.

2. Write structures for propane, *n*-butane, isobutane, and *n*-pentane which illustrate the three-dimensional aspects of the molecules.

3. Write structures for the nine isomers of heptane (C_7H_{16}), and, using the information available in Appendix C, name each isomer.

4. Use equations to illustrate the mechanism of the reaction of chloroform ($CHCl_3$) with Cl_2 under the influence of light.

5. The production of HCl by the reaction of hydrogen gas with chlorine is known to be a free-radical chain reaction. The Cl—Cl bond is easier to break than the H—H bond. Postulate a mechanism for the reaction.

6. Free-radical chain reactions, such as the chlorination of methane, are not 100% efficient. The formation of one chlorine atom is not enough to bring about complete reaction of Cl_2 and CH_4. In other words, the chain must be broken. Can you think of any reactions of the species involved in the chlorination of methane that would break the chain? [*Hint:* Any reaction of a free radical that does not produce another free radical will terminate the chain.]

7. Sketch diagrams which illustrate the three-dimensional aspects of the molecules for ethylene, propylene, and 1-butene.

8. There are three isomeric alkenes with the formula $C_2H_2Cl_2$. Write structures for these, and predict the direction of the dipole moment for each structure.

9. Write equations that illustrate the following conversions.

   a) CH_3CH_2Br to ethylene

   b) $CH_3CH_2CH_2CH_2CH_2Br$ to 1-pentene

   c) $CH_3-\overset{\overset{\displaystyle CH_3}{|}}{\underset{\underset{\displaystyle Br}{|}}{C}}-CH_3$ to isobutylene

10. Predict the products of the following reactions.

   a) $CH_3-\overset{\overset{\displaystyle H}{|}}{\underset{\underset{\displaystyle OH}{|}}{C}}-CH_3 \xrightarrow{H_2SO_4}$

   b) $CH_3-\overset{\overset{\displaystyle CH_3}{|}}{\underset{\underset{\displaystyle CH_3}{|}}{C}}-CH_2CH_2Br \xrightarrow{KOH}$

   c) —OH $\xrightarrow{H_2SO_4}$

d) 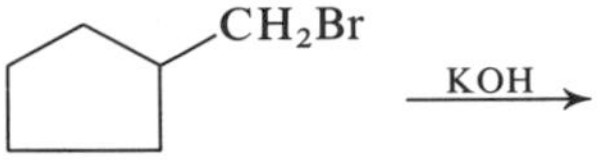 CH₂Br → KOH →

e) 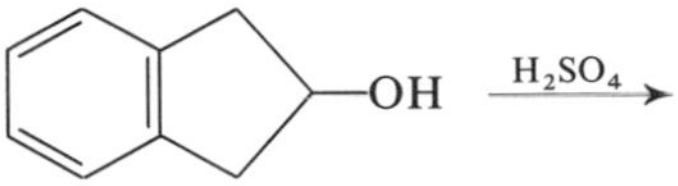 —OH → H₂SO₄ →

f) Products of parts (b), (d), and (e) + H_2 —Pt→

11. The product of the addition of hydrogen halides and other acids to an alkene can be predicted by *Markovnikov's rule*, which states that the proton will add to the carbon of the double bond which has the greater number of hydrogens attached to it. Based on Markovnikov's rule, predict the products of the following.

a) Propene + H_2SO_4 →
b) Isobutylene + HCl →
c) 1-butene + HI →
d) Cyclohexene + HBr →

e) 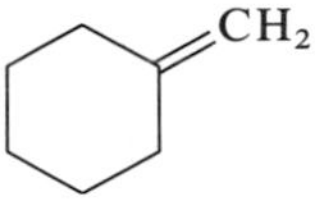 =CH₂ + HI →

12. How many resonance structures of the Kckulé type shown on page 461 can you write for naphthalene, phenanthrene, anthracene, and azulene?

13. Calculate the hydrogen ion concentrations in a liter each of phenol, water, and ethanol, using the dissociation constants given on page 464.

14. Which of the following would you expect to react rapidly with $ZnCl_2/HCl$?

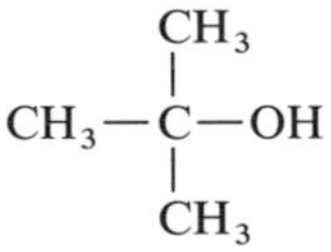

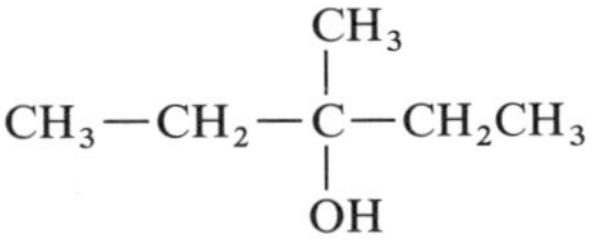

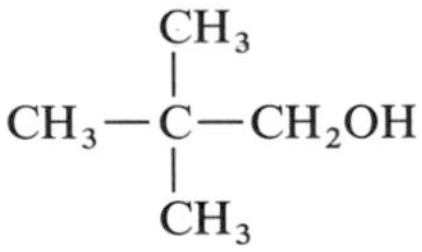

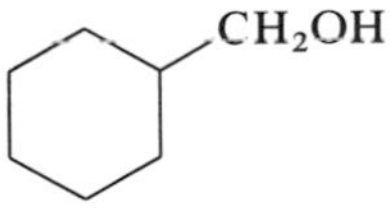

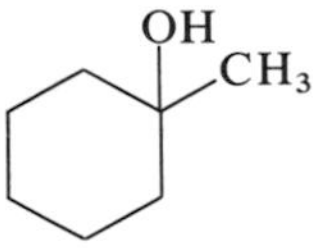

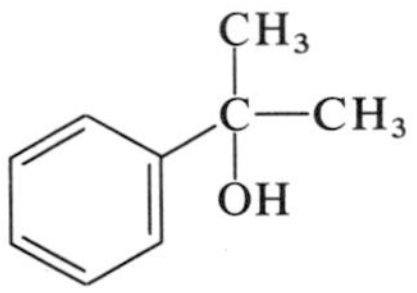

$$CH_3-\underset{\underset{CH_3}{|}}{\overset{\overset{CH_3}{|}}{C}}-OH$$

$$CH_3-CH_2-\underset{\underset{OH}{|}}{\overset{\overset{CH_3}{|}}{C}}-CH_2CH_3$$

$$CH_3-\underset{\underset{CH_3}{|}}{\overset{\overset{CH_3}{|}}{C}}-CH_2OH$$

CH₂OH

OH CH₃

C—CH₃, CH₃, OH

15. Predict the organic products of the following reactions.

a) $CH_3CH_2OH + H_2SO_4$ —Heat→

b) 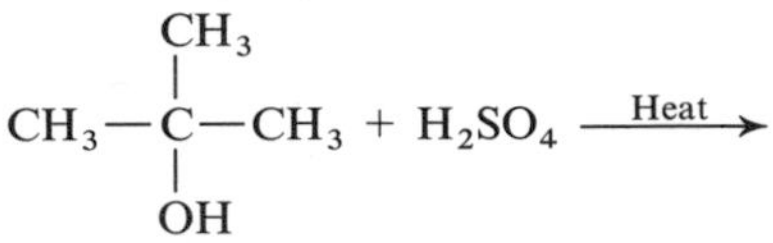 $CH_3-\underset{\underset{OH}{|}}{\overset{\overset{CH_3}{|}}{C}}-CH_3 + H_2SO_4$ —Heat→

c) 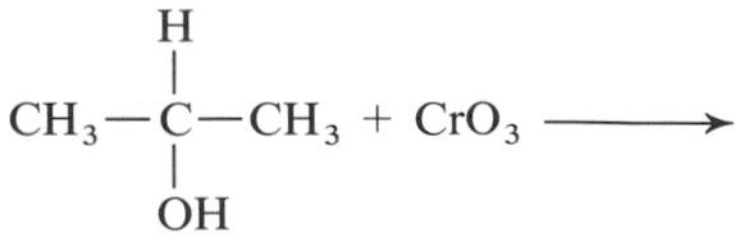 $CH_3-\overset{\displaystyle H}{\underset{\displaystyle OH}{C}}-CH_3 + CrO_3 \longrightarrow$

d) $CH_3CH_2CH_2CH_2OH + CrO_3 \longrightarrow$

e) 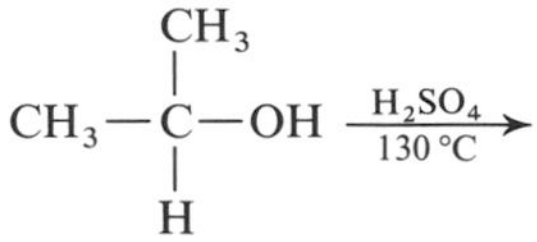 $CH_3-\overset{\displaystyle CH_3}{\underset{\displaystyle H}{C}}-OH \xrightarrow[130\ °C]{H_2SO_4}$

f) 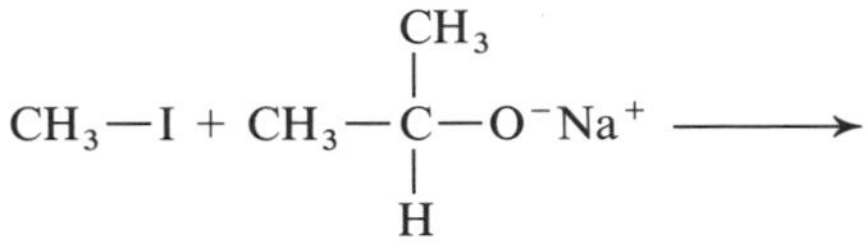 $CH_3-I + CH_3-\overset{\displaystyle CH_3}{\underset{\displaystyle H}{C}}-O^-Na^+ \longrightarrow$

16. Write equations for a plausible mechanism for the conversion of acetone to its ketal with ethanol and anhydrous HCl.

17. Compound A decolorizes solutions of Br_2 and $KMnO_4$. The hydration of compound A with aqueous H_2SO_4 gives compound B. Oxidation of compound B under carefully controlled conditions gives compound C, which reduces Benedict's and Tollen's reagents. Oxidation of compound C or vigorous oxidation of compound B gives compound D. Write structures for A, B, C, and D.

18. Would you expect

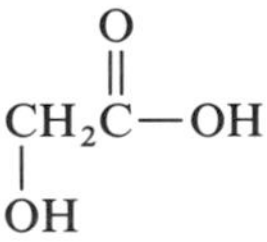

$$CH_2\overset{\displaystyle O}{\overset{\displaystyle \|}{C}}-OH$$
$$\underset{\displaystyle OH}{|}$$

to be a stronger or weaker acid than acetic acid? Explain your answer.

19. Write structures for the organic products of the following reactions.

a) 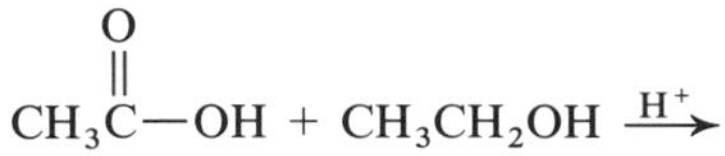 $CH_3\overset{\displaystyle O}{\overset{\displaystyle \|}{C}}-OH + CH_3CH_2OH \xrightarrow{H^+}$

b) $CH_3CH_2\overset{\displaystyle O}{\overset{\displaystyle \|}{C}}-OH + CH_3OH \xrightarrow{H^+}$

c) $CH_3\overset{\displaystyle O}{\overset{\displaystyle \|}{C}}-O-\overset{\displaystyle O}{\overset{\displaystyle \|}{C}}CH_3 + CH_3CH_2OH \longrightarrow$

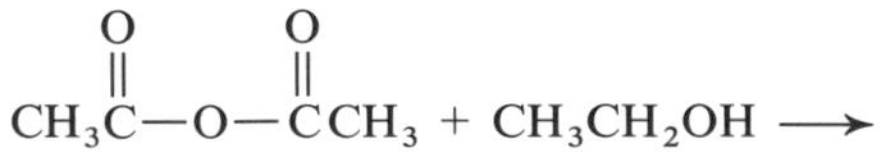

d) 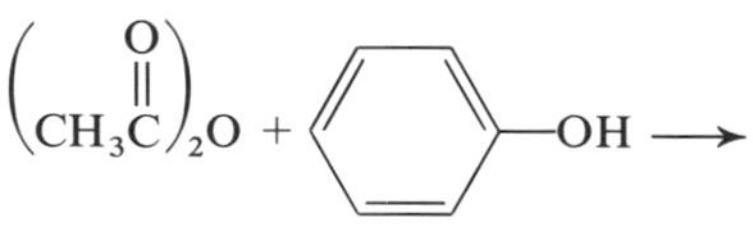 $\left(CH_3\overset{\displaystyle O}{\overset{\displaystyle \|}{C}}\right)_2O + $ ⬡—OH $\longrightarrow$

e) $CH_3-\overset{O}{\overset{||}{C}}-O-\overset{O}{\overset{||}{C}}-CH_3 + H_2O \longrightarrow$

f) $CH_3O-\overset{O}{\overset{||}{C}}CH_2CH_2CH_3 + H_2O \xrightarrow{H^+}$

g) $-\overset{O}{\overset{||}{C}}-OCH_2CH_3 + H_2O \xrightarrow{H^+}$

h) $-NH_2 + \left(CH_3\overset{O}{\overset{||}{C}}-\right)_2 O \longrightarrow$

i) $(CH_3)_2N-$ $+ CH_3I \longrightarrow$

20. Arrange the following sets of amines in order of increasing basicity.

a) CH_3NH_2, NH_3, $(CH_3)_2NH$, $(CH_3)_3N$

b) NH_3, CH_3NH_2, $-NH_2$

17. STEREOCHEMISTRY AND REACTION MECHANISMS

17-1 INTRODUCTION

We can explain the chemical properties of some compounds by considering electronic factors. For example, the —OH group in secondary alcohols introduces a dipole into the molecule, which makes it possible for the molecule to undergo certain reactions. The acid-catalyzed dehydration of a secondary alcohol to an alkene (Fig. 17–1) is due in part to the attraction which the oxygen atom in the —OH group has for a proton. This is an electronic factor; that is, the oxygen with its high electronegativity has a slight negative charge, and thus readily attracts a proton.

Other compounds may exhibit properties in various reactions which cannot be entirely explained by electronic influences. Careful experimentation has shown that many properties of compounds may be determined instead by the *spatial arrangement* of the atoms in the molecule, the *geometric orientation*. The study of the spatial arrangements of atoms in a molecule and the effect of this geometry on the properties of a compound is called *stereochemistry*. In this chapter we shall study various compounds, with the emphasis on the three-dimensional nature of the molecules.

17-2 CARBON–CARBON SINGLE-BOND ROTATION AND CONFORMATION

The simplest compound containing a carbon–carbon single bond is ethane. We can consider ethane as two tetrahedra joined at the points. To help us visualize the

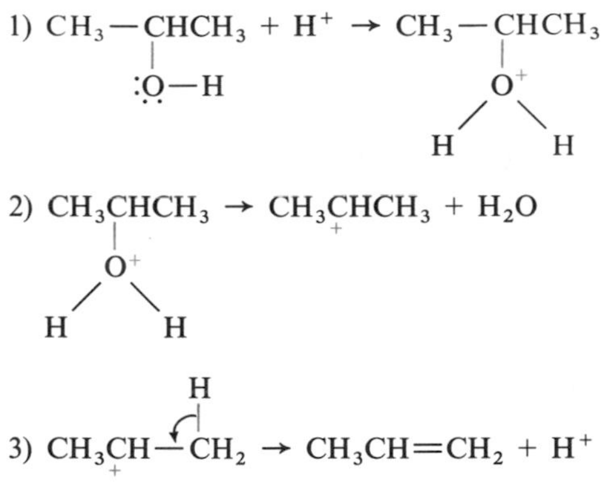

1) $CH_3-CHCH_3 + H^+ \rightarrow CH_3-CHCH_3$

  $:\underset{\cdot\cdot}{O}-H$ $\overset{+}{O}$
                                                   H H

2) $CH_3CHCH_3 \rightarrow CH_3\underset{+}{C}HCH_3 + H_2O$

  $\overset{+}{O}$
  H H

                          H
3) $CH_3CH\overset{}{\underset{+}{-}}CH_2 \rightarrow CH_3CH{=}CH_2 + H^+$

Fig. 17–1 Mechanism of the acid-catalyzed dehydration of an alcohol.

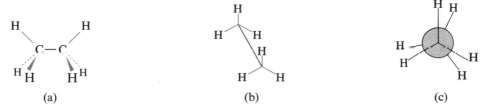

(a)   (b)   (c)

Fig. 17–2 Three-dimensional representations of the ethane molecule.

geometrical relationships between the atoms and the bonding electrons in the ethane molecule, we can refer to various kinds of three-dimensional diagrams.*

Figure 17–2 shows three of these. In Fig. 17–2a, the dashed lines represent bonds sticking out behind the paper and the wedges represent bonds in front of the paper. The solid lines are in the plane of the paper.

Figure 17–2b is a *sawhorse* representation of the ethane molecule. The long line represents the carbon–carbon single bond (lengthened for visual convenience); the carbon atoms are considered to be at the points where the bonds from the hydrogens converge, but the symbol for carbon is omitted for simplicity. In the sawhorse projection, it is easier to see the angular relationship of one set of hydrogen atoms to the other.

Figure 17–2c shows a *Newman projection*, in which the viewer is looking along the direction of a carbon–carbon bond; the circle represents the electrons in this bond. The three bonds which extend to the center of the circle represent the hydrogen–carbon bonds of the methyl group closer to the viewer. The point of convergence of these three bonds indicates the position of the carbon atom of this methyl (CH_3)

*   A set of molecular models, if available, will show the three-dimensional orientation of these molecules even more clearly.

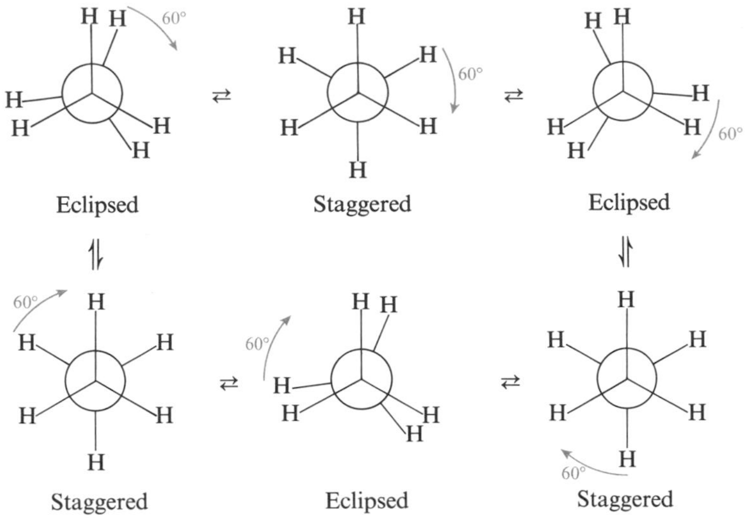

Fig. 17–3 Conformations of the ethane molecule, illustrated with Newman projections.

group. The bonds from the three hydrogens to the perimeter of the circle belong to the methyl group away from the viewer; the carbon atom is obscured by the electrons in the C—C bond. Figure 17–2c is meant to represent an ethane molecule in which the hydrogens of the rear methyl group are directly behind those of the front methyl group; it is necessary to show the rear hydrogens at a slight angle so they may be seen. Newman projections make it very easy to see the angles between the C—H bonds in the first methyl group and the C—H bonds in the second methyl group. The orientation that one methyl group assumes with respect to the second is called a *conformation*. In principle, ethane can have an infinite number of conformations, but we need consider only two.

Figure 17–3 shows different conformations which illustrate the rotation of one methyl group relative to the other; that is, rotation about the carbon–carbon single bond. In the Newman projection in Fig. 17–2c the hydrogens in each methyl group are exactly opposite one another. If the forward methyl group is held stationary and the rear methyl group is rotated by 60°, a new conformation is obtained, with the hydrogens no longer in line with respect to one another. After six of these 60° rotations, the original conformation is restored. In the course of this rotation the molecule passes through three conformations, in which the hydrogen atoms and the bonding electrons of one methyl group are at a maximum distance from the hydrogen atoms and bonding electrons of the other methyl group. These are called *staggered conformations*, and are identical in energy. The other three conformations (in which the hydrogens and their bonding electrons are as close as possible) are also equal in energy, and are called *eclipsed conformations*.

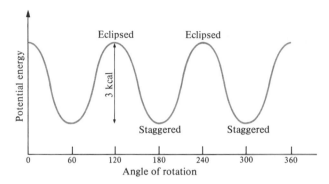

Fig. 17–4 Variation of potential energy with rotation about the carbon-carbon bond in the ethane molecule.

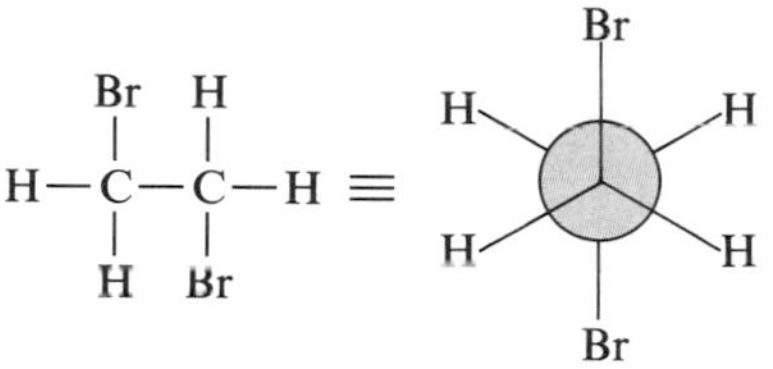

$$
\begin{array}{ccc}
\text{Br} & \text{H} \\
| & | \\
\text{H}-\text{C}-\text{C}-\text{H} \\
| & | \\
\text{H} & \text{Br}
\end{array}
\equiv
$$

Fig. 17–5 The most stable conformation of 1-2-dibromoethane.

Chemists used to think that the rotation about the carbon–carbon bond in ethane was "free," i.e., that there was no energy barrier to rotation. This is equivalent to saying that the energies of the staggered and eclipsed forms of ethane are equal. Calculations have shown, however, that there *is* an energy barrier to rotation about the C—C bond in ethane, a barrier of about 3 kcal/mole. That is, the eclipsed and staggered conformations are not equal in energy. But this is not a great amount of energy, and at ordinary temperatures the rotation can occur readily. As Fig. 17–4 shows, the staggered conformation is more stable than the eclipsed one.

▶ The precise reason for the higher energy of the eclipsed conformation in ethane is not known. Presumably, electron-electron repulsions between the C—H bonding electrons in the methyl groups are primarily responsible. ◀

If the hydrogens in ethane are replaced by larger atoms, the barrier to rotation increases. In 1,2-dibromoethane, for example, x-ray data show that in the solid state only one conformation is present: that in which the Br atoms are as far apart as possible (Fig. 17–5).

Hindered Rotation. The alkenes exhibit much greater barriers to rotation (25–65 kcal/mole) than the alkanes. This is to be expected, as we can see if we consider

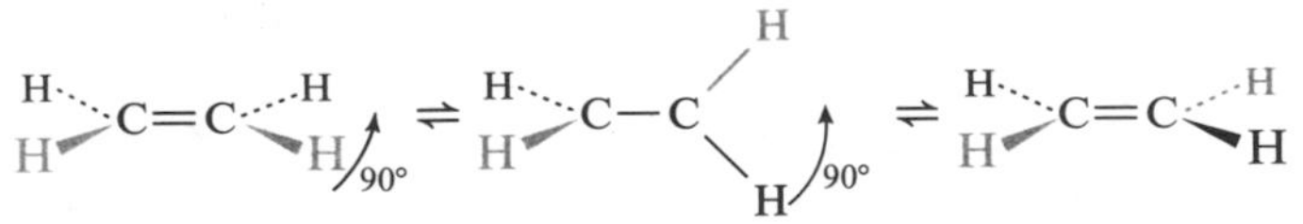

Fig. 17-6 Schematic representation of π-bond cleavage during rotation about carbon-carbon single bond.

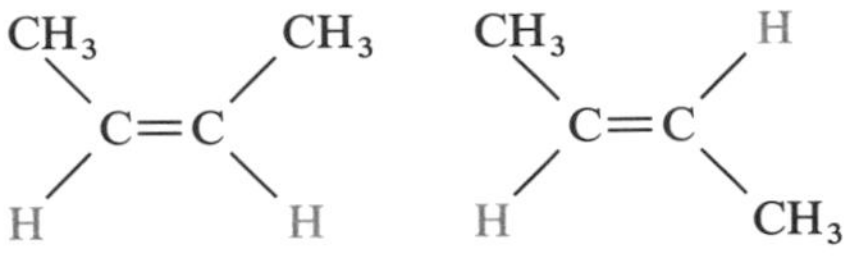

Fig. 17-7 Geometric isomers of 2-butene.

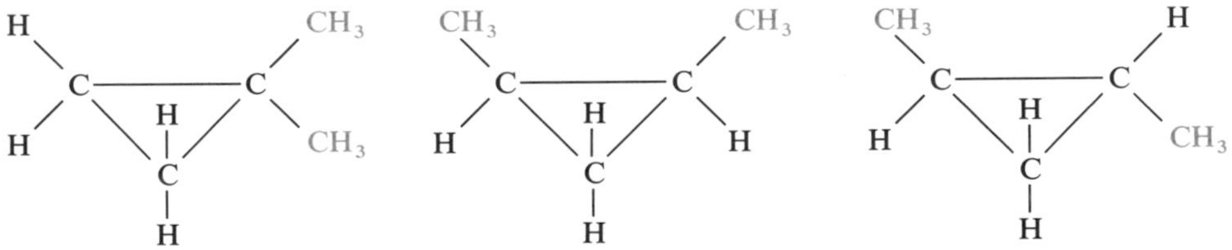

cis-2-butene   trans-2-butene

1,1-dimethylcyclopropane   cis-1,2-dimethylcyclopropane   trans-1,2-dimethylcyclo-
propane

Fig. 17-8 Three isomeric dimethylcyclopropanes.

the case of ethylene. Figure 17–6 shows that rotation about the double bond in ethylene requires the breaking of the π-bond rather than a simple crowding of atoms. This process does not take place at room temperature, since it is a relatively high-energy process.

The consequences of the high barrier to rotation can be illustrated with 2-butene, $CH_3CH=CHCH_3$. We can write the structure of 2-butene in two ways: with the methyl groups on the same side of the double bond or on opposite sides of the double bond (Fig. 17–7). If, as we have mentioned, the geometry of atoms in a molecule influences the properties of the molecule, these two structures should exhibit different properties. In other words, the lack of rotation about the double bond in 2-butene is an example of a new type of isomerism; this is called *geometric isomerism*. Both 2-butenes are known compounds. The butene with both methyl groups on the same side of the double bond is called *cis*, while the compound with opposed methyl groups is called *trans* (Fig. 17–7).

Just as the addition of a second bond between two atoms prevents rotation of the atoms about the bonds, incorporation of a bond into a ring of atoms can effectively prevent rotation about the bond. Thus geometric isomerism is found in the cycloalkane series as well as in the alkene series. For example, Fig. 17–8 shows three different dimethyl cyclopropanes. Table 17–1 lists these and other examples of geometric isomers.

Table 17–1. Structures and properties of representative geometric isomers

Name	Formula	Melting point, °C	Boiling point, °C
cis-2-butene	CH_3 ⟍ CH_3 ╱ C=C ╱ H ⟍ H	−139	4
trans-2-butene	CH_3 ⟍ H ╱ C=C ╱ H ⟍ CH_3	−106	1
cis-2-pentene	CH_3 ⟍ CH_2CH_3 ╱ C=C ╱ H ⟍ H	−151	37
trans-2-pentene	CH_3 ⟍ H ╱ C=C ╱ H ⟍ CH_2CH_3		36
cis-1,2-dichloroethene	Cl ⟍ Cl ╱ C=C ╱ H ⟍ H	−80	60
trans-1,2-dichloroethene	Cl ⟍ H ╱ C=C ╱ H ⟍ Cl	−50	48

(continued)

Table 17-1 (*continued*)

Name	Formula	Melting point, °C	Boiling point, °C
cis-1,2-dimethylcyclopentane	CH₃ ... CH₃ H ... H	− 62	99
trans-1,2-dimethylcyclopentane	CH₃ ... H H ... CH₃	− 120	92
cis-1,2-cyclopentanediol	OH ... OH H ... H		30
trans-1,2-cyclopentanediol	OH ... H H ... OH		55

17-3 STEREOCHEMISTRY AND REACTION MECHANISMS

Knowledge of stereochemistry such as the above is useful in interpreting the mechanism of certain reactions. For example, when we add one mole of bromine to acetylene, we obtain *trans*-1,2-dibromoethylene as the major product:

$$H-C{\equiv}C-H \xrightarrow{\ Br_2\ } \quad \begin{matrix} Br \\ \diagdown \\ \end{matrix}\, C{=}C \,\begin{matrix} H \\ \diagup \\ \end{matrix}$$

It is assumed that the addition of Br_2 to a double bond involves the attachment of one bromine atom followed by the attachment of the second bromine atom on the opposite side of the molecule. This reaction is a general one for carbon–carbon double bonds.

We can illustrate this reaction mechanism, and show that it is consistent with the stereochemical data, by considering the reaction between bromine and cyclo-pentene. The current interpretation of the mechanism of this reaction proposes the initial formation of a complex between a bromine molecule and the π-electrons of

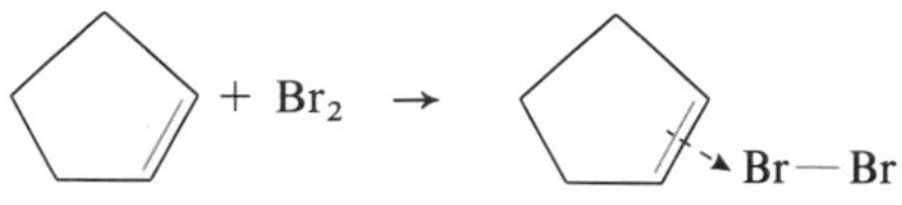

Cyclopentene Complex

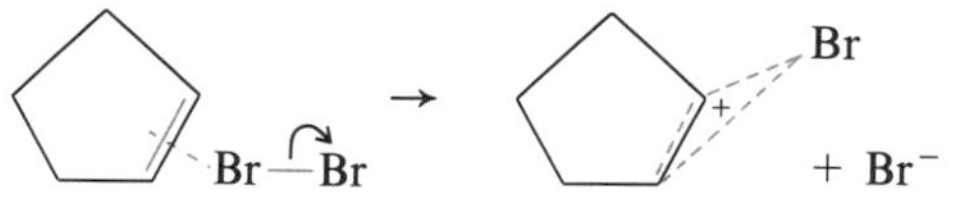

Cyclic bromonium ion

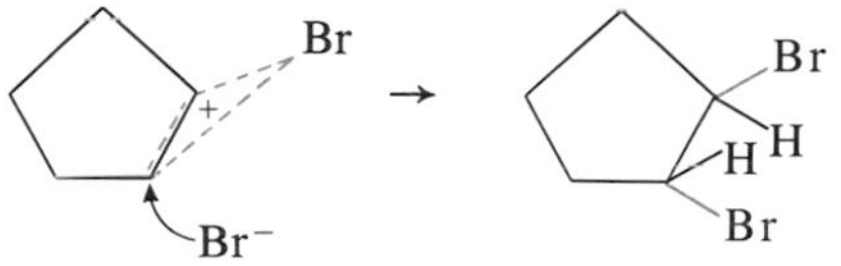

trans-1,2-dibromocyclopentane

Fig. 17-9 Mechanism for bromination of an alkene.

the double bond in cyclopentene. This complex collapses to give a bromide ion and an unstable species in which the bromine is bonded in a three-membered ring to the carbon atoms. Although an ion of this type (called a *cyclic* bromonium ion) has never been isolated, chemists have postulated its transient existence in order to explain the phenomenon of *trans* addition. Attack of bromide ion is blocked on one side by the presence of the cyclic bromonium ion and must proceed from the other side of the molecule, resulting in the *trans* product, as shown in Fig. 17-9.

▶ We often speak of one reagent attacking another in a chemical reaction. In this context the word *attacking* is meant to describe the entire process in which one reagent approaches another, the electron clouds begin to mix and become distorted, and finally bonds are broken and formed. ◀

Most addition reactions of alkenes give products resulting from *trans* addition or a mixture of *cis* and *trans* products, with the *trans* product predominating:

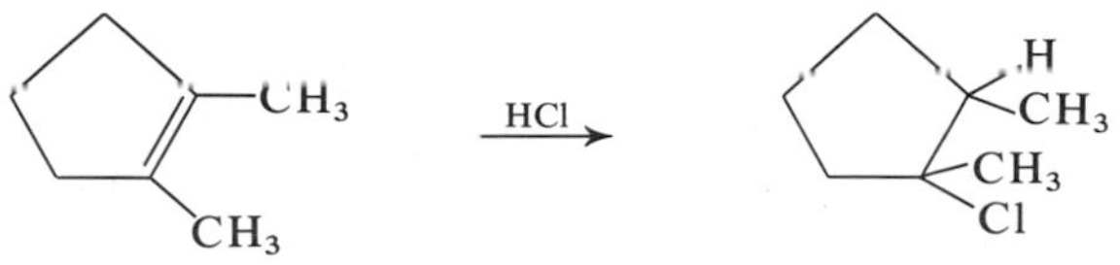

1,2-dimethylcyclopentene *trans*-1,2-dimethylchlorocyclopentane

A few addition reactions of alkenes give products resulting from specific *cis* (from the same side) addition. Such reactions must therefore proceed by a mechanism quite different from that postulated above for *trans* addition reactions. In many cases

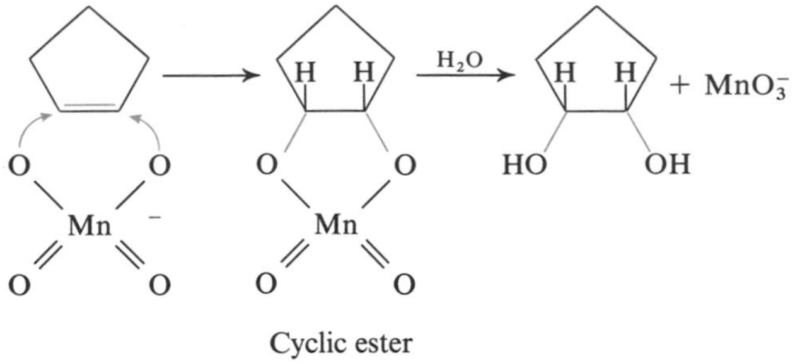

Fig. 17–10 Proposed mechanism for the *cis*-hydroxylation of a double bond.

Cyclic ester

there may be cyclic intermediates. For example, the permanganate test for alkenes, described in Chapter 16, proceeds via a *cis* addition exclusively:

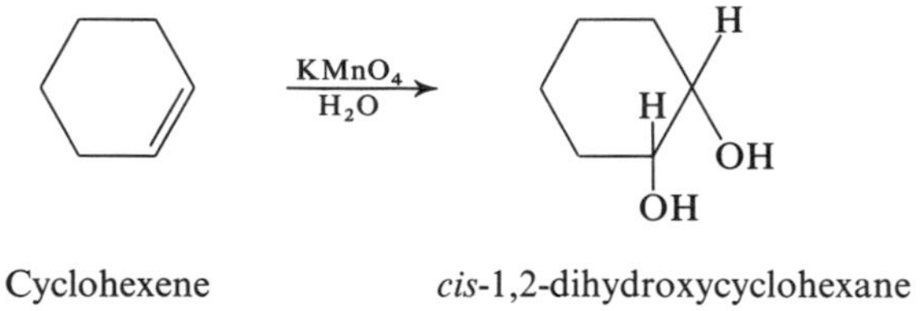

$$\xrightarrow[\text{H}_2\text{O}]{\text{KMnO}_4}$$

Cyclohexene   *cis*-1,2-dihydroxycyclohexane

The mechanism for this reaction must take this into account. The currently accepted view is that initially there is formed a cyclic manganese ester, which is unstable in water and which is hydrolyzed to the *cis* addition product. The cyclic manganese ester can form only with both oxygens bonded on the same side of the cyclopentane ring, as shown in Fig. 17–10. Bonding of one oxygen below the ring and the other above the ring in a *trans* manner would require inordinate stretching of the bonds involved. This illustrates an important point: We can use knowledge of the mechanism of a reaction to predict the stereochemistry of the product of the reaction. We shall talk more about the application of stereochemical principles to reaction mechanisms in Section 17–5.

17–4 OPTICAL ISOMERISM

Plane Polarized Light. Materials which have the capability of rotating plane polarized light are called *optically active*. For many biologically important materials there seems to be a connection between optical activity and biological activity. Therefore a prerequisite to the study of biochemistry is a discussion of optical activity. And before we can discuss optical activity, we must briefly describe plane polarized light.

Ordinary light is composed of electromagnetic radiation of various wavelengths. If we pass such light through a prism (Fig. 17–11), the different wavelengths are spread apart and we can obtain light of one particular wavelength (recall Chapter 5).

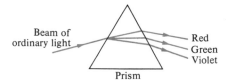

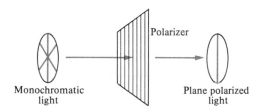

Fig. 17–11 A monochromator separating light into component wavelengths.

Fig. 17–12 The formation of plane polarized light.

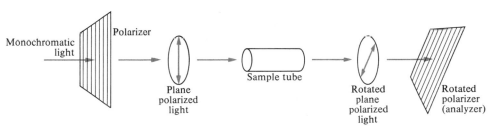

Fig. 17–13 A polarimeter.

Light of a uniform wavelength is called *monochromatic light*, and the device used to obtain it is called a *monochromator*. Monochromatic light may be considered to be vibrating in all possible directions, but when this light is passed through a polarizer (Fig. 17–12), only the light vibrating in one plane passes through the polarizer. (Ordinary light is also polarized in this way. Polaroid sunglasses do this.)

▶ This description of light vibrating in one plane and being polarized in one plane is an oversimplification. The actual description of polarized light is much more complex, and very difficult to understand. But the simplified description of plane polarized light will suit our purposes. ◀

This *plane polarized light* will pass through a second polarizer only if the second polarizer is lined up parallel to the plane of vibration of the light waves. With this fact in mind, we can construct a simple instrument called a *polarimeter* (Fig. 17–13), which measures the extent to which a given optically active material rotates plane polarized light.

In the polarimeter, monochromatic light, often obtained from a sodium or mercury lamp, is passed through a polarizer and then through the sample to be measured. The degree and direction of rotation of the emerging light is determined by measuring the extent to which a second polarizer (called the *analyzer*) must be rotated in order to allow the maximum amount of light to pass.

Requirements for Optical Activity

In order to have optical activity, it is necessary to have *dissymmetry*. An object which is dissymmetric cannot be superimposed on its mirror image. For example, the left hand can be considered to be a crude mirror image of the right hand. If you place your left hand palm down on the back of your right hand, the hands do not superimpose. Thus the hands are dissymmetric.

On the molecular level, in order to predict whether a compound is optically active or not, we apply this same criterion of dissymmetry. We construct a model of the molecule and its mirror image, and try to superimpose them. If the mirror images are nonsuperimposable, then the molecules represented by the models cannot be identical, and must be capable of independent existence.

If we were to isolate separate examples of the compounds represented by the molecular models, we would find that a sample of one compound would rotate plane polarized light a certain number of degrees to the right and the second compound would rotate the light *an equal amount in the opposite direction*, to the left.

The first compound is called *dextrorotatory* and its optical isomer is called *levorotatory*. Optical isomers of this type are called *enantiomers*.

If we take a certain amount of the dextrorotatory compound and mix it with an equal amount of the levorotatory enantiomer, the rotations cancel and no effect on plane polarized light will be seen. Such a mixture is referred to as a *racemic mixture* or a *dl pair*.

Optical Activity Due to Restricted Rotation

We can apply the criterion for optical activity to the study of rotational energy barriers. Figure 17–14 shows Newman projections of *n*-butane, illustrating rotation about the central carbon–carbon single bond. Only the staggered forms are shown, in their three different conformations. The conformations in which the methyl groups are adjacent are called *gauche* conformations; the conformation in which the methyl groups are as far apart as possible is called *anti*.

Figure 17–15a depicts "sawhorse" projections of the two *gauche* forms, showing that they are nonsuperimposable mirror images, while the mirror image of the *anti* form *is* superimposable.

Thus it should theoretically be possible to separate the gauche forms into optically active compounds. In practice this has not been possible. The energy barrier to rotation about the central C—C bond in butane is so low that one *gauche* form is readily converted to the other (giving a *dl* pair) or to the optically inactive *anti* form. However, the principle is sound. If the energy barrier to rotation could be increased

CH₃CH₂CH₂CH₃

n-butane

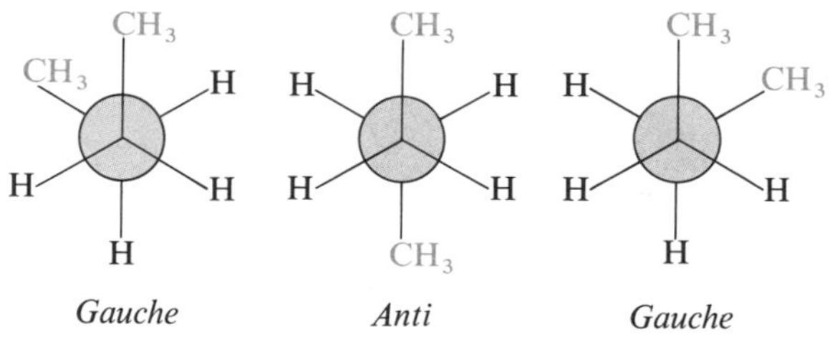

Fig. 17–14 Newman projections of rotational conformers of *n*-butane.

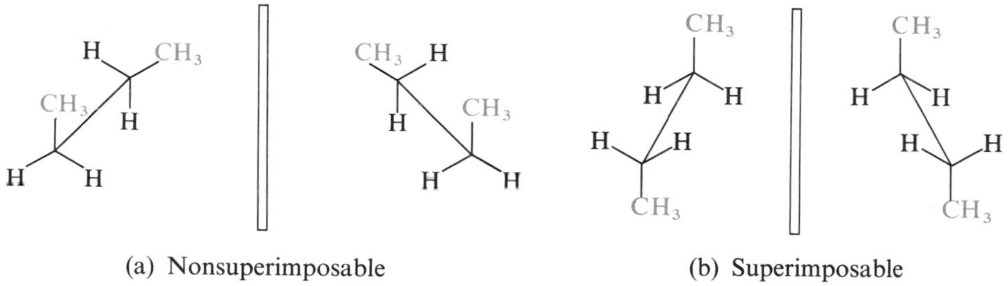

Gauche   *Anti*   *Gauche*

(a) Nonsuperimposable   (b) Superimposable

Fig. 17–15 Mirror-image representations of the *gauche* and *anti* conformers of *n*-butane.

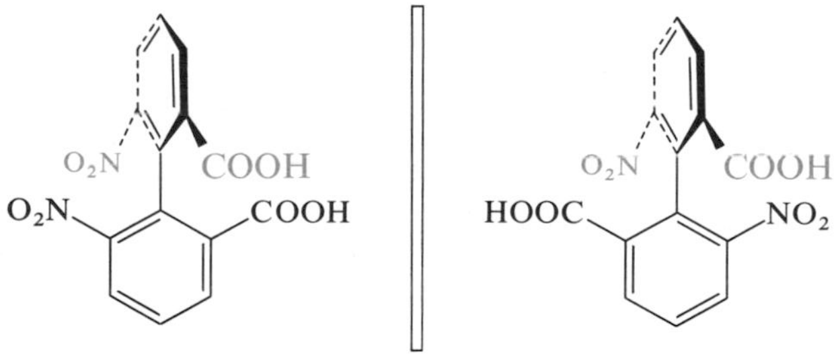

Fig. 17–16 Nonsuperimposable mirror images of a biphenyl derivative (o,o-dinitrodiphenic acid).

enough to prevent the atoms from rotating about a carbon–carbon single bond and if the rotational conformers were dissymmetric, it should be possible to isolate the optically active enantiomers. This has been done in the biphenyl series (see Fig. 17–16). The benzene rings in each isomer are mutually perpendicular and are prevented from becoming coplanar by the bulk of the nitro ($-NO_2$) and carboxyl ($-COOH$) groups. This effectively prevents rotation about the C—C bond connecting the two rings and makes possible the separation of the optical isomers. If either

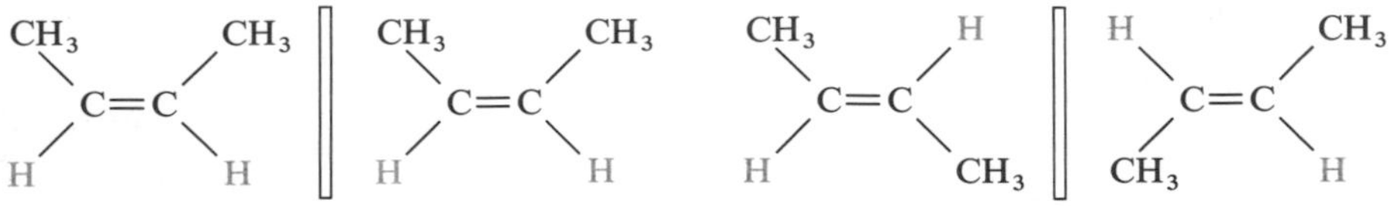

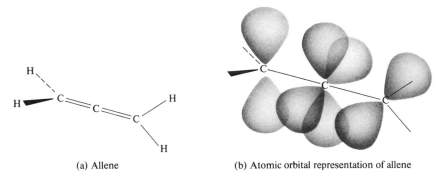

(a) *cis*-2-butene   (b) *trans*-2-butene

Fig. 17–17 Superimposable mirror images of *cis*- and *trans*-2-butene.

isomer is heated sufficiently, rotation about the bond can occur and a *dl* mixture is formed. We can determine the rate at which this loss of optical activity takes place by measuring the change of rotation of plane polarized light with time. From such measurements we may calculate the rotational energy barrier.

Since we know that in alkenes the double bond between the two carbon atoms prevents rotation at ordinary temperatures, it is logical to ask whether optical isomerism is found in such compounds. Recalling that the sp² hybrid orbitals require planar geometry, we can see that simple alkenes cannot be dissymmetric. In Fig. 17–17, for example, the *cis* and *trans* 2-butenes are shown along with the corresponding mirror images. Since the mirror images of *cis* and *trans* 2-butene are superimposable, optical isomerism is not possible.

(a) Allene   (b) Atomic orbital representation of allene

Fig. 17–18 Structures illustrating the nonplanar geometry of allene.

In the rather unusual compounds called *allenes*, we have a different situation. As Fig. 17–18a shows, allene has a carbon atom participating in two different double bonds (like the bonding in carbon dioxide, $O=C=O$). The orbital diagram (Fig. 17–18b) shows that these double bonds must be at right angles to each other, since the p-orbitals which overlap to form the π-bonds must be mutually perpendicular. The carbon atoms at the ends are ordinary sp² hybridized alkene carbon atoms, while the central carbon atom exhibits alkyne-like sp-hybridization.

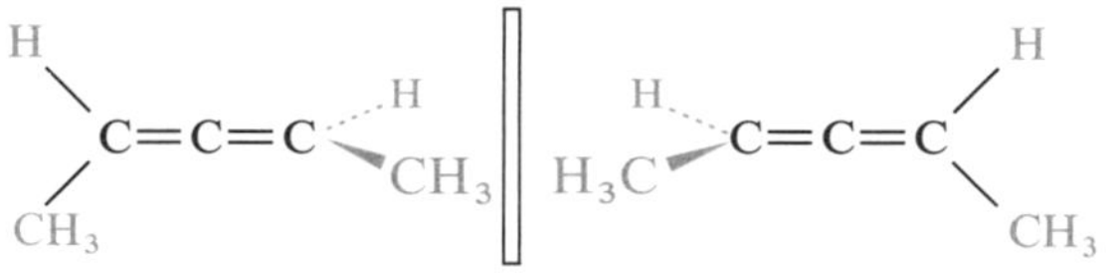

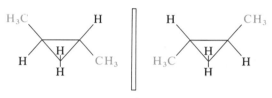

Fig. 17-19 Nonsuperimposable mirror images of 1, 3-dimethylallene.

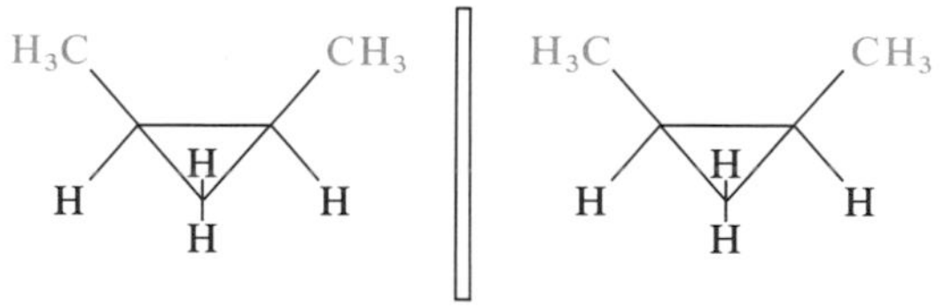

(a) *cis*-1,2-dimethylcyclopropane

(b) *trans*-1,2-dimethylcyclopropane

Fig. 17-20 Mirror-image representations of *cis*- and *trans*-1, 2-dimethylcyclopropane.

Figure 17–19 illustrates the structure of 1,3-dimethylallene. The bonding involved requires that the end groups be in a mutually perpendicular orientation like that observed in biphenyl isomerism (Fig. 17–16). As a consequence, 1,3-dimethylallene can be separated into two optically active isomers.

By the same token, the restricted bond rotations in cyclic systems can also introduce dissymmetry into the molecule. As illustrated with 1,2-dimethylcyclopropane in Fig. 17–20, the *cis* isomer can be superimposed on its mirror image, while the *trans* isomer cannot. Consequently, *trans*-1,2-dimethylcyclopropane must exist in *d* and *l* forms.

Optical Activity Due to Asymmetry

The most common and most important type of optical isomerism is that in saturated, sp³-hybridized, carbon atoms. As illustrated in Fig. 17–21, a carbon atom with *four different groups attached* is not superimposable on its mirror image and should be separable into dextrorotatory and levorotatory isomers.

A carbon atom with four different groups attached* is called *asymmetric*. Many optically active compounds occur in nature because they contain one or more asymmetric carbon atoms. The optical activity of asymmetric carbon atoms distinguishes

---

*   You should satisfy yourself that a single carbon atom containing any two identical groups cannot be optically active.

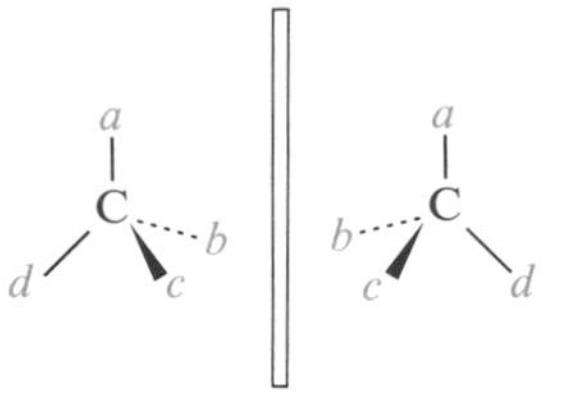

Fig. 17-21 Nonsuperimposable mirror image of a carbon atom containing four different groups.

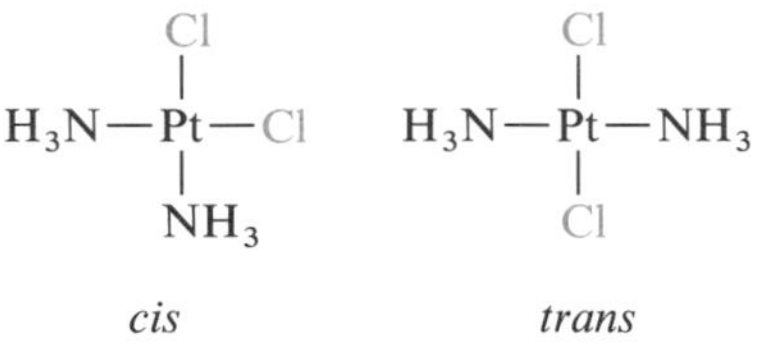

Fig. 17-22 Isomeric *cis*- and *trans*-dichlorodiammine platinum (II) complexes, illustrating square-planar geometry.

$$
\begin{array}{cc}
\text{Cl} & \text{Cl} \\
| & | \\
\text{H}_3\text{N}-\text{Pt}-\text{Cl} & \text{H}_3\text{N}-\text{Pt}-\text{NH}_3 \\
| & | \\
\text{NH}_3 & \text{Cl} \\
cis & trans
\end{array}
$$

between the two possible geometric orientations for a carbon atom: square planar or tetrahedral (recall Chapter 16). If a carbon atom had a square-planar arrangement, with the bonding electrons directed to the corners of a square, there would not be dissymmetry as a consequence of four different groups being attached to the carbon atom. That is, the mirror images would be superimposable. Thus the tetrahedral shape proposed by van't Hoff and LeBel explained the puzzling phenomenon of optical activity and successfully predicted the existence of optically active materials unknown at the time.

A similar stereochemical argument has been used to demonstrate the square-planar arrangement of certain transition metal complexes. The fact that a Pt(II) complex containing four different ligands has not been resolved into optical antipodes is negative evidence for the square-planar arrangement. Positive stereochemical evidence is found in the existence of the *two* dichlorodiammine platinum(II) complexes shown in Fig. 17-22. Geometric isomerism of this type would not be possible with a tetrahedral orientation.

The Nomenclature of Optically Active Compounds

Figure 17-23 shows two representative *dl* pairs. The minus sign indicates the levorotatory isomer and the plus sign the dextrorotatory isomer. An example of the way living organisms distinguish between optical isomers is *lactic acid*, which is present in soured milk, as well as in muscle tissue after strenuous exercise. Investigation shows that the lactic acid in the milk is exclusively the dextrorotatory isomer, while that found in fatigued muscles is the levorotatory isomer.

Figure 17-23 distinguishes between the optical isomers of lactic acid and of alanine by indicating the sign of the rotation of plane polarized light. However, there is no obvious correlation between the sign of rotation and the configuration of the asymmetric carbon atom. Since in many cases we need to know the configuration of

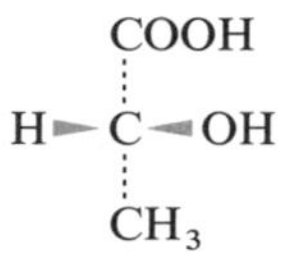

COOH   COOH

H►C◄OH   HO►C◄H

CH₃   CH₃

(−)-lactic acid 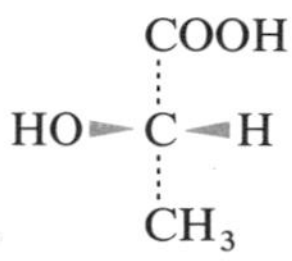   (+)-lactic acid

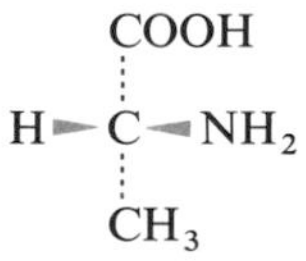

COOH   COOH

H►C◄NH₂   H₂N►C◄H

CH₃   CH₃

(−)-alanine 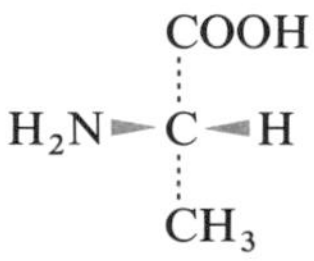   (+)-alanine

Fig. 17–23 Common optically active substances containing an asymmetric carbon atom.

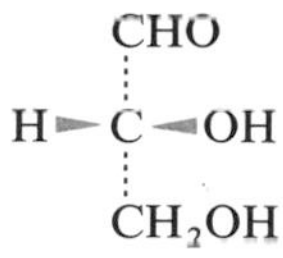

CHO   CHO

H►C◄OH   HO►C◄H

CH₂OH   CH₂OH

D(+)-glyceraldehyde 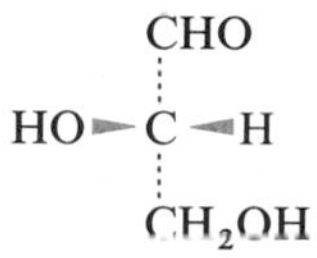   L(−)-glyceraldehyde

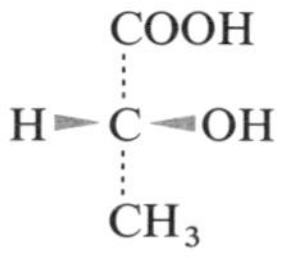

COOH   COOH

H►C◄OH   HO►C◄H

CH₃   CH₃

D(−)-lactic acid 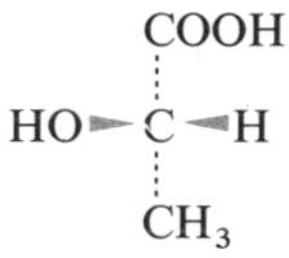   L(+)-lactic acid

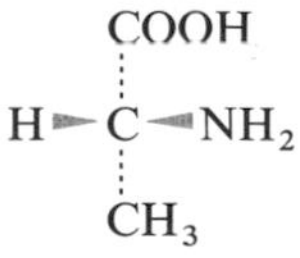

COOH   COOH

H►C◄NH₂   H₂N►C◄H

CH₃   CH₃

D(−)-alanine 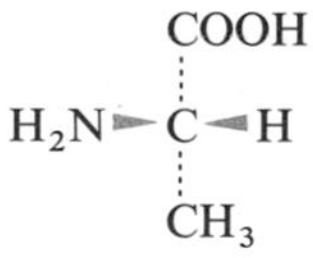   L(+)-alanine

Fig. 17–24 Glyceraldehyde-based nomenclature of optically active compounds.

a given optically active compound relative to a second similar compound, arbitrary methods for designating configurations have been devised.

One method is based on *glyceraldehyde*. Figure 17–24 shows the glyceraldehyde enantiomer in which the hydroxyl group is on the right while the more oxidized (aldehyde) group is at the top. This is called the D-*configuration*. The other enantiomer represents the L-*configuration*. Note that the designation of relative configuration says nothing about the sign of rotation.

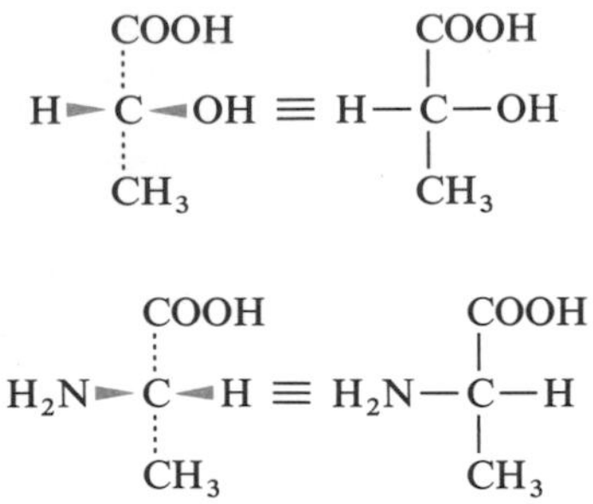

$$
\begin{array}{ccc}
\text{COOH} & & \text{COOH} \\
\text{H} \blacktriangleright \overset{\vdots}{\underset{\vdots}{\text{C}}} \blacktriangleleft \text{OH} & \equiv & \text{H} - \overset{|}{\underset{|}{\text{C}}} - \text{OH} \\
\text{CH}_3 & & \text{CH}_3
\end{array}
$$

$$
\begin{array}{ccc}
\text{COOH} & & \text{COOH} \\
\text{H}_2\text{N} \blacktriangleright \overset{\vdots}{\underset{\vdots}{\text{C}}} \blacktriangleleft \text{H} & \equiv & \text{H}_2\text{N} - \overset{|}{\underset{|}{\text{C}}} - \text{H} \\
\text{CH}_3 & & \text{CH}_3
\end{array}
$$

Fig. 17-25 Conventional planar representation of asymmetric carbon atoms.

We may also adopt a second convention when we deal with optically active compounds of the type under discussion. To avoid having to use wedges and dashed lines to indicate the tetrahedral nature of an asymmetric carbon atom, we can write the structure in planar form and *assume* that the groups on the top and the bottom project behind the plane of the paper, while those on the sides project in front of the plane of the paper. Even though we use these convenient planar representations, we must be careful to *think* of these structures in three dimensions (see Fig. 17-25).

Although the D and L nomenclature for D-glyceraldehyde is useful, particularly when it comes to similarly substituted compounds such as the α-hydroxy acids and the α-amino acids, its use is rather limited. More widely applicable systems of nomenclature have been devised, but they are often cumbersome. The safest way to indicate the configuration of asymmetric atoms in a molecule is to use a model or a drawing indicating the three-dimensional orientation of the molecule.

Compounds with More than One Asymmetric Carbon Atom

So far we have confined our discussion to molecules containing only one asymmetric carbon. When there is more than one in a molecule, the maximum number of different isomers is 2^n, where n is the number of asymmetric carbon atoms.

For example, we can write four different structures for 3-bromo-2-butanol (Fig. 17-26); there are two mirror-image pairs. The four isomers are all known as separate compounds and together satisfy the 2^n rule ($2^2 = 4$).

We can also write four structures for tartaric acid (Fig. 17-27), but when we analyze the situation we see that the first two structures represent a *dl* pair and the second two represent only one compound. That is, the mirror images are superimposable.

A single molecule which contains more than one asymmetric carbon atom can, if these asymmetric atoms are substituted with the same groups, have the tendency to rotate light in one direction canceled by an equal tendency to rotate light in the opposite direction. Such compounds, even though they contain asymmetric carbon

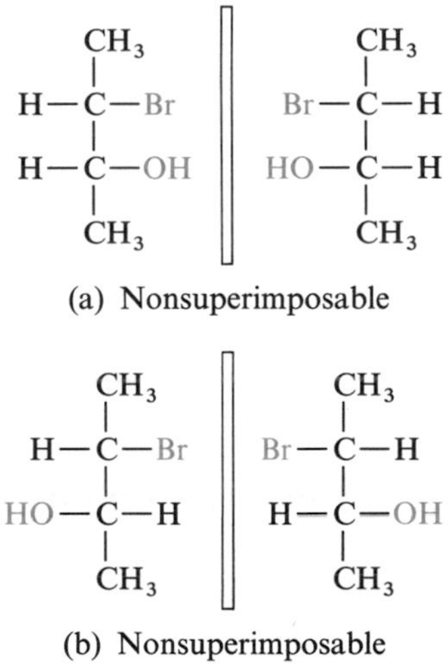

$$CH_3 \qquad\qquad\qquad CH_3$$
$$H-C-Br \qquad\qquad Br-C-H$$
$$H-C-OH \qquad\qquad HO-C-H$$
$$CH_3 \qquad\qquad\qquad CH_3$$

(a) Nonsuperimposable

$$CH_3 \qquad\qquad\qquad CH_3$$
$$H-C-Br \qquad\qquad Br-C-H$$
$$HO-C-H \qquad\qquad H-C-OH$$
$$CH_3 \qquad\qquad\qquad CH_3$$

(b) Nonsuperimposable

Fig. 17–26 Four isomeric 3-bromo-2-butanols represented as two *dl* pairs.

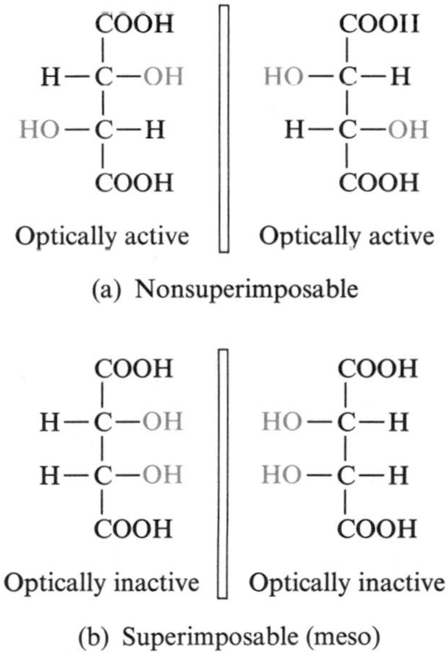

$$COOH \qquad\qquad\qquad COOH$$
$$H-C-OH \qquad\qquad HO-C-H$$
$$HO-C-H \qquad\qquad H-C-OH$$
$$COOH \qquad\qquad\qquad COOH$$

Optically active   Optically active

(a) Nonsuperimposable

$$COOH \qquad\qquad\qquad COOH$$
$$H-C-OH \qquad\qquad HO-C-H$$
$$H-C-OH \qquad\qquad HO-C-H$$
$$COOH \qquad\qquad\qquad COOH$$

Optically inactive   Optically inactive

(b) Superimposable (meso)

Fig. 17–27 Three isomeric tartaric acids, including one *dl* pair and a *meso* compound.

atoms, are not dissymmetric and therefore do not rotate plane polarized light. A compound which is symmetrical and which contains two or more asymmetric carbon atoms is called *meso*. Thus the maximum number (2^n) of optical isomers can be obtained only when no two of the asymmetric carbon atoms bear the same four substituents.

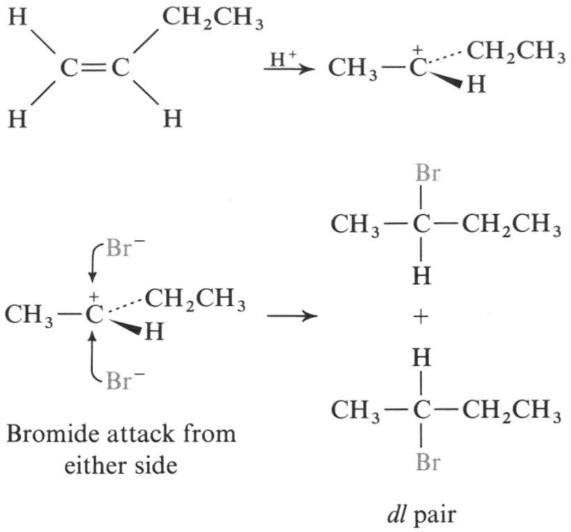

Fig. 17-28 Mechanism of the addition of HBr to an alkene.

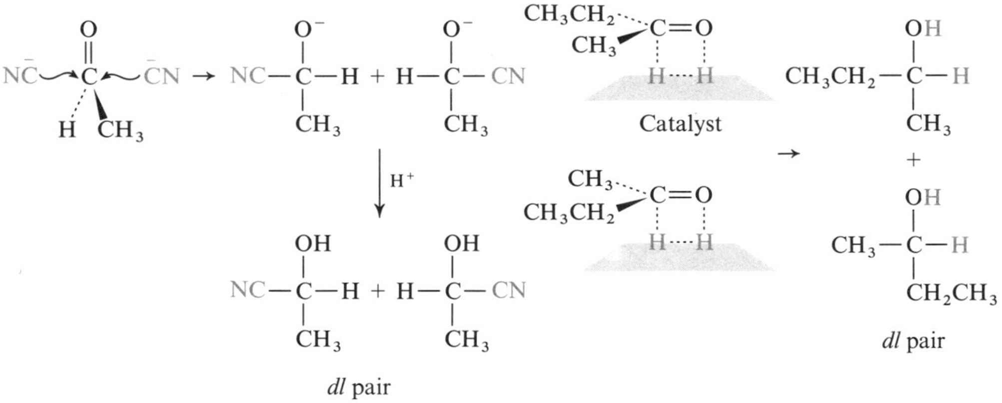

Bromide attack from
either side

dl pair

Catalyst

dl pair

dl pair

Fig. 17-29 Reactions illustrating the formation of *dl* pairs from reactions involving symmetrical materials.

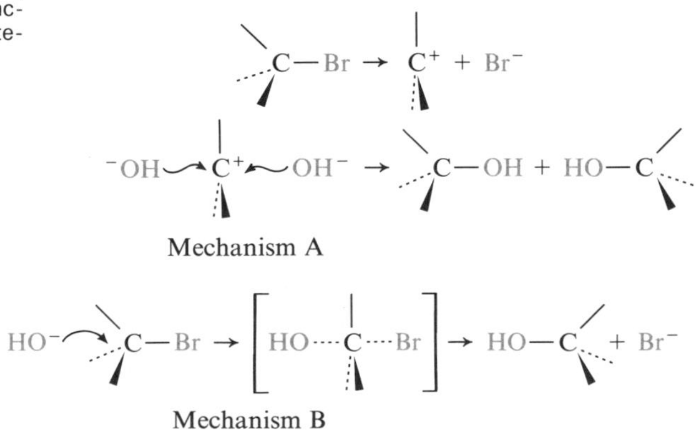

Mechanism A

Fig. 17-30 Two likely mechanisms for the reaction of an alkyl bromide with hydroxide to form an alcohol.

Mechanism B

17-5 OPTICAL ISOMERISM AND REACTION MECHANISMS

Preparation of Optically Active Compounds. Any time we synthesize an asymmetric carbon atom using starting materials and conditions (reagents, solvent, catalysts, etc.) that are not optically active, we obtain a *dl* mixture. If 1-butene is treated with HBr, for example, an asymmetric carbon atom is introduced into the molecule, but the product does not rotate plane polarized light:

$$
\begin{array}{cc}
\text{H} \diagdown & \diagup \text{CH}_2\text{CH}_3 \\
 & \text{C}{=}\text{C} \\
\text{H} \diagup & \diagdown \text{H}
\end{array}
\xrightarrow{\text{HBr}}
\begin{array}{c}
\text{H} \quad \text{CH}_2\text{CH}_3 \\
\text{H}{-}\overset{|}{\underset{|}{\text{C}}}{-}\overset{|}{\underset{|}{\text{C}}}{-}\text{Br} \\
\text{H} \quad \text{H}
\end{array}
$$

dl-2-bromobutane

This can only mean that the optical isomers are produced in equal amounts. This result is a reasonable one when we consider the mechanism of the addition of HBr to an olefin. The first step in the reaction is the attack of the acid proton at the double bond, forming a carbonium ion intermediate (Fig. 17–28). The carbonium ion carbon atom is sp² hybridized, and therefore planar. Since the intermediate is symmetrical, attack by a bromide ion is equally probable from either side of the carbonium ion, and, since we are dealing with millions of molecules, the product will contain equal amounts of the dextro- and levorotatory isomers.

We obtain similar results when we study the addition of hydrogen cyanide (HCN) to an aldehyde, or when we study the formation of an alcohol by catalytic hydrogenation of a ketone to an alcohol, as illustrated in Fig. 17–29. In the HCN addition, attack by cyanide ion is equally probable on either side of the planar carbonyl group. In the catalytic reduction, hydrogen will add with the same equal probability. In fact, innumerable experiments of this type have demonstrated that any time an optically inactive starting material is treated with optically inactive reagents, an optically inactive product must result. In other words, an optically active compound can be prepared only by using optically active starting materials.

▶ Since, as we shall see in subsequent chapters, optical activity is intimately associated with life processes, an interesting evolutionary question arises. How did the first optically active compound come to be? ◀

Optical Activity and Reaction Mechanisms

To see how we can use optical activity in the study of reaction mechanisms, let us consider a simple substitution reaction. When we treat an alkyl bromide with hydroxide, an alcohol is formed. We can write two reasonable mechanisms for this reaction. Mechanism A, shown in Fig. 17–30, is a two-step process. The first step involves cleavage of the C—Br bond so that a carbonium ion and a bromide ion are formed. The carbonium ion is sp² hybridized and must be planar. Its symmetrical nature makes possible, in the second step, the formation of a bond between the carbon and oxygen atoms, at either the side vacated by the bromide ion or the opposite side, with equal probability.

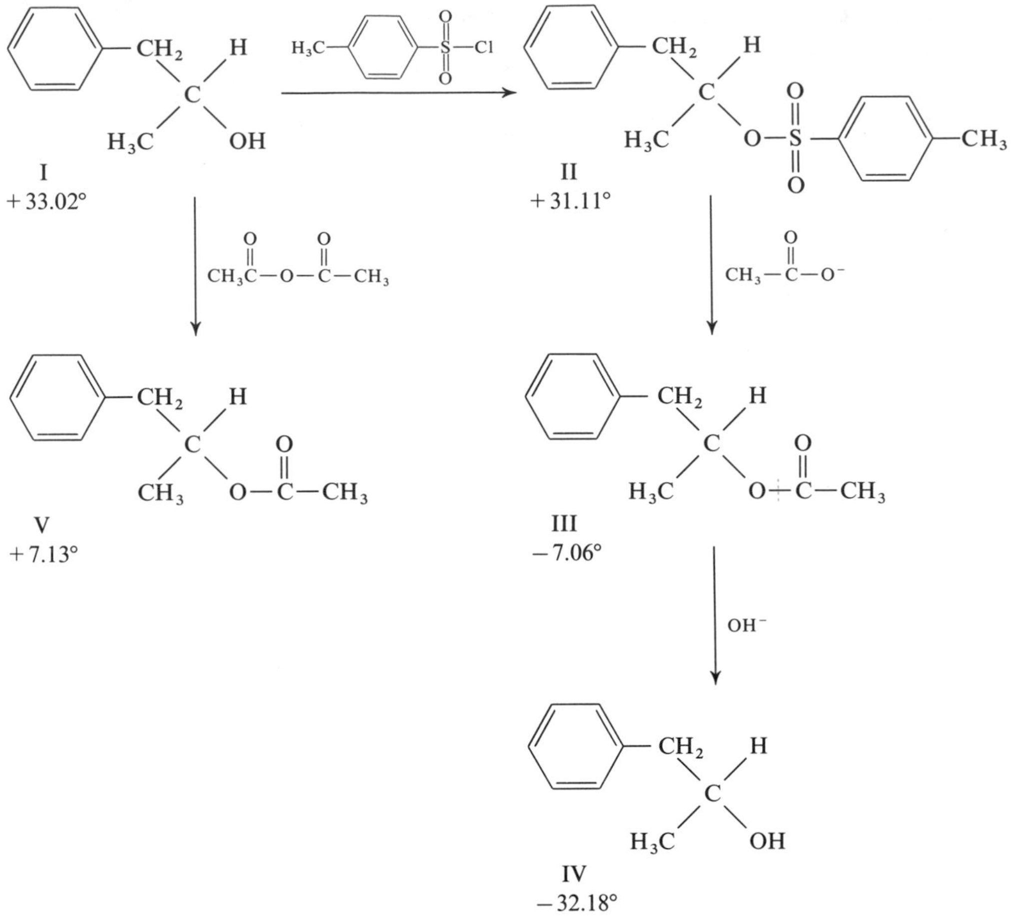

Fig. 17-31 Reaction sequence involving inversion of configuration. The numbers refer to optical rotations.

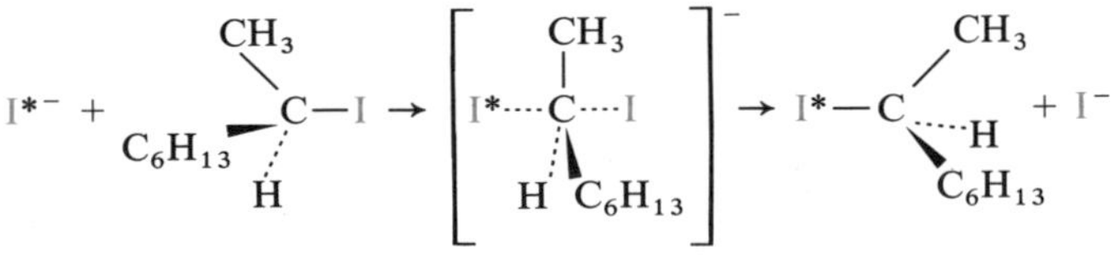

Fig. 17-32 Racemization of optically active 2-iodooctane by means of radioactive iodide ions (I*).

The second mechanism, Mechanism B, is a one-step process. At the same time the carbon–bromine bond is breaking, the hydroxide ion is forming a bond with carbon. As a consequence of this mechanism, the hydroxyl group in the product must occupy a position opposite to that previously held by the bromine. Any reaction that proceeds by this second mechanism must have the configuration of the carbon atom inverted.

A reaction of this type, exemplified in Fig. 17–31, is called a *Walden inversion*. The starting alcohol I is converted to its enantiomer IV by a series of three reactions. One of these reactions must have proceeded by a mechanism which required that the configuration of the asymmetric atom be inverted. The first reaction (forming II) involves only the oxygen–hydrogen bond of the alcohol, and therefore could not bring about inversion of the carbon atom. The third reaction (forming IV)—base-catalyzed hydrolysis of an ester to the corresponding alcohol and acid salt—proceeds by a mechanism involving attack of hydroxide ion at the carbon atom doubly bonded to oxygen, followed by cleavage of the bond indicated by the dashed line. Thus the formation of IV from III cannot involve inversion of configuration; the second reaction (forming III) must be a Walden inversion. Treatment of the original alcohol I with acetic anhydride yields the ester V. This ester has a configuration opposite to that of III, which further supports the thesis that the inversion occurs during attack by acetate ion.

An ingenious experiment (see Fig. 17–32) has been devised to prove the mechanism postulated for the Walden inversion: Optically active 2-iodooctane is treated with radioactive iodide ions. Each time a radioactive iodide ion displaces a nonradioactive iodide ion, 2-iodooctane (which has the inverted configuration) is formed. After half the original 2-iodooctane has undergone substitution, the two enantiomeric forms of 2-iodooctane are found in equal amounts; a *dl* mixture has been formed and no optical activity remains. When the reaction is studied, the rate of incorporation of radioactivity is measured by means of radioactive counting equipment (see Chapter 21), and the rate of formation of the *dl* mixture is measured by studying the change of rotation of plane polarized light with time. The results prove the mechanism proposed for the Walden inversion.

▶ Strictly speaking, ordinary iodine and radioactive iodine should be considered different groups and a true *dl* mixture should not be formed. However, the distinction between the optical behavior of the isotopic iodides is too small to be measured with today's equipment. ◀

All substitution reactions do not proceed by a Walden inversion mechanism, however. In many reactions, such as the reaction of the alcohol (1-phenylethanol) in Fig. 17–33 with hydrochloric acid, Mechanism A (Fig. 17–30) is more consistent with the data. When optically active 1-phenylethanol is treated with concentrated hydrochloric acid, a *dl* mixture of the chloro compound is obtained. We can explain this by postulating the formation of a *symmetrical* carbonium ion intermediate. On

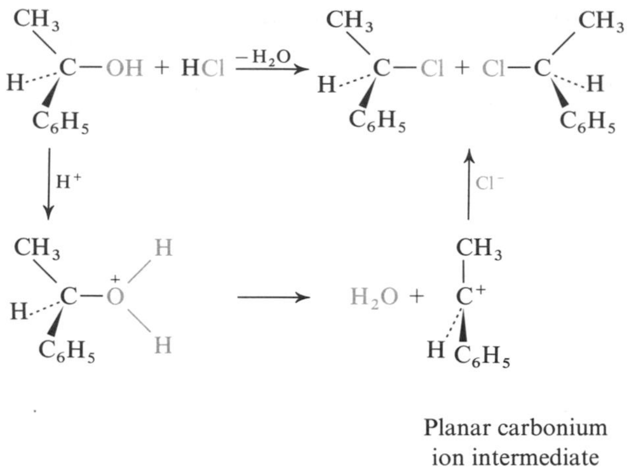

Fig. 17-33 Mechanism explaining the formation of 1-phenylethylchloride as a dl pair from optically active 1-phenylethanol.

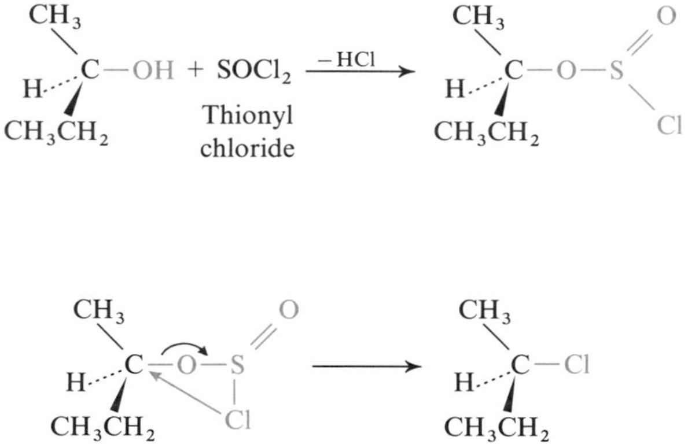

Planar carbonium ion intermediate

Fig. 17-34 Example of a reaction proceeding with retention of configuration: chlorination of 2-butanol with thionyl chloride.

the other hand, when optically active 2-butanol is treated with thionyl chloride, the configuration of the product, 2-chlorobutane, remains the same. Yet retention of configuration is not consistent with either mechanism A or B discussed above. Figure 17-34 postulates a mechanism to explain this rather unusual reaction.

17-6 SUMMARY

We have presented stereochemistry as a prerequisite for understanding the relation between structure and reaction, and also as a tool for the investigation of reaction mechanisms. We have described—albeit briefly—a few of the studies which have been performed in this field.

Optical activity is not necessarily confined to carbon compounds. Any molecule which is dissymmetric can exist in optically active forms. Figure 17-35 presents a few examples.

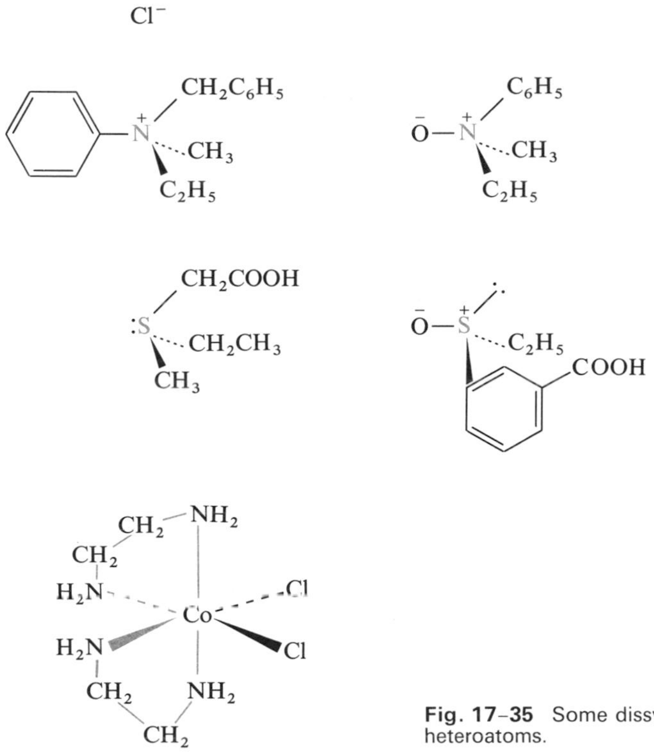

Cl^-

Fig. 17-35 Some dissymmetric compounds containing heteroatoms.

As we shall see in later chapters, the shape of a molecule may also be influenced by factors weaker than those discussed in this chapter. Weak bonding and attractive forces within a molecule can determine the shape of the molecule and consequently its properties.

The following references contain more detailed discussions of stereochemistry.

SUGGESTIONS FOR FURTHER READING

Bent, R. L., "Stereoisomerism," *J. Chem. Ed.* **30**, 328 (1953)

Conrow, K., and R. N. McDonald, *Deductive Organic Chemistry: A Short Course*, Reading, Mass.: Addison-Wesley, 1966

Cram, D. J., and G. S. Hammond, *Organic Chemistry*, second edition, New York: McGraw-Hill, 1964

Eliel, E. L., *Stereochemistry of Carbon Compounds*, New York: McGraw-Hill, 1962

Mislow, K., *Introduction to Stereochemistry*, New York: W. A. Benjamin, 1966

Roberts, J. D., and M. C. Caserio, *Basic Principles of Organic Chemistry*, New York: W. A. Benjamin, 1965

Morrison, R. T., and R. N. Boyd, *Organic Chemistry*, second edition, Boston: Allyn and Bacon, 1966

QUESTIONS

1. Using dashed lines and wedges, as in Fig. 17-2, sketch a three-dimensional structure for propane. Using the same technique, draw another three-dimensional structure for butane, which represents a 180° rotation about the middle C—C bond of the first structure.

2. Using Newman projections, show all the staggered and eclipsed forms for rotation about the C_1—C_2 bond in butane, as well as for rotation about the C_2—C_3 (middle) bond.

3. Draw all staggered and eclipsed forms of 1,2-dibromoethane. Are all the eclipsed forms the same? Are the staggered forms the same?

4. Which of the following would exist in *cis* and *trans* forms? Draw the structures.

$$CH_3CH{=}CHCH_3, \quad CH_3CH{=}CH_2, \quad CH_3CH{=}CHCH_2CH_3,$$

$$\underset{\displaystyle CH_2{=}CCH_2CH_3,}{\overset{\displaystyle CH_3}{|}} \quad BrCH{=}CHBr, \quad Cl_2C{=}CH_2, \quad CH_3CH{=}CHBr,$$

$$\underset{\displaystyle \underset{\displaystyle CH_3}{\overset{\displaystyle |}{}}}{\overset{\displaystyle CH_3}{\underset{\displaystyle |}{}}}$$
$$CH_3{-}C{=}CCH_3$$

5. Using the diagrams from question 1 in Chapter 16, predict the number of isomers expected for $CHCl_3$ and CH_2Cl_2 in the various geometric arrangements. Use an arrow to indicate the expected direction of the dipole moment for each structure.

6. Which of the following would you predict to exist in enantiomeric (*dl*) forms?

$$\underset{\displaystyle D}{\overset{\displaystyle |}{CH_3{-}CHCl,}} \quad \underset{\displaystyle OH}{\overset{\displaystyle |}{C_6H_5{-}CHCH_2NH_2,}} \quad trans\text{-}1,3\text{-dimethylcyclobutane,}$$

$$BrCH{=}C{=}CHBr, \quad CH_3CH{=}C{=}C{=}CHCH_3, \quad trans\text{-}1,2\text{-dimethylcyclobutane,}$$

$$\underset{\displaystyle NH_2}{\overset{\displaystyle |}{CH_3CHCOOH}}$$

7. Write out structures for all the isomers of the following:

$$\underset{\displaystyle OH\,OH}{\overset{\displaystyle |\;\;|}{CH_3CHCHCOOH,}} \quad \underset{\displaystyle OH\,OH}{\overset{\displaystyle |\;\;|}{CH_3CHCHCH_3,}} \quad \underset{\displaystyle OH\,OH\,OH}{\overset{\displaystyle |\;\;|\;\;|}{CH_3CHCHCH\overset{\displaystyle O}{\overset{\displaystyle \|}{C}}{-}OH,}} \quad \text{and}$$

$$\underset{\displaystyle Br\;Br\;Br}{\overset{\displaystyle |\;\;|\;\;|}{CH_3CHCHCHCH_3}}$$

Indicate which are *dl* pairs and which are meso forms.

8. Write a mechanism for the addition of HBr to optically active

$$\underset{\displaystyle Br}{\overset{\displaystyle |}{CH_3CH{-}CH{=}CH_2}}$$

(refer to Fig. 17–28). Is the intermediate symmetrical?

9. The reaction in question 8 gives two products: one optically active and one inactive. Draw structures for these products.

10. The reaction in question 8 involves a carbonium ion intermediate. Newman projections of the carbonium ion are:

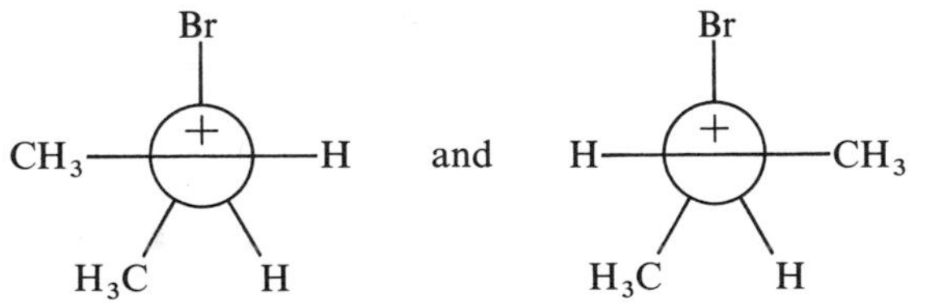

Which of the two conformations of the carbonium ion is the more stable? If the Br^- attacks the carbonium ion from the side farthest away from the Br already in the molecule, which of the two products in question 9 will predominate?

18. BIOCHEMISTRY: AN INTRODUCTION

In man's attempt to understand the physical universe, the complexity of living systems has always presented a challenge. Biochemistry—the chemistry of living systems—is a particularly difficult area of study. In an investigation of the chemistry of the brain, for example, a person can take a portion of the brain, place it in a flask, and study composition, properties, and reactions. He cannot, however, directly relate the experimentation in the flask to the functioning of the living system. Nor can he exactly reproduce, in a controlled manner, the conditions of the living system.

This difficulty in studying the chemistry of living systems caused biochemistry to develop at a relatively slow pace for a long time. Recently, however, new techniques have accelerated research activity in biochemistry, and the resulting knowledge is stimulating ever-more-sophisticated research. Although the study of biochemistry is usually reserved for advanced students, the importance and breadth of this area of study are sufficient reasons for including it in an introductory textbook. It is safe to predict that many of the outstanding scientific findings of the near future will be in biochemistry.

In this and subsequent chapters, we shall discuss the chemical structure of biologically important materials, as well as some of the chemical reactions related to life processes.

18–1 LIPIDS

Most of the organic matter of living cells is comprised of proteins, carbohydrates, and lipids. Classifying a material as a lipid does not necessarily depend on its chemical structure or reactivity, but on its solubility. Lipids are generally insoluble in water, but soluble in less-polar organic solvents such as ether, benzene, and chloroform. Lipids are divided into three main classifications: (1) *Simple lipids*, including fats and waxes, (2) *complex lipids*, including phospholipids and glycolipids, and (3) *steroids*.

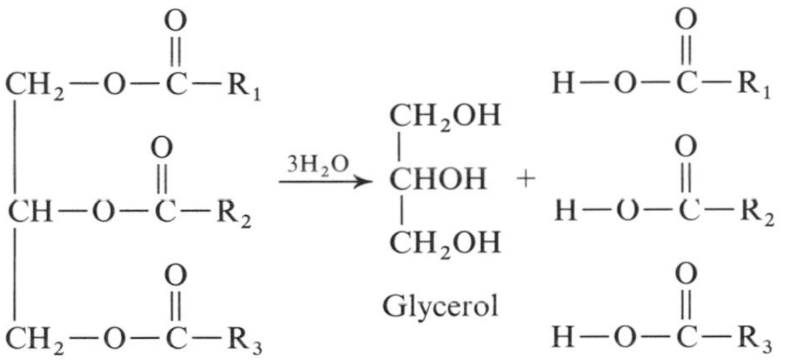

Fig. 18–1 Hydrolysis of a fat.

18–2 FATS

Fats are esters which can be hydrolyzed (Fig. 18–1) to an alcohol fragment and carboxylic acid fragments. The alcohol portion of a fat is the trihydroxy alcohol, glycerol. The acid portions, called fatty acids, depend on the source of the fat. In animals, including man, fats form the fatty tissue of the body and represent the main storehouse of energy. Fats also exist in the plant kingdom, particularly in seeds. The distinguishing characteristics of a fat depend on the particular fatty acids that are incorporated in the ester. Table 18–1 presents a list of common fatty acids. Note that some acids are saturated and others unsaturated.

Table 18–1. Fatty acids

Name	Formula	Source
Butyric acid	$CH_3(CH_2)_2COOH$	Butter
Palmitic acid	$CH_3(CH_2)_{14}COOH$	Palm oil
Stearic acid	$CH_3(CH_2)_{16}COOH$	Plant and animal fat
Oleic acid	$CH_3(CH_2)_7CH{=}CH(CH_2)_7COOH$	Plant and animal fat
Linoleic acid	$CH_3(CH_2)_4CH{=}CHCH_2CH{=}CH$ $\qquad\qquad\qquad\qquad(CH_2)_7COOH$	Linseed oil
Linolenic acid	$CH_3CH_2CH{=}CHCH_2CH{=}CHCH_2CH{=}CH$ $\qquad\qquad\qquad\qquad\quad HOOC(CH_2)_7$	Fish-liver oil
Arachadonic acid	$C_{19}H_3COOH$ (4 double bonds)	Butter

18-3 SOME PROPERTIES OF FATS

When a fat contains many double bonds, it is often a liquid at room temperature, and is called an *oil*. Oils are typically isolated from the seeds of plants. Animal fat, however, is a semi-solid material in which the acid portion of the ester contains few double bonds. Until recently, semi-solid fat was preferred for cooking purposes and the polyunsaturated oils were converted to saturated fats by hydrogenation of the double bonds, using a powdered-nickel catalyst:

$$\begin{matrix} \diagdown \\ \diagup \end{matrix}C=C\begin{matrix} \diagup \\ \diagdown \end{matrix} + H_2 \xrightarrow[\text{nickel}]{\text{Powdered}} \begin{matrix} | \\ -C- \\ | \\ H \end{matrix}\begin{matrix} | \\ C- \\ | \\ H \end{matrix}$$

Portion of a fat

In recent years, however, excessive animal fats in the diet have been linked to high levels of cholesterol in the blood. Since deposits of cholesterol clog the aorta and other arteries when a person has arteriosclerosis, people have begun to use poly-unsaturated vegetable oils more and more in cooking. One can determine the degree of unsaturation in a fat by measuring the amount of iodine that will react with the double bonds in an addition reaction. When this is expressed as the number of grams of iodine that will react with 100 g of fat, it is called the *iodine number*:

$$\begin{matrix} \diagdown \\ \diagup \end{matrix}C=C\begin{matrix} \diagup \\ \diagdown \end{matrix} + I_2 \rightarrow \begin{matrix} | \\ -C- \\ | \\ I \end{matrix}\begin{matrix} | \\ C- \\ | \\ I \end{matrix}$$

Portion of a fat

Oxidation. The double bond in fats can also react with the oxygen in air. In paints containing linseed oil, this leads to polymerization of the fat molecules into very large interlocked molecules which hold the paint pigment. As the reaction proceeds, catalyzed by light, the molecules become even larger and more brittle, leading to peeling and cracking of the paint. The reaction is exothermic (i.e., accompanied by the giving off of heat) and that is why paint-soaked rags, when left bunched up so that the heat of the reaction cannot be dissipated, can spontaneously burst into flames.

In the case of cooking fats, oxidation at elevated temperatures is generally fol-lowed by cleavage of the double bonds. This yields carboxylic acids and aldehydes which have a strong taste and odor. The fat is then said to be *rancid*. Since this oxidative rancidification takes place only in the presence of double bonds, it is pri-marily a problem with polyunsaturated fats and oils. The rancidification process can be slowed by adding antioxidants, such as hydroquinone, vitamin C, or vitamin E (Fig. 18–2). All these substances are readily oxidized by the oxygen in air, and thus retard the oxidation of the unsaturated fat.

Rancidity can also occur as a result of hydrolysis of fats, particularly fats con-taining low-molecular-weight fatty acids. Rancidity of butter is due primarily to the hydrolytic formation of butyric acid. At high temperatures the glycerol produced by the hydrolysis of the fat is unstable and is dehydrated to acrolein, a pungent-

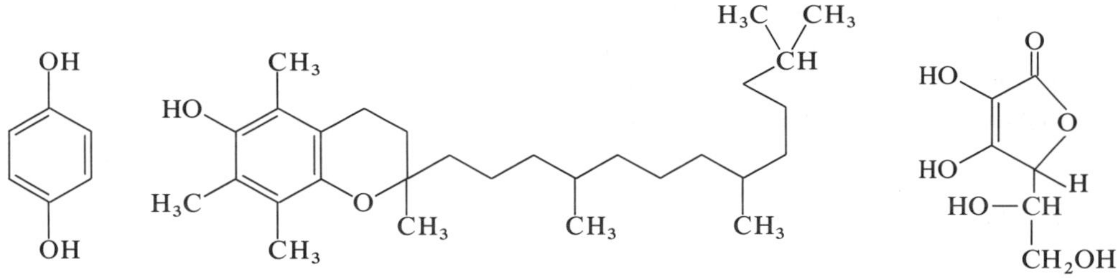

Fig. 18-2 Antioxidants.

Hydroquinone   α-tocopherol (vitamin E)   L-ascorbic acid (vitamin C)

smelling substance that irritates the eyes:

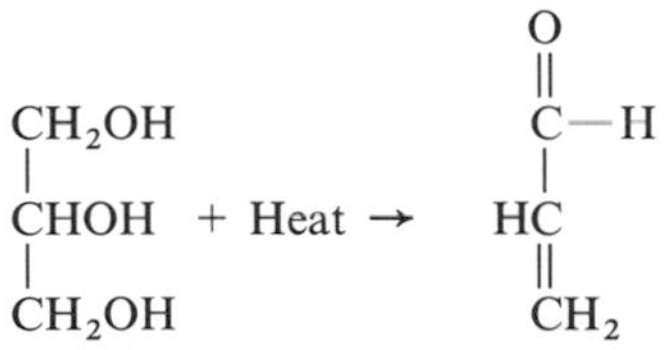

$$
\begin{array}{ccc}
\text{CH}_2\text{OH} & & \overset{\displaystyle O}{\overset{\|}{\text{C}}}-\text{H} \\
| & & | \\
\text{CHOH} & + \text{Heat} \rightarrow & \text{HC} \\
| & & \| \\
\text{CH}_2\text{OH} & & \text{CH}_2
\end{array}
$$

Acrolein

Saponification. When the hydrolysis of a fat occurs under alkaline conditions, glycerol and a mixture of fatty acid salts are formed. The mixture of fatty acid salts is a soap, and the process (alkaline hydrolysis of an ester) is called *saponification* (soap-making):

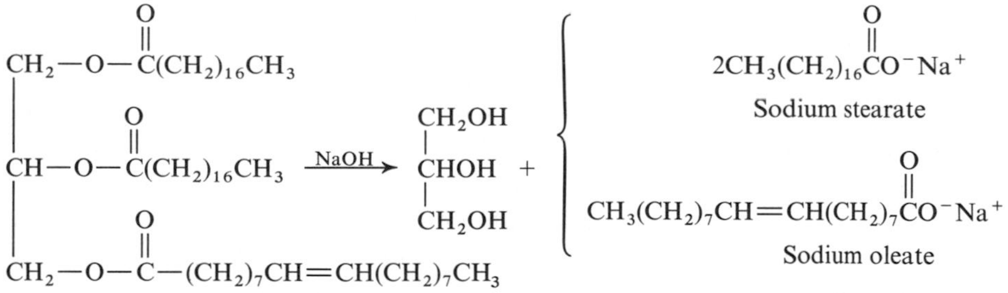

$$
\begin{array}{l}
\text{CH}_2-\text{O}-\overset{O}{\overset{\|}{\text{C}}}(\text{CH}_2)_{16}\text{CH}_3 \\
| \\
\text{CH}-\text{O}-\overset{O}{\overset{\|}{\text{C}}}(\text{CH}_2)_{16}\text{CH}_3 \xrightarrow{\text{NaOH}} \\
| \\
\text{CH}_2-\text{O}-\overset{O}{\overset{\|}{\text{C}}}-(\text{CH}_2)_7\text{CH}=\text{CH}(\text{CH}_2)_7\text{CH}_3
\end{array}
$$

$$
\begin{array}{l}
\text{CH}_2\text{OH} \\
| \\
\text{CHOH} \quad + \\
| \\
\text{CH}_2\text{OH}
\end{array}
$$

$$
\left\{
\begin{array}{l}
2\text{CH}_3(\text{CH}_2)_{16}\overset{O}{\overset{\|}{\text{C}}}\text{O}^-\text{Na}^+ \\
\text{Sodium stearate} \\
\\
\text{CH}_3(\text{CH}_2)_7\text{CH}=\text{CH}(\text{CH}_2)_7\overset{O}{\overset{\|}{\text{C}}}\text{O}^-\text{Na}^+ \\
\text{Sodium oleate}
\end{array}
\right.
$$

Soap

Most soaps are mixtures of sodium salts of fatty acids, but for special purposes (shaving cream, liquid soaps), the softer and more soluble potassium salts are used.

The cleaning action of soaps is a complex phenomenon, but in general terms we can say that it is the result of a molecule having a water-soluble end and a fat-soluble end. In a soap molecule, the carboxylate salt has an affinity for water, and the long hydrocarbon end is water insoluble, but has an affinity for nonpolar materials. As a consequence, the hydrocarbon ends join together in a cluster called a *micelle*, which attracts and dissolves grease and the like. The carboxyl ion end allows the grease to be dispersed (emulsified) in water and ultimately washed away.

The limitations of soap restrict its use in many cases; it is unstable in acid and many of its salts are insoluble. Hard water, for example, contains calcium and magnesium ions which form insoluble salts of the fatty acid molecules and result in precipitation. The problem can be solved by using water softeners to remove the unwanted metal ions, but in many cases the expense of water-softening makes it impractical, and synthetic detergents are employed. Alkyl sulfonates, prepared from petroleum hydrocarbons, are common detergents which are not precipitated in hard water.

$$RH + SO_2 + Cl_2 \xrightarrow{\text{Light}} RSO_2Cl + HCl$$

Petroleum   Alkyl sulfonyl
hydro-   chloride
carbon

$$RSO_2Cl + 2NaOH \longrightarrow RSO_3^-Na^+ + NaCl + H_2O$$

Sodium alkyl
sulfonate

▶ The history of synthetic detergents has been a story of profit and controversy. There is no question that the discovery of synthetic detergents in the 1930's filled a need. The detergents won widespread acceptance in homes and factories throughout the world, and, in the United States, became the basis of an industry with annual sales of $1.5 billion. As the use of these detergents increased, a problem surfaced. The detergents, which were highly branched-chain alkyl benzene sulfonates, were not biodegradable. The enzymes present in water bacteria, which could break up soap molecules to carbon dioxide and water, were ineffective on these synthetic detergents. The result was a foam which appeared in streams, lakes, and wells, and clogged sewage-treatment plants. The cry of an outraged public was heard by some in Congress. The detergent companies felt the pressure, replaced the branched-chain alkyl groups with straight-chain groups, and marketed, after 1965, detergents which were biodegradable.

A new problem presented itself, however. The detergent makers found that the cleansing action was improved by the addition of large amounts of sodium phosphate to the detergents. About 500 million pounds of phosphorus per year are consumed for this purpose. Phosphorus contains small amounts of arsenic as an impurity. The results of increasing the arsenic content of our waters are not yet known. But the results of phosphate addition are known.

Strictly speaking, phosphate is not a pollutant, since it is a natural fertilizer. In such high quantities, however, it stimulates an overproduction of algae. The plant life in the water ultimately dies, and, in the process of decay, consumes oxygen. This oxygen depletion results in the death of fish, and, ultimately, in the death of the lakes. The problem is generally recognized, and the pressure of public opinion will require the detergent makers to find a substitute for phosphates.

What will be the long-range effects of these new additives (or of the enzymes now being used) remains to be seen. But we must be prepared to recognize them and do something about them. ◀

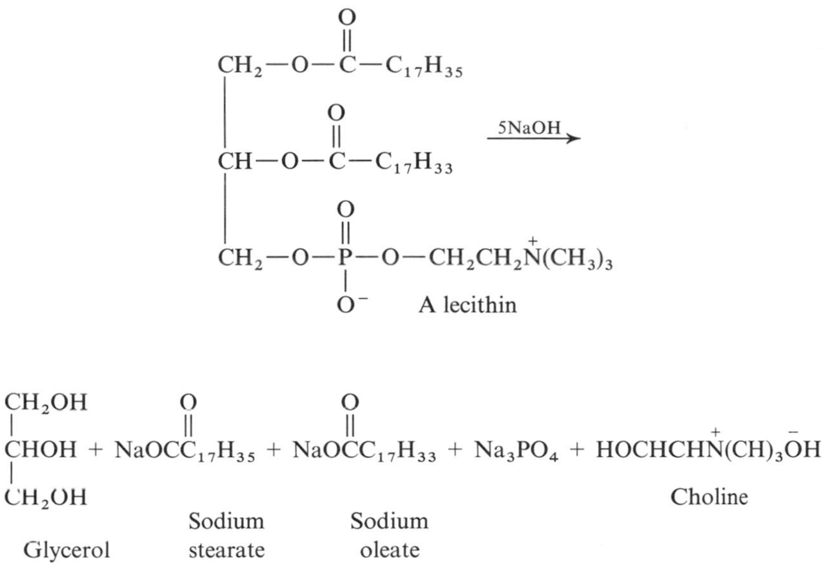

$$CH_2-O-\overset{\overset{\textstyle O}{\|}}{C}-C_{17}H_{35}$$

$$CH-O-\overset{\overset{\textstyle O}{\|}}{C}-C_{17}H_{33} \qquad \xrightarrow{\text{5NaOH}}$$

$$CH_2-O-\underset{\underset{\textstyle O^-}{|}}{\overset{\overset{\textstyle O}{\|}}{P}}-O-CH_2CH_2\overset{+}{N}(CH_3)_3 \qquad \text{A lecithin}$$

$$\begin{array}{l} CH_2OH \\ | \\ CHOH \\ | \\ CH_2OH \end{array} + NaO\overset{\overset{\textstyle O}{\|}}{C}C_{17}H_{35} + NaO\overset{\overset{\textstyle O}{\|}}{C}C_{17}H_{33} + Na_3PO_4 + HOCHCH\overset{+}{N}(CH)_3\overset{-}{O}H$$

Glycerol   Sodium stearate   Sodium oleate   Choline

Fig. 18–3 Results of saponification of a lecithin.

Saponification Number. We can estimate the molecular weight of a fat by measuring the amount of sodium hydroxide required to saponify a given weight of the fat. Since one mole of sodium hydroxide reacts with one mole of an ester, one mole of a simple fat requires three moles of sodium hydroxide for complete saponification. Fats isolated from natural sources are generally mixtures, and quantitative saponification gives only the average molecular weight of the fat mixture. This average molecular weight is often listed using a relative term, *saponification number*, i.e., the number of grams of sodium hydroxide needed to saponify 100 g of fat.

18-4 COMPLEX LIPIDS

Phospholipids. Phospholipids, which are integral parts of cell membranes, are found in abundance in the tissue of the brain and nervous system, as well as in egg yolk and plant seeds. Saponification of a phospholipid yields glycerol, a mixture of sodium salts of fatty acids, sodium phosphate, and a nitrogen-containing fragment which is usually basic.

Phospholipids are classified according to the particular nitrogen-containing fragment found in the molecule. *Lecithins*, first isolated from egg yolk, contain the quaternary ammonium hydroxide choline (Fig. 18–3).

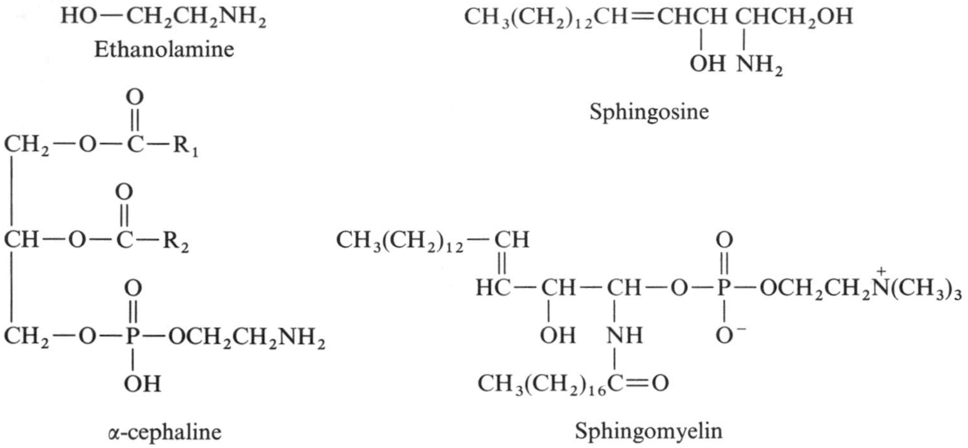

$$HO-CH_2CH_2NH_2$$

Ethanolamine

$$CH_3(CH_2)_{12}CH=CHCH\ CHCH_2OH$$
$$\qquad\qquad\qquad\qquad\ OH\ NH_2$$

Sphingosine

$$CH_2-O-\overset{\overset{O}{\parallel}}{C}-R_1$$

$$CH-O-\overset{\overset{O}{\parallel}}{C}-R_2$$

$$CH_3(CH_2)_{12}-CH$$
$$\qquad\qquad\qquad\parallel$$
$$HC-CH-CH-O-\overset{\overset{O}{\parallel}}{P}-OCH_2CH_2\overset{+}{N}(CH_3)_3$$

$$CH_2-O-\overset{}{\underset{OH}{P}}-OCH_2CH_2NH_2$$

$$\qquad\qquad OH\quad NH\qquad\ O^-$$
$$\qquad\qquad CH_3(CH_2)_{16}C=O$$

α-cephaline

Sphingomyelin

Fig. 18-4 Complex lipids.

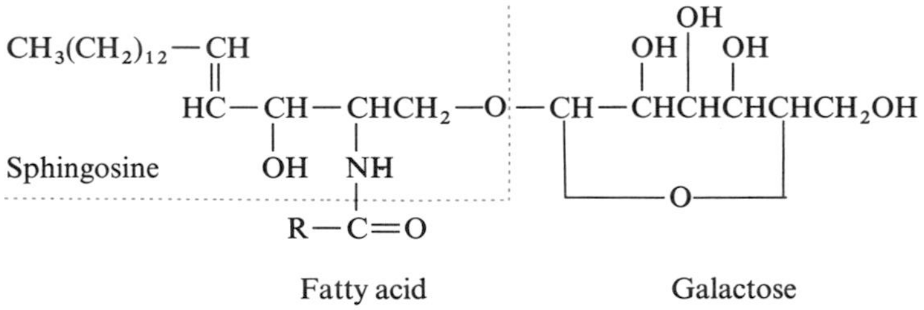

$$CH_3(CH_2)_{12}-CH$$
$$\qquad\qquad\qquad\parallel$$
$$HC-CH-CHCH_2-O-CH-\overset{\overset{OH}{\overset{OH\ \ OH}{|\ \ |}}}{CHCHCHCHCH_2OH}$$

Sphingosine   OH   NH

R—C=O

Fatty acid   Galactose

Fig. 18-5 A cerebroside.

Cephalins (brain tissue) and *sphingo lipids* (nervous tissue) contain the bases ethanolamine and sphingosine, respectively, as shown in Fig. 18-4.

Glycolipids. Certain lipids which are found mainly in the brain contain fatty acids, sphingosine, and the sugar galactose. These are called *glycolipids*, indicating the sugar portion (*glyco* from the Greek word meaning sweet) or *cerebrosides*, indicating the origin in the brain. Figure 18-5 depicts a glycolipid, with dashed lines indicating the component parts.

The common bases of the phospholipids—choline, ethanolamine, and sphingosine—all have an amino group separated by two carbons from a hydroxyl group. Since the phospholipids exist in nerve tissue, pharmaceutical chemists investigated many compounds containing structural units similar to ethanolamine, i.e., com-

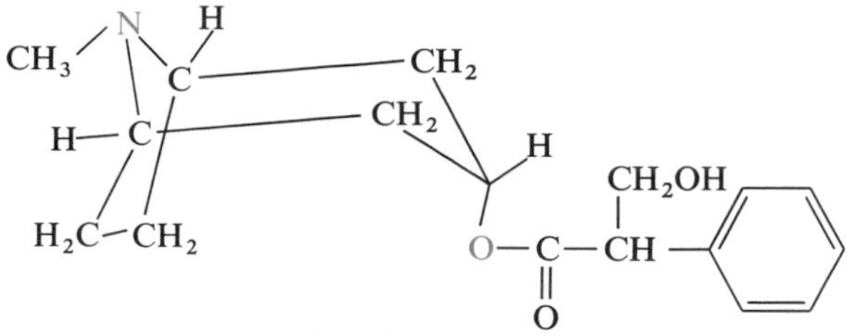

Atropine

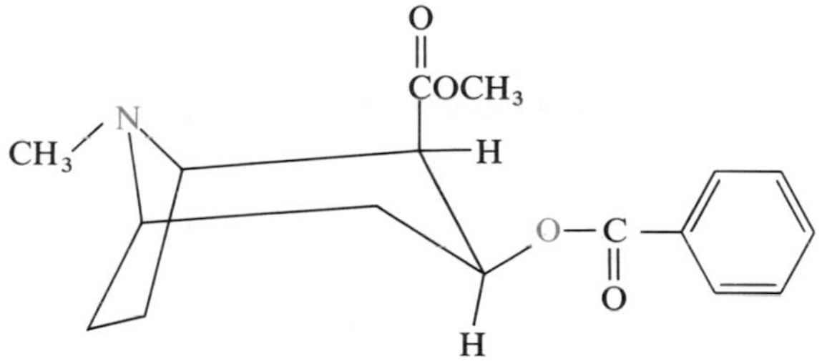

Cocaine

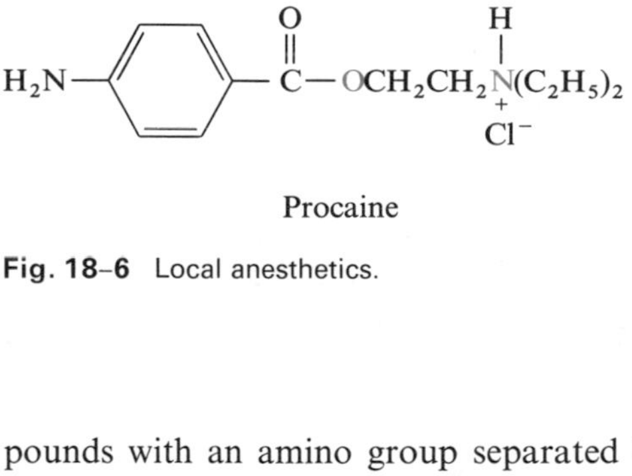

Procaine 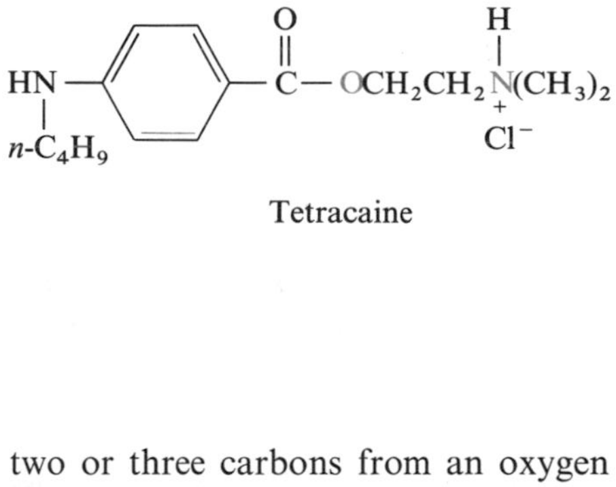   Tetracaine

Fig. 18-6 Local anesthetics.

pounds with an amino group separated by two or three carbons from an oxygen atom. Many compounds of this type have been found to have capabilities as local anesthetics. Figure 18-6 lists a few of them.

18-5 STEROIDS

The steroids are a large group of compounds with structural similarities that exhibit a wide variety of biological activity. The sterols, bile acids, vitamin D, and various hormones belong to this class of compounds. Naturally occurring steroids and synthetically produced analogs are used in the treatment of heart disease and certain hormonal imbalances.

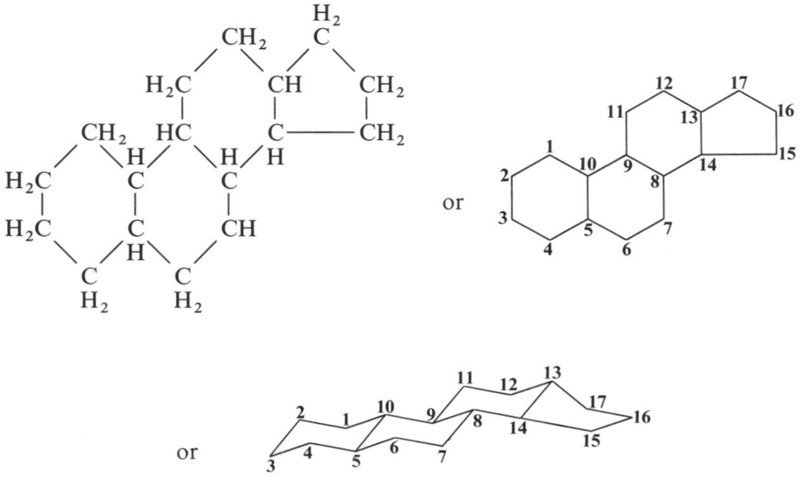

Fig. 18–7 Three structural representations of the steroid ring system.

The steroid nucleus, illustrated in Fig. 18–7, consists of three fused* six-membered rings and a fused five-membered ring. In most cases methyl substituents are found at positions 10 and 13 and a branched hydrocarbon chain is found at the 17 position.

Sterols. Certain steroid alcohols—sterols—are found in lipid-soluble portions of plants and animals. These belong to a group of compounds called *nonsaponifiable lipids*. When brain tissue is saponified, the fats, phospholipids, and other complex lipids are converted to water-soluble, ether-insoluble materials. A certain portion of the original lipid fraction, however, remains water insoluble and ether soluble, and is termed nonsaponifiable. The main constituent sterol of this nonsaponifiable lipid is *cholesterol* (Fig. 18–8). Cholesterol is found in all body tissue; the blood normally contains about 200 mg of cholesterol per ml.

---

*   Two rings are said to be fused when they share two adjacent carbon atoms so that the rings are joined along one edge.

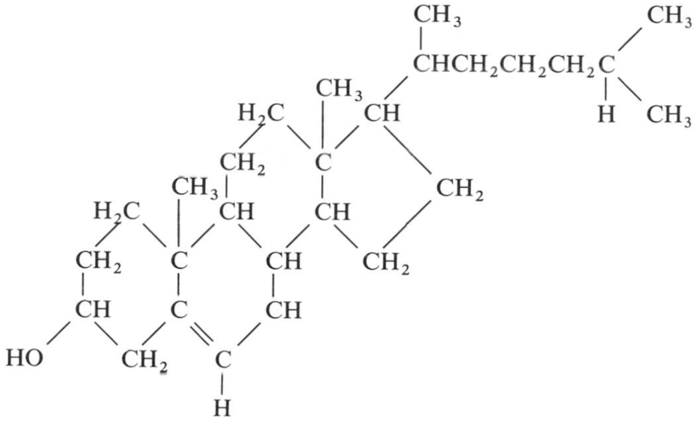

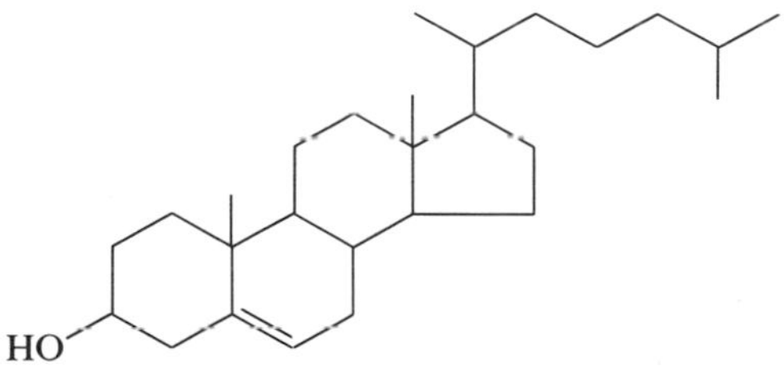

Fig. 18-8 Cholesterol.

Cholesterol was first isolated from gallstones, which consist of about 80% of this sterol. Gallstones, dropping from the gall bladder into the bile duct, can cause biliary blockage, preventing the flow of bile required for digestion. This, along with the role that cholesterol plays in arteriosclerosis, overshadows its beneficial role as a source of bile acids.

Studies conducted via radioactive isotope tracers have shown that cholesterol is the biosynthetic precursor of bile acids. Bile acids, as we shall see in the following chapter, are necessary for the digestion of fats. Although cholesterol may be absorbed during the digestion of animal fat, the body is capable of synthesizing cholesterol from acetate which is readily available in the body. The sequence of this synthetic route, determined by means of radioactive tracer techniques, is shown in Fig. 18-9.

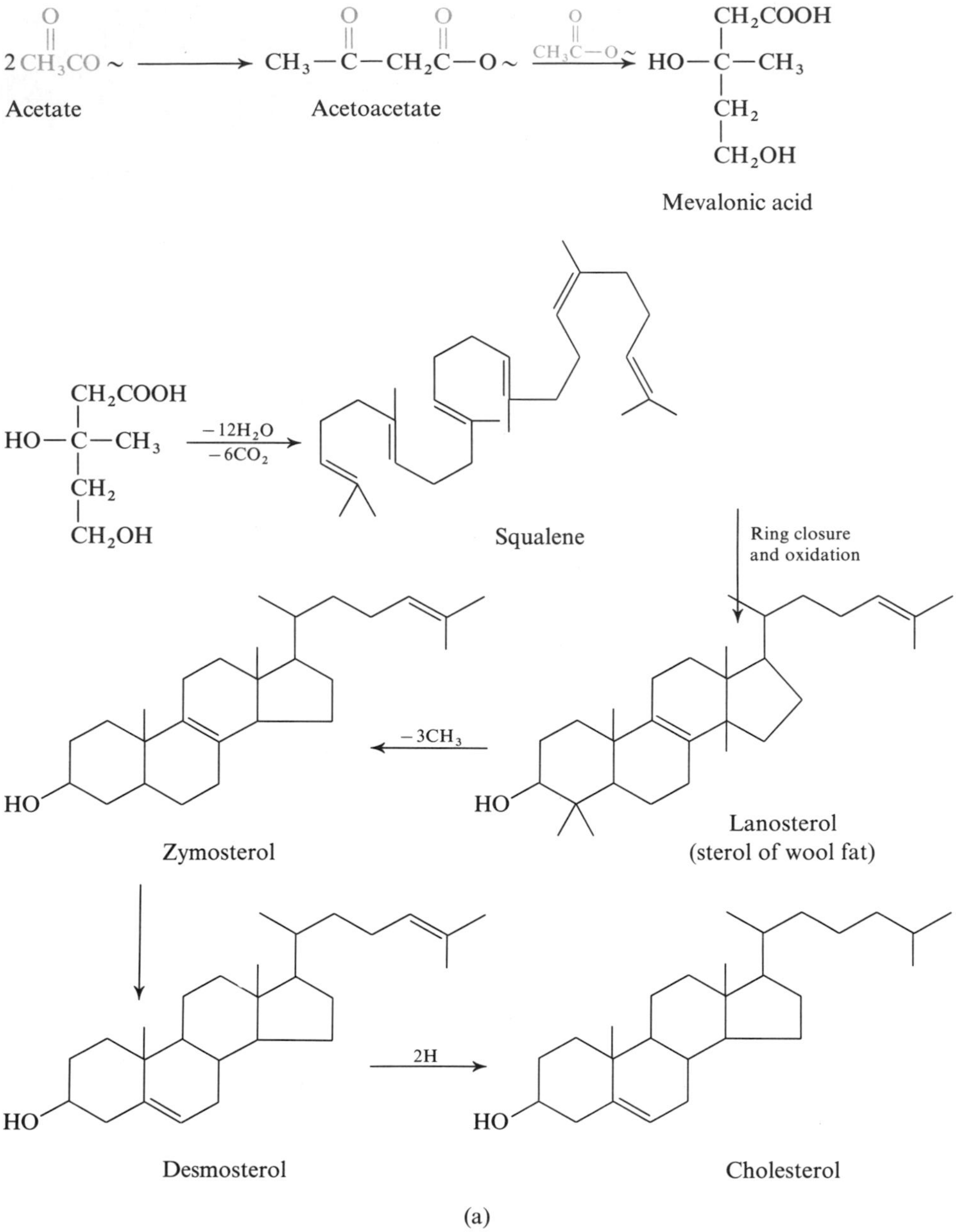

Fig. 18-9 Biosynthesis of cholesterol.

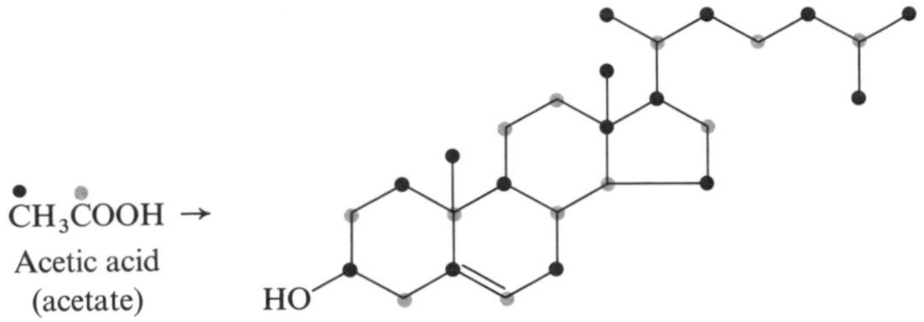

CH₃COOH →
Acetic acid
(acetate)   HO

(b)

Fig. 18-9 (*continued*)

Part (a) of Fig. 18–9 shows how three acetate units form mevalonic acid and how six mevalonic acid groups are joined together to form squalene, and, ultimately, cholesterol. Chemists have been able to follow each step in the process by using radioactive carbon in the molecules. Part (b) of Fig. 18–9 indicates where the two different types of carbons in the acetate molecule wind up in cholesterol. The black atoms in cholesterol are carbon atoms which were originally in a methyl group in the acetate; the colored atoms were originally carboxyl carbon atoms.

Many other steroids are found in plants and animals, and the isolation and structural elucidation of these materials have been one of the success stories of organic chemistry. We shall discuss the hormonal nature of some of these substances in Chapter 20.

CARBOHYDRATES

Carbohydrates are named for the general formula $(C \cdot H_2O)_n$, even though they are not hydrates in the usual sense. The carbohydrates, which are major components of most living systems, act as the main source of energy for biological processes as well as a structural component in plants. Carbohydrates are the end products of photosynthesis in plants, and may be found as low-molecular-weight, sweet-tasting, water-soluble sugars or as high-molecular-weight, less-soluble *polysaccharides*.

In this section we shall discuss the organic chemistry of those carbohydrates containing six-carbon fragments, the *hexoses*. When the hexose unit stands alone it is called a *monosaccharide*. When two hexose units are joined together, the molecule is called a *disaccharide*, and many hexose units joined together comprise a *polysaccharide*.

18-6 MONOSACCHARIDES

Disaccharides and polysaccharides may be converted to monosaccharides by hydrolysis in the presence of an acid. Therefore we may define a monosaccharide as a carbohydrate which cannot be broken down by acid hydrolysis into a simpler carbo-

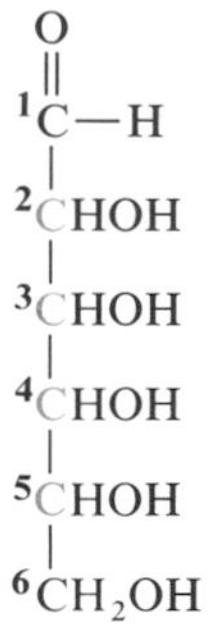

$$
\begin{array}{c}
O \\
\parallel \\
{}^{1}C\text{--}H \\
| \\
{}^{2}CHOH \\
| \\
{}^{3}CHOH \\
| \\
{}^{4}CHOH \\
| \\
{}^{5}CHOH \\
| \\
{}^{6}CH_2OH
\end{array}
$$

Fig. 18–10 An aldohexose.

hydrate. Hexoses are polyhydroxy aldehydes (aldohexoses) or polyhydroxy ketones (ketohexoses).

Figure 18–10 illustrates the general formula for an aldohexose. The carbon atoms numbered 2, 3, 4, and 5 all have four different groups attached and are therefore asymmetric. The presence of four different asymmetric carbon atoms in the molecule enables us to predict $(2)^4$ or 16 different isomers of this aldohexose, 8 dl pairs. All the compounds are known; some exist in nature and others have been synthesized.

Glucose. By far the most abundant naturally occurring monosaccharide is glucose. Glucose is found in honey, ripe grapes and other sweet fruits, and in the blood and urine of man. When treated with acetic anhydride, glucose forms a pentaacetate, indicating the presence of five hydroxy groups in the molecule. The fact that glucose reduces Cu^{2+} in Benedict's solution—coupled with other evidence—demonstrates the presence of an aldehyde group. Glucose is therefore an aldohexose.

Figure 18–11 indicates the actual configuration of naturally occurring glucose. As the structure shows, the configuration at carbon-5 in glucose is identical to that of D-glyceraldehyde (page 503), and by convention D-glucose is the correct name.

When glucose is crystallized from methanol solution and the crystals are dissolved in water, plane polarized light from a sodium lamp is rotated by the solution $+112°$ under standard conditions of measurement. When the solution is allowed to stand,

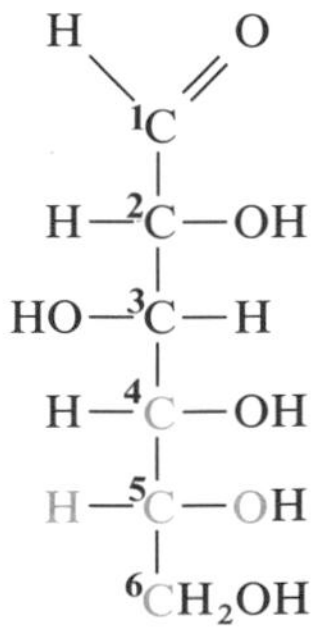

$$
\begin{array}{c}
H \qquad O \\
\diagdown \;\; \diagup\!\!\diagup \\
{}^{1}C \\
\parallel \\
H\text{--}{}^{2}C\text{--}OH \\
| \\
HO\text{--}{}^{3}C\text{--}H \\
| \\
H\text{--}{}^{4}C\text{--}OH \\
| \\
H\text{--}{}^{5}C\text{--}OH \\
| \\
{}^{6}CH_2OH
\end{array}
$$

Fig. 18–11 D-glucose (open-chain form).

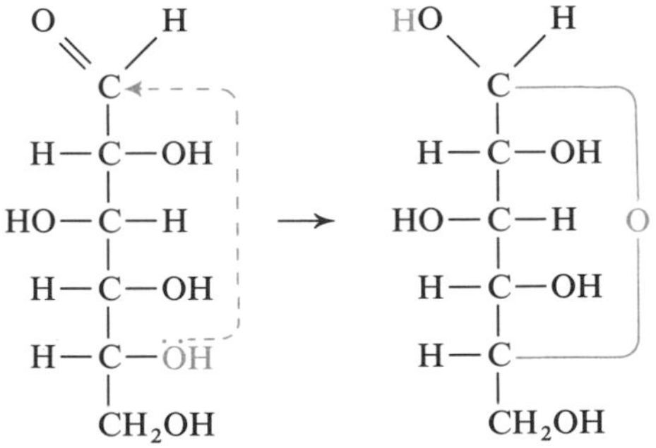

Fig. 18-12 Glucose cyclic hemiacetal formation.

however, the rotation gradually changes, a phenomenon known as *mutarotation*. The change, which can be accelerated by adding a trace of acid, will stop at $+52°$.

When glucose is crystallized from hot acetic acid and the crystals are dissolved in water, the initial rotation, under conditions identical to those used previously, is $+19°$. This glucose solution also exhibits mutarotation; the rotation changes gradually until a final value of $+52°$ is reached.

We can explain this phenomenon by examining the structure of glucose. In Chapter 16 we discussed the reaction of aldehydes with alcohols to form hemiacetals:

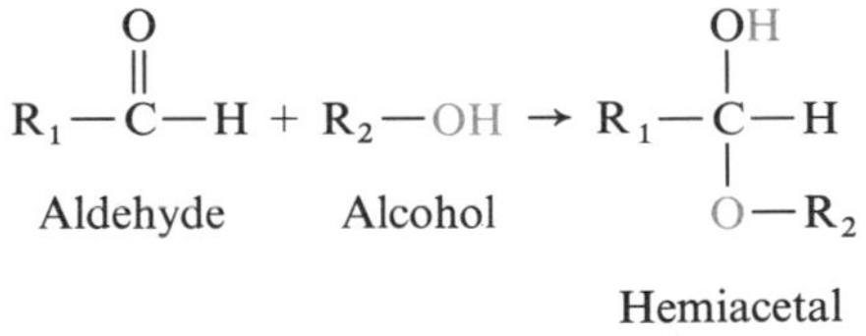

$$R_1-\overset{\overset{O}{\|}}{C}-H + R_2-OH \rightarrow R_1-\overset{\overset{OH}{|}}{\underset{\underset{O-R_2}{|}}{C}}-H$$

Aldehyde   Alcohol   Hemiacetal

The mechanism of the reaction involves an initial attack by a nonbonded pair of electrons from the alcohol oxygen atom on the carbonyl carbon of the aldehyde group. The glucose structure contains an aldehyde group and five hydroxyl groups, and it is possible for a cyclic hemiacetal to form by intramolecular reaction.

We can apply our knowledge of stereochemistry to this problem in order to predict which of the five hydroxyl groups will participate in the reaction. Since we know that 5- and 6-membered rings are more stable than rings containing 3, 4, or 7 atoms, we would predict that hydroxyl groups at positions 4 and 5 in the glucose molecule are likely to react to form the cyclic hemiacetal. The 6-membered ring system contains less bond angle strain than the 5-membered ring, and hence it is probable that the hydroxyl at carbon-5 will react with the aldehyde.

Figure 18–12 illustrates the reaction. The long carbon–oxygen–carbon bonds merely emphasize the points of attachment, but it is more instructive to show the correct stereochemistry of the system.

The fact of hemiacetal formation cannot by itself explain the phenomenon of mutarotation. We can get closer to the answer, however, if we consider the stereochemistry of the glucose molecule just before cyclization occurs.

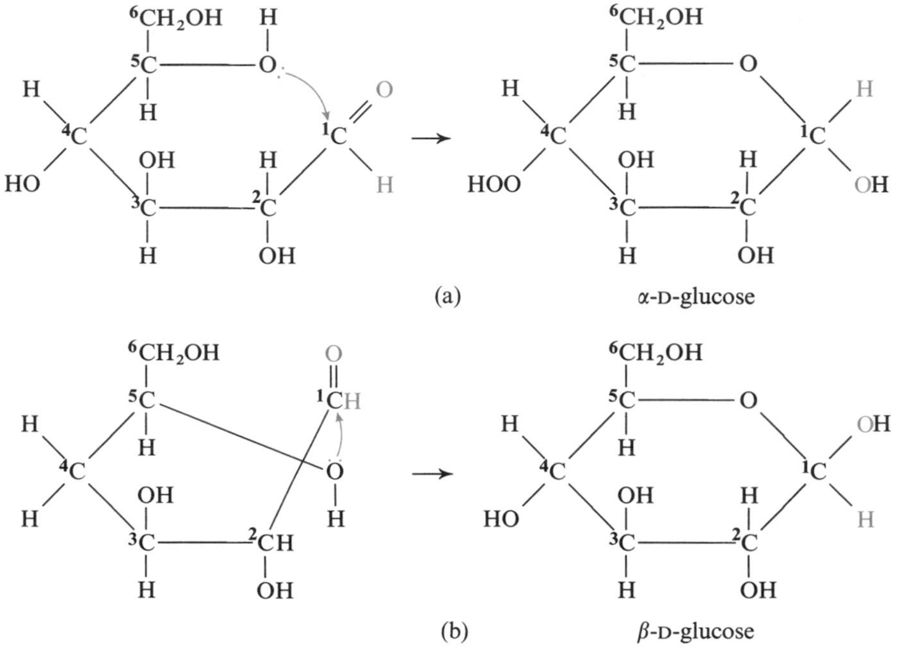

Fig. 18-13 Hemiacetal formation. (a) Attack from above. (b) Attack from below

Figure 18–13 shows that the product of cyclization from above yields the cyclic compound α-D-glucose, with the hydroxyl group at carbon-1 *trans* to the $-CH_2OH$ group at carbon-5. Cyclization from below yields β-D-glucose, with these groups in the *cis* orientation. When cyclization occurs, a new asymmetric atom is introduced into the molecule. The aldehyde, when converted to the hemiacetal by attack of the carbon-5 hydroxyl group from above, forms a new asymmetric carbon atom opposite in configuration to that resulting from hydroxyl attack from below. Thus α-D-glucose has a configuration at carbon-1 opposite to the configuration at carbon-1 in β-D-glucose. These compounds are not enantiomers, since the configurations at carbon-2, carbon-3, carbon-4, and carbon-5 are the same in both compounds. Isomers which contain more than one asymmetric atom but differ only in the configuration at one atom are called *epimers*.

Mutarotation of glucose, therefore, depends on an equilibrium between the α and β forms in water solution. As Fig. 18–14 shows, α-D-glucose has a rotation of $+112°$ and β-D-glucose $+19°$, and when either isomer is placed in water solution it slowly forms the other isomer via the open-chain compound, until an equilibrium mixture containing 37% α and 63% β (with less than 1% of the open-chain compound) is formed.

α-D-glucose β-D-glucose

37% α-D-glucose 63% β-D-glucose

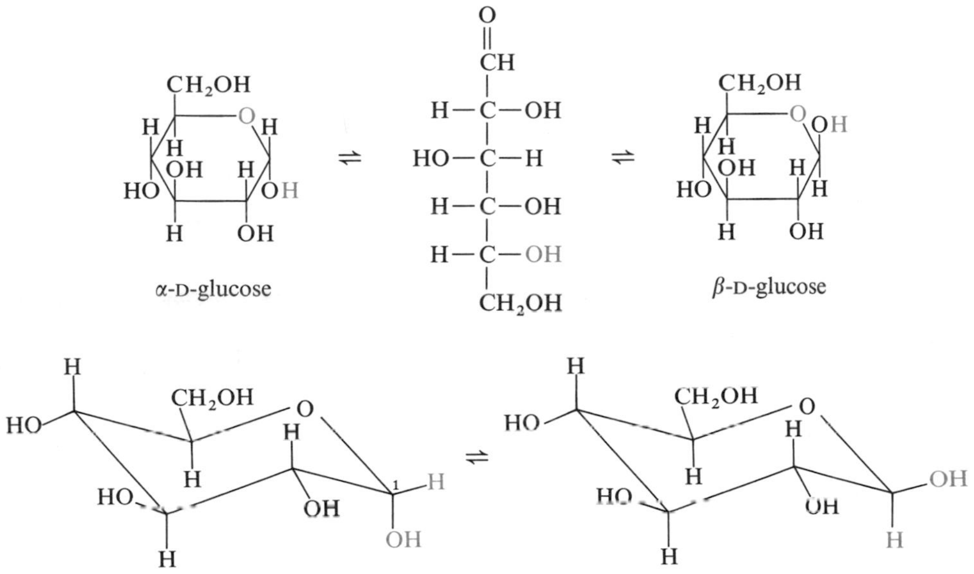

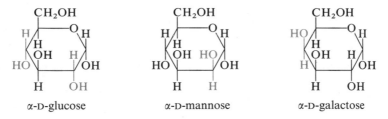

Fig. 18–14 Glucose forms two isomeric cyclic structures.

α-D-glucose α-D-mannose α-D-galactose

Fig. 18–15 α-D-mannose and α-D-galactose are epimers of glucose, differing in configuration at only one carbon atom.

The conformational structures in Fig. 18–14 show that the α-isomer must have one substituent in an axial position (the carbon-1 hydroxyl), while the β-isomer can have all substituents in the more stable equatorial position (see section 16–3), and is therefore formed in excess.

Other monosaccharides also exist in ring form and can form α and β isomers. Figure 18–15 depicts α-D isomers of common monosaccharides, to show that they differ from glucose only in the configuration at one carbon atom.

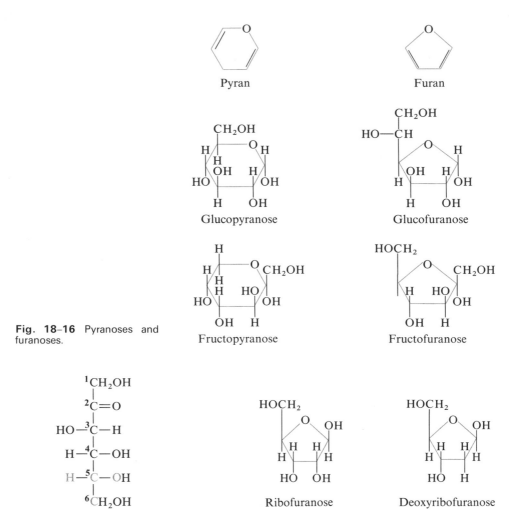

Fig. 18–16 Pyranoses and furanoses.

Pyran

Furan

Glucopyranose

Glucofuranose

Fructopyranose

Fructofuranose

Fig. 18–17 D-fructose.

Ribofuranose

Deoxyribofuranose

Fig. 18–18 Two important pentoses.

In the preceding discussion all the cyclic sugars were written in the more stable six-membered-ring forms. Sugars in this six-membered ring are called *pyranoses*, after the six-membered oxygen heterocycle pyran. Monosaccharides are also found in nature as the less-stable five-membered cyclic systems called *furanoses*, after furan (see Fig. 18–16).

Fructose. Fructose is a ketohexose rather than an aldohexose, and as a result has one less asymmetric carbon atom. Fructose in the open-chain form (Fig. 18–17) has a configuration at carbon-5 similar to D-glyceraldehyde, and is called D-*fructose*, even though it is strongly levorotatory. (Fructose is also known as *levulose*, due to

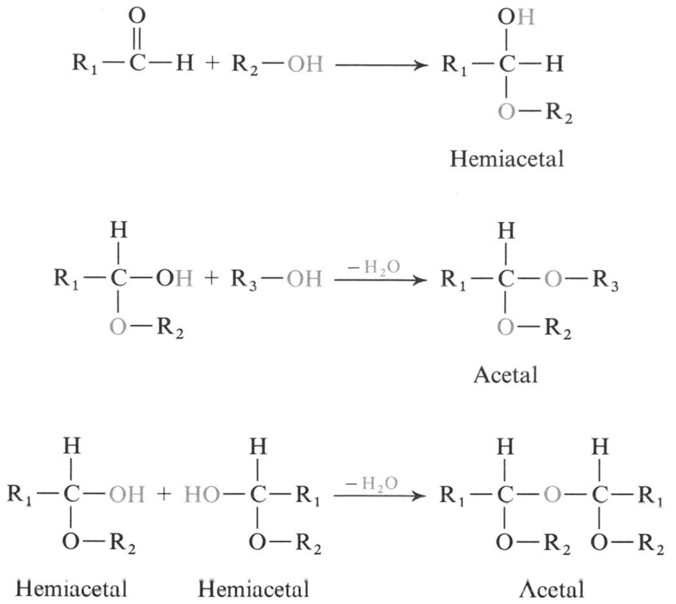

$$R_1-\underset{\displaystyle \overset{\displaystyle \|}{O}}{C}-H \; + \; R_2-OH \;\longrightarrow\; R_1-\underset{\displaystyle O-R_2}{\overset{\displaystyle OH}{C}}-H$$

Hemiacetal

$$R_1-\underset{\displaystyle O-R_2}{\overset{\displaystyle H}{C}}-OH \; + \; R_3-OH \;\xrightarrow{-H_2O}\; R_1-\underset{\displaystyle O-R_2}{\overset{\displaystyle H}{C}}-O-R_3$$

Acetal

$$R_1-\underset{\displaystyle O-R_2}{\overset{\displaystyle H}{C}}-OH \; + \; HO-\underset{\displaystyle O-R_2}{\overset{\displaystyle H}{C}}-R_1 \;\xrightarrow{-H_2O}\; R_1-\underset{\displaystyle O-R_2}{\overset{\displaystyle H}{C}}-O-\underset{\displaystyle O-R_2}{\overset{\displaystyle H}{C}}-R_1$$

Hemiacetal   Hemiacetal   Acetal

Fig. 18–19 Acetal formation can take place between two hemiacetals.

its strong levorotation.) When fructose exists uncombined in various fruits, it has a pyranose structure resulting from hemiketal formation by the ketone at carbon-2 and the hydroxyl at carbon-6. When fructose exists combined, as it does in sucrose, it has the furanose structure which results from reaction at carbon-2 and carbon-5 to yield the hemiketal.

Pentoses. Monosaccharides containing only five carbon atoms are called *pentoses*. Pentoses are commonly found in plants, and a few pentoses play important roles as constituents in nucleic acids, the genetic information carriers. Figure 18–18 shows the two most important pentoses, ribose and deoxyribose, in the furanose forms.

18–7 DISACCHARIDES

The three most important disaccharides are *sucrose*, *maltose*, and *lactose*. They are called disaccharides because on hydrolysis they yield two monosaccharide molecules.

Let us first discuss the structure of the molecules on a general basis. In Chapter 16 we mentioned that hemiacetals, formed by the reaction of an aldehyde with an alcohol, were not ordinarily stable, but could react with a second alcohol molecule to form the more stable acetal. As Fig. 18–19 shows, acetal (ketal) formation may result from the reaction of one hemiacetal with another.

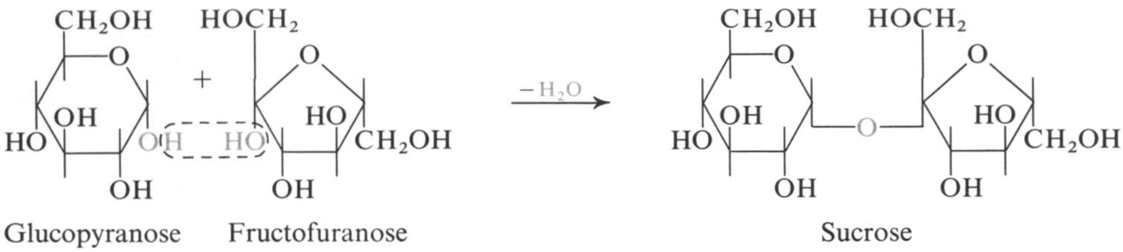

Glucopyranose   Fructofuranose   Sucrose

Fig. 18-20 Sucrose formed from a hemiacetal (glucose) and a hemiketal (fructose).

Since monosaccharides are cyclic hemiacetals, acetal can form between two monosaccharide molecules to yield a disaccharide held together by an acetal linkage.

Sucrose. Sucrose is ordinary sugar, found in high concentration in sugar cane and sugar beet. On hydrolysis sucrose yields equal amounts of glucose and fructose. Sucrose does not reduce the Cu^{2+} ion in Benedict's solution, as do glucose and fructose in the ring forms, and hence the hemiacetal function of glucose and the hemiketal function of fructose must no longer be available.* Thus sucrose is a disaccharide formed by the condensation of the hemiacetal hydroxyl group of glucose and the hemiketal hydroxyl group to form the acetal (Fig. 18-20).

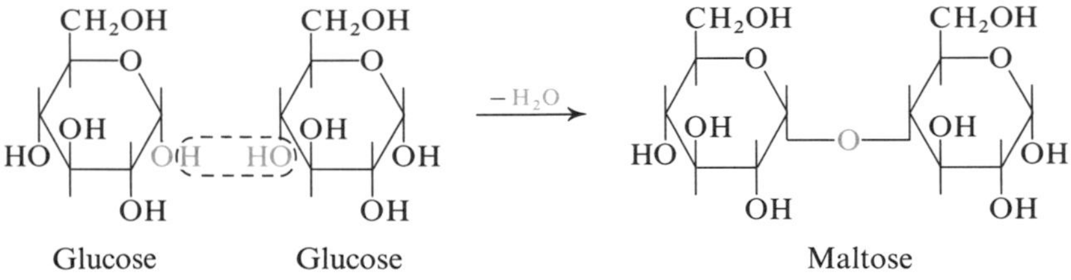

Glucose   Glucose   Maltose

Fig. 18-21 Maltose, an acetal formed from two hemiacetals.

Maltose. When starch is partially hydrolized under natural conditions, *maltose* or malt sugar is formed. On further hydrolysis, maltose yields two molecules of glucose. Since maltose is a reducing sugar (it reduces Cu^{2+} in Benedict's solution), the hemiacetal hydroxyl groups of both glucose molecules cannot be tied up in an acetal linkage. At least one of the two hemiacetal units must be free in order to form the aldehyde group (by opening the ring) needed to react with the Benedict's solution.

The structure of maltose shown in Fig. 18-21 is consistent with these data. One glucose molecule forms an acetal linkage with a second glucose molecule by reacting with the hydroxyl group at carbon-4 rather than the hydroxyl group at carbon-1 of the second molecule. Thus the maltose molecule contains one acetal unit and one hemiacetal unit.

---

* Glucose in the ring form has no aldehyde group, but the hemiacetal group, on hydrolysis, gives an aldehyde which can reduce Cu^{2+} to Cu_2O. Fructose, in the open-chain form, also reduces Cu^{2+}, even though it has no aldehyde function.

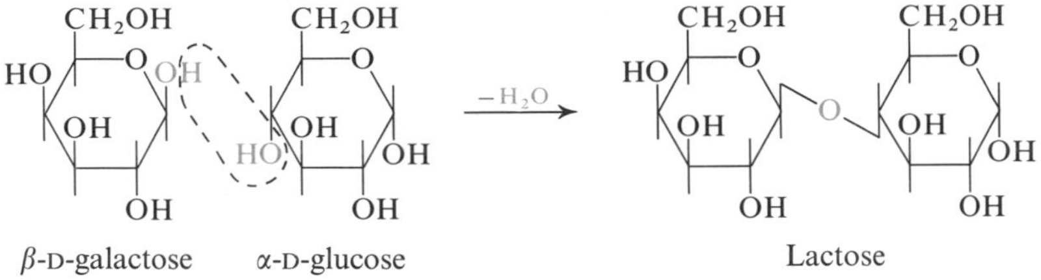

Fig. 18–22 Lactose, a β-1,4-linkage.

Lactose. Lactose, the sugar found in the milk of mammals and called *milk sugar*, is also a reducing sugar, and must therefore possess a hemiacetal function. Hydrolysis of lactose gives glucose and galactose, and further structural studies indicate that lactose is a disaccharide formed by acetal formation between the hemiacetal hydroxyl of β-D-galactose and the hydroxyl at carbon-4 of α-D-glucose (Fig. 18–22).

Sucrose and maltose form acetal linkages between hydroxyl groups that are on the same side of the molecule relative to the hydroxyl-methylene ($-CH_2OH$) groups; such a bond is termed an α-linkage. The linkage between galactose and glucose to form lactose results from acetal formation between hydroxyl groups of opposite orientation relative to the hydroxyl methylene group, a β-linkage. Thus we describe the linkage in maltose as being α-1,4 in order to indicate an α acetal linkage between the hydroxyl group at carbon-1 and the hydroxyl group at carbon-4 of a second glucose molecule. The geometry of the linkage between monosaccharide units has important consequences for man, as we shall see in the next section and in Chapter 19.

18–8 POLYSACCHARIDES

Polysaccharides are high-molecular-weight carbohydrate molecules formed by the linkage of many monosaccharide units. The principle of bond formation between the monosaccharide units to give polysaccharides is identical to that discussed above for disaccharides: acetal formation.

Although there are many different polysaccharides found in nature, we shall confine our discussion to only three types: starch, cellulose, and glycogen. All three have glucose as the monosaccharide unit, and differ only in the way the glucose molecules are linked together.

Starch. The principal dietary source of carbohydrate for man is starch. Starch, found in high concentration in plants such as peas, beans, potatoes, and corn, is a mixture of two types: *amylose* and *amylopectin*. Both give α-D-glucose when they are subjected to acid hydrolysis, but differ in the way they are bonded. Amylose has a straight-chain structure,* in which approximately 1800 glucose units are joined in a manner analogous to that found in maltose. Figure 18–23 gives a partial structure of amylose.

---

*   Actually the molecule is not straight; it is coiled into a spiral arrangement called a *helix*.

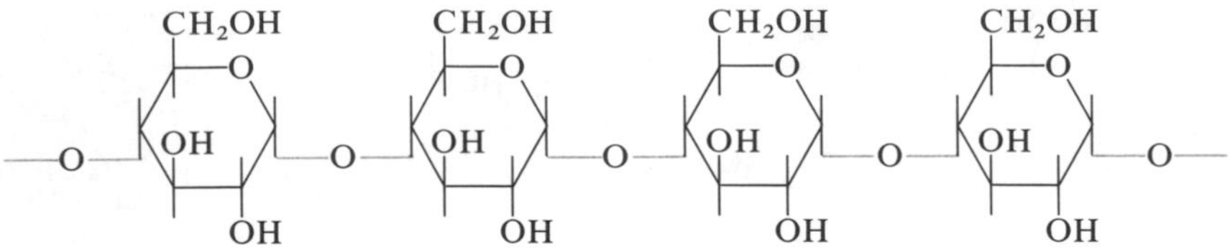

Fig. **18–23** Amylose (α-1,4-glucose links).

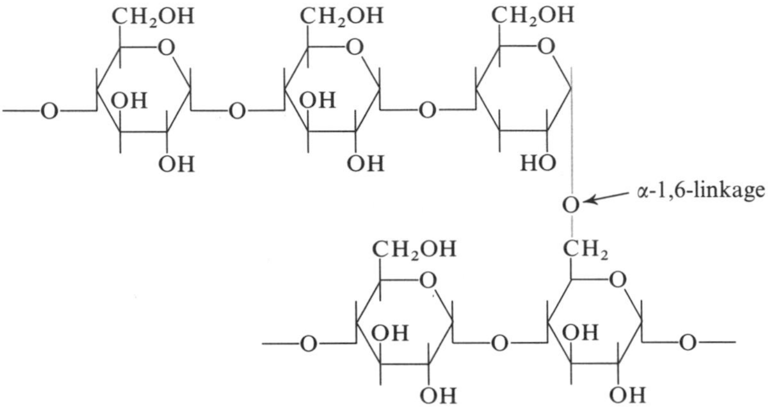

Fig. **18–24** Partial structure of amylopectin.

Amylopectin, instead of having the simple straight-chain structure of amylose, has a branched structure. It contains the α-1,4 linkage, but also an α-1,6 linkage, with the α-1,4 linkage predominant (∼96%). Figure 18–24 gives a partial picture of the structure of amylopectin, which shows that amylose-like α-1,4-linked chains are terminated by an α-1,6 linkage to a second amylose-like chain. This creates a branching effect in the molecule, which rules out the helical structure found in amylose.

Cellulose. Cellulose, by far the most abundant polysaccharide, is the major structural constituent of plants; it is found in plants' cell walls and woody tissue. Complete acid hydrolysis of cellulose yields glucose, but if the reaction is not allowed to proceed to completion, a disaccharide may be isolated. The disaccharide, *cellubiose*, contains β-1,4-linked glucose molecules indicative of the structure of cellulose (Fig. 18–25). Cellulose is a long (300 to 2500 glucose units) polymer of glucose joined by β-1,4 links (Fig. 18–26). The β-1,4 linkage makes it impossible for the cellulose molecule to coil into a helical structure. This is in part responsible for the differences between the properties of cellulose and those of amylose.

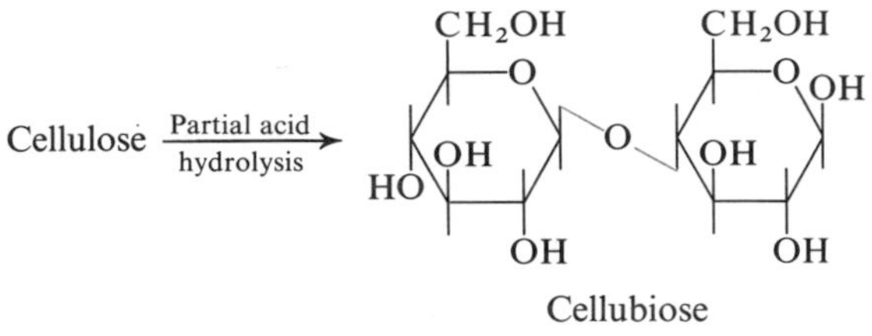

Cellulose $\xrightarrow[\text{hydrolysis}]{\text{Partial acid}}$

Cellubiose

Fig. 18–25 Cellubiose with its β-1,4-link.

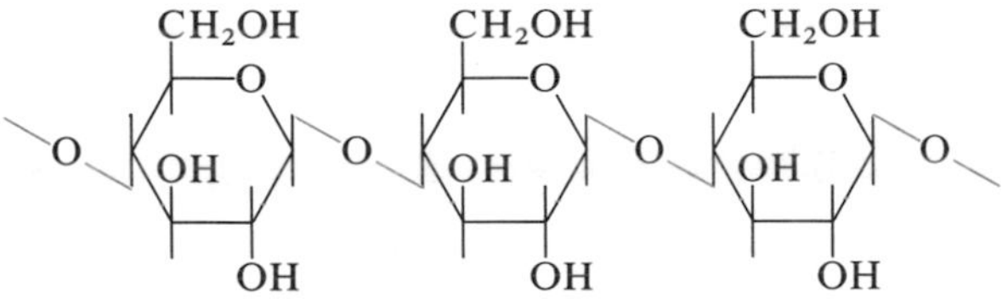

Fig. 18–26 Cellulose (partial structure).

Glycogen. The principal source of energy in living systems is glucose. In plants the glucose is stored, ready for use, in the form of starch, while in animals glucose is stored as *glycogen*. The structure of glycogen is very similar to that of amylopectin, containing 1700 to 22,000 glucose units bonded in a branched-chain fashion. The linkages are α-1,4 and α-1,6, but the branching is somewhat more extensive than that of amylopectin (91% α-1,4 compared with 9% α-1,6). We shall discuss the function of glycogen as an energy storehouse further in Chapter 19.

PROTEINS

The structure of proteins is a much more complex problem than the structure of lipids or carbohydrates. The number of different proteins active in living systems seems to be without limit, and the detailed description of the structures of any one of these represents an enormous problem. However, interpreting the importance of proteins to growth, reproduction, energy transfer, and nearly all other biochemical processes demands a knowledge of their structure.

The first information relating to the structure of proteins comes from studies of hydrolysis experiments. When any protein is hydrolyzed (usually in acid), the product is a mixture of amino acids, with the occasional presence of other compounds. Thus the protein molecule must be made up of amino acid residues joined together.

18–9 AMINO ACIDS

Studies of the products of protein hydrolysis indicate that proteins are made up largely of just twenty amino acids. Their structures are shown in Fig. 18–27.

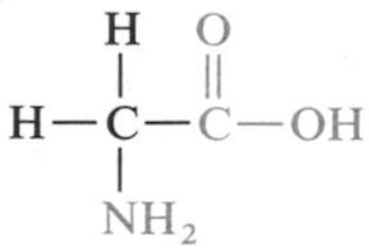

$$
\begin{array}{c}
H \quad O \\
\mid \quad \parallel \\
H-C-C-OH \\
\mid \\
NH_2
\end{array}
$$

Glycine (gly)

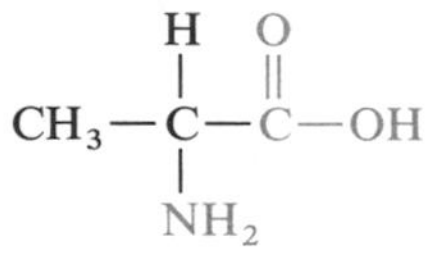

$$
\begin{array}{c}
H \quad O \\
\mid \quad \parallel \\
CH_3-C-C-OH \\
\mid \\
NH_2
\end{array}
$$

Alanine (ala)

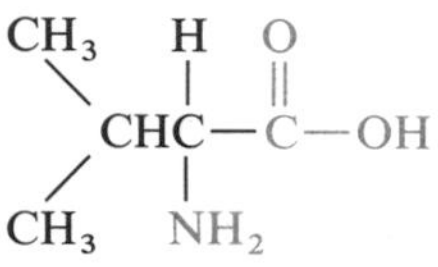

$$
\begin{array}{c}
CH_3 \quad H \quad O \\
\diagdown \quad \mid \quad \parallel \\
CHC-C-OH \\
\diagup \quad \mid \\
CH_3 \quad NH_2
\end{array}
$$

Valine

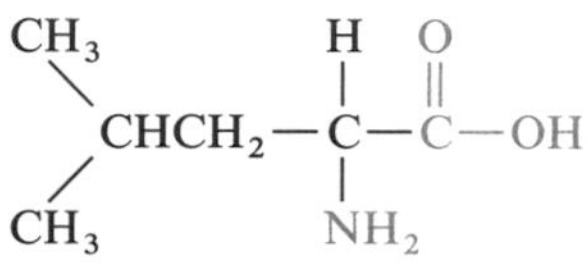

$$
\begin{array}{c}
CH_3 \qquad H \quad O \\
\diagdown \qquad \mid \quad \parallel \\
CHCH_2-C-C-OH \\
\diagup \qquad \mid \\
CH_3 \qquad NH_2
\end{array}
$$

Leucine (leu)

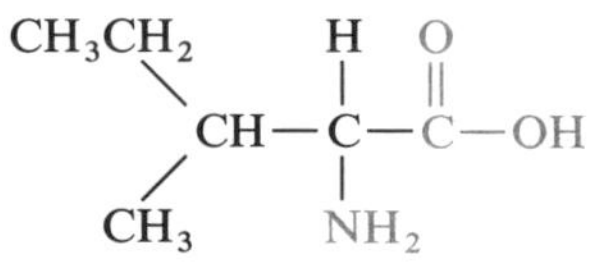

$$
\begin{array}{c}
CH_3CH_2 \qquad H \quad O \\
\diagdown \qquad \mid \quad \parallel \\
CH-C-C-OH \\
\diagup \qquad \mid \\
CH_3 \qquad NH_2
\end{array}
$$

Isoleucine (ileu)

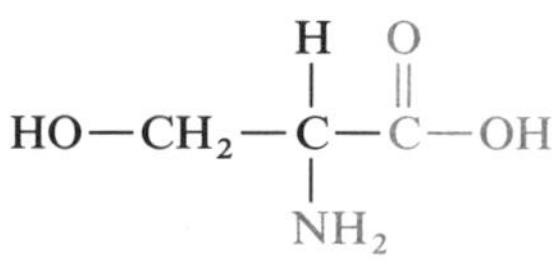

$$
\begin{array}{c}
H \quad O \\
\mid \quad \parallel \\
HO-CH_2-C-C-OH \\
\mid \\
NH_2
\end{array}
$$

Serine (ser)

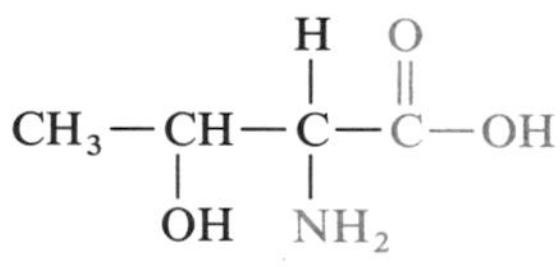

$$
\begin{array}{c}
H \quad O \\
\mid \quad \parallel \\
CH_3-CH-C-C-OH \\
\mid \qquad \mid \\
OH \quad NH_2
\end{array}
$$

Threonine (thr)

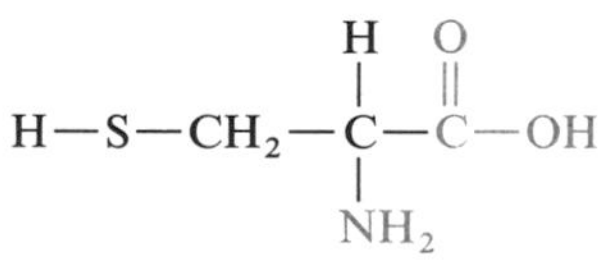

$$
\begin{array}{c}
H \quad O \\
\mid \quad \parallel \\
H-S-CH_2-C-C-OH \\
\mid \\
NH_2
\end{array}
$$

Cysteine (cySH)

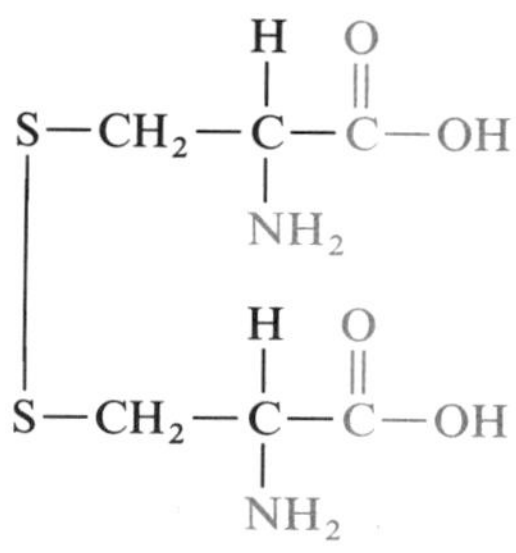

$$
\begin{array}{c}
H \quad O \\
\mid \quad \parallel \\
S-CH_2-C-C-OH \\
\mid \qquad \mid \\
\quad \qquad NH_2 \\
\mid \\
H \quad O \\
\mid \quad \parallel \\
S-CH_2-C-C-OH \\
\mid \\
NH_2
\end{array}
$$

Cystine (cyS)$_2$

$$
\begin{array}{c}
H \quad O \\
\mid \quad \parallel \\
CH_3-S-CH_2CH_2-C-C-OH \\
\mid \\
NH_2
\end{array}
$$

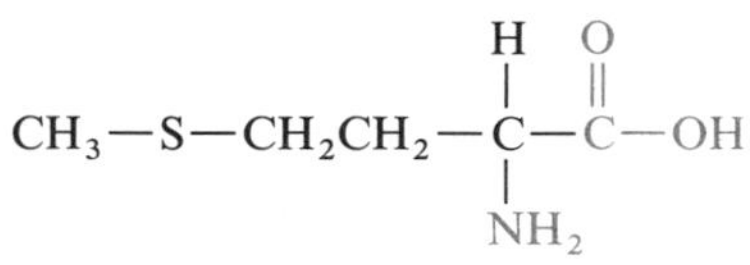

Methionine (met)

Fig. 18–27 Structures of amino acids obtained from the hydrolysis of proteins.

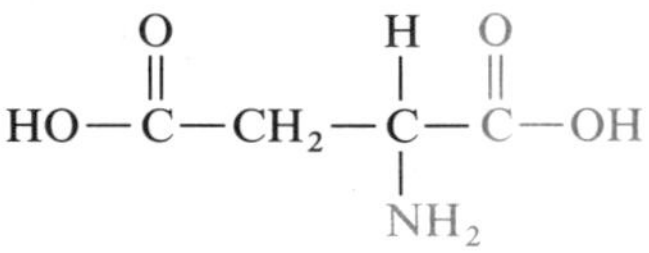

Aspartic acid (asp)

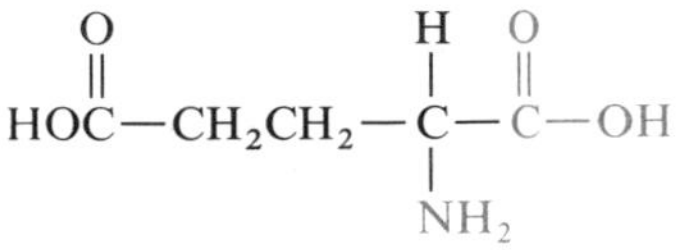

Glutamic acid (glu)

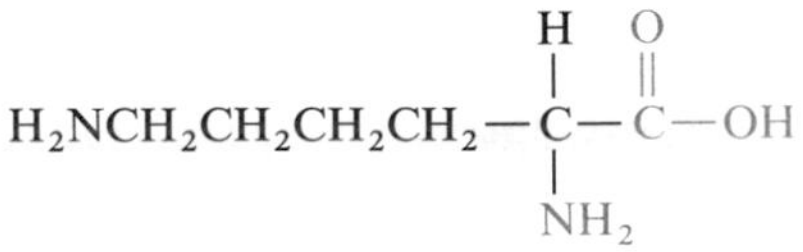

Lysine (lys)

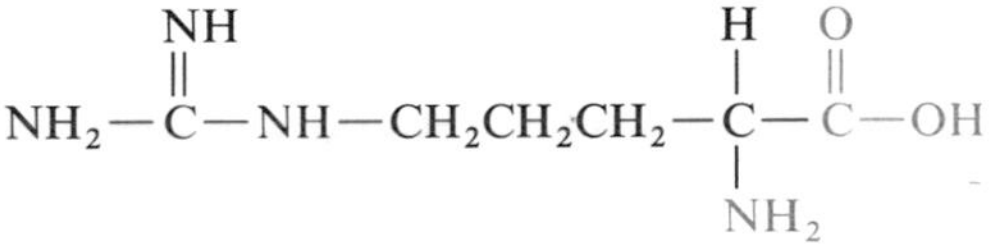

Arginine (arg)

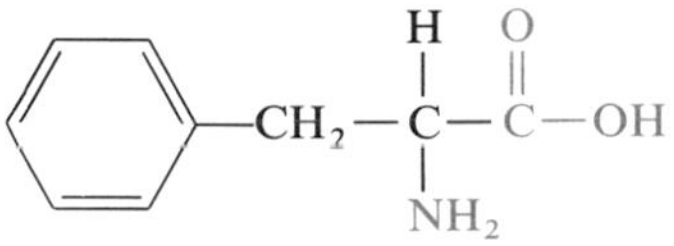

Phenylalanine (phe)

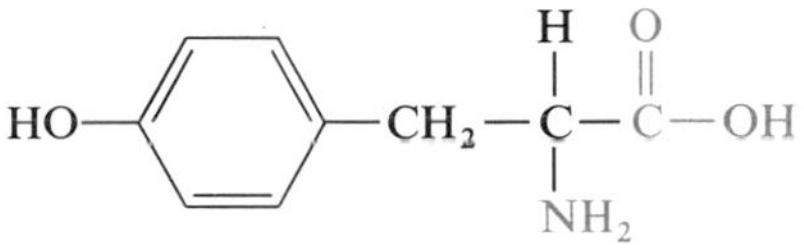

Tyrosine (tyr)

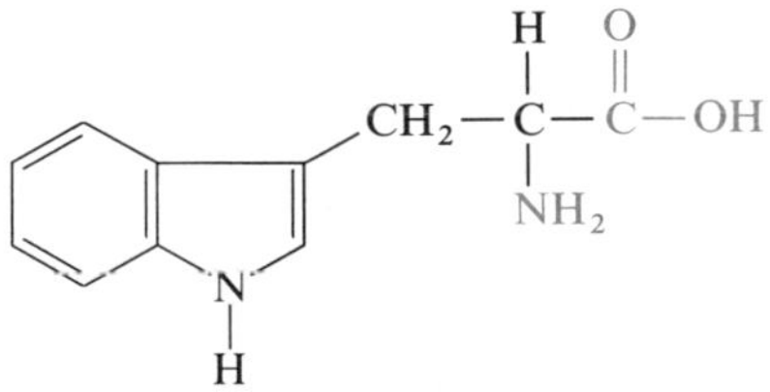

Tryptophan (try)

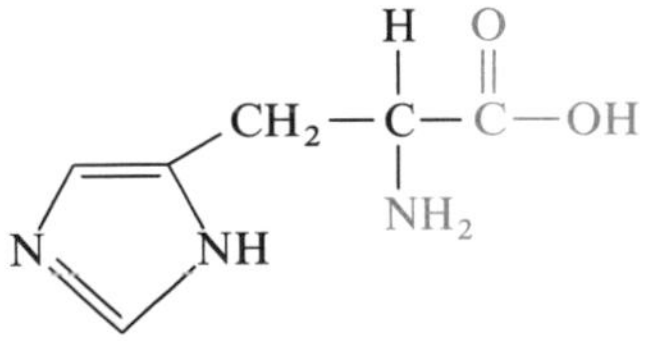

Histidine (his)

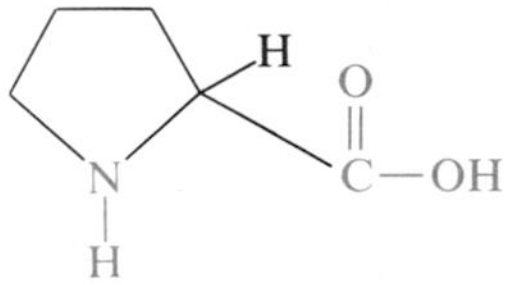

Proline (pro)

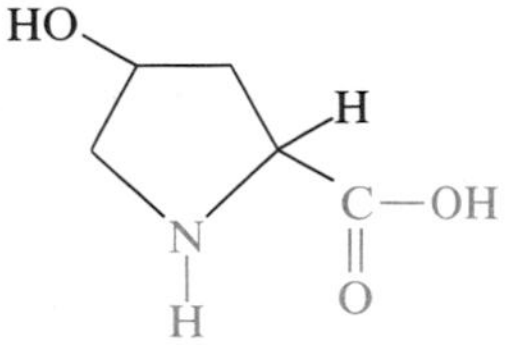

Hydroxyproline (hypro)

Fig. 18-27 (*continued*)

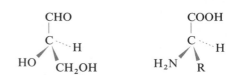

Fig. 18-28 The relationship between essential
amino acids and L-glyceraldehyde.

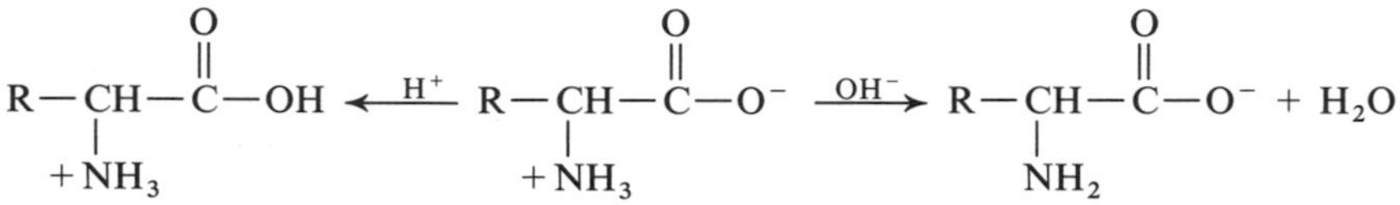

L-glyceraldehyde   L-amino acid

$$R-CH-\overset{\overset{\displaystyle O}{\|}}{C}-OH \xleftarrow{H^+} R-CH-\overset{\overset{\displaystyle O}{\|}}{C}-O^- \xrightarrow{OH^-} R-CH-\overset{\overset{\displaystyle O}{\|}}{C}-O^- + H_2O$$
$$\quad\ \ \ \overset{|}{+NH_3} \qquad\qquad\qquad \overset{|}{+NH_3} \qquad\qquad\qquad \overset{|}{NH_2}$$

Fig. 18-29 Amino acids in the zwitterionic form can act as buffers, neutralizing excess acid and
excess base.

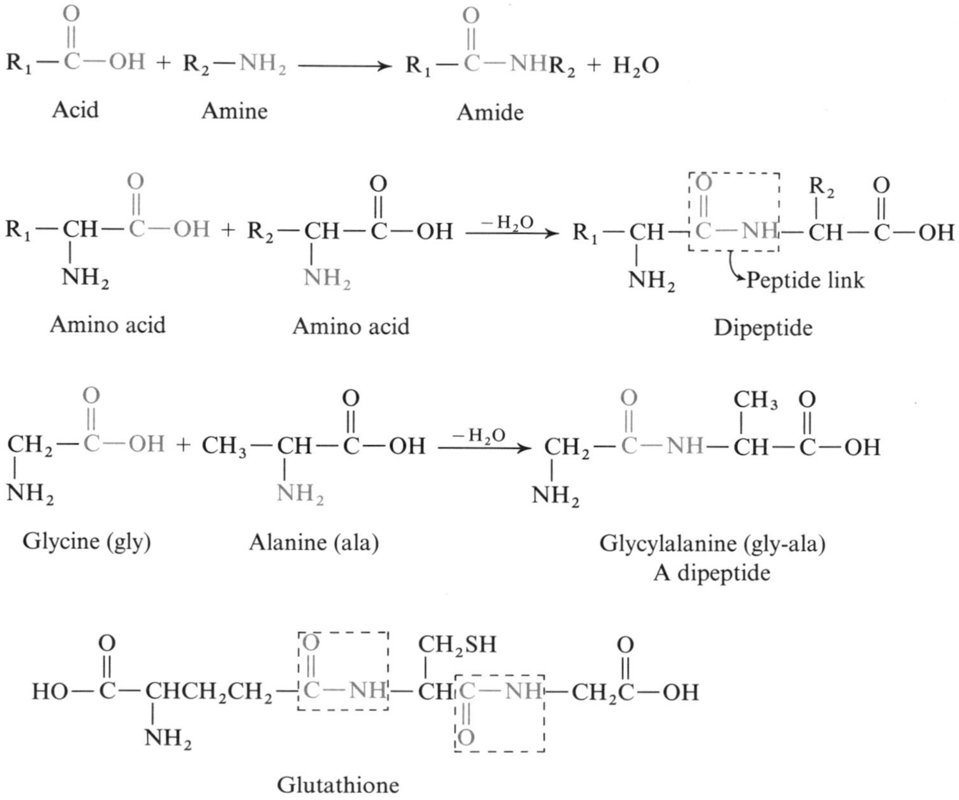

$$R_1-\overset{\overset{\displaystyle O}{\|}}{C}-OH + R_2-NH_2 \longrightarrow R_1-\overset{\overset{\displaystyle O}{\|}}{C}-NHR_2 + H_2O$$

  Acid   Amine   Amide

$$R_1-CH-\overset{\overset{\displaystyle O}{\|}}{C}-OH + R_2-CH-\overset{\overset{\displaystyle O}{\|}}{C}-OH \xrightarrow{-H_2O} R_1-CH-\overset{\overset{\displaystyle O}{\|}}{C}-NH-CH-\overset{\overset{\displaystyle O}{\|}}{C}-OH$$
$$\qquad\ \overset{|}{NH_2} \qquad\qquad\quad \overset{|}{NH_2} \qquad\qquad\qquad\ \overset{|}{NH_2} \qquad \overset{|}{}\ \overset{|}{R_2}$$

  Amino acid   Amino acid   Dipeptide

Peptide link

$$CH_2-\overset{\overset{\displaystyle O}{\|}}{C}-OH + CH_3-CH-\overset{\overset{\displaystyle O}{\|}}{C}-OH \xrightarrow{-H_2O} CH_2-\overset{\overset{\displaystyle O}{\|}}{C}-NH-CH-\overset{\overset{\displaystyle O}{\|}}{C}-OH$$
$$\overset{|}{NH_2} \qquad\qquad\quad \overset{|}{NH_2} \qquad\qquad\qquad\ \overset{|}{NH_2} \qquad\quad \overset{|}{CH_3}$$

  Glycine (gly)   Alanine (ala)   Glycylalanine (gly-ala)
                                                          A dipeptide

$$HO-\overset{\overset{\displaystyle O}{\|}}{C}-CHCH_2CH_2-\overset{\overset{\displaystyle O}{\|}}{C}-NH-CH-\overset{\overset{\displaystyle CH_2SH}{|}}{C}-NH-CH_2\overset{\overset{\displaystyle O}{\|}}{C}-OH$$
$$\qquad\qquad\ \overset{|}{NH_2}$$

  Glutathione

Fig. 18-30 Some peptide links.

Although there exists a wide variation in structural types among the common amino acids, there are also certain similarities. They all contain an amino group attached to the same carbon atom as the carboxyl group, and are designated α-amino acids. Further, the α-carbon atom is asymmetrically substituted in all cases except glycine. The optically active natural amino acids obtained from proteins have a configuration at the α-carbon atom like that of L-glyceraldehyde, and are therefore designated as L-amino acids (Fig. 18–28).

All amino acids have a basic group ($-NH_2$) and an acidic group ($-COOH$), which can react with each other or with other sources of acid or base. In solution, there is a pH value at which the amino acid is electrically neutral (the *isoelectric point*) and exists as a dipolar ion called *zwitterion* (Fig. 18–29). The zwitterion reacts with excess acid or with excess base in such a fashion as to prevent a large increase in acidity or basicity of a solution. This is an important buffering mechanism for the biological control of pH.

The most important reaction of amino acids is that involving amide formation. You will recall that an acid reacts with an amine to give an amide. This reaction may also take place between the acid group of one amino acid and the amine group of a second to yield a *dipeptide*. A dipeptide, as we see in Fig. 18–30, is a molecule formed from two amino acids; the amide group is termed a *peptide link*. Note that the dipeptide still contains an acid group and an amino group and is capable of reaction at either end of the molecule with another amino acid to form a *tripeptide*. Glutathione is a naturally occurring tripeptide containing glutamic acid, cysteine, and glycine residues held together by peptide links. The tripeptides must also have free amino and acid groups in the molecule which are capable of further condensation with amino acids to form a polymeric *polypeptide*.

18-10 PRIMARY STRUCTURE OF PROTEINS

Proteins are polypeptide molecules containing from several hundred to several thousand amino acid residues held together by peptide links. Since there are only 20 amino acids in proteins, the same amino acid residue must occur in several positions in the protein chain. We can see that such an arrangement easily accounts for the huge number of naturally occurring proteins. Consider a polypeptide comprised of only 20 amino acids, with each of the common amino acids appearing once. The number of possible arrangements such a system can have is approximately 2×10^{18}. The number of possible arrangements a protein molecule can have is therefore nearly limitless.

The order in which the various amino acids are joined in a protein molecule— the *amino acid sequence*—is called the *primary protein structure*. It is very hard to work out the amino acid sequence for a protein molecule, and only recently has it been possible to attack the problem. After chemists carry out the difficult experimental processes of isolation, purification, and determination of molecular weight, they subject the protein to hydrolysis, using a strong acid or base, and analyze the

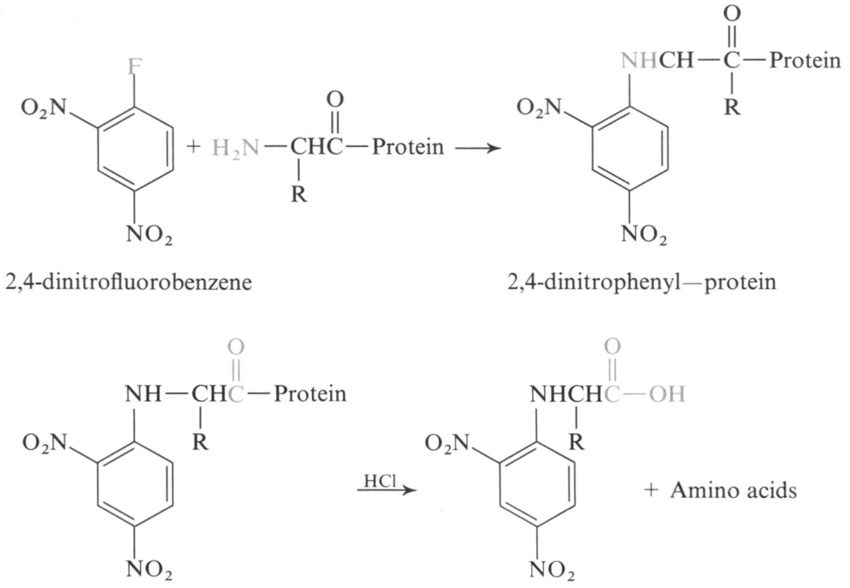

Fig. 18–31   End group analysis of protein.

hydrolysate for the constituent amino acids. Once they have determined the number and kind of amino acid residues, the really hard part of the problem begins.

When amino acids condense to form polypeptides, one end of the molecule must have a free amino group and the other end must have a free carboxyl group. These are called the N-terminal and C-terminal positions, respectively. In 1945 the British chemist Frederick Sanger elucidated the structure of the protein-hormone insulin by a process that involved, in part, the sequential analysis of the N-terminal groups.

Although insulin is a very small protein (molecular weight = 6000), the procedure worked out by Sanger may in principle be applied to larger protein molecules. He found that 2,4-dinitrofluorobenzene would combine with the terminal α-amino group to give a dinitrophenyl (DNP) derivative. Subsequent hydrolysis of the protien on analysis of the mixture would identify the amino acid containing the DNP group as the N-terminal group in the original protein (see Fig. 18–31).

Similar methods of analysis are available for C-terminal groups. Techniques have been developed which allow selective hydrolysis of the end groups, leaving the rest of the molecule intact, and with a new end group that can be analyzed. It is also possible to partially hydrolyze the protein molecule into small polypeptide fragments which can be analyzed separately. And by using a combination of all these techniques, Sanger and his fellow workers were able to explain the primary structure of insulin, illustrated in Fig. 18–32. There are two polypetide chains linked by disulfide bridges formed between two cysteine residues.

Using techniques similar to those applied to insulin, chemists have determined the amino acid sequence for other proteins, including ribonuclease (124 amino acids), hemoglobin (67,000 molecular weight), and myoglobin (17,000 molecular weight).

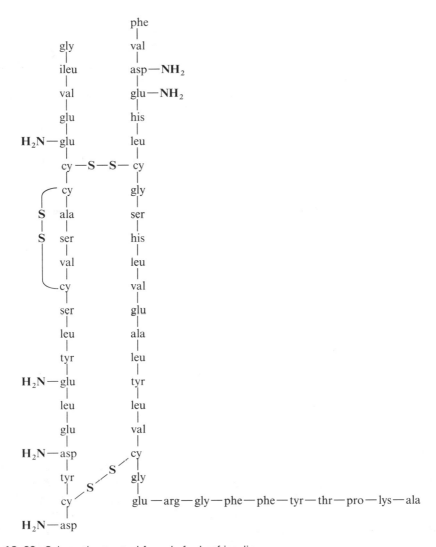

Fig. 18–32 Schematic structual formula for beef insulin.

18-11 SECONDARY AND TERTIARY STRUCTURE OF PROTEINS

It might be expected that the long-chain polypeptide would be arranged in a random fashion and have a completely flexible and variable conformation. The conformations of protein molecules, however, appear to be fixed; they are neither random nor variable. It is the specific conformation of the protein molecule that is called the *secondary protein structure*.

In 1951, Linus Pauling and R. B. Corey studied x-ray data of amino acids and small peptides and found the bond angles and distances shown in Fig. 18–33. The length of the bond between the nitrogen and the carbonyl carbon is shorter than

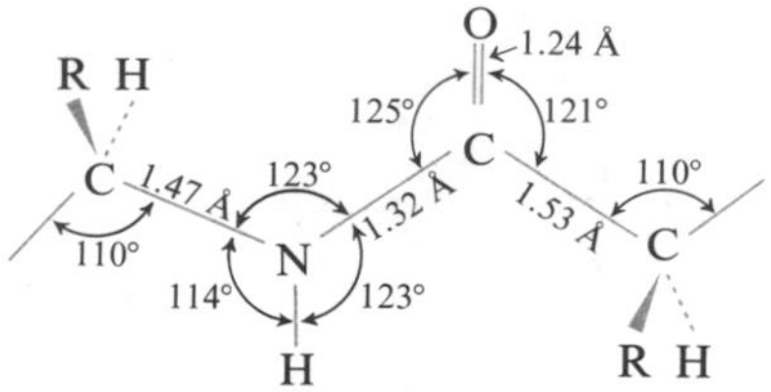

Fig. 18-33 The dimensions of the peptide link in proteins.

that expected for a carbon–nitrogen single bond. It is assumed that this bond has considerable double-bond character; i.e., the peptide link may be considered a resonance hybrid of the structures shown in Fig. 18–34.

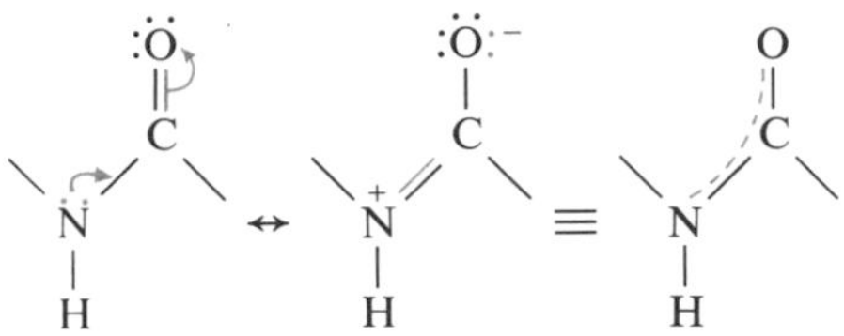

Fig. 18-34 Resonance in the amide group.

The double-bond character of the carbon–nitrogen bond presents a barrier to rotation about this bond, and suggests that the peptide-amide linkage probably has a planar structure, with the attached groups in a *trans* orientation, as depicted in Fig. 18–34. Since each nitrogen in the molecule contains a hydrogen which should be capable of hydrogen bonding to a nearby carbonyl oxygen (N—H---O=C), Pauling and Corey assumed that the most stable conformation of a protein molecule would be that in which hydrogen bonding could occur to the greatest extent.

On the basis of this assumption, they proposed a structure in which the protein molecule was arranged in a spiral conformation called an *α-helix* (Fig. 18–35). The α-helix is so constructed that there are 3.6 amino acid residues for each turn of the helix (each amino acid takes up a 100° turn). Each peptide link forms a hydrogen bond with the third peptide link along the chain, such that each turn of the helix is held by hydrogen bonds. Although hydrogen bonds are weak, the cumulative effect of several hundred hydrogen bonds in a protein molecule is enough to ensure struc-

Fig. 18-35 Schematic representation of the protein α-helix.

tural integrity. This structural integrity is essential to the biological function of the protein, and if it is irreversibly destroyed the protein can no longer function.

The helical structure of proteins has been demonstrated by x-ray analysis in a few cases, but it is not true that all proteins have this structure. The important point is that stereochemical factors and weak interactions give proteins a discrete, non-random secondary structure.

Finally, let us note that the secondary structure of proteins may be folded and bent and wrapped around to give what is known as the *tertiary* structure of the protein. This tertiary structure does not appear to be a random orientation either, but a fixed and specific orientation. Of all the nearly infinite orientations that such a molecule could take, one arrangement seems to be favored. Presumably the ultimate arrangement represents a low energy system determined by the primary and secondary structures of the molecule. It has not been possible to take all the factors that determine tertiary structure and construct a theory which would explain it. As of this writing, however, the tertiary structure of several proteins has been deduced from x-ray studies.

18-12 CLASSIFICATION OF PROTEINS

In the previous discussion we considered those proteins which on hydrolysis yield only amino acids: the *simple proteins*. A second class of proteins, the *conjugated proteins,* yield, on hydrolysis, materials other than amino acids.

Simple Proteins

a) Albumins—from blood serum and egg white
b) Globulins—from blood serum
c) Glutelins—from cereal grains: wheat, corn, and barley
d) Scleroproteins—such as the *elastin* of connective tissue, the *keratin* of hair, horn, hoofs, and feathers, and *collagen*, the most abundant animal protein
e) Protamines—low-molecular-weight basic proteins usually associated with nucleic acids
f) Histones—such as *globin*, the protein portion of hemoglobin

Conjugated Proteins

a) Phosphoproteins—with one or more phosphate group attached usually to the —OH group of a serine molecule, such as *pepsin*, found in the stomach
b) Lipoproteins—with a lipid unit bonded in some fashion to the protein; approximately 70% of the lipid found in blood plasma is bound to protein
c) Mucoproteins—with large amounts of carbohydrates
d) Glycoproteins—with small ($\sim 4\%$) amounts of bound carbohydrate
e) Chromoproteins—with a group attached affording a visible color, including hemoglobin and other respiratory pigments

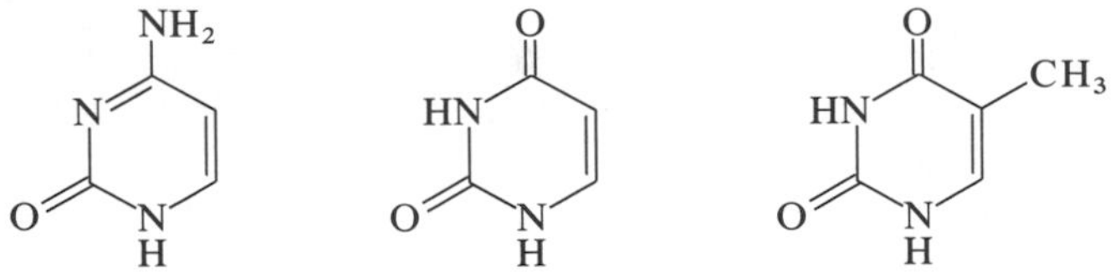

Cytosine Uracil Thymine

Pyrimidines

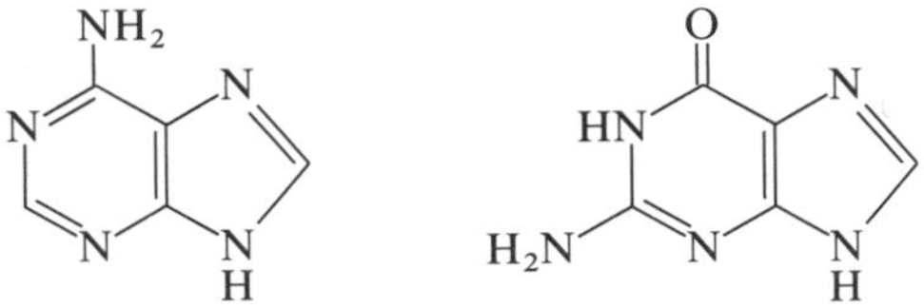

Adenine Guanine

Purines

Fig. 18–36 Bases common in DNA and RNA molecules.

NUCLEIC ACIDS

No area of scientific inquiry in the twentieth century has generated more interest and speculation than the study of nucleic acids. It has been demonstrated that nucleic acids are responsible for the transfer of genetic information in most, if not all, instances. If scientists can unlock the secrets of the structure and function of nucleic acids, in the future they may be able to control genetics. The medical, sociological, and moral implications are staggering.

Since nucleic acids have the property of replication—that is, under certain conditions they can reproduce themselves—there is no question that an understanding of nucleic acids is required before we can understand the molecular basis of life. The recently reported synthesis of an active DNA molecule from nonliving agents was considered important enough to warrant front-page headlines in the newspapers. Although this remarkable feat will probably not be accepted as a synthesis of life, it certainly makes it seem probable that scientists may soon be able to produce simple living systems synthetically.

The nucleic acid macromolecules, found in all living cells and in all viruses, are of two major types: *deoxyribose nucleic acids* (DNA) and *ribose nucleic acids* (RNA).

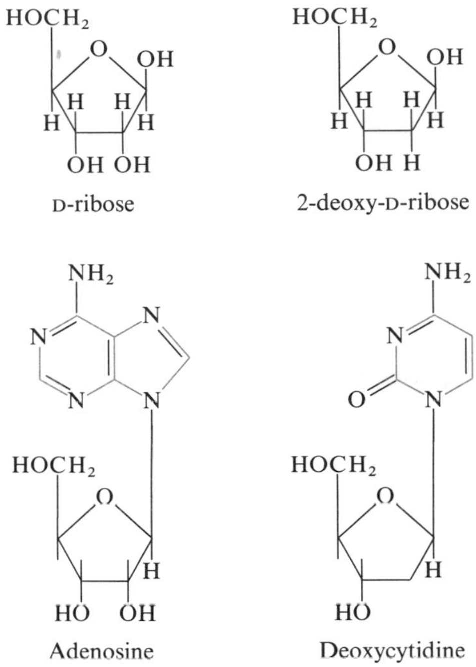

Fig. 18–37 Typical nucleosides.

DNA is found in the nucleus of the cell, while RNA is found in the cellular fluid outside the nucleus, the *cytoplasm*. Both are found bonded to proteins as nucleoproteins, but techniques are available for separating the pure nucleic acid from the protein.

18–13 POLYNUCLEOTIDE STRUCTURE

Nucleic acids are very large polymeric molecules comprised of repeating units called *nucleotides*. The nucleotide unit contains a phosphate residue, a sugar residue, and a base residue. The base residues are organic heterocyclic bases belonging to the pyrimidine class and the purine class. Figure 18–36 shows the pyrimidines (cytosine, uracil, and thymine) and purines (adenine and guanine) which are commonly found in DNA and RNA molecules.

   The nucleotide units of DNA and RNA molecules have the purine and pyrimidine bases attached to the deoxyribose or ribose molecules to give nucleosides. Figure 18–37 shows structures of the pentoses (ribose and deoxyribose), as well as typical nucleoside structures.

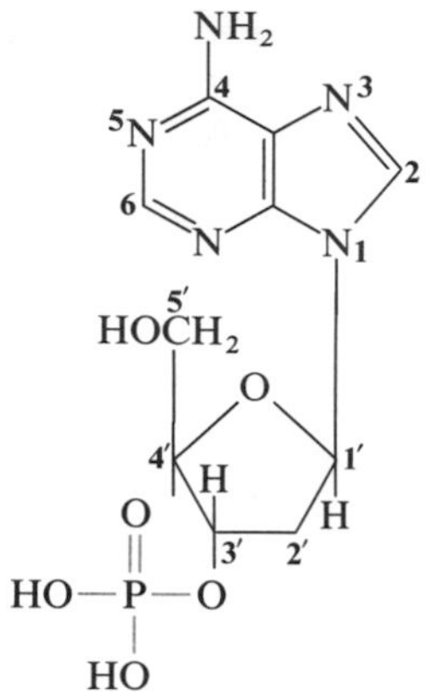

Deoxy-3′-adenylic acid
(deoxyadenosine-3′-phosphate)

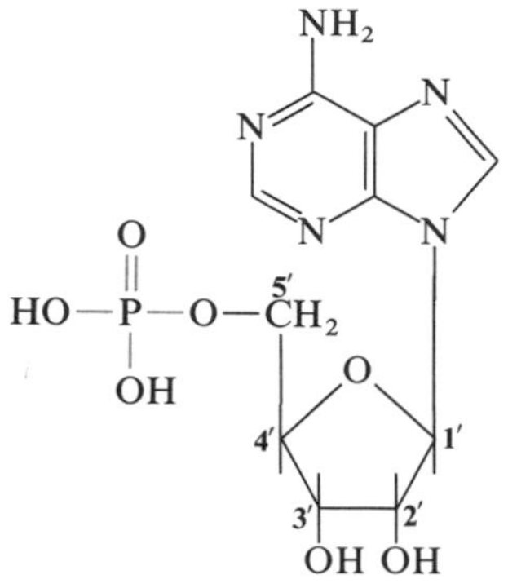

5′-adenylic acid
(adenosine-5′-phosphate)

Fig. 18–38 Typical nucleotides.

Nucleotides are phosphate esters of nucleosides, with a phosphate group attached to the pentose residue. The base (purine or pyrimidine) is attached to the carbon-1′ position* of the pentose and the phosphate is attached to the carbon-3′ or carbon-5′ position. Figure 18–38 shows two typical nucleotides.

Nucleotides represent the basic unit of nucleic acids; that is, DNA and RNA molecules are composed of repeating nucleotide units. The nucleotides are joined through the phosphate residues by the formation of phosphate esters. A phosphate on the 5′ position of the pentose reacts with the free hydroxyl on the 3′ position of a second nucleotide molecule, forming an ester linkage. The phosphate on the 5′ position of the second nucleotide can link with the 3′ position of a third nucleotide, and the process can continue until a polynucleotide is formed.

Figure 18–39 depicts a partial structure of DNA, along with a schematic representation of a polynucleotide. The schematic representation shows that all DNA molecules are essentially the same with respect to the sugar-phosphate backbone; the same is true for RNA molecules. The differences in nucleic acids, and therefore the differences in action, must be due to differences in the relative number of bases,

---

*   Since the positions in the purine and pyrimidine rings are numbered (1, 2, 3, etc.), the carbons of the ribose rings are indicated with primed numbers (1′, 2′, 3′, etc.) to avoid confusion.

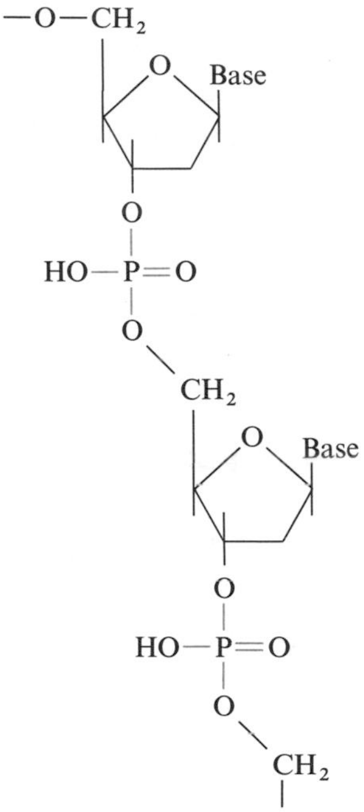

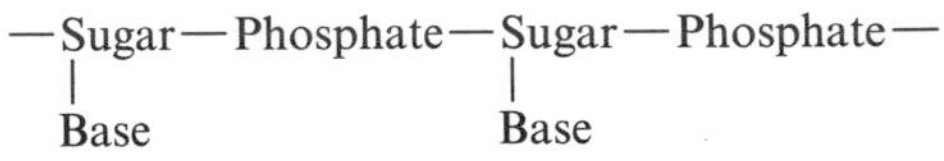

$-O-CH_2$

O Base

O

$HO-P=O$

O

CH_2

O Base

O

$HO-P=O$

O

CH_2

—Sugar—Phosphate—Sugar—Phosphate—

Base   Base

Fig. 18–39 Schematic partial structures of DNA.

and—more important—to differences in the sequence of attachment of the bases. Even though there are only a few bases to consider, the number of possible sequences is enormous when one considers that a single polynucleotide can contain more than 20,000 nucleotide units.

18–14 STRUCTURE OF DNA

Studies of the basic constitution of several DNA preparations indicated that the number of adenine molecules always equals the number of thymine molecules (A = T) and the number of guanine molecules equals the number of cytosine molecules (G = C). This information, along with x-ray data which indicated a helical structure for DNA, served as a basis for understanding the structure of DNA.

In 1953, American chemist James Watson and British chemist Francis Crick proposed a structure for DNA in which two parallel polynucleotide chains were twisted in a helix. The attached bases were pointed toward the middle of the helix and served to hold one polynucleotide strand to the other by virtue of hydrogen bonding. Adenine in one chain would always be opposite thymine in the second chain, and guanine would always be opposite cytosine, thus explaining the A = T and G = C phenomenon.

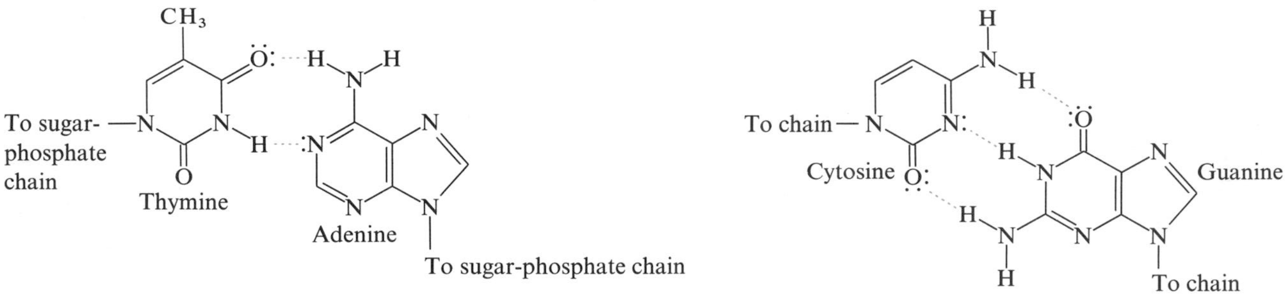

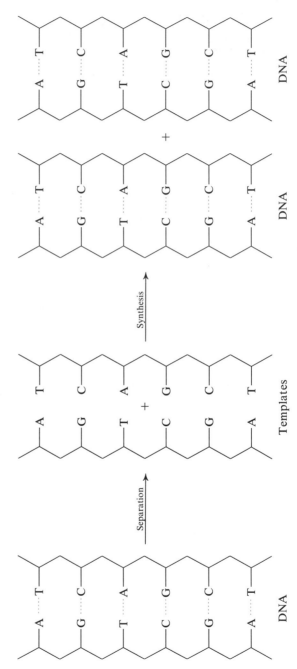

Fig. 18–40 Hydrogen bonding between adenine–thymine and guanine–cytosine.

Fig. 18–41 Self-replication of DNA.

The A–T pairing is held by two hydrogen bonds and the C–G pairing by three hydrogen bonds, as shown in Fig. 18–40.

The Watson-Crick structure for DNA has been widely accepted by the scientific community because it seems to make possible an explanation of the action of DNA and RNA. In brief, two processes must be explained: the self-replication of DNA and the synthesis of protein. Although the details of these processes are not well understood and are the subject of intensive investigation, both processes require that the strands of the DNA double helix be separated.

Fig. 18–42   Schematic drawing of DNA double helix.

In the instance of self-replication, the DNA can be thought of as splitting into two polynucleotide strands. Each strand serves as a surface or template for the synthesis of a new polynucleotide. Since—according to Watson and Crick—adenine must be H-bonded to thymine and guanine to cytosine, the base sequence of the new polynucleotide is determined by the base sequence of the template, and must be identical to the nucleotide to which the template was bonded in the original DNA molecule. Since this process can occur on both strands of the original DNA molecules, two identical molecules are produced from the original molecule. Figure 18–41 gives a schematic representation. Figure 18–42 is a schematic representation of the DNA double helix held together by H-bonding.

Protein synthesis is known to be determined by RNA rather than DNA. The RNA base sequence, however, is determined by DNA. In other words, the RNA is said to act as a messenger for the transmittal of genetic information to the synthesis of a particular protein. The DNA molecule again separates and acts as a template for the synthesis of RNA; the base sequence in the RNA is dependent on the base sequence in the DNA. The RNA molecule acts as a template for protein synthesis. Presumably amino acids are held on the surface of the RNA in a sequence determined by the base sequence of the RNA, and then linked together to form the protein. Thus the amino-acid sequence of the protein is determined by the base sequence of RNA, which is in turn determined by the base sequence of DNA.

Let us emphasize that the information given in this section is speculative and should not be taken as absolute fact. Intensive investigation currently being conducted by chemists, biologists, physicists, and mathematicians may change the picture overnight. There can be no question that the area of molecular biology is attracting some of the world's best scientists and generating strong interaction among the various scientific disciplines. Doubtless the efforts of these researchers will soon give us much detailed information about basic life processes. It is not surprising that molecular biology is one of the most exciting areas of study on the scientific scene.

19. DIGESTION AND METABOLISM

When food is ingested into the human body, it starts a remarkable series of chemical reactions. The food particles are degraded by the process of digestion to a form suitable for absorption into the body through the walls of the small intestine. Once the digestive process is completed, the absorbed materials are converted into muscle and fat and hair and bone and blood and energy and all the materials the human mechanism needs to sustain itself. Since all the matter of the body is constantly undergoing degradation, a build-up process must proceed at a rate equal to the degradation process for adults; for growing children, the build-up process must surpass the degradation.

An investigation of the complex chemical processes of digestion and metabolism discloses the remarkable efficiency of the body. Reactions which would be difficult, if not impossible, to carry out in even the best-equipped laboratories take place readily in the body under very mild conditions to give high yields of the desired products. The speed with which many reactions occur in the body is much greater than similar reactions in the laboratory. The increase in speed and the ability of the body to perform complex chemical transformations in an efficient manner is due to the presence in the body of special catalysts called *enzymes*.

19–1 ENZYMES

Enzymes are complex organic catalysts that are present in all living organisms and are required for the chemical activity in living systems. It has been known for centuries that glucose could be converted to alcohol and carbon dioxide by the process

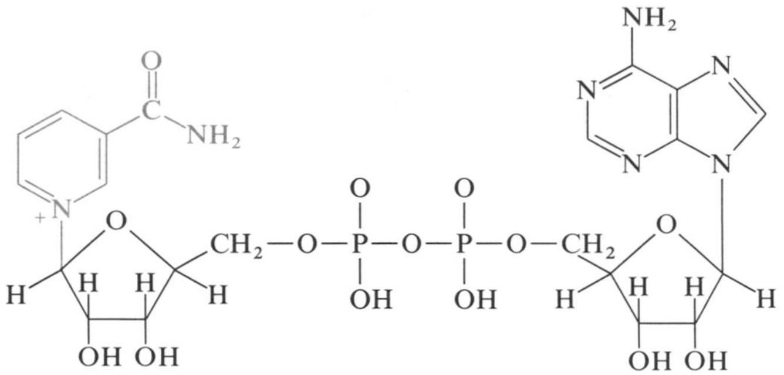

Fig. 19-1 Nicotinamide adenine dinucleotide (NAD).

of fermentation. In the late nineteenth century, however, Pasteur demonstrated that a solution of pure glucose in water was stable until living yeast cells were added. It was later found that, when the yeast cells were ground and pressed, a cell-free extract could be obtained which would efficiently catalyze the fermentation of glucose.

Thus the first enzyme was separated from its source; many more were to follow. Chemists developed techniques for isolating and purifying enzymes, and it became possible to study the enzymes' chemical constitution. That enzymes were proteins was soon realized. Some were found to be pure proteins, while others were found to contain a non-protein portion which was essential for the catalytic action of the enzyme. The non-protein portion of the enzyme is called a *coenzyme* and the protein portion the *apoenzyme*. Neither has separate catalytic activity.

19-2 COENZYMES

The coenzyme portion of an enzyme is non-protein and non-catalytic in nature. Generally speaking, coenzymes participate directly in a given reaction. In some instances, the role of the coenzyme in catalytic processes is understood in detail; in other cases very little is known. NAD (Fig. 19–1) is a coenzyme that participates in the catalysis of some oxidation reactions. In these reactions it is the nicotinamide portion (in color) of the NAD molecule that participates directly in the reaction. The pyridinium ion in the molecule is readily reduced to NADH and, in the process, an oxidation occurs.

An example of this process is the cellular oxidation of lactic acid to pyruvic acid (Fig. 19–2). The NAD is reduced to NADH and the secondary alcohol group in lactic acid is oxidized to a ketone. The overall reaction is catalyzed by an enzyme. Since the reaction is an oxidation reaction, the enzyme may be referred to as an

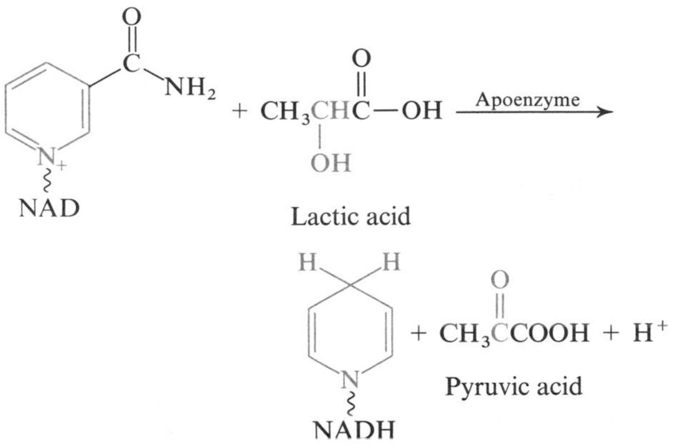

Fig. 19-2 NAD is reduced to NADH as the alcohol is oxidized to a ketone.

*oxidase.** More particularly, the type of oxidation involves the removal of hydrogen from lactic acid and the enzyme is called a *dehydrogenase*. The particular enzyme which catalyzes this reaction has been isolated, and is called *lactic acid dehydrogenase*.

The NADH which is formed in the oxidation of lactic acid is readily reoxidized to NAD and, in the presence of a suitable apoenzyme, can act as a coenzyme in reduction reactions. Such an enzyme system is called a *reductase* or a *hydrogenase*.

Other coenzyme systems are known and more will undoubtedly be discovered. Determining the chemical details of these systems will be one of the most exciting results of current and future research.

19-3 ENZYME REACTIONS

The reactions catalyzed by enzymes are too numerous to mention. In fact, one of the remarkable features of enzymes is their specificity. A given enzyme may act as a catalyst for a single specific reaction!

Therefore it is not reasonable for us to try to discuss the specific mechanisms of an enzyme reaction. Also the complexity of the enzyme molecule makes mechanistic interpretations even more difficult. Many research groups are concentrating on understanding the details of the catalytic functioning of enzymes. It seems clear that the catalytic action of enzymes is due, in part, to their ability to hold the reactants in the precise position required for reaction.

Of the various enzyme-catalyzed reactions, we shall be particularly interested in hydrolysis reactions and oxidation reactions.

---

* In naming enzymes, biochemists often tack the ending *ase* onto the function. Thus *oxidase* is an enzyme which catalyzes oxidation, *hydrolysase* is an enzyme which catalyzes hydrolysis, etc.

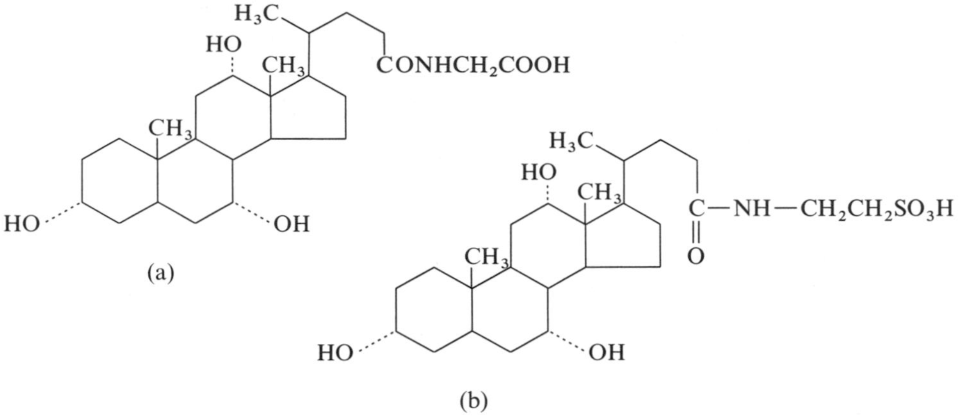

Fig. 19-3 (a) Glycholic acid. (b) Taurocholic acid. (The dashed bonds indicate groups which are behind the plane of the paper.)

Table 19-1

Enzyme	Type	Source	Substrate
Pepsin	Peptidase	Gastric juice	Proteins
Trypsin	Peptidase	Pancreatic juice	Proteins
Chymotrypsin	Peptidase	Pancreatic juice	Proteins
Amylase	Carbohydrase	Saliva, pancreatic juice, intestinal juice	Carbohydrates
Maltase		Pancreatic juice Intestinal juice	Maltose
Lactase		Intestinal juice	Lactose
Sucrase		Intestinal juice	Sucrose
Lipase	Esterase	Gastric juice Pancreatic juice Intestinal juice	Lipids

19-4 DIGESTION

Before man can utilize the compounds present in food, these compounds must pass through the wall of the small intestine. Often this is not a problem: water, some vitamins, soluble salts, monosaccharides, and certain lipids can pass through the walls of the small intestine unaided. Larger molecules, including polysaccharides, proteins, and *lipids* (the biochemist's term for *fats*) must be digested before they can be absorbed.

The digestion process involves hydrolysis reactions catalyzed by various *hydrolysases*. If the enzymes required for hydrolysis are not present, the material cannot be absorbed. This is why, for example, cows can graze and people cannot. Man does not have the enzymes required to hydrolyze cellulose. The enzymes which catalyze hydrolysis reactions are often further classified according to the substrate they work on. For example, *peptidases* hydrolyze the peptide link in proteins. Table 19-1 lists the various digestive enzymes, along with their place of action and function.

The digestive process begins in the mouth when the amylase that is present in saliva begins to hydrolyze any starch present. The action of the amylase continues so long as the optimum pH range of the saliva (6.4 to 6.9) is maintained. But when it becomes mixed with the more acidic gastric juice, the amylase loses its effectiveness.

When the food enters the stomach, it is mixed with gastric juice containing pepsinogen, which is converted to pepsin by the action of the hydrochloric acid in the stomach. From the stomach, the food passes into the small intestine and is mixed with the juice secreted by the pancreas, along with intestinal juice secreted by the walls of the small intestine. It is in the small intestine that the rest of the digestive process occurs.

19-5 DIGESTION OF LIPIDS

Part of the difficulty involved in the digestion of fats arises from the fact that the fat molecule is insoluble in water. Because of this insolubility, the fat must be emulsified—dispersed into very small droplets—so that the surface contact of the fat molecules with the lipases in the small intestine may be increased.

The body accomplishes this emulsification by mixing the fats with bile from the gall bladder. The bile contains salts of glycholic and taurocholic acid which are formed by the metabolism of cholesterol (Fig. 19-3). It is these *bile salts* which act as emulsifying agents to finely disperse the fat in the water system of the small intestine.

After the fat is emulsified, part of it is hydrolyzed with the aid of lipase, forming some diglyceride, monoglyceride, and glycerol, plus the corresponding fatty acids

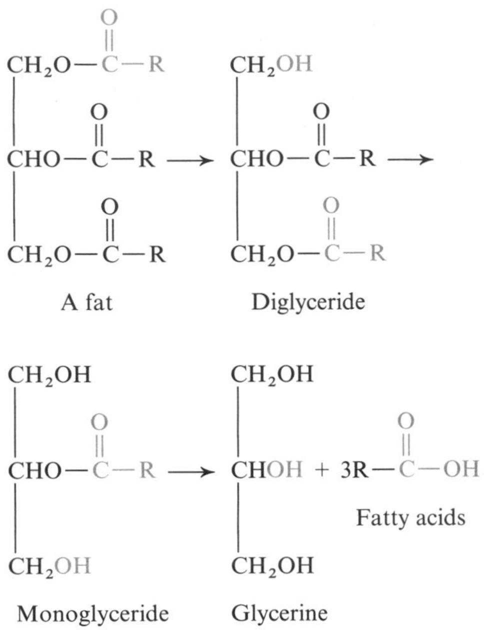

$$CH_2O-\overset{\overset{\displaystyle O}{\|}}{C}-R \qquad CH_2OH$$

$$CHO-\overset{\overset{\displaystyle O}{\|}}{C}-R \longrightarrow CHO-\overset{\overset{\displaystyle O}{\|}}{C}-R \longrightarrow$$

$$CH_2O-\overset{\overset{\displaystyle O}{\|}}{C}-R \qquad CH_2O-\overset{\overset{\displaystyle O}{\|}}{C}-R$$

A fat   Diglyceride

$$CH_2OH \qquad\qquad CH_2OH$$

$$CHO-\overset{\overset{\displaystyle O}{\|}}{C}-R \longrightarrow CHOH + 3R-\overset{\overset{\displaystyle O}{\|}}{C}-OH$$

Fatty acids

$$CH_2OH \qquad\qquad CH_2OH$$

Monoglyceride   Glycerine

Fig. 19-4 Stepwise hydrolysis of a fat.

(Fig. 19–4). The rest of the fat is absorbed, without hydrolysis; it passes through the walls of the small intestine into the blood stream via the lymphatic system.

19-6 METABOLISM OF FATS

Once in the blood stream the fat can be either stored in various *fat depots* or oxidized. The extent of storage of fat in the fat depots around the muscles and organs is determined by many different factors, and effective ways of controlling the extent of fat deposition may vary for each individual. Even fat which is stored, however, does not remain in a static state. Radioactive labeling experiments have demonstrated that depot fat is continuously being removed and new fat continuously being deposited. The type of fat that is deposited (though not the amount of it) appears to be characteristic of the animal in question and largely independent of the composition of the diet.

The oxidation of the fat in the body proceeds by a complex process. Many of the details of the process have been worked out, but there are still unsolved mysteries. The glycerol portion of the fat undergoes an enzyme-catalyzed oxidation to glyceraldehyde, and then phosphate esters are formed. Further oxidation to pyruvic acid links the metabolism (i.e., biochemical change) of glycerols to the metabolism of carbohydrates.

Oxidation of the fatty-acid portion of fat is a process that yields a significant amount of energy. Thus stored fat is a source of energy for the body. When it needs to do so, the body can oxidize the stored fat to produce energy.

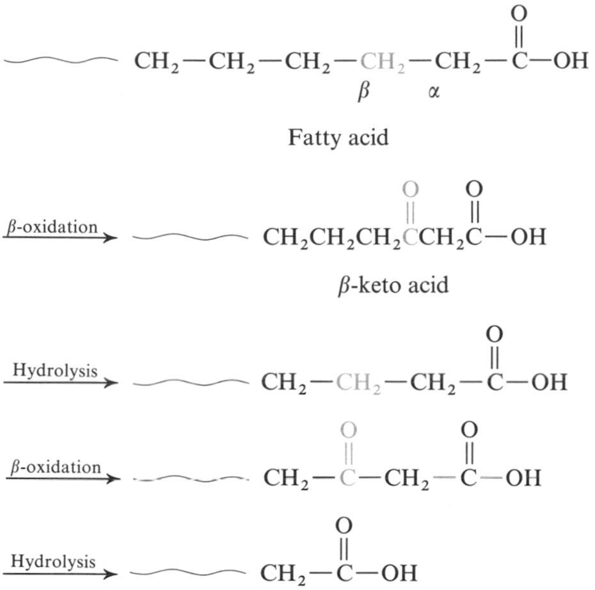

$$\text{---} \quad CH_2-CH_2-CH_2-CH_2-CH_2-\overset{\overset{\displaystyle O}{\|}}{C}-OH$$
$$\quad\quad\quad\quad\quad\quad\quad\quad\quad\beta\quad\quad\alpha$$

Fatty acid

β-oxidation $\longrightarrow$ --- $CH_2CH_2CH_2\overset{\overset{\displaystyle O}{\|}}{C}CH_2\overset{\overset{\displaystyle O}{\|}}{C}-OH$

β-keto acid

Hydrolysis $\longrightarrow$ --- $CH_2-CH_2-CH_2-\overset{\overset{\displaystyle O}{\|}}{C}-OH$

β-oxidation $\longrightarrow$ --- $CH_2-\overset{\overset{\displaystyle O}{\|}}{C}-CH_2-\overset{\overset{\displaystyle O}{\|}}{C}-OH$

Hydrolysis $\longrightarrow$ --- $CH_2-\overset{\overset{\displaystyle O}{\|}}{C}-OH$

Fig. 19-5 Oxidative metabolism of fatty acids.

The usual pathway for the metabolism of the fatty acids is described as *β-oxidation*. By an enzyme-catalyzed reaction, a carbon atom in a position *β* to the carboxyl group is oxidized to a ketone (Fig. 19-5).

The ketone undergoes hydrolysis, which results in cleavage of a two-carbon fragment and the formation of a new fatty acid. The new acid can eliminate a second two-carbon fragment by the same process, and repeated steps result in complete conversion of the fatty acid to carbon dioxide, water, and energy. The oxidation step is not a simple one, but involves at least four separate enzyme-catalyzed reactions, plus coenzyme participation.

19-7 SYNTHESIS OF FATS

The body is capable of synthesizing fat. It is well known, for example, that fat can be deposited even when a person is on a low-fat diet. Carbohydrates or proteins—when eaten to excess—wind up being deposited as fat. Studies on the biosynthesis of fats have thrown some light on the details of this process. For example, the starting point of the synthesis involves the formation of acetyl coenzyme A. Since acetyl coenzyme A is one of the end products of the metabolism of proteins and carbohydrates (see Fig. 19-13 and Section 19-11), we can see how the body can convert protein and carbohydrate to fat.

19-8 DIGESTION OF CARBOHYDRATES

Most of the carbohydrate material in the diet of man is in the form of starch and sucrose, with smaller amounts of glucose and fructose. In order for the body to be able to utilize carbohydrate, the digestion process must hydrolyze the starch and other polysaccharides to monosaccharides, which can be absorbed into the bloodstream through the walls of the small intestine.

This digestive hydrolysis of the polysaccharide chain begins in the mouth with the aid of the saliva's amylase enzyme, *ptyalin* (pronounced tie'-ah-lynn). Ptyalin catalyzes the hydrolysis of the 1,4-α links in the carbohydrate chain. This hydrolytic cleavage continues to break up the polysaccharide molecule into smaller and smaller fragments, until the low pH of the stomach renders the salivary enzymes inactive.

Little further digestion occurs until the mixture enters the small intestine and mixes with the enzymes in the pancreatic and intestinal juices. These enzymes catalyze the hydrolysis of the remaining 1,4-α links as well as 1,6-α links; they also help to convert disaccharides to monosaccharides. The overall process is very efficient, and ordinarily about 98% of the carbohydrate we eat is absorbed into the bloodstream.

19-9 ABSORPTION AND STORAGE OF CARBOHYDRATES

Since the primary monosaccharide resulting from digestion is glucose, and since the other common monosaccharides—fructose, galactose, and mannose—can be converted into glucose by enzymatic reactions in the liver, the rest of our discussion of the metabolism of carbohydrates will deal only with glucose.

The mechanism by which glucose passes through the wall of the small intestine is not entirely understood. After glucose enters the portal vein (which flows from the intestine to the liver), however, we know that the glucose is transported to the liver, where part is removed and part is allowed to pass into the general circulation of the bloodstream.

The glucose removed by the liver is converted to glycogen, a highly branched polysaccharide similar to amylopectin, which is stored in the liver. Although its capacity for storing glycogen is relatively small, the liver rapidly synthesises glycogen from fructose, galactose, mannose, lactic acid, and metabolic products of fats and amino acids, as well as from glucose itself.

The liver's glycogen is important in that it acts as a source of glucose for the bloodstream. Glucose in the bloodstream is kept at a very constant level: An average adult has about 80 mg of glucose/100 ml of blood. Shortly after the person eats, this value may rise to about 150 mg/100 ml, but then it rapidly declines to the original value, as shown in Fig. 19-6.

In a diabetic person, the level of glucose in the blood may start higher (though it need not). The shape of the diabetic person's glucose tolerance curve differs from that of the average in that it is slower to reach a maximum and slower to

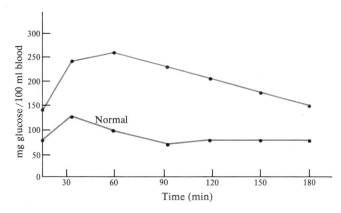

Fig. 19-6 Glucose tolerance curves for normal and for diabetic people.

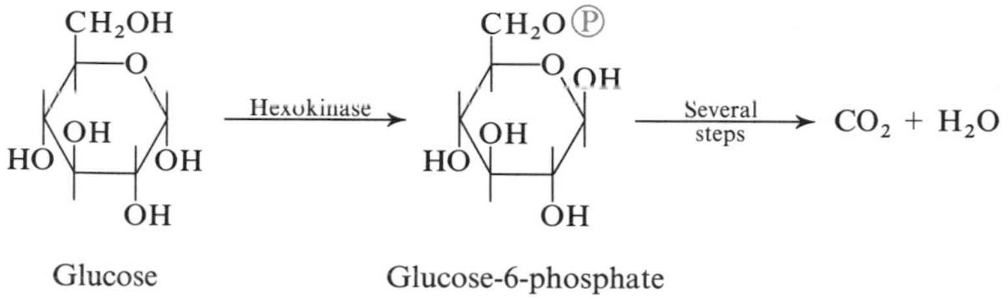

Hexokinase

Several steps

$CO_2 + H_2O$

Glucose

Glucose-6-phosphate

Fig. 19-7 Process by which the blood-sugar level is lowered.

decline. If the level of glucose in a person's blood exceeds a value of about 160 mg/100 ml, glucose is excreted and appears in the urine.

In order for the level of glucose in the blood to be maintained at a constant value, a balance must be achieved between the amount of glucose being stored as glycogen, the amount of glycogen being converted into blood glucose, and the amount of blood glucose being further metabolized. Glucose, in the presence of the enzyme hexokinase, is converted to glucose-6-phosphate (Fig. 19-7). The glucose-6-phosphate is then metabolized by a series of reactions to carbon dioxide and water.

The end result of this process is a lowering of the level of glucose in the blood. The extent to which this process takes place is determined by the presence of the hormone *insulin*. Hormones are present in secretions of the endocrine glands. The hormone insulin is secreted in the pancreas, and is necessary for the activation of the enzyme *hexokinase*. A deficiency of insulin results in a lowering of the level of glycogen in the liver and a less-efficient lowering of the level of sugar in the blood after a meal. An excess of insulin can lead to a marked decrease in the level of sugar in the blood, with dizziness, coma, and even ultimate death.

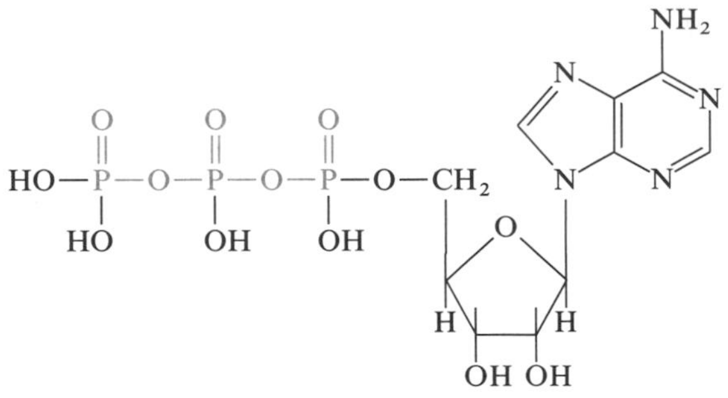

Fig. 19-8 Adenosine triphosphate (ATP).

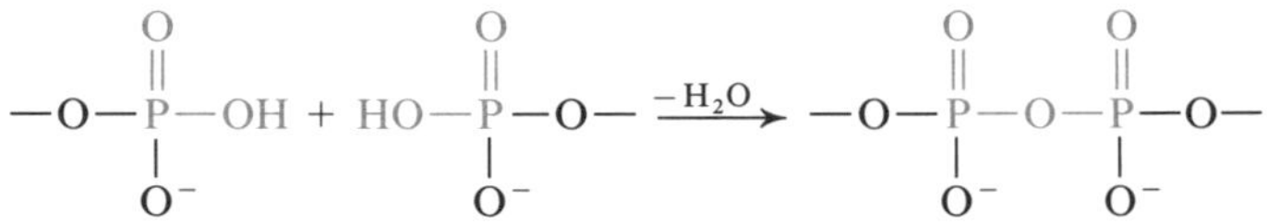

Fig. 19-9 Formation of a high-energy phosphate bond.

Among the hormones that control the level of glucose in the blood when the carbohydrate entering the bloodstream is insufficient, the fastest acting is *adrenaline* (epinephrine). Adrenaline is required to activate the enzyme which catalyzes the conversion of liver glycogen to glucose-1-phosphate, which is ultimately converted to carbon dioxide and water. Diabetics often exhibit an excess of adrenaline or a deficiency of insulin.

19-10 ENERGY PRODUCTION

The ultimate fate of glucose in the body is its oxidation to carbon dioxide and water. The overall process is an exothermic one, and the energy liberated is available for use by the body:

$$C_6H_{12}O_6 + 6O_2 \rightarrow 6CO_2 + 6H_2O + 686 \text{ kcal/mole.}$$

Although some of this energy can be used to maintain body temperature, the requirements are much more diverse. We need energy for muscle action, for nerve response, for chemical synthesis, and for innumerable cellular functions that occur at various times in various places. Thus the body must be able to take the energy available in food, transport it to the various parts of the body, and store it in a readily available fashion.

The body does this very efficiently by a process called *oxidative phosphorylation*. One of the keys to this process is the molecule adenosine triphosphate (ATP) shown

in Fig. 19–8. The most important part of the molecule, so far as energy transfer is concerned, is the polyphosphate portion. The phosphorus-oxygen bond formed by the condensation of two phosphate groups (Fig. 19–9) is energetically unstable toward hydrolysis. The hydrolysis reaction is a slow one, however, and, to proceed at a reasonable rate, requires a suitable catalyst. Thus, in the presence of an appropriate enzyme, one of the phosphate groups in ATP can be hydrolyzed, with the production of adenosine diphosphate (ADP) and the release of about 8 kcal/mole of energy:

$$ATP + H_2O \xrightarrow{\text{Enzyme}} ADP + \text{energy.}$$

The ATP, then, can act as a storehouse for the energy released by the metabolism of food:

$$ADP + \text{phosphate} + \text{energy} \xrightarrow{\text{Phosphorylation}} ATP$$
$$\text{Oxidation}\uparrow$$
$$\text{Food}$$

The ATP is formed from ADP and inorganic phosphate, by *oxidative phosphorylation*. The details of this complex process are under investigation by several research groups.

19–11 THE METABOLISM OF GLUCOSE

In the body, glucose is metabolized in two stages: The first involves the conversion of glucose-6-phosphate to pyruvic acid by the Embden-Meyerhof pathway (glycolysis) as shown in Fig. 19–10. The second (Fig. 19–11) involves the oxidation of pyruvic acid to carbon dioxide and water by the Krebs citric acid cycle. (Figures 19–10 and 19–11 appear on pages 562 and 563.)

In the Embden-Meyerhof pathway, two molecules of ATP are consumed in steps 1 and 3 for each molecule of glucose, but in steps 6 and 9 two molecules of ATP are formed from a glyceraldehyde fragment. Since each glucose can afford two glyceraldehyde fragments, the end result of one glucose molecule being converted to pyruvic acid is the formation of two energy-rich ATP molecules.

We can see the necessity of both oxidation and phosphorylation by looking at steps 5 and 6 in Fig. 19–10. Although the details of the process are still not known, we can see that the 1,3-diphosphoglyceric acid has two types of phosphate groups in the molecule. One of these has the ℗ —O— attached to a saturated carbon atom (℗ —O—CH$_2$—) and the second has the carbon atom in a higher oxidation state by virtue of step 5:

$$-O-\overset{\overset{\displaystyle O}{\|}}{C}-℗$$

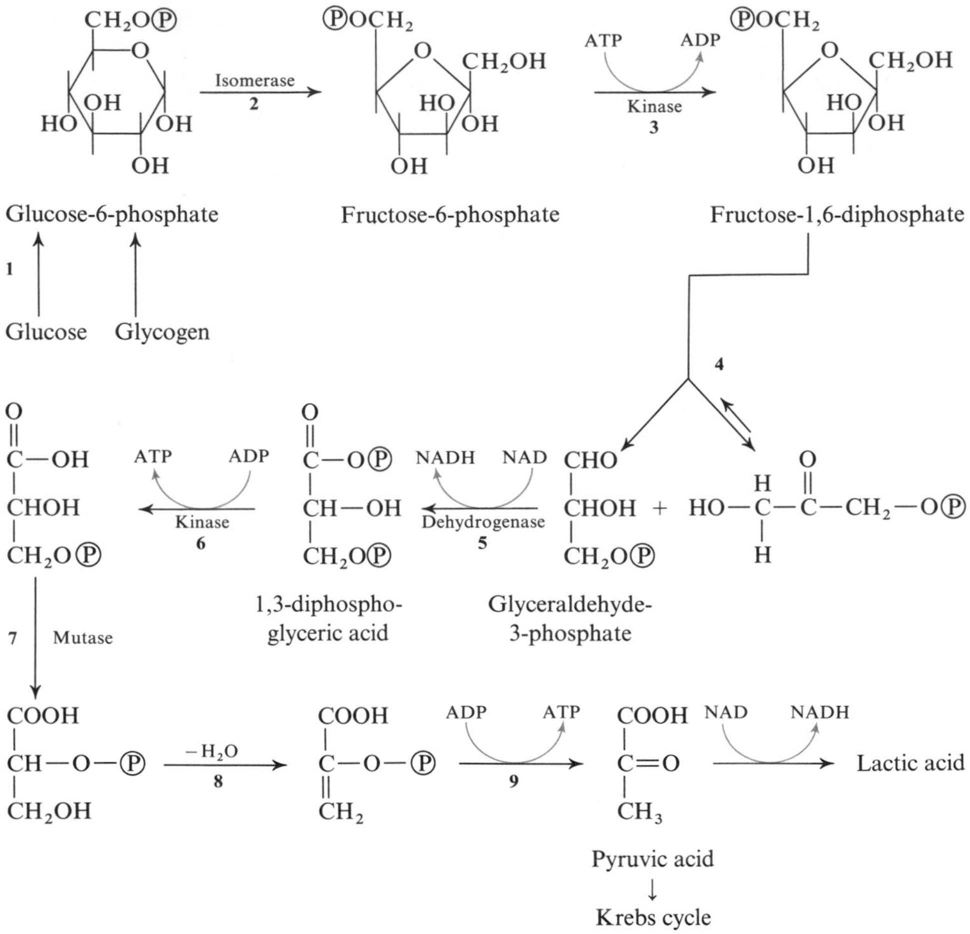

CH₂O℗ ... Glucose-6-phosphate → (Isomerase 2) → ℗OCH₂ ... Fructose-6-phosphate → (ATP ADP, Kinase 3) → ℗OCH₂ ... Fructose-1,6-diphosphate

$CH_2O℗$

Glucose-6-phosphate

$℗OCH_2$ ... CH_2OH

Fructose-6-phosphate

$℗OCH_2$ ... CH_2OH

Fructose-1,6-diphosphate

Isomerase 2

ATP   ADP
Kinase 3

1

Glucose   Glycogen

4

$$
\begin{array}{l}
O \\
\| \\
C-OH \\
| \\
CHOH \\
| \\
CH_2O℗
\end{array}
\quad
\begin{array}{c}
ATP \qquad ADP \\
\xleftarrow{\hspace{2cm}} \\
Kinase \\
6
\end{array}
\quad
\begin{array}{l}
O \\
\| \\
C-O℗ \\
| \\
CH-OH \\
| \\
CH_2O℗
\end{array}
\quad
\begin{array}{c}
NADH \qquad NAD \\
\xleftarrow{\hspace{2cm}} \\
Dehydrogenase \\
5
\end{array}
\quad
\begin{array}{l}
CHO \\
| \\
CHOH \\
| \\
CH_2O℗
\end{array}
+
\begin{array}{l}
\quad\quad\quad O \\
\quad H \quad \| \\
HO-C-C-CH_2-O℗ \\
\quad | \\
\quad H
\end{array}
$$

1,3-diphospho-
glyceric acid

Glyceraldehyde-
3-phosphate

7 | Mutase

$$
\begin{array}{l}
COOH \\
| \\
CH-O-℗ \\
| \\
CH_2OH
\end{array}
\quad
\begin{array}{c}
-H_2O \\
\xrightarrow{\hspace{1.5cm}} \\
8
\end{array}
\quad
\begin{array}{l}
COOH \\
| \\
C-O-℗ \\
\| \\
CH_2
\end{array}
\quad
\begin{array}{c}
ADP \qquad ATP \\
\xrightarrow{\hspace{2cm}} \\
9
\end{array}
\quad
\begin{array}{l}
COOH \\
| \\
C=O \\
| \\
CH_3
\end{array}
\quad
\begin{array}{c}
NAD \qquad NADH \\
\xrightarrow{\hspace{2cm}}
\end{array}
\quad \text{Lactic acid}
$$

Pyruvic acid
↓
Krebs cycle

Fig. 19–10 The Embden-Meyerhof pathway of glycolysis. The symbol ℗ stands for the phosphate group —PO₃H.

Hydrolysis of the second phosphate bond is apparently exothermic enough to cause the phosphorylation of ADP in step 6, whereas the first phosphate bond is stable. We can see a certain similarity between the unstable bonding arrangement in ATP, on the left below, and in the 1,3-diphosphoglyceric acid, on the right:

$$
\begin{array}{cc}
O \quad\quad O \\
\| \quad\quad \| \\
O-P-O-P- \\
| \quad\quad | \\
O
\end{array}
\qquad\qquad
\begin{array}{cc}
O \quad\quad O \\
\| \quad\quad \| \\
O-P-O-C- \\
| \\
O
\end{array}
$$

and are tempted to predict that a phosphate group attached to an atom doubly bonded to something else will be energetically unstable. The phosphorylation in step

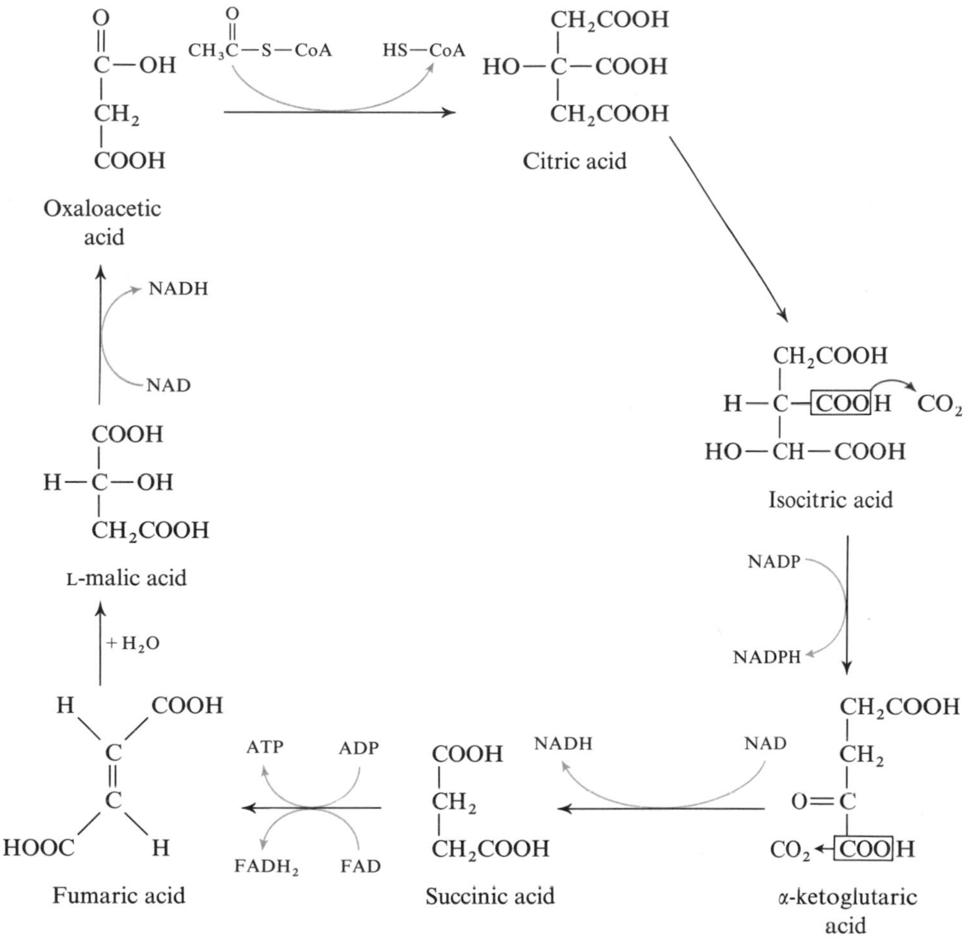

Fig. 19–11 The Krebs citric acid cycle.

9 is further evidence, since it involves the necessary bonding arrangement:

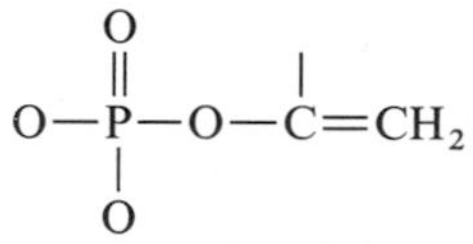

$$O-\overset{\overset{\displaystyle O}{\|}}{\underset{\underset{\displaystyle O}{|}}{P}}-O-\overset{|}{C}=CH_2$$

    The pyruvic acid, formed as the end product in the glycolysis process, in the absence of oxygen is reduced to lactic acid. This is a condition that occurs in the muscles after violent exercise. The cells are depleted of oxygen and lactic acid is produced; this causes the symptoms of muscle fatigue. Under ordinary circumstances, however, the pyruvic acid enters the Krebs citric acid cycle, as illustrated in Fig. 19–11. In order for this to occur, there must be prior enzyme-catalyzed con-

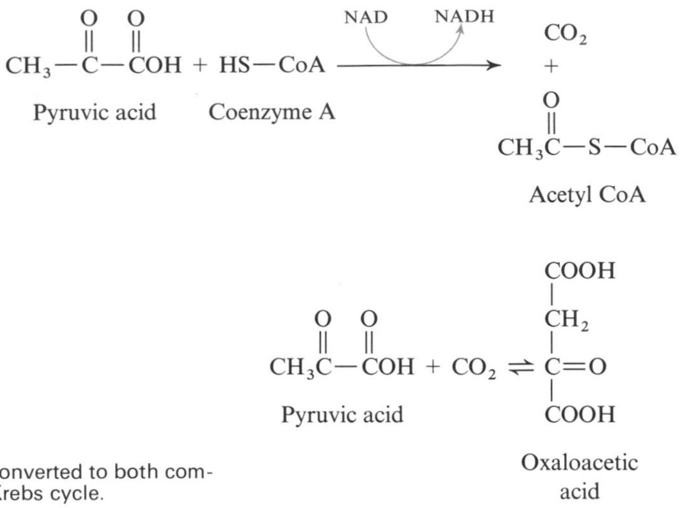

$$CH_3-\overset{\overset{O}{\|}}{C}-\overset{\overset{O}{\|}}{C}OH + HS-CoA \xrightarrow{\quad NAD \quad NADH \quad} \begin{array}{c} CO_2 \\ + \\ \overset{O}{\overset{\|}{}} \\ CH_3C-S-CoA \end{array}$$

Pyruvic acid   Coenzyme A   Acetyl CoA

$$CH_3\overset{\overset{O}{\|}}{C}-\overset{\overset{O}{\|}}{C}OH + CO_2 \rightleftharpoons \begin{array}{c} COOH \\ | \\ CH_2 \\ | \\ C=O \\ | \\ COOH \end{array}$$

Pyruvic acid   Oxaloacetic acid

Fig. 19-12 Pyruvic acid is converted to both components needed to start the Krebs cycle.

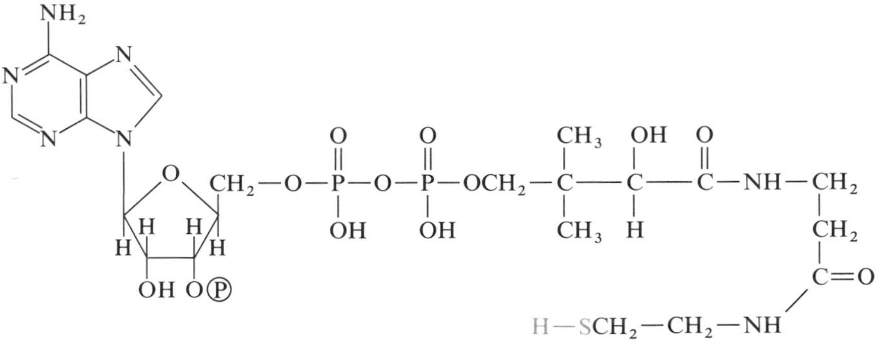

Fig. 19-13 Coenzyme A.

version of some of the pyruvic acid to acetyl coenzyme A and other pyruvic acid to oxaloacetic acid (Fig. 19-12).

The oxaloacetic acid then combines with acetyl CoA to form citric acid. The structure of coenzyme A is shown in Fig. 19-13. The important part of the molecule for our purposes is the —SH group which is used to form acetyl coenzyme A.

The Krebs citric acid cycle includes four oxidation steps. Three of these involve the reduction of NAD to NADH or of phosphorylated NADP to NADPH, and the fourth involves another oxidative nucleotide FAD (flavin-adenine-dinucleotide) which is reduced to $FADH_2$. The FAD oxidation produces one ATP molecule. In addition, each NADH (or NADPH) molecule formed is reoxidized by a complex pathway to NAD (or NADP) with the formation of three ATP molecules, and

$FADH_2$ is reoxidized to FAD with the formation of two ATP molecules. The formation of acetyl CoA also produces NADH and subsequently three ATP molecules. The grand total, then, for one pass through the Krebs cycle (i.e., from oxaloacetic acid through the cycle and back to oxaloacetic acid) is 15 high-energy ATP molecules.

When acetyl CoA is formed, the by-product is carbon dioxide. Carbon dioxide is also formed at two other points in the Krebs cycle. Since the end of the cycle gives us another oxaloacetic acid molecule ready to combine with a new acetyl CoA molecule, the net result of the Krebs cycle is the oxidation of pyruvic acid to carbon dioxide and water, with the energy being used to form 15 ATP molecules:

$$C_3H_4O_3 + \tfrac{5}{2}O_2 + 15ADP + 15H_3PO_4 \rightarrow 3CO_2 + 2H_2O + 15ATP \cdot H_2O$$

Since the end result of the β-oxidation of fatty acids previously discussed is acetyl CoA, we can see how fats can also contribute to the energy produced in the Krebs cycle.

19–12 THE DIGESTION OF PROTEINS

Although the body is capable of synthesizing glucose from other carbohydrates and even from lipides and can also synthesize fats from carbohydrates and proteins, there are ten amino acids that the body cannot synthesize. Since these amino acids (*essential amino acids*) are necessary for proper protein synthesis in the body, they must be included in the diet for proper nutrition. The best source of proteins containing these amino acids is animals (meat or fish) or animal by-products (milk and eggs). Although vegetables and grain contain proteins, the protein content is usually lower and often one or more of the essential amino acids is missing. That is why relying on a diet of rice—which is the situation in much of the world—often leads to malnutrition.

In order for the body to be able to utilize protein, the protein must be hydrolyzed to the constituent amino acids, which are readily absorbed. The digestion occurs primarily in the stomach (gastric juices) and in the small intestine (pancreatic and intestinal juices).

The main enzyme of the gastric juice is pepsin. Pepsin is an *endopeptidase*, which means that it catalyzes the hydrolysis of peptide (amide) links in the middle portion of the molecule. The result of pepsin action is the cleavage of protein molecules into smaller polypeptide fragments.

Passing from the stomach into the small intestine, the polypeptide fragments come into contact with trypsin, chymotrypsin, and carboxypeptidase produced in active form in the pancreatic juice, as well as aminopeptidase and dipeptidase found in the intestinal juices. Like pepsin, chymotrypsin and trypsin are endopeptidases which catalyze hydrolysis in the inner portion of the polypeptide chain.

On the other hand, aminopeptidase and carboxypeptidase are *exopeptidases*. Aminopeptidase catalyzes the hydrolysis of the amide link at the end of the polypeptide chain containing the free amino group. Carboxypeptidase catalyzes

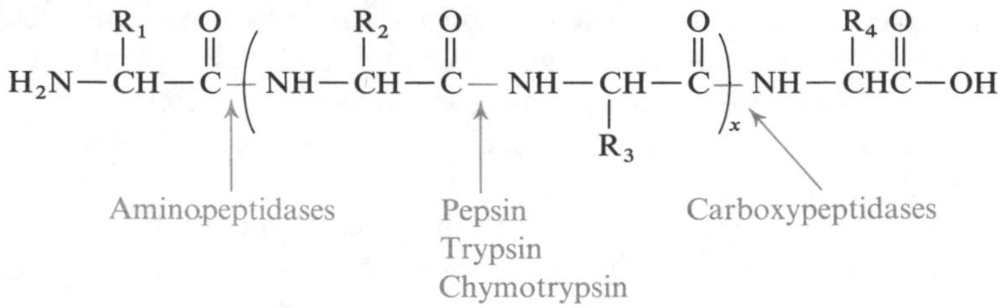

$$\underset{\text{Aminopeptidases}}{\uparrow} \quad H_2N-\underset{\underset{R_1}{|}}{CH}-\underset{\underset{O}{\|}}{C}\left(NH-\underset{\underset{R_2}{|}}{CH}-\underset{\underset{O}{\|}}{C}-\underset{\underset{\text{Pepsin}}{\uparrow}}{NH}-\underset{\underset{R_3}{|}}{CH}-\underset{\underset{O}{\|}}{C}\right)_x NH-\underset{\underset{R_4}{|}}{CH}\underset{\underset{O}{\|}}{C}-OH$$

Aminopeptidases   Pepsin   Carboxypeptidases
Trypsin
Chymotrypsin

Fig. 19-14 Sites of catalytic action for various exo- and endopeptidases.

the hydrolysis of the amide grouping at the end bearing the free carboxyl (Fig. 19–14). The combined effect of these enzymes is to convert protein into its constituent amino acids very efficiently.

19-13 METABOLISM OF AMINO ACIDS

The amino acid fragments resulting from digestion pass through the walls of the small intestine and into the bloodstream via the portal vein. The liver cells remove the amino acids from the bloodstream and direct their further utilization along three general pathways:

1. Degradation to carbon dioxide and urea.

2. Synthesis of proteins.

3. Synthesis of other compounds.

Loss of Amino Nitrogen. Amino acids are converted to α-keto acids with the loss of the amino group by virtue of enzyme-catalyzed oxidation (Fig. 19–15). A more important reaction, both for the removal of an amino group and the synthesis of new amino acids, is *transamination*.

The reaction requires a *transaminase* enzyme and pyridoxal phosphate as a coenzyme (Fig. 19–16). The nitrogen in the amino acids is utilized for the synthesis of other amino acids or is excreted as urea:

$$H_2N-\underset{\underset{O}{\|}}{C}-NH_2$$

The fate of the rest of the molecule varies for each amino acid. Alanine and glutamic acid, for example, can be converted by transamination or oxidative deamination to pyruvic acid and α-ketoglutaric acid, respectively, and thereby enter the Krebs cycle.

Another—although less common—method of degradation of amino acids is by enzyme-catalyzed decarboxylation (Fig. 19–17). Such a process is believed to be responsible for the formation of histamine from histidine and serotonin from 5-hydroxy-tryptophan, as well as other amines (Fig. 19–18).

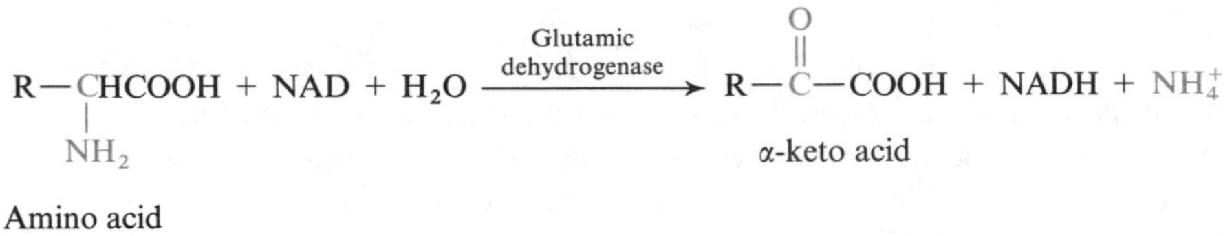

$$R-\underset{\underset{NH_2}{|}}{CH}COOH + NAD + H_2O \xrightarrow[\text{dehydrogenase}]{\text{Glutamic}} R-\overset{\overset{O}{\|}}{C}-COOH + NADH + NH_4^+$$

Amino acid α-keto acid

Fig. 19-15 Oxidative deamination.

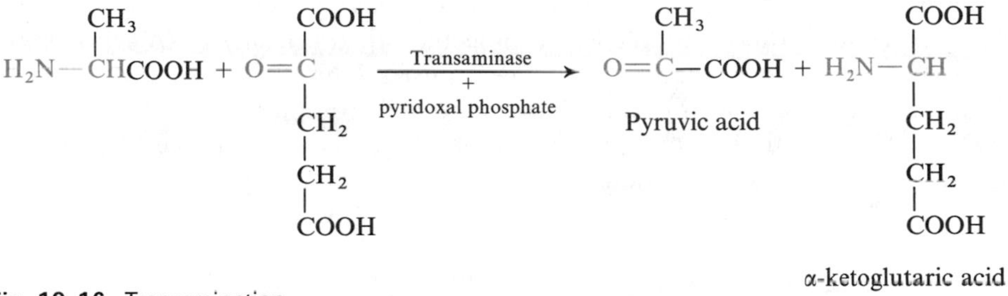

$$H_2N-\underset{}{CH}COOH + \underset{\underset{\underset{\underset{COOH}{|}}{CH_2}}{|}}{\underset{CH_2}{|}}\overset{\overset{CH_3}{|}}{\underset{}{O=C}} \xrightarrow[\text{pyridoxal phosphate}]{\underset{+}{\text{Transaminase}}} \overset{\overset{CH_3}{|}}{O=C}-COOH + H_2N-\underset{\underset{\underset{\underset{COOH}{|}}{CH_2}}{|}}{\underset{CH_2}{|}}\overset{\overset{COOH}{|}}{CH}$$

                                                        Pyruvic acid

                                                                        α-ketoglutaric acid

Fig. 19-16 Transamination.

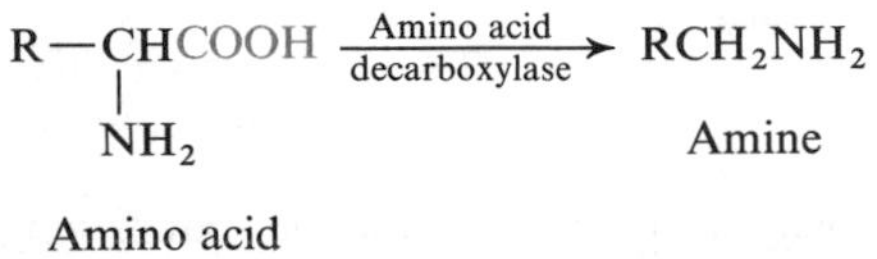

$$R-\underset{\underset{NH_2}{|}}{CH}COOH \xrightarrow[\text{decarboxylase}]{\text{Amino acid}} RCH_2NH_2$$

Amino acid   Amine

Fig. 19-17 Amino acid decarboxylation.

Histidine   Histamine

5-hydroxytryptophan   Serotonin

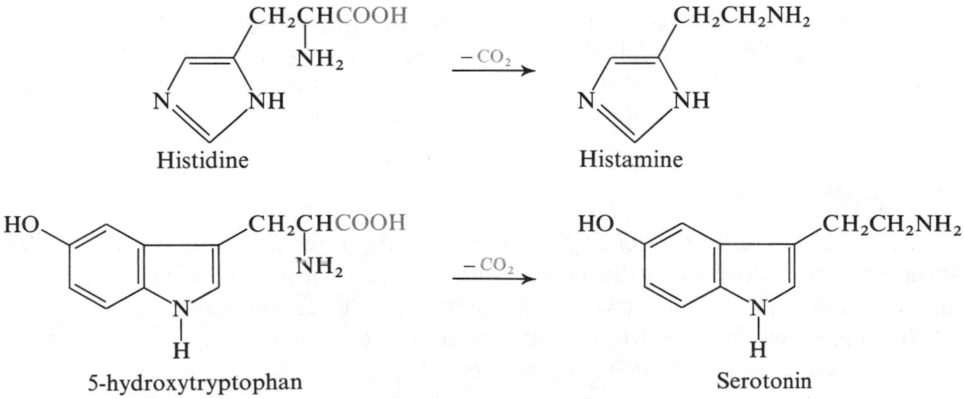

Fig. 19-18 Formation of two amines by amino acid decarboxylation.

Synthesis of Proteins.   Radioactive labeling techniques have shown that the protein content of the body is constantly undergoing change. Old protein is continuously being degraded to amino acids, which are metabolized, and new protein is continuously being synthesized. Thus, for an adult to be in proper health, the proteins of the muscles, bones, brains, blood, enzymes, etc., must be synthesized at a rate equal to the metabolism of protein. In the child, of course, for growth to take place, the rate of protein synthesis must exceed the rate of protein metabolism.

As mentioned in Chapter 18, protein synthesis is controlled by DNA and RNA. This remarkably efficient process is very complex, but some of the details are known. The base sequence in the DNA molecule determines the protein structure. This information is transferred to an RNA molecule called a *messenger RNA*. In the meantime, another RNA molecule called a *transfer RNA* brings the amino acids to the messenger RNA. The transfer RNA with the amino acid is held there while a second transfer RNA amino acid complex attaches itself to the m-DNA. The two amino acids are then joined together. This process is repeated until the proper protein is synthesized.

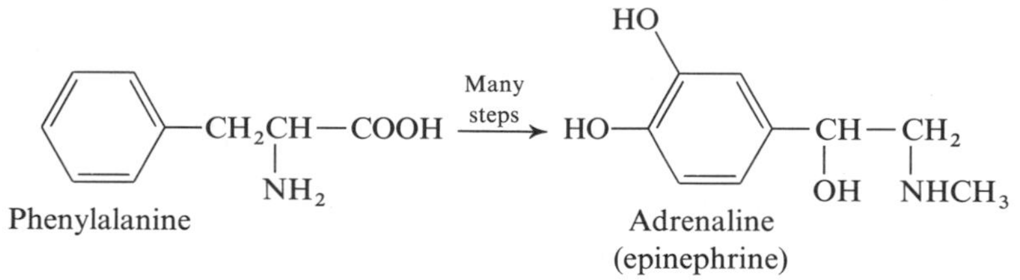

Fig. 19-19 The hormone adrenaline is synthesized in the body, starting from the amino acid phenylalanine.

Further reactions of amino acids vary widely, depending on the nature of the particular amino acid. Phenylalanine, for example, serves as a biosynthetic precursor of adrenaline (Fig. 19–19), as well as of phenylacetic acid, tyrosine, thyroxine, the skin pigment melanin, and many other compounds. Glycine finds itself converted to serine, creatine, hippuric acid, oxalic acid, acetylcholine, etc. Reactions of other amino acids are likewise varied and complex.

19-14 CONCLUSION

This chapter has only scratched the surface of the information currently known about metabolic processes, and what is currently known is only a start toward an understanding of the chemistry of living systems. The day is undoubtedly coming when chemists will be able to penetrate the mysteries of disease, genetic alteration, and the physical and mental development of mankind.

20. VITAMINS, HORMONES, AND DRUGS

In addition to the proteins, carbohydrates, and fats previously considered, there are numerous chemical compounds that exhibit a wide variety of physiological properties. Some of these compounds are found naturally in the body or in the diet and are required for normal functioning, while others are not usually found in the body, but are ingested to yield a desired physiological effect. This effect may be to overcome a deficiency in the body chemistry, to control some abnormality, to interrupt the ordinary body functions, or to produce some desired sensation.

The chemical mode of action of these compounds is often unknown both in the case of the compounds natural to the body and those artificially produced. In the latter case, the long-term biological effects of the compounds are also open to question. In view of the remarkable strides made by the pharmaceutical industry in recent years as well as the general public's overwhelming acceptance of pill-taking, the details of both the long-term and short-term effects of biologically active compounds need to be elucidated. Although time and space prevent us from in-depth consideration, we shall discuss in this chapter the structure and properties of some of the vitamins, hormones, and drugs.

20-1 VITAMINS

To qualify as a *vitamin*, an organic compound—with the exception of carbohydrates, fats, and proteins—must be found in small quantities in natural foods,

be essential for normal growth and health, and not be synthesized in the body. In addition, the absence of that vitamin from the diet must cause a specific disease.

Although evidence for the presence of special nutritive factors in certain foods had been available for centuries, vitamins fulfilling the above requirements were not recognized until the early part of the twentieth century. In 1912 the Polish chemist Casimir Funk proposed the existence of four amines necessary for life, and coined the word *vitamine*. One, he proposed, was an antiscurvy factor, one an antipellagra factor, and the other two antiberiberi and antirickets factors.

In the same year, at Cambridge University, Sir Frederick Hopkins reported that milk contains in minute quantities compounds essential for growth, which he termed *growth factors*. Subsequent to Hopkins' work (awarded the Nobel Prize in medicine in 1929), it was shown that the growth factors could be separated into the fat-soluble "growth factor A" and the water-soluble "growth factor B," and further, that "growth factor B" was identical to the "antiberiberi vitamine." Since the A factor contained no amine in the molecule, the "e" was dropped from the name and the factors were called vitamin A and vitamin B. As new vitamins were discovered, new letters were used to designate them; for example, vitamins C, D, E, K. Refinements in techniques of isolation and purification soon demonstrated that the antiberiberi vitamin B was in fact a mixture, the components of which were designated by number, such as vitamins B_1, B_6, and B_{12}.

20–2 THE FUNCTION AND SOURCE OF VITAMINS

Details of the biochemistry of most vitamins are still obscure. There is increasing evidence, however, that vitamins affect the metabolism in a catalytic fashion, that they are coenzymes or are biochemically transformed into coenzymes. Thus a vitamin deficiency in the diet can have dramatic effects on health and growth. Some of the common vitamins are discussed below in terms of their chemical structure, dietary source, and biochemical function.

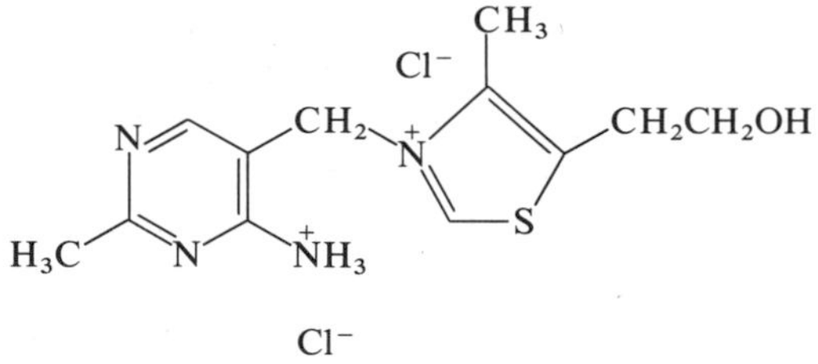

Fig. 20–1 Thiamine hydrochloride (vitamin B_1).

Vitamin B_1

In 1890 symptoms similar to beriberi in man were found in hens fed exclusively on polished rice. It was further found that the symptoms could be cured by adding the rice polishings (bran) to the feed. Extensive studies finally resulted in the isolation of *thiamine*, or vitamin B_1, in the form of the hydrochloride salt. Thiamine was later

isolated from yeast and wheat germ, and is known to be present in eggs, meat, beans, and peas. The isolation technique was improved to such a degree that chemists were able to obtain 5 grams of thiamine hydrochloride from a ton of rice bran. The structure of thiamine hydrochloride (Fig. 20–1) was demonstrated and the synthesis accomplished in the 1930's.

In the body, thiamine is converted to the pyrophosphate ester, cocarboxylase (Fig. 20–2), which is important to the utilization of pyruvic acid in carbohydrate metabolism.

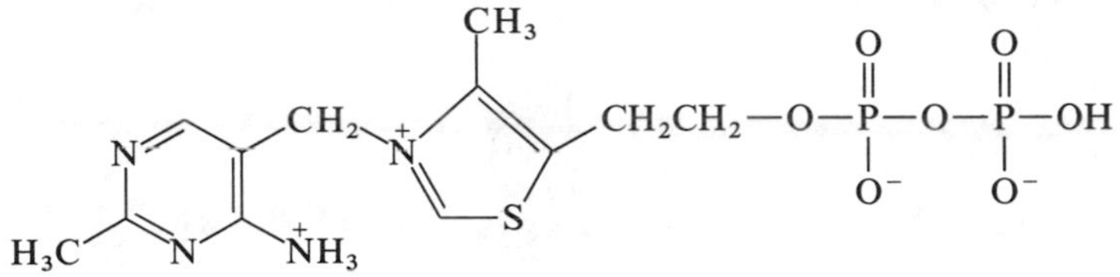

Fig. 20-2 Cocarboxylase.

Thiamine deficiency, still found in the Far East where diets of polished rice are common, leads to beriberi. Thiamine deficiency also exists in the Western world under conditions which put great stress on the metabolism of carbohydrates, such as pregnancy and alcoholism.

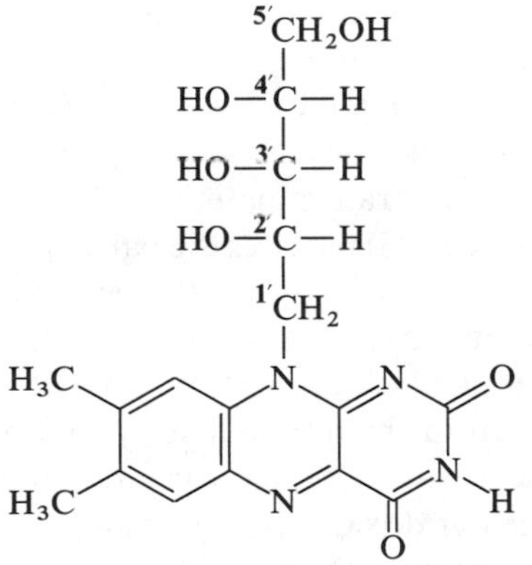

Fig. 20-3 Riboflavin.

Vitamin B₂ (Riboflavin)

Riboflavin (Fig. 20–3) is available in eggs, milk, and green vegetables, as well as in animal organs such as the heart, liver, and kidneys. The symptoms of riboflavin deficiency are sometimes vague, but inhibited growth and vision, as well as skin sores, are often present.

Riboflavin is a biological precursor of two coenzymes, flavine mononucleotide with a phosphate ester group in the 5′-position, and flavine-adenine-dinucleotide (FAD) described in Chapter 19. Both coenzymes participate in oxidative processes, the former in amino acid and monosaccharide oxidation and the latter in the Krebs cycle.

Vitamin B_6

Vitamin B_6 is not a single organic compound, but a group of closely related and biochemically interconvertible compounds including pyridoxine, pyridoxal, and pyridoxamine (Fig. 20–4).

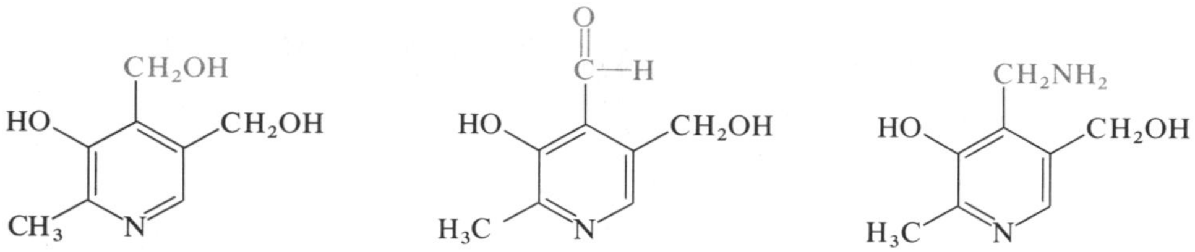
CH₂OH
HO
CH₂OH
CH₃
N
O
C—H
HO
CH₂OH
H₃C
N
CH₂NH₂
HO
CH₂OH
H₃C
N

Pyridoxine Pyridoxal Pyridoxamine

Fig. 20–4 Compounds of the vitamin B_6 complex.

Originally isolated from rice bran as well as brewer's yeast, vitamin B_6 exists in many foods, particularly vegetables, whole-grain cereals, and liver. Deficiency of vitamin B_6 in the diet retards growth, as well as being a factor in many diverse symptoms such as infant convulsions and increased dental cavities.

The coenzyme form of vitamin B_6 is a pyridoxal phosphate called *codecarboxylase*. It is strongly implicated in the metabolism of amino acids, since it plays a catalytic role in—among other reactions—decarboxylation, transmination, and racemization. The transamination reaction (Fig. 20–5 on page 573) requires the participation of a metal ion such as Cu^{2+}, Fe^{3+}, or Al^{3+}. A complex is formed by the combination of pyridoxal plus an amino acid, pyridoxamine plus an α-keto acid, and the metal ion. This complex is converted to a second complex, which decomposes. The result is that the pyridoxal accepts an amino group from the amino acid, yielding pyridoxamine and an α-keto acid, while pyridoxamine donates an amino group to the keto acid to yield pyridoxal and an amino acid. Since pyridoxal, pyridoxamine, and the metal ion are all regenerated, the net result is transamination (Fig. 20–6).

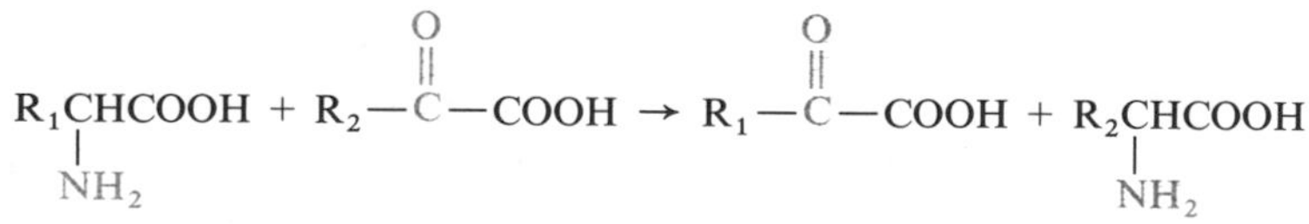
O
O
R₁CHCOOH + R₂—C—COOH → R₁—C—COOH + R₂CHCOOH
NH₂
NH₂

Fig. 20–6 Transamination.

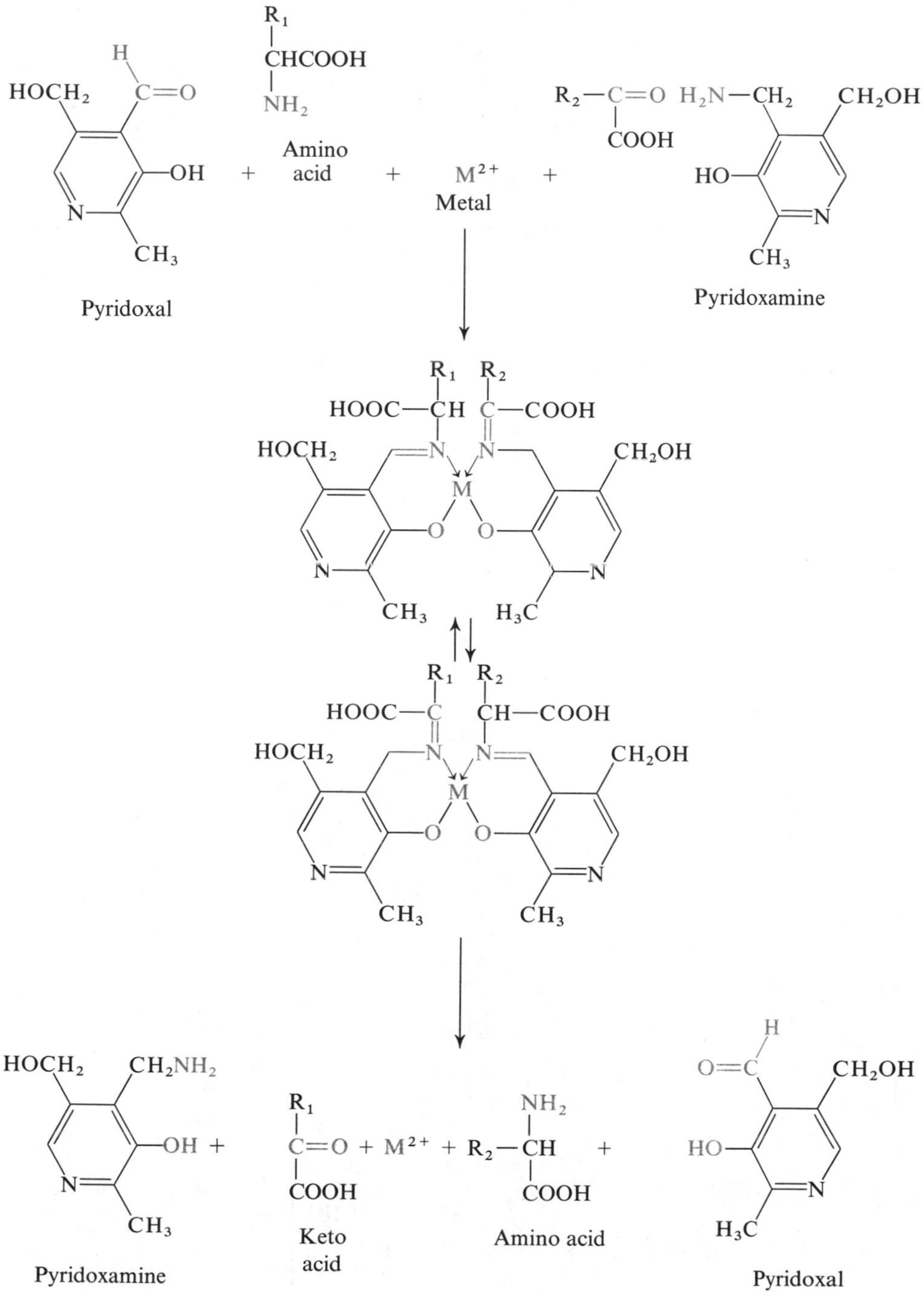

Fig. 20-5 Role of pyridoxal in transamination.

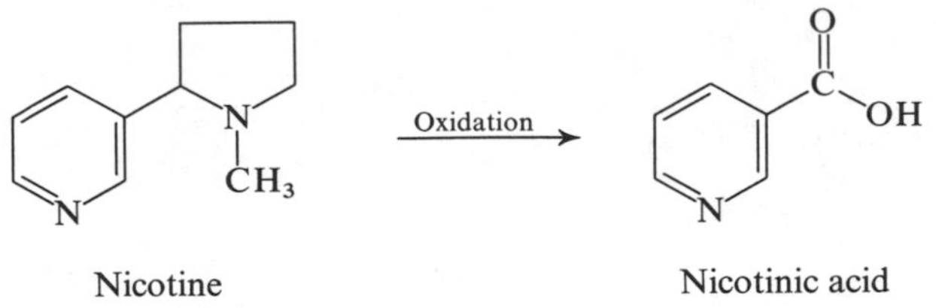

Nicotine   Nicotinic acid

Fig. 20-7 The structure of nicotinic acid was determined during a study of the structure of nicotine.

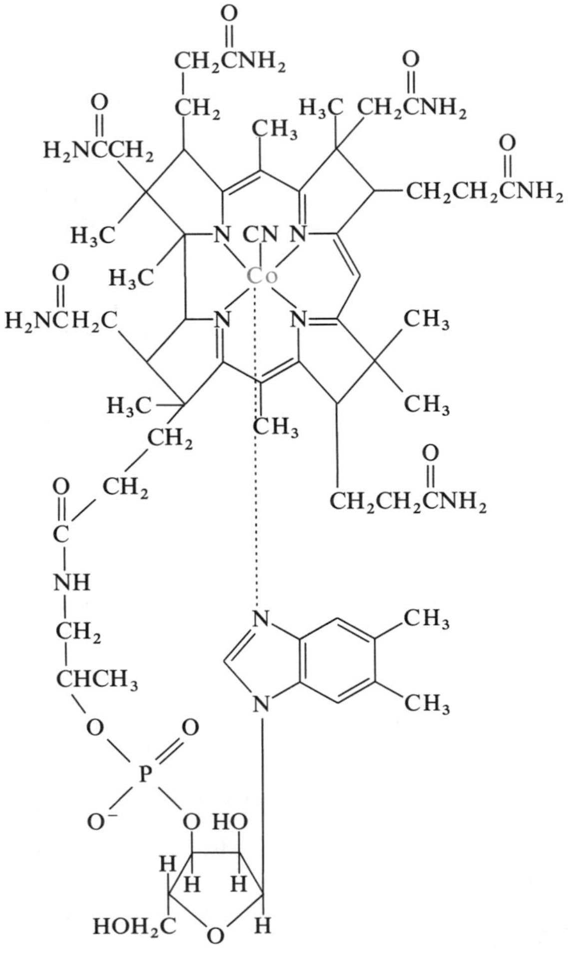

Fig. 20-8 Structure of vitamin B_{12} (cyanocobalamin).

Nicotinic Acid

Nicotinic acid (Fig. 20–7) has been known for a century as one of the oxidation products of nicotine, but the recognition of nicotinic acid and nicotinamide (niacin) as the antipellagra vitamins present in liver was delayed until the late 1930's. The metabolic function of nicotinic acid and amide is a result of the conversion to NAD and phosphorylated NAD, with consequent participation in metabolism of the carbohydrates and in oxidative phosphorylation.

Vitamin B₁₂

The structure of vitamin B_{12} (Fig. 20–8) is very complex. This vitamin was first isolated from liver containing about one part per million. Therefore it is not surprising that vitamin B_{12} was one of the last of the known vitamins to be discovered.

In 1926 it was recognized that pernicious anemia could be successfully treated by feeding patients whole liver. This initiated studies into the isolation of the active constituent. In 1948 vitamin B_{12} was isolated as a red crystalline compound which was capable of combating pernicious anemia in dosages as small as 3×10^{-6} g. Studies of molecular weight, plus analytical data, indicated a formula of $C_{61\ 64}H_{86\ 92}N_{14}O_{13}PCo$. The cobalt forms a hexavalent complex with a tightly bound but replaceable CN group. Measurements of magnetic susceptibility indicated that the vitamin was diamagnetic, and therefore cobalt was trivalent. Within a ten-year period the detailed structure of the molecule was determined, by a combination of chemical degradation and x-ray crystallographic studies.

Vitamin B_{12} is required for normal growth, blood formation, and proper function of the nervous system. The metabolic role of the vitamin is not clear. It is known to be converted to a coenzyme when the cyano group is replaced by an adenyl-sugar residue. Chemists recognize the importance of the coenzyme in various stages of the biosynthesis of amino acids, but they still lack details.

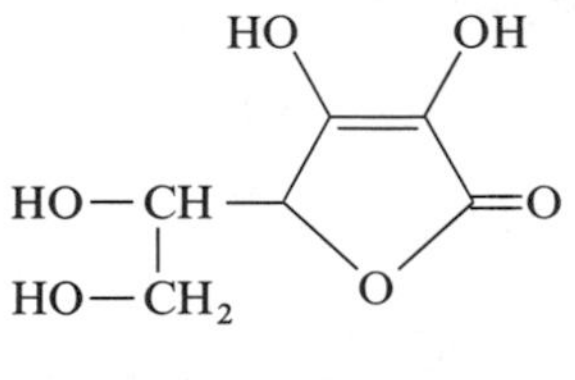

Fig. 20–9 L-ascorbic acid (vitamin C).

Vitamin C (Ascorbic Acid)

In the early 1700's it was discovered that scurvy could be prevented among crews of the English ships of the East India Company by including supplies of lemons and oranges in the diet. About 200 years later, the antiscurvy factor, vitamin C, was isolated from oranges and lemons and the structure was shown to be that of L-ascorbic acid (Fig. 20–9).

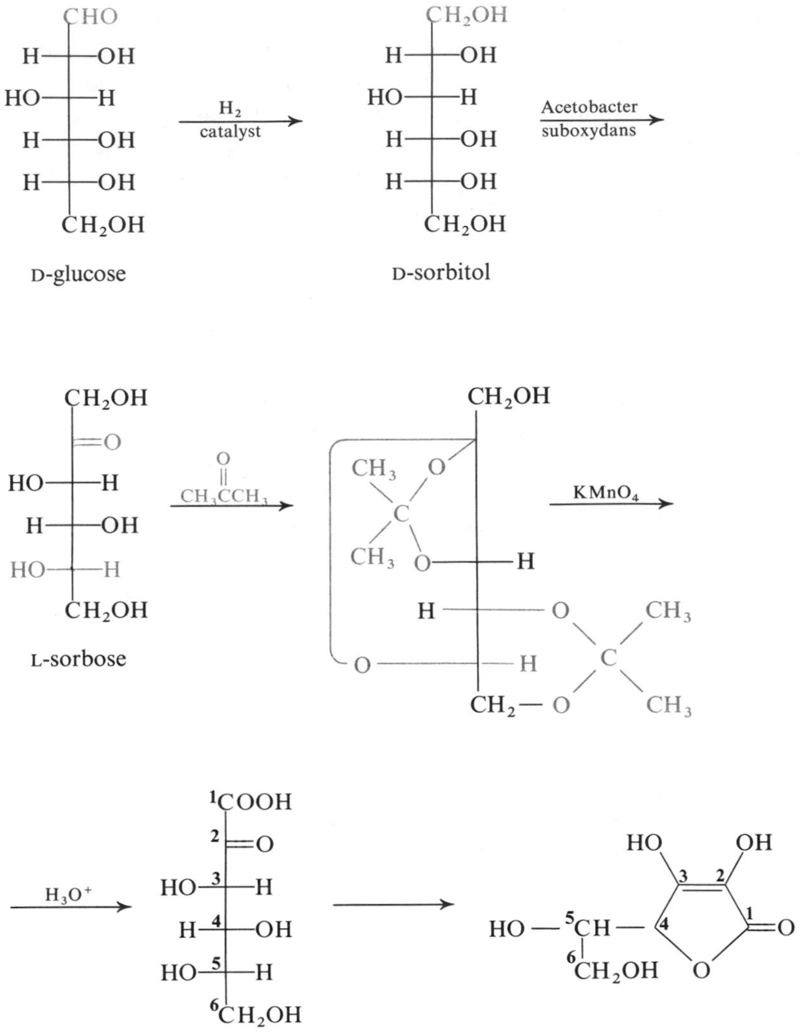

Fig. 20-10 A synthetic route to L-ascorbic acid.

Vitamin C is widely distributed in nature. It is concentrated in liver, milk, fruits, and green vegetables. Ascorbic acid can be commercially synthesized by a route described in Fig. 20–10. The D-sorbital formed from the catalytic hydrogenation of D-glucose is microbiologically oxidized and isomerized to L-sorbose by *Acetobacter suboxydans* cultures. All the reactive centers except one can be protected from oxidation by ketal formation with acetone. The unprotected alcohol can then be oxidized to an acid group. Acid-catalyzed hydrolysis of the ketal groups gives simultaneous conversion to the L-ascorbic acid.

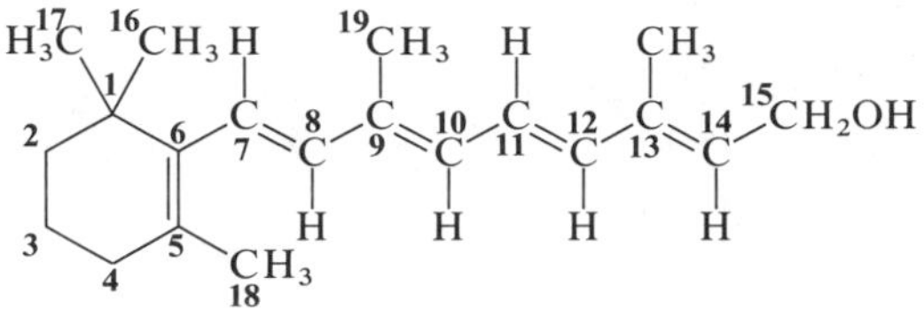

Fig. 20-11 Retinol (vitamin A).

Vitamin C appears to play an important metabolic role in the synthesis of collagen, the main protein of connective tissue, and is therefore necessary for growth and tissue repair. The mechanism of the process and the coenzyme functions of the vitamin are still unproven, as is its proposed cold-prevention.

Vitamin A

Nutritional studies indicated that a person's body needed certain fat-soluble materials in the diet to attain normal growth and development. Children existing principally on vegetable diets or on skimmed milk were found to develop eye problems. Investigations led to the discovery of vitamin A. The photosensitive pigments in the retina of the eye were later shown to be complexes of vitamin A with proteins. Vitamin A activity is found in plants containing certain yellow-red pigments. The pigments (carotenes) behave as biosynthetic precursors (provitamins) to the vitamin A group, being converted into vitamin A in the body.

Retinol, the first A-vitamin isolated, has the structure shown in Fig. 20-11. It exists in high concentration as an ester in fish-liver oil and can be isolated by saponification.

Various *cis-trans* isomers of retinol are possible. Four of these are found in nature and exhibit vitamin A activity. Another member of the group, vitamin A$_2$ (dehydroretinol) has one extra double bond in the 3,4-position.

There are also vitamin A types of molecules in which the alcohol group at C-15 has been oxidized to the aldehyde (retinal) or the acid (retinoic acid).

Today the role of vitamin A compounds in the visual process is quite well understood. The cones (for bright light) and the rods (for dim light) of the retina contain photosensitive pigments which are protein complexes of the A vitamins. The mechanism by which a person sees is too complex a subject to be discussed here. Let us say only that, in essence, it involves a bleaching of the pigment by light—which detaches the vitamin A from the protein—followed by isomerization and oxidation-reduction reactions of the vitamin A material. The net result is the conversion of light energy to chemical energy, which in turn is converted to an electrical impulse which goes to the optic nerve. We know that A vitamins are also implicated in other metabolic processes, but the details are more obscure.

The A vitamins are found in meat and meat products such as milk, cheese, eggs, liver, kidney, and heart, but not in vegetables. Vegetables such as carrots, spinach, and broccoli do contain carotene pigments which act as provitamin A, and are converted into vitamin A in the body.

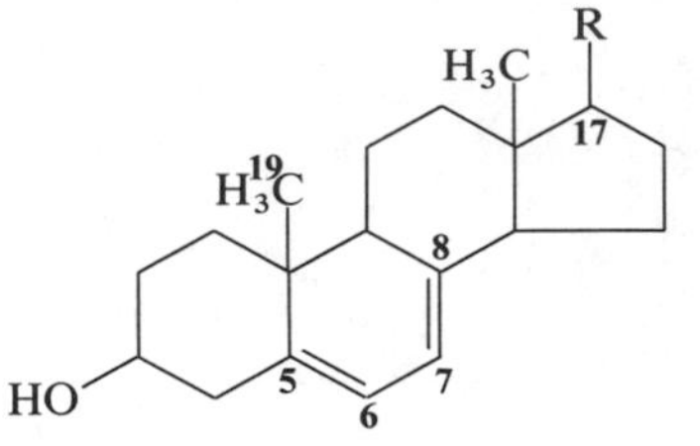

Provitamin D

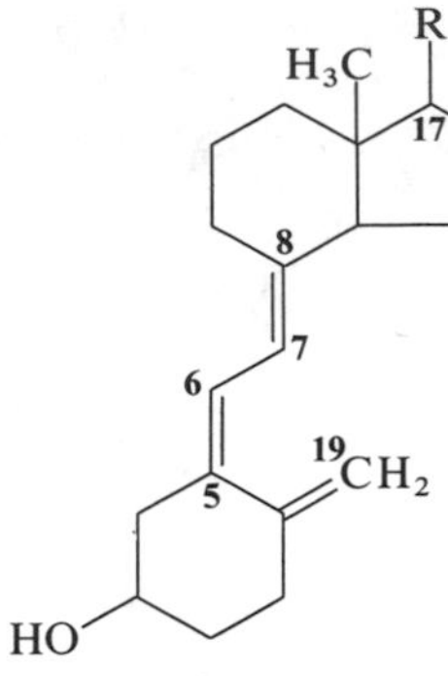

Fig. 20-12 Structures of provitamin D, for example 7-dehydro-cholesterol, differ in the substituent on C-17. They are converted into different D vitamins.

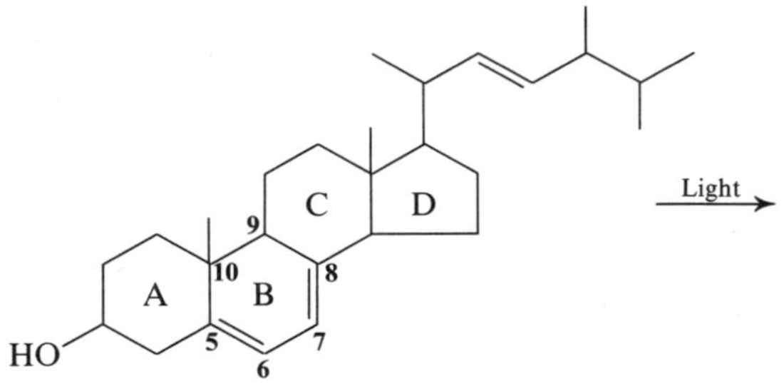

Vitamin D

Ergosterol
(provitamin D_2)

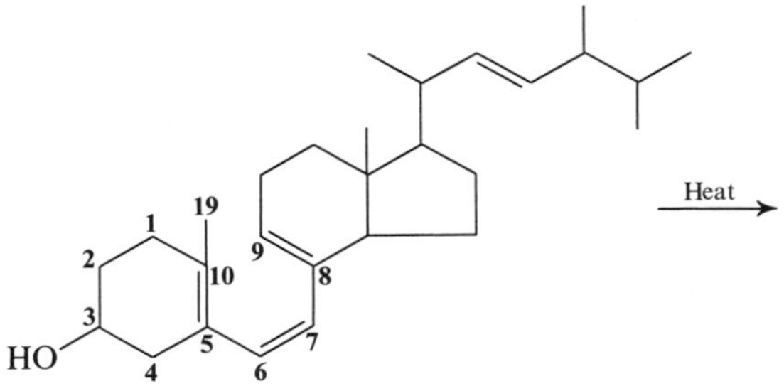

Pre-ergocalciferol
(previtamin D_2)

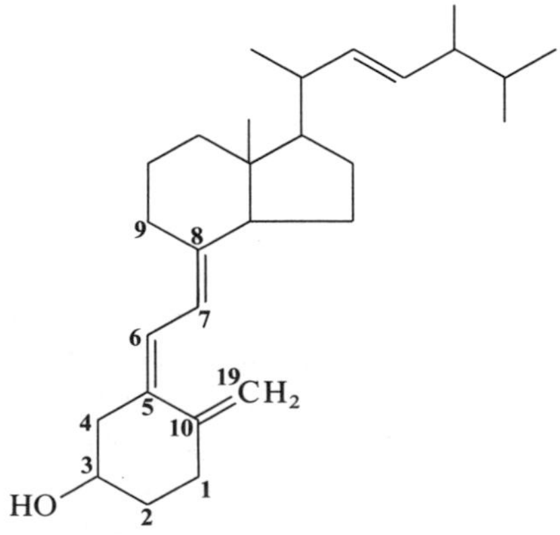

Ergocalciferol
(vitamin D_2)

Fig. 20-13 Formation of D vitamins from bio-chemical precursors.

The Vitamin D Group

The D vitamins are also represented by a group of closely related and interconvertible molecules. The first evidence of their existence was a recognition of the curative effects of cod liver oil and sunlight on rickets, although the reasons were not known. Vitamin D materials also have precursor molecules—called previtamins and provitamins (see Figs. 20–12 and 20–13).

The relationship of a provitamin to the vitamin can be illustrated with ergosterol. The first step involves the cleavage of the 9,10-bond of the B ring in ergosterol (Fig. 20–13). Since the process is catalyzed by light, it explains the role of sunlight in curing rickets. It also explains the practice of irradiating milk with ultraviolet light to increase the vitamin D content. The second step in the sequence involves the thermal isomerization of the previtamin D_2 to vitamin D_2.

The main biochemical role of the vitamin D group is to control the deposition of calcium and inorganic phosphate required for bone formation, but the details of the process remain to be discovered.

The Vitamin E Group

The factors that control full-term pregnancy in rats as well as the fertility of male and female rats have been designated as the E vitamins. They are fat-soluble, and are found in natural oils (soybean, corn, etc.) as well as in yeast, eggs, butter, liver, and peas. The structure of one of the E-vitamins, α-tocopherol, is shown in Fig. 20–14. The others differ only in the length and extent of unsaturation in the side chain or in the number of methyl groups on the aromatic ring.

The E vitamins are easily oxidized to quinones (Fig. 20–15), and therefore are used as antioxidants in the preservation of foods. Much of the metabolic function

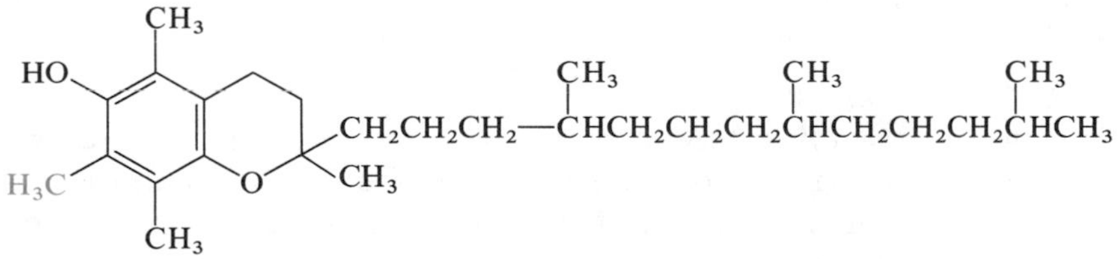

Fig. 20–14 Vitamin E (α-tocopherol). Other E vitamins differ in the length or unsaturation in the side chain or in the ring substitution; for example, β-tocopherol has the colored methyl group missing.

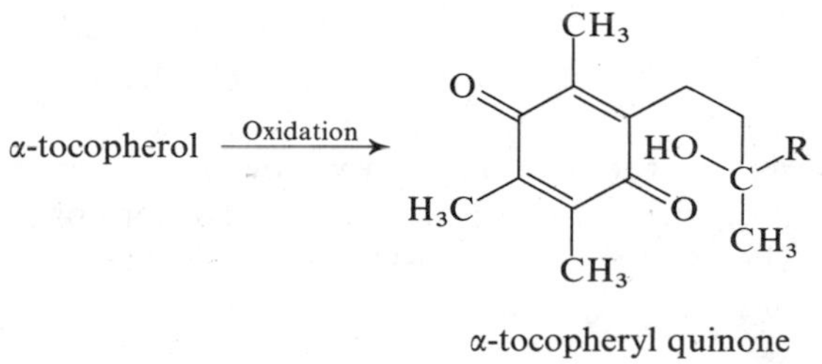

α-tocopheryl quinone

Fig. 20–15 Oxidation of vitamin E.

of vitamin E materials has been ascribed to this general antioxidant behavior rather than to any specific coenzyme function.

The role of vitamin E in human metabolism has not been demonstrated. It has been used with varying degrees of success in the treatment of habitual auto-abortion, sterility, and other disorders.

The Vitamin K Group

Another group of fat-soluble vitamins of common function is the vitamin K group: the vitamin required for proper clotting of the blood. Figure 20–16 shows the structure of vitamin K_1. The naphthoquinone ring and the methyl group at C-2 are common to all the K vitamins. The length of the side chain and the extent of unsaturation may vary.

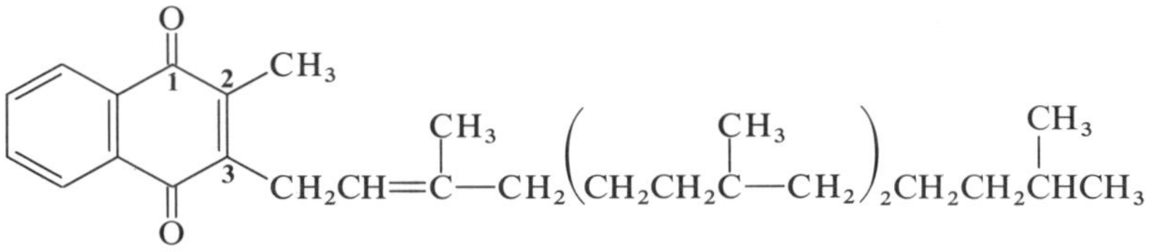

Fig. 20–16 Vitamin K_1 (phylloquinone).

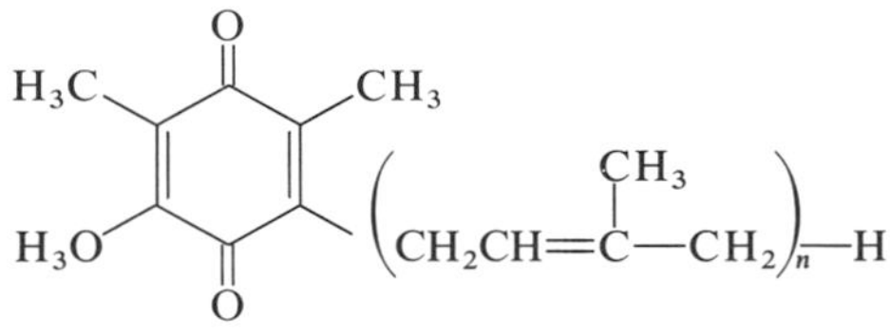

Fig. 20–17 General structure for coenzyme Q; n varies from 6 to 10.

The role of vitamin K in the blood-clotting process has long been recognized. The interruption of oxidative phosphorylation (the process whereby the energy from the oxidation of food is made available for the body's use) in cellfree bacteria devoid of vitamin K—and its restoration when vitamin K is added—indicates that vitamin K plays a much more fundamental role in the body's metabolism. The structural similarity of vitamin-K compounds to the coenzyme-Q group (Fig. 20–17), which is known to participate in this process of energy conversion, is striking.

20–3 HORMONES

Another group of compounds that exert a profound effect on various processes in the body are the *hormones*. The main distinction between vitamins and hormones is that the former must be ingested, whereas the latter are synthesized in the body. They are produced in the endocrine glands.

Hormones may be proteins (or polypeptides such as insulin), steroids, or other organic compounds, and the action of one hormone is usually intertwined with the action of others.

Thyroid Hormones. The thyroid gland is closely associated with the regulation of the body's rate of metabolism. The amino acid tyrosine is iodinated to diiodotyrosine (Fig. 20–18). Two diiodotyrosines then combine to give the hormone thyroxine, which has recently been involved in the control of Parkinson's disease. Or diiodotyrosine combines with monoiodotyrosine to yield the even more active hormone 3,5,3′-triiodothyronine. The synthesis of hormones is the main reason why iodine is required in the diet. In spite of the importance of the thyroid hormones, no one has yet been able to explain how they operate.

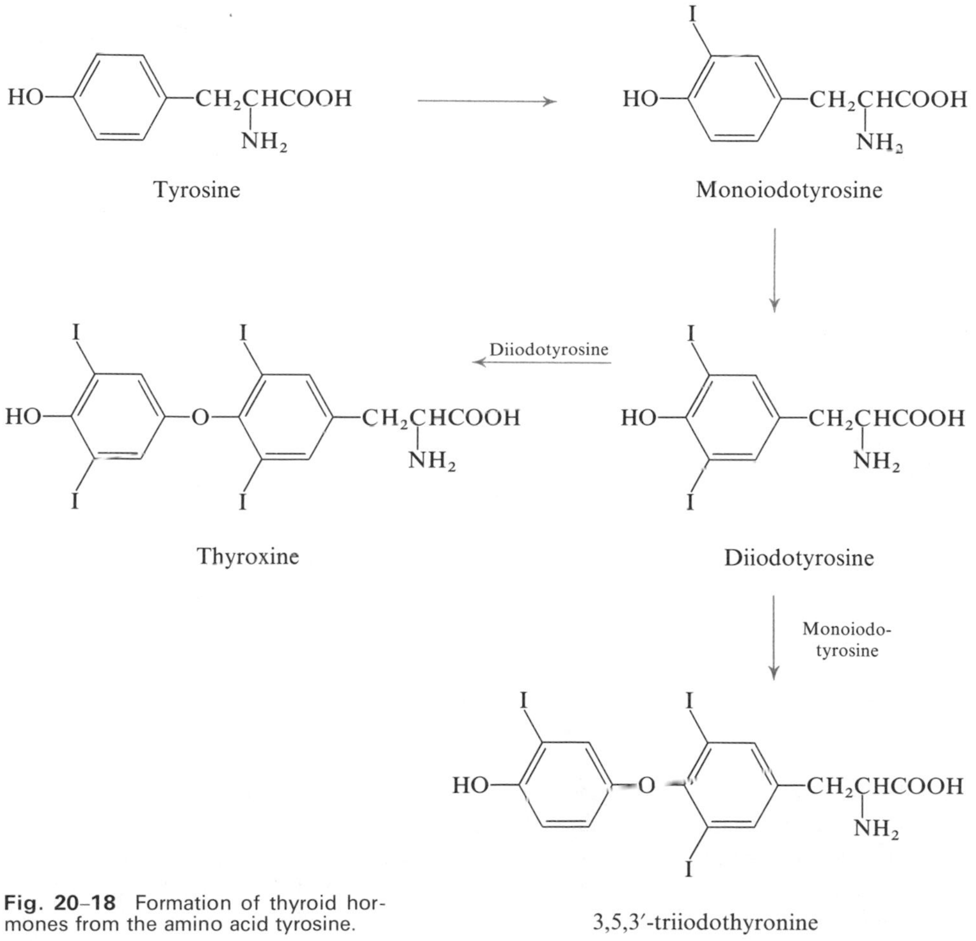

Fig. 20–18 Formation of thyroid hormones from the amino acid tyrosine.

Parathyroid Hormones. The parathyroid glands produce a hormone which is protein in nature, and is intimately involved in the metabolism of calcium and phosphorus, and therefore in the formation of bone.

Pituitary Hormones. The pituitary gland, which is divided into three sections, produces three kinds of hormones: The growth hormone (*somatotrophin*) controls growth and sexual development. The thyroid-stimulating hormone (TSH) is called *thyrotropin*. It controls the activity of the thyroid gland and thereby the metabolic rate. Similarly the *adrenocorticotropic hormone* (ACTH) stimulates the adrenal gland's production of steroid hormones. All these hormones, as well as others produced in the pituitary gland, are polypeptides.

The Adrenal Hormones. The inner portion of the adrenal gland—*medulla*—secretes the hormones *epinephrine* (adrenaline) and *norepinephrine* (noradrenaline), whose structures are shown in Fig. 20–19.

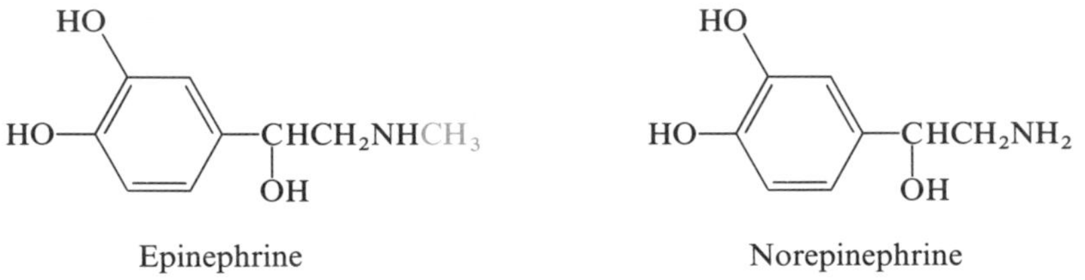

Epinephrine   Norepinephrine

Fig. 20-19 Adrenal hormones.

Chapter 19 described the role of epinephrine in controlling the levels of glucose in the blood. Norepinephrine functions in the same way. In addition, both hormones increase the rate of heartbeat and raise the blood pressure, although not to the same extent.

The outer portion of the adrenal gland—the *cortex*—produces the so-called adrenocortical hormones. These are steroidal in nature, and, as indicated in Fig. 20–20, have similar structures.

These hormones are important in the synthesis of liver glycogen and in carbohydrate metabolism in general, although the mechanism of the hormone action is unknown. In addition, there is a relation between these hormones and the sodium/potassium ion ratio in the blood. These hormones, and synthetic analogs of them, have found widespread use as drugs.

Sex Hormones. The masculinizing hormones, produced primarily in the testes but also in the ovaries and adrenal cortex, are also steroidal in nature (Fig. 20–21).

The *androgens* are hormones that control the secondary sex characteristics of the male, as well as exerting effects on the formation of sperm and the metabolism of protein.

Corticosterone

Cortisol

Cortisone

Aldosterone

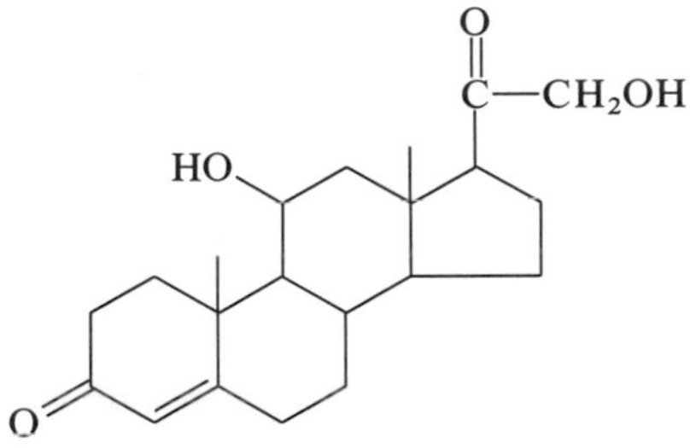

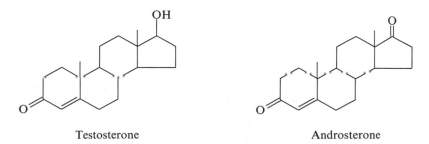

Fig. 20–20 Structures of four of the adrenocortical hormones. Several other structurally similar steroids are also produced in the adrenal cortex.

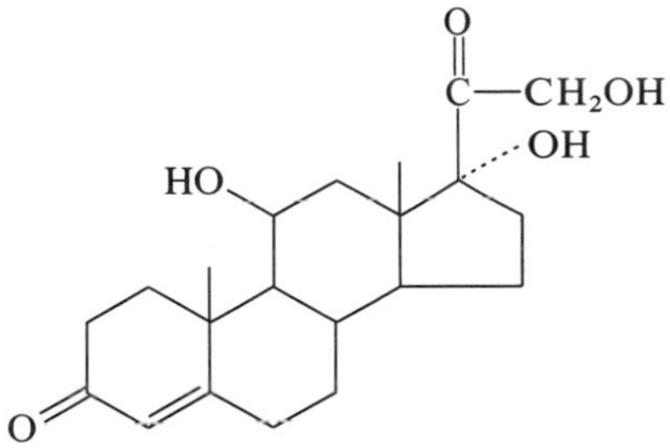

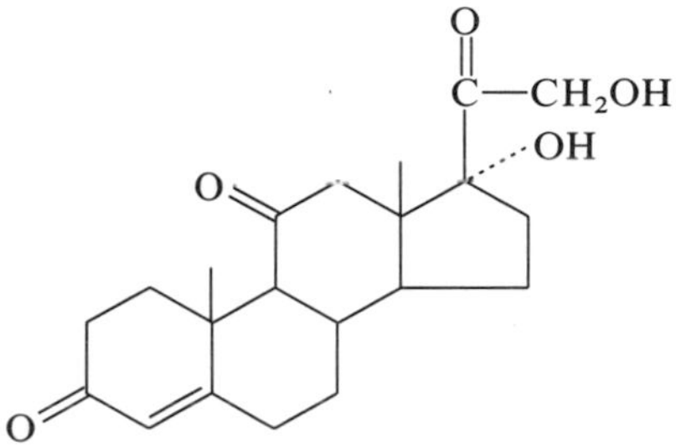

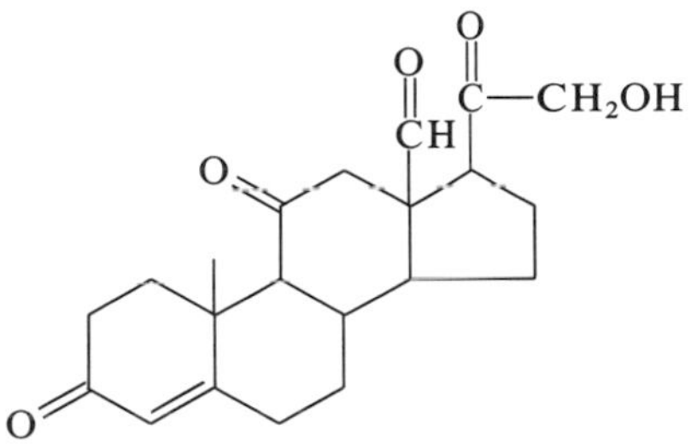

Testosterone

Androsterone

Fig. 20–21 The principal androgens, testosterone and androsterone, are found in the blood of both men and women, although the level in men is usually about 5 times greater than in women.

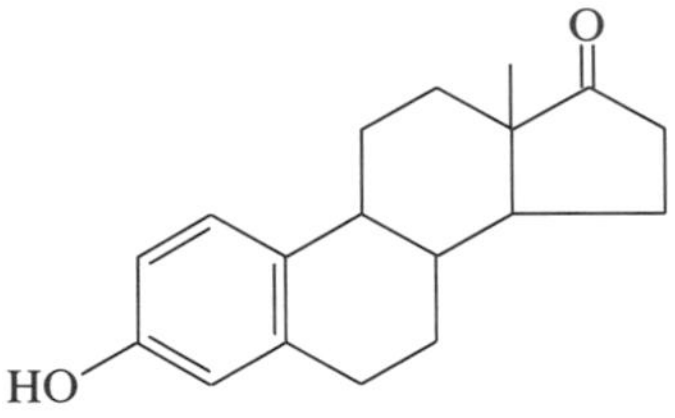

Estrone

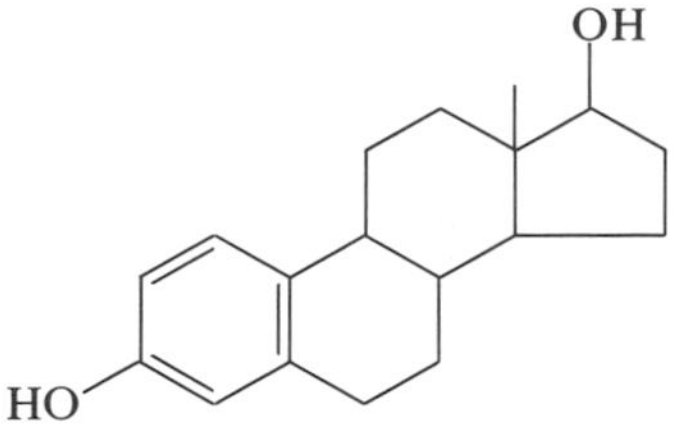

Estradiol

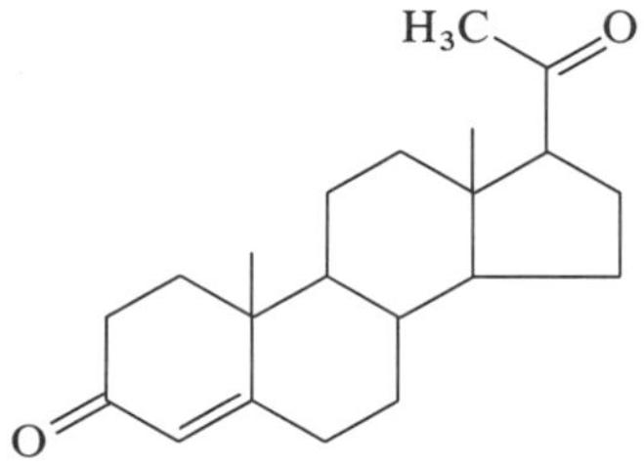

Fig. 20-22 Female sex hormones such as estrone and
estradiol (estrogens) are found in males also, but in much
lower concentration.

Progesterone

The *estrogens* are hormones that determine the secondary sex characteristics of
the female, as well as various reproductive and metabolic functions. Produced primarily in the ovaries, estrogens are also found in the adrenal gland, and in the testes.
Several estrogens have been found in nature, the two most important for humans
being estrone and estradiol (Fig. 20–22). Another steroid hormone, progesterone,
helps to maintain pregnancy, in that it prevents spontaneous abortion of the fetus.

Although the effects of these various hormones are well documented, the detailed
biochemical reactions are still unknown.

20-4 DRUGS

The area of chemistry that is concerned with drugs is often called *pharmaceutical* or
medicinal chemistry. In this classification there are a large number of chemists conducting a wide variety of studies. Generally, however, these people are concerned
with the isolation and identification of compounds which can be used in the treatment
of disease (drugs), the synthesis of such compounds, and an understanding of the
mode of action of each drug. Over the years the first two efforts have been remarkably
successful, while the last has been largely ineffective, even though extensive studies
have been made.

The first drugs, of course, were those herbs and other concoctions of plant or
animal origin that were used in ancient times to treat disease.

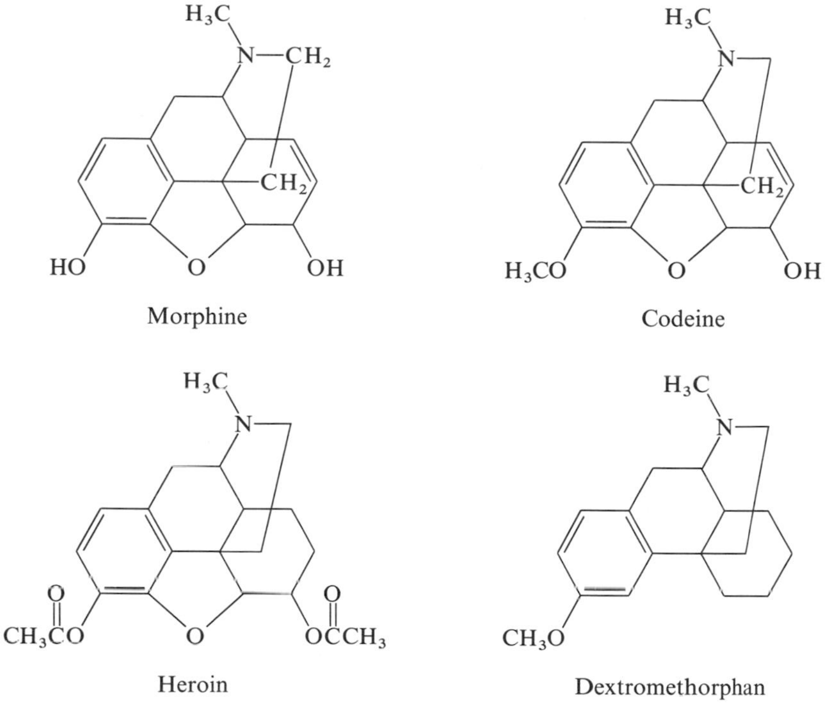

Fig. 20-23 Analgesics of the morphine type.

Many herbs used over 5000 years ago in China have been shown to contain alkaloids of medicinal importance. The use by South American Indians of the cinchona bark led to the isolation of the antimalarial drug, quinine. Modern investigation of many of the naturally occurring remedies discovered in the folk lore of ancient as well as in modern times has led to the isolation of new drugs.

People in early times and in the Middle Ages were filled with superstitions (not unlike today!) about drugs, and only very recently have efforts been made to apply a rational, scientific approach to the problem.

One of the oldest and most varied areas of drug studies involves pain-killing *analgesics*. One of the first analgesics used was opium, whose active agents are alkaloids. The most important of these active agents are morphine and codeine, which act on the central nervous system.

The use of morphine as an analgesic has been restricted because of the addictive properties of the drug. The same is true of the morphine derivatives—codeine and heroin—shown in Fig. 20–23. The fact that these compounds have a euphoric effect, coupled with their addictive properties, has led to a serious social problem.

Chemists who were investigating the structural relationship between heroin, codeine, and morphine found that heroin is a stronger analgesic than morphine and codeine, and also that minor structural changes in the molecule of morphine can alter the drug's effects. In fact, codeine is about one-sixth as active as an analgesic

as morphine, and is less addicting as well. Thus it seems reasonable to hope that a suitable alteration in the molecule would result in a compound which had pain-killing properties similar to those of morphine but which was not addictive.

To test this hypothesis, many compounds have been synthesized. One of the most promising groups of compounds are derivatives of morphinan (Fig. 20–24).

Dextromethorphan (Fig. 20–23), for example, is totally nonaddicting, and, although not an analgesic, has cough-suppressant (antitussive) properties so that it can supplant codeine in cough medicines. More recent research has developed analgesics that seem to approach the ideal properties of potency without addiction.

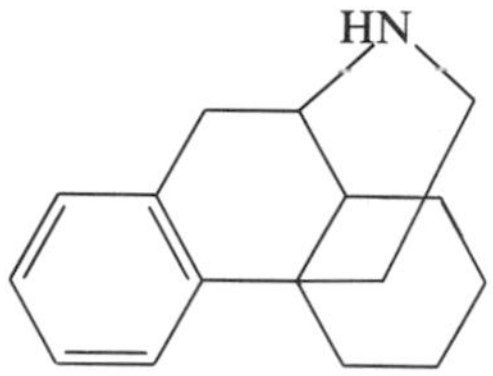

Fig. 20–24 Morphinan.   Morphinan

20–5 DRUG DESIGN

The development of local anesthetics followed a pathway which is a classic demonstration of the rationale behind drug design.

Cocaine was used during the Middle Ages as a euphoric and a hallucinogen, but it was not until much later that the local-anesthetic properties of cocaine were noted. The fact that cocaine was both toxic and addictive—along with its other undesirable side effects—prevented its widespread use as a local anesthetic. The problem, then, was to find a compound with suitable anesthetic activity but no undesirable side effects.

Conversion of cocaine to tropacocaine (Fig. 20–25) created a drug which retained cocaine's anesthetic properties, but was highly toxic. The discovery of the local-anesthetic properties of α-eucaine—and later the less-toxic β-eucaine—demonstrated that the cocaine ring system was not a prerequisite for anesthetic activity. Similar studies showed that ethyl-p-aminobenzoate and its derivatives also had value as local anesthetics.

It was postulated that the activity of cocaine was dependent on having a tertiary amino group attached by 2 or 3 carbon atoms to an alcohol group which was esterified by an aromatic acid,

$$\begin{matrix} R \\ | \\ R-N-(CH_2)_{2 \text{ or } 3} \end{matrix} \qquad \begin{matrix} O \\ \| \\ O-CC_6H_5 \end{matrix}$$

By making the aromatic acid p-amino benzoic acid, chemists hoped to combine the positive properties of both types. The result was the synthesis of procaine (Fig. 20–25), one of the most widely used local anesthetics.

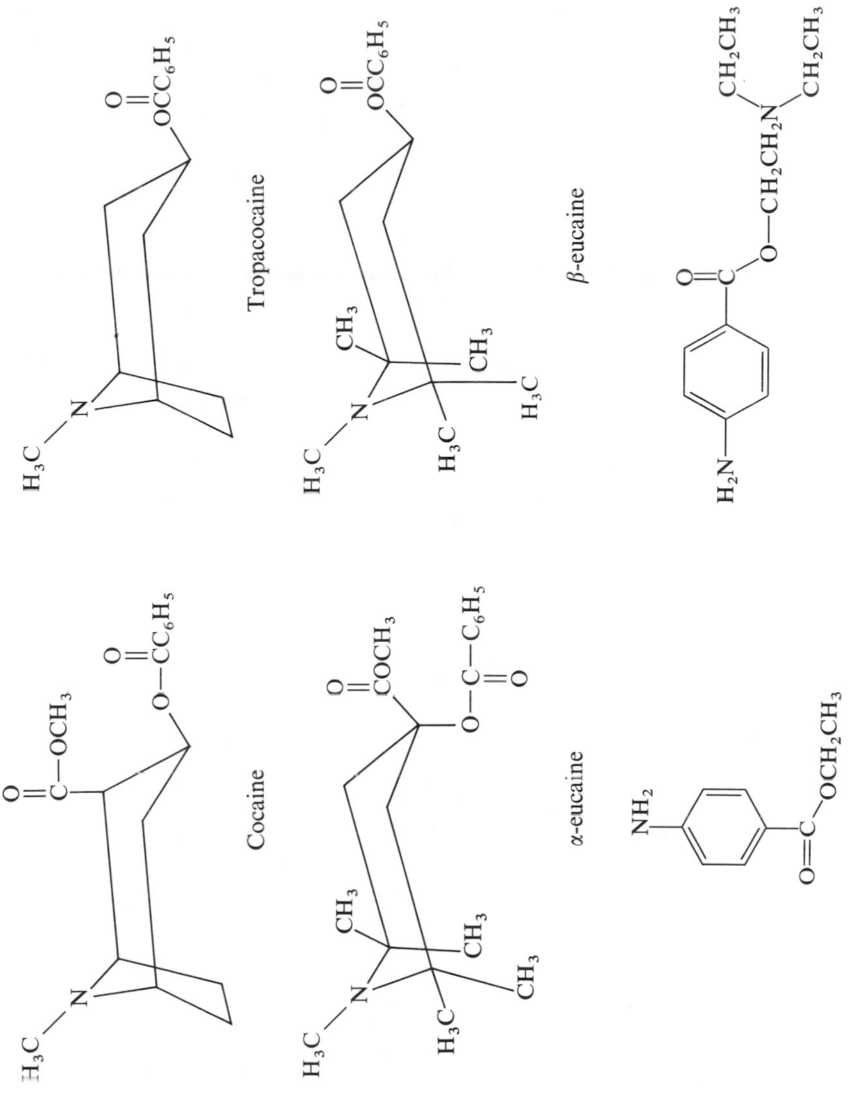

Fig. 20–25 Some local anesthetics.

20-6 CONCLUSION

The chemical world has had great success in producing compounds which have physiological activity. We have mentioned only a few of these in this chapter, but many more are known: The ability to chemically control ovulation in females, via The Pill, is well known. The so-called wonder drugs have had their share of headlines. Chemicals are being used more and more as food additives (sweeteners, tenderizers, preservatives, flavor enhancers), although this practice is now under scrutiny by the Food and Drug Administration.

The chemist today has the ability to produce compounds which have highly specific action, although not without occasional tragic failures in application (such as happened in the sixties in the case of thalidomide). This ability has not only been demonstrated but has received wide acceptance, and this acceptance in turn has created an atmosphere of danger. All of us seem to be too ready to choose a chemical solution to our difficulties. Chemicals not only cure illness and conquer pain, but they control weight gain and loss, stimulate, relax, cause happiness and euphoria, sadness and depression, and countless other physiological and psychological events. That the chemist can do this has been demonstrated; that he will continue to produce compounds with even more specific activity is beyond doubt. What is lacking—and what is needed not only by the chemist but also by the doctor, the government official, the legislator, and the layman—is the wisdom to use the compounds wisely.

21. NUCLEAR AND RADIOCHEMISTRY

21-1 RADIOACTIVITY

The chemical and physical properties of substances are determined almost exclusively by the electronic configuration of the atoms. The nucleus is important only insofar as it influences the size and reactivity of the electron cloud. Consequently, the interpretation and explanation of most chemical phenomena require only a detailed understanding of electronic configurations.

The phenomenon of radioactivity is one property, however, that cannot be explained by electron structure; it is solely a result of the structure of the nucleus. Henri Becquerel in 1896 discovered that uranium compounds emitted radiation that could darken a photographic plate after the radiation passed through a sheet of paper or a thin sheet of metal which stopped ordinary light. After intensive investigation, Pierre and Marie Curie discovered two new radioactive elements, polonium and radium.

If a sample of a radium compound is placed in a lead block, as indicated in Fig. 21-1, a stream of radiation can be directed between the poles of a powerful magnet. The radium sample actually emits three different types of radiation: One of these (α) behaves like a positive particle when it is passed through the magnetic field; the second (γ) is unaffected by the field; the third (β) behaves like a negative particle. Further investigation revealed that the properties of these particles are as outlined in Table 21-1.

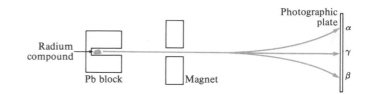

Fig. 21-1 Effect of a magnetic field on alpha, beta, and gamma radiations.

The alpha particle, the least penetrating of the three, is a helium nucleus which has high kinetic energy. The gamma ray—the most penetrating—is electromagnetic radiation of higher energy (shorter wavelength) than x-rays. The beta particle is a high-energy electron moving at various speeds, some close to the speed of light. When either an alpha or a beta particle is emitted from a nucleus, an atom of another element must be formed. Since both these particles are charged, the charge in the nucleus must change and the product of the reaction must have a different atomic number and be a different element.

Table 21-1. Properties of alpha, beta, and gamma radiations

Radiation	Charge	Mass	Nature
α	$+2$	4.0026 amu	Helium nucleus
β	-1	0.000548 amu	Electron
γ	0	0	Electromagnetic radiation

Table 21-2. Uranium-238 disintegration series

$$^{238}_{92}U \xrightarrow{-\,^4_2He} {}^{234}_{90}Th \xrightarrow{-\,^0_{-1}e} {}^{234}_{91}Pa \xrightarrow{-\,^0_{-1}e} {}^{234}_{92}U \xrightarrow{-\,^4_2He} {}^{230}_{90}Th \xrightarrow{-\,^4_2He}$$

$$^{226}_{88}Ra \xrightarrow{-\,^4_2He} {}^{222}_{86}Rn \xrightarrow{-\,^4_2He} {}^{218}_{84}Po \xrightarrow{-\,^4_2He} {}^{214}_{82}Pb \xrightarrow{-\,^0_{-1}e} {}^{214}_{83}Bi \xrightarrow{-\,^0_{-1}e}$$

$$^{214}_{84}Po \xrightarrow{-\,^4_2He} {}^{210}_{82}Pb \xrightarrow{-\,^0_{-1}e} {}^{210}_{83}Bi \xrightarrow{-\,^0_{-1}e} {}^{210}_{84}Po \xrightarrow{-\,^4_2He} {}^{206}_{82}Pb$$
$$\text{Stable}$$

For example, the uranium isotope U-238 decomposes spontaneously to lead-206 by the pathway illustrated in Table 21-2. The first step in this process involves the loss of a helium nucleus from the uranium-238 nucleus:

$$^{238}_{92}U \rightarrow {}^4_2He + {}^{234}_{90}Th.$$

The second product of the reaction must, therefore, weigh 4 mass units less and have only 90 protons in the nucleus. The element whose atomic number is 90 is thorium, and thus the other product of the radioactive disintegration of uranium-238 must be thorium-234.

The second step in the disintegration involves the loss of a beta particle from thorium. The beta particle appears to come from the *nucleus*, and thus the nucleus must lose a negative charge:

$$^{234}_{90}\text{Th} \rightarrow \,_{-1}^{0}\text{e} + \,^{234}_{91}\text{Pa}$$

This can be envisioned as the conversion of a neutron into a proton and an electron:

$$^{1}_{0}\text{n} \rightarrow \,^{1}_{1}\text{H} + \,_{-1}^{0}\text{e}$$

The end result of beta emission is the formation of an element of the same atomic weight as the parent element, but whose atomic number is *increased* by 1. All the steps involved in the conversion of uranium-238 to lead-206 can be explained as either an alpha or beta emission from the nucleus.

The product nuclei formed by emission of an α or β particle are often in an excited energy state. The collapse of these nuclei to lower energy states is accompanied by the emission of discrete amounts of energy as electromagnetic radiation, γ-rays.

21–2 NUCLEAR STABILITY

The lead-206 nucleus formed from the series of reactions illustrated in Table 21–2 is stable to further nuclear decomposition. When we consider various isotopes of the elements, we see that some have nuclear stability, while others do not. For example, the hydrogen nucleus ($^{1}_{1}\text{H}$) and the deuterium nucleus ($^{2}_{1}\text{H}$) are stable, but the tritium nucleus ($^{3}_{1}\text{H}$) is unstable and emits a beta particle. In the same fashion, carbon-12 is stable and carbon-14 is unstable. In the case of the heavier elements, no stable isotopes are known above atomic number 84.

Although there is no simple explanation of why certain isotopes are stable and others are not, one factor which seems to be important is the neutron-to-proton ratio. Naturally occurring radioactive materials apparently have too many neutrons for the number of protons. To correct this situation, they can eject an alpha or a beta particle. Either process results in a decrease in the neutron-to-proton ratio. To see why a high neutron-to-proton ratio means instability requires an understanding of the structure of the nucleus that we don't have today. It is an area of research that nuclear physicists are enthusiastically pursuing.

21–3 SYNTHESIS OF ISOTOPES

The discovery of radioactivity made available to the scientist a new tool. The alpha particles that were emitted from the nucleus could be focused with the aid of magnets, and could be used to bombard various materials. One such reaction carried out involved the bombardment of nitrogen:

$$^{14}_{7}\text{N} + \,^{4}_{2}\text{He} \rightarrow \,^{17}_{8}\text{O} + \,^{1}_{1}\text{H}.$$

The alpha particle must be of high energy in order to overcome the electrostatic repulsion between the positive nitrogen nucleus and the positive helium nucleus.

In 1932, the English physicist Sir James Chadwick bombarded beryllium with alpha particles and discovered the neutron:

$$^{9}_{4}Be + ^{4}_{2}He \rightarrow ^{12}_{6}C + ^{1}_{0}n.$$

The neutron formed in this reaction has a large amount of kinetic energy and can also be used as a bombarding particle. In this respect the neutron has an advantage over the alpha particle, in that it is uncharged and does not experience electrostatic repulsion as it approaches the nucleus. Energetic neutrons, available from the alpha-particle bombardment of beryllium or from other sources to be mentioned, represent very efficient bombarding agents. Some representative reactions are:

$$^{16}_{8}O + ^{1}_{0}n \rightarrow ^{13}_{6}C + ^{4}_{2}He$$
$$^{40}_{20}Ca + ^{1}_{0}n \rightarrow ^{40}_{19}K + ^{1}_{1}H$$
$$^{35}_{17}Cl + ^{1}_{0}n \rightarrow ^{35}_{16}S + ^{1}_{1}H$$
$$^{14}_{7}N + ^{1}_{0}n \rightarrow ^{14}_{6}C + ^{1}_{1}H$$
$$^{238}_{92}U + ^{1}_{0}n \rightarrow ^{239}_{92}U$$

The last reaction, the neutron bombardment, of uranium-238, is particularly significant. The product of the reaction, $^{239}_{92}U$, is unstable, and rapidly emits a beta particle to form neptunium-239. The neptunium-239 is also unstable and emits a beta particle, forming plutonium-239:

$$^{239}_{92}U \rightarrow ^{0}_{-1}e + ^{239}_{93}Np, \qquad ^{239}_{93}Np \rightarrow ^{0}_{-1}e + ^{239}_{94}Pu.$$

Neither neptunium nor plutonium occur in nature. Thus they represent man-made elements. Plutonium-239 is unstable and decays with the emission of an alpha particle. The reaction is very slow, however, with a half-life of 24,000 years.* For this reason, it has been possible to prepare and store large quantities of plutonium-239 for use as fissionable fuel in nuclear weapons, atomic power plants, and reactors.

The synthesis of new elements and the study of nuclear reactions in general was advanced considerably by the discovery of the cyclotron in 1929 by Ernest O. Lawrence at the Berkeley campus of the University of California. The cyclotron and other types of particle accelerators take advantage of alternating positive and negative fields to push and pull light, positively charged nuclei until they attain very high velocities. These high-speed particles can then be used in studies of bombardment of various materials. Some of the reactions are the following.

$$^{39}_{19}K + ^{1}_{1}H \rightarrow ^{36}_{18}Ar + ^{4}_{2}He \qquad ^{239}_{94}Pu + ^{4}_{2}He \rightarrow ^{242}_{96}Cm + ^{1}_{0}n$$
$$^{12}_{6}C + ^{1}_{1}H \rightarrow ^{13}_{7}N + ^{0}_{0}\gamma \qquad ^{238}_{92}U + ^{12}_{6}C \rightarrow ^{246}_{98}Cf + 4^{1}_{0}n$$
$$^{238}_{92}U + ^{4}_{2}He \rightarrow ^{241}_{94}Pu + ^{1}_{0}n \qquad ^{207}_{82}Pb + ^{2}_{1}H \rightarrow ^{208}_{82}Pb + ^{1}_{1}H$$

---

*   The *half-life* is the time required for one-half of a given amount of sample to decompose.

▶ There has been a good deal of activity recently in the synthesis of heavy elements. Workers at the Institute for Nuclear Research at Dubna, in the Soviet Union, announced the synthesis of element 104 and proposed the name kurchatovium (Kh or Ku?). Workers in the United States, at the University of California at Berkeley, however, were unable to repeat the Russian work. But they did synthesize a different isotope of element 104 by the reaction

$$^{249}_{98}Cf + {}^{12}_{6}C \rightarrow {}^{257}_{104}? + 4{}^{1}_{0}n.$$

The Berkeley workers have suggested the name rutherfordium (Rd?) for the new element. These studies were carried out using the heavy ion linear accelerator (HILAC) at Berkeley. The same system was used for the synthesis of element 105— proposed name: hahnium (Ha)—announced by Albert Ghiorso of Berkeley's Lawrence Radiation Laboratory on April 27, 1970. The reaction used was

$$^{249}_{98}Cf + {}^{15}_{7}N \rightarrow {}^{260}_{105}Ha + 4{}^{1}_{0}n.$$

The Berkeley group say that the synthesis of elements 106 and 107 is likely in the near future. The HILAC is to be shut down and modified so that it may be used to accelerate ions as heavy as those of uranium. The production of many more synthetic elements is the likely result. ◀

NUCLEAR ENERGY

21-4 BINDING ENERGY

If we carefully determine the masses of the elements, we see that the sum of the masses of the separate component particles of an atom does not equal the total mass of the atom. For example, the mass of two protons (2 × 1.0073) plus the mass of two neutrons (2 × 1.0087) equals 4.0320 amu, but the mass of a helium nucleus is only 4.0017. Therefore 0.0303 amu disappears when we take 2 protons and 2 neutrons and combine them into a helium nucleus. We account for this loss of mass by saying that this amount of mass *has been converted into energy.*

According to the Einstein equation,*

$$E = mc^2,$$

energy (E) in ergs and mass (m) are interconvertible, and, since the velocity of light (c) is large (3 × 10^{10} cm/sec), a small mass is equivalent to a large energy. The 0.0303-amu loss of mass for helium is equivalent to a large amount of energy, approximately 28 million electron volts. This energy is called the *binding energy*, and can be thought of as the energy liberated when the four nuclear particles combine to form a helium nucleus. Or it can be thought of as the energy required to separate a helium nucleus into its components.

---

*   Named after Albert Einstein (1879–1955), the German-born American physicist, who was responsible for the *theory of relativity*, which this equation summarizes.

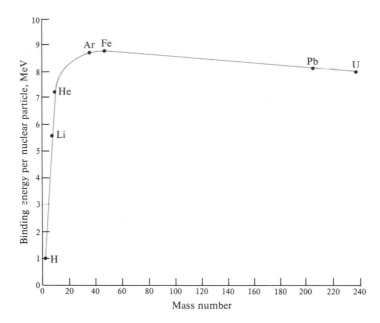

Fig. 21–2 Binding energy per nuclear particle, in million electron volts, versus mass number.

Figure 21–2 shows a plot of binding energy per nuclear particle, in million electron volts (MeV), versus mass number. The curve rises very steeply for the light elements, reaches a maximum around mass number 55, and then gradually decreases. This curve tells us that the most stable nuclei are those around mass number 55. Thus, if very heavy atoms could be split into particles in this more-stable-mass range, a significant amount of mass would be lost and *very* large amounts of energy would be liberated. If the very light elements could be fused into heavier particles, much larger amounts of energy would become available.

21–5 NUCLEAR WEAPONS

Although the Einstein equation had been known for years, there did not seem to be any way in which the source of energy could be tapped until, in 1939, German physicists noted the formation of barium and other large fragments in samples of uranium. It appeared that the splitting (fission) of uranium atoms into large fragments was an occasional natural phenomenon.

It soon became apparent that this fission process, if it could be caused to occur to a greater extent, represented a huge source of energy. Since this was a time of war or near-war over most of the world, a period of frenzied activity was initiated among scientists. In the United States the powers in the government were convinced of the importance of atomic energy in weaponry, and in 1940 a great and diverse collection

of scientists began working on the key to atomic fission. The collection and direction of so many scientific minds toward one end had never before been seen, and probably never will again. The sociological and psychological implications of this collective effort are still being argued.

On August 6, 1945, the United States dropped an atomic bomb on Hiroshima, Japan, culminating five years of extraordinary scientific effort and plunging the world into a new era: The atomic age had begun.

The difficulties which had to be overcome during these exploratory years were enormous. In the early stages of the investigations, it became apparent that uranium-238 underwent fission as a result of bombardment by high-speed neutrons, and that it split into fragments with the production of more neutrons. The neutrons which were produced, however, usually did not have enough energy to cause further fission of uranium-238. The process was very slow and inefficient. In order for the energy available in fission processes to be used in a bomb, it had to be released in a very short period of time.

It was finally discovered that uranium-235 and plutonium-239 would do the job. Both of these would undergo fission by slow neutrons. The fission process would produce nuclei of lower mass plus *excess neutrons*, which were capable of causing further fission reactions and even more neutrons. Thus the reaction would accelerate very rapidly, with a tremendous burst of energy.

In order for the reaction to be self-sustaining, however, it had to have a certain amount of fissionable material, called the *critical mass*. For example, if only a small amount of uranium-235 is available when one nucleus undergoes fission, the neutrons which are formed are unlikely to strike another uranium-235 nucleus. In order to increase the probability of a neutron colliding with a fissionable nucleus, a greater mass of the material must be present. Therefore, in order to construct an atom bomb, we needed subcritical masses of fissionable material. The subcritical masses had then to be compressed at the appropriate time into a supercritical mass for explosion.

This is not the whole story, however. Uranium-235 occurs as only 0.7% of natural uranium. Most of the uranium is the 238 isotope; plutonium-239 does not occur at all in nature to any appreciable extent. Thus getting hold of enough fuel was a crucial problem. The methods of separation and purification of materials available at the time depended on differences in chemical and physical properties that were gross differences. But since the two isotopes of uranium (U-235 and U-238) have identical electronic configurations, their properties are nearly identical. It was necessary to develop separation techniques that could take advantage of the slight difference in mass of the two isotopes. That chemists succeeded in doing this is a tribute to the people involved, and the techniques they developed have since been applied to the solution of many other important—though less dramatic—problems.

With the end of the war, activity did not cease. Scientists refined techniques and undertook the synthesis of large amounts of plutonium-239 by bombarding $^{238}_{92}U$ with neutrons. In addition, they were aware that the tremendous amounts of energy that were released by fission processes could even be surpassed if they could only manage to *fuse* light elements into heavier ones (recall Table 21–2).

The fusion of hydrogen nuclei to form a helium nucleus and two *positrons* (a particle with the mass of an electron, but positively charged) had been postulated as the reaction that occurs on the surface of the sun, and that is responsible for the energy released by the sun:

$$4^1_1H \rightarrow {}^4_2He + 2^0_1e + energy.$$

This reaction occurs only at the very high temperatures that exist on the surface of the sun. It was only after scientists had figured out how to set off self-sustaining fission reactions that the probability of their being able to produce fusion reactions increased.

The heat generated during the explosion of a plutonium bomb is great enough to trigger a fusion reaction. The advantage of a fusion bomb (if indeed any bomb can be said to have an advantage) is that greater energy is produced from smaller quantities of material, i.e., more mass is converted to energy. In 1952 lithium deuteride ($^6_3Li^2_1H$) was placed in an "ordinary" fission bomb and the first fusion reaction occurred. In the ten years or so that followed, most scientific effort was concentrated on a race between the United States and the Soviet Union to construct bigger and better bombs. At the present time there seems to be a lull in the competition, and signs that both nations are genuinely concerned with control of the development, production, and spread of nuclear weapons.

21–6 CONTROLLED NUCLEAR ENERGY

Since fusion reactions require temperatures on the order of a million degrees, and since we know of no materials that can withstand temperatures anywhere near that range, it will require a major breakthrough if fusion reactions are to be carried out in a controlled manner.*

This is not true for fission reactions. Since the rate of a fission reaction is proportional to the number of neutrons that strike a fissionable nucleus, if we control the number of neutrons, we can control the rate of the reaction. This is essentially what is done in a nuclear reactor. A typical reactor utilizes uranium-235 or plutonium-239 as fuel. The fuel is surrounded by a moderator—such as graphite or heavy water—which slows down the neutrons. Rods of cadmium or other neutron-absorbing materials are inserted. The reactor can be controlled by inserting or withdrawing the cadmium rods as necessary.

The reactor may be used as a source of neutrons for studies of neutron bombardment, and for the synthesis of isotopes not available from other sources. The heat from the reactor is usually transferred and used as a source of energy to drive turbines, supply electricity, propel ships, etc. There is little doubt that nuclear reactors will continue to increase in importance as sources of energy.

---

*   Very recent evidence suggests that such a breakthrough may be imminent. Extremely high temperatures have been contained by very strong magnetic fields.

▶ Nuclear energy should not be thought of as the ideal solution to our energy needs, though. The water that is used as a coolant in these reactors, if released into our rivers and lakes, can significantly increase the temperature of the natural waters. The effects of this *thermal pollution* are not known with certainty, but it is a problem that must be solved.

Two other serious problems must be overcome as well. The first of these involves the small amounts of radioactivity that are released from the nuclear reactors. We must study the long-term effects on the ecology of continued exposure to low-level radioactivity.

The second problem involves the disposal of waste materials. The waste materials are radioactive and must be disposed of with care. The problem is even more acute when we realize that the half-lives of these waste materials may be thousands of years long. This may well be the most serious drawback to large-scale production of nuclear energy. ◀

21–7 ISOTOPIC TRACERS

With the availability of reactors for the synthesis of new isotopes, it became possible to study the mechanism of reactions in a new manner. This has been particularly useful in following the pathway of biochemical reactions. For example, methionine, in which a very small percentage of the CH_3—S— carbon atoms are radioactive carbon-14, has been used in the investigation of the biosynthesis of nicotine:

$$^{14}CH_3—S—CH_2—CH_2—CHCOOH \rightarrow$$

Labeled methionine   Labeled nicotine

When this radioactively labeled methionine is fed to tobacco or tomato plants and the nicotine is isolated from these plants, the N-methyl group is found to be radioactive. Thus the methyl group in methionine, by some process or other, becomes the methyl group in nicotine.

Similar studies have shown that methionine is a common biochemical source of methyl groups. Such information would be very difficult to obtain without the use of isotopic tracers. Tracer techniques have been used in a variety of complex studies directed toward the understanding of biochemical processes. Studies of photosynthesis by Melvin Calvin at the University of California and the biosynthesis of cholesterol by Konrad Bloch at Harvard were judged worthy of Nobel prizes. Both these investigations relied heavily on isotope-tracing techniques.

21–8 OTHER APPLICATIONS OF RADIOACTIVITY

New applications of radioactivity are being discovered very rapidly. The treatment of cancer by irradiation—either from an external source such as cobalt-60 or by

implantation of the radioactive material directly into the tumor—is a commonly employed technique. The effect of radiation on the properties of various materials such as wood and a wide variety of foods is currently under investigation.

A knowledge of the rates of radioactive decay can be applied to the dating of objects. For example, the half-life of carbon-14 is 5668 years. In other words, one-half of any sample of carbon-14 will decompose in 5668 years. The carbon-14 content of all living things is a constant.* Bombardments by cosmic rays in the outer atmosphere result in the steady-state production of carbon-14. That is, the rate of formation of the isotope carbon-14 is exactly equal to the rate of decomposition, so that the concentration does not change. The result is that the CO_2 in the air contains a constant percentage of carbon-14. Likewise the plants that utilize CO_2 through photosynthesis, and the animals that eat the plants, contain the same constant ratio of carbon-14 to carbon-12 in the organic compounds present. If the plant or the animal is killed, the cycle is broken and the carbon-14/carbon-12 ratio decreases as the carbon-14 decays.

By measuring the natural radioactivity of a given organic material (leather, parchment, fabric, coal, etc.), we can calculate the amount of carbon-14 present and, knowing the half-life, we can determine the age of the material. The method is accurate to within about 5% and is limited to objects less than about 30,000 years old. It has been used to date ancient art objects and other materials of archeological interest. For older objects of geological interest, it is necessary to measure an isotope of longer half-life. Scientists use several isotopes by measuring the ratio of the isotope to its known decay product. For example, the half-life of U-238 is 4.5×10^9 years, and the end product of decomposition is Pb-206. Thus the ratio of U-238 to Pb-206 in old rocks can be used to date the earth as more than three billion years old.

21–9 CONCLUSION

The applications of nuclear and radiochemisty in all areas of science and technology are only now beginning to unfold. We can only hope that mankind will control the tremendous potential for destruction, and allow the equally tremendous potential for good to be developed.

QUESTIONS

1. Explain why different isotopes of the same element—for example, $^{12}_{6}C$ and $^{13}_{6}C$—have nearly identical chemical properties.

2. How did the discovery of radioactivity change atomic theory?

3. What product is formed when $^{232}_{90}Th$ decays with the stepwise loss of the following particles: α, β, β, α, α, α, α, β? Write the equations for each step.

---

*   The ^{14}C content is actually no longer constant. The $^{14}C/^{12}C$ ratio is increasing slightly, due to the production of ^{14}C in nuclear reactors.

4. Several artificially produced isotopes have neutron-to-proton ratios which are too low rather than too high. This situation is corrected by the emission of a positron, $_1^0e$. (Recall that the positron is a particle with the mass of an electron but with opposite charge.) When $_6^{11}C$ decays with the loss of a positron, what isotope is formed?

5. Complete the equations:

a) $_4^9Be + ? \rightarrow 3_2^4He + _0^1n$

b) $_{15}^{31}P + _1^2H \rightarrow ? + _1^1H$

c) $_{92}^{238}U + _1^2H \rightarrow 2_0^1n + ?$

d) $? \rightarrow _{94}^{238}Pu + _{-1}^0e$

e) $_8^{16}O + _0^1n \rightarrow ? + _2^4He$

f) $_3^7Li + _1^1H \rightarrow 2?$

g) $_{94}^{240}Pu \rightarrow ? + _{-1}^0e$

h) $_{95}^{242}Am \rightarrow ? + _{-1}^0e$

i) $_{92}^{238}U + _6^{12}C \rightarrow ? + 4_0^1n$

j) $_{94}^{244}Pu + _{10}^{20}Ne \rightarrow ? + 2_0^1n$

6. Why is more energy available from the fusion of one gram of a light element than from the fission of one gram of a heavy element?

7. Why are neutrons considered more efficient bombarding agents than protons?

8. Explain what is meant by the term *critical mass*.

9. How many calories of energy would be produced by the conversion of one gram of matter into energy?

$$\left(1 \text{ erg} = \frac{1 \text{ g-cm}^2}{\text{sec}^2} = 0.239 \times 10^{-7} \text{ cal}\right)$$

10. The combustion of natural gas (mostly methane) liberates 212,800 calories for each mole of methane burned. How many grams of methane must be burned to liberate the same amount of energy as in question 9?

11. In the saponification of ethyl acetate, two mechanisms are possible:

a) $CH_3\overset{\overset{\displaystyle O}{\|}}{C}-O-CH_3 + OH^- \rightarrow CH_3\overset{\overset{\displaystyle O}{\|}}{C}-O^- + CH_3OH$

b) $CH_3\overset{\overset{\displaystyle O}{\|}}{C}-O-CH_3 + OH^- \rightleftharpoons CH_3-\overset{\overset{\displaystyle :\ddot{O}:^-}{|}}{\underset{\underset{\displaystyle O-H}{|}}{C}}-O-CH_3$

$CH_3-\overset{\overset{\displaystyle :\ddot{O}:^-}{|}}{\underset{\underset{\displaystyle O-H}{|}}{C}}-O-CH_3 \rightleftharpoons CH_3-\overset{\overset{\displaystyle O}{\|}}{\underset{\underset{\displaystyle OH}{}}{C}} + {}^-O-CH_3$

$CH_3-\overset{\overset{\displaystyle O}{\|}}{C}-OH + CH_3-O^- \rightleftharpoons CH_3-\overset{\overset{\displaystyle O}{\|}}{C}-O^- + CH_3OH$

How could the oxygen-18 isotope be used to distinguish between these two mechanisms?

12. The half-life of radium is 1590 years. How long will it take for 75% of a radium sample to disappear?

13. The decomposition of carbon-14 in atmospheric carbon dioxide gives 15.3 counts per minute in a Geiger counter. A sample of CO_2 obtained by burning material found in some archeological diggings gives 1.9 counts per minute. Assume that the half-life of carbon-14 is 5670 years. How old is the material?

Appendix A: MATHEMATICS

EXPONENTIALS

Chemists often have to deal with very large or very small numbers: The number of molecules in one mole of a substance—Avogadro's number—is 602,000,000,000,000,000,000,000, and the equilibrium constant for the acid dissociation of HCN is 0.0000000005. In order to avoid writing out all the zeros needed to deal with such numbers, we use exponential notation. For example, the number 100 is equal to 10×10 or 10^2. We can say that $1 \times 10^2 = 100$, where the exponent, 2, indicates the number of times that 1 must be multiplied by 10 to yield 100. Similarly, 200 is equal to $2 \times 10 \times 10$ or 2×10^2, where the exponent again tells us the number of times we must multiply 2 by 10 to get 200. We can express any number we wish to in this fashion:

$$4{,}000 = 4 \times 10 \times 10 \times 10 = 4 \times 10^3,$$
$$13{,}909 = 1.3909 \times 10 \times 10 \times 10 \times 10 = 1.3909 \times 10^4,$$
$$3{,}000{,}000 = 3 \times 10^6,$$
$$602{,}000{,}000{,}000{,}000{,}000{,}000{,}000 = 6.02 \times 10^{23}.$$

Exponential notation can be applied to numbers less than 1 as well:

$$0.1 = \frac{1}{10} = 1 \times 10^{-1},$$

$$0.0001 = \frac{1}{10{,}000} = \frac{1}{10 \times 10 \times 10 \times 10} = \frac{1}{10^4} = 1 \times 10^{-4},$$

$$0.0006 = 6 \times 10^{-4},$$
$$0.000018 = 1.8 \times 10^{-5},$$
$$0.0000000005 = 5 \times 10^{-10}.$$

In many of the problems encountered in chemistry, large and small numbers must be multiplied or divided, and again the most convenient method is to use exponential notation. To multiply exponential numbers, we add the exponents:

$$10^3 \times 10^5 = 10^8,$$
$$10^6 \times 10^{-5} = 10^1,$$
$$10^{-5} \times 10^2 = 10^{-3}.$$

To divide exponential numbers, we subtract the exponents:

$$10^5 \div 10^3 = 10^2,$$
$$\frac{10^5}{10^{-3}} = 10^8,$$
$$\frac{10^{-5}}{10^3} = 10^{-2}.$$

Thus, to find the product of two numbers written exponentially, we multiply the numbers and add the exponents:

$$(4 \times 10^4)(3 \times 10^6) = (4 \times 3)(10^4 \times 10^6) = 12 \times 10^{10};$$
$$(6.5 \times 10^{13})(4.3 \times 10^{-9}) = (6.5 \times 4.3)(10^{13} \times 10^{-9}) = 28 \times 10^4 = 2.8 \times 10^5.$$

Division follows a similar route:

$$\frac{4.4 \times 10^6}{2.2 \times 10^4} = \frac{4.4}{2.2} \times \frac{10^6}{10^4} = 2.0 \times 10^2 = 200;$$
$$\frac{3.26 \times 10^{-3}}{2.0 \times 10^4} = \frac{3.26}{2.0} \times \frac{10^{-3}}{10^4} = 1.6 \times 10^{-7}.$$

SIGNIFICANT FIGURES

Many of the numbers that we deal with in chemistry are not exact. We may speak of exactly *two* atoms or *one* molecule, but when we speak of Avogadro's number of molecules, do we mean that we have exactly 602,000,000,000,000,000,000,000 molecules, or could it be that we have 602,000,000,000,000,000,000,001? There is no way that we can measure the number accurately enough to answer the question, and the most precise calculations give a value for Avogadro's number of 6.02252×10^{23}. We say, then, that the number is precise to six significant figures.

Similarly, assume that we weigh one sample on three balances, of increasing precision, and get the values 13.2 g, 13.22 g, and 13.2176 g. The first value is accurate to three significant figures. This serves to tell us that the weight of the sample is closer to 13.2 g than either 13.3 g or 13.1 g, but no more. The increased precision of the second and third balances is reflected by the four and six significant figures given by the weighings.

Thus the number of digits used in writing a number reflects the accuracy to which the number is known, or else reflects the accuracy required by the particular situation.

When we place zeros after the decimal place but before the digits, they are not significant; but zeros placed after a similar number *are* significant. For example, 0.00046 has *two* significant figures; the zeros merely locate the decimal place. But 0.4600 has *four* significant figures. For numbers greater than 1, there is some ambiguity. The number 25,000 could have 2, 3, 4, or 5 significant figures. We can clearly indicate the number of significant digits here by writing the number using exponents. Thus 2.50×10^4 would show three significant figures. Some other examples follow:

47.63	four significant figures
5.00	three significant figures
0.00001	one significant figure
5.7×10^3	two significant figures
57,000.00	seven significant figures
1.67×10^{-8}	three significant figures

If we carry out a mathematical operation, the number of significant figures in the answer is determined by the least accurately known number used in the operation. For example, if we want to know the weight of one mole of calcium sulfate ($CaSO_4$), we add the weights of the

individual atoms:

$$Ca \text{ atomic weight} = 40.08$$
$$S \ \text{ atomic weight} = 32.06$$
$$O \ \text{ atomic weight} = 15.9994$$
$$15.9994$$
$$15.9994$$
$$15.9994$$
$$\overline{}$$
$$136.1376$$

Thus 1 mole of $CaSO_4$ weighs 136.14 g. It would be incorrect for us to list the formula weight of $CaSO_4$ to the fourth decimal place, since the atomic weight of calcium is given only to the second decimal place. There is no reason for us to use the more precise weight value of oxygen; we can round it off to the second decimal place:

$$Ca = 40.08 \text{ amu}$$
$$S \ = 32.06 \text{ amu}$$
$$O \ = 16.00 \text{ amu}$$
$$O \ = 16.00 \text{ amu}$$
$$O \ = 16.00 \text{ amu}$$
$$O \ = 16.00 \text{ amu}$$
$$\overline{}$$
$$136.14 \text{ amu}$$

In addition or subtraction, then, we find that the answer is written to reflect the accuracy of the least precisely known number involved.

In multiplication or division, we follow the rule which states that the number of significant figures in the answer is equal to the number of significant figures in the least precisely known number used in the calculations. For example, $4.0 \times 1.37 \times 10.012 = 54.865760$ if we do not round off. It is, of course, wrong to write an answer containing eight significant figures when we are performing a multiplication problem involving a number (4.0) with only two significant figures. If we round off the answer to two significant figures, we get 55. There is no reason, then, for us to go through the tedium of carrying out the multiplication to eight figures. Why not round off all the numbers to two significant figures? Thus the problem becomes $4.0 \times 1.4 \times 10 = 56$. It appears that we have introduced an error here, but this only reflects our doubt in the last figure in each case. The last significant figure is considered to be in doubt. This means that 4.0 is really 4.0 ± 0.1, and thus the answers become 56 ± 1 and 55 ± 1, which can turn out to be the same.

Division follows the same rules as multiplication:

$$2426 \div 1.64 = 1479.27 = 1480 = 1.48 \times 10^3$$

or

$$243 \div 1.64 = 1481 = 1.48 \times 10^3.$$

UNITS

Most of the mathematical operations required of a chemist do not involve pure numbers, but rather utilize numbers which measure a particular property. If we are concerned about the mass of an object, it is meaningless to describe the mass as 12 unless we indicate the units; that is, 12 grams, 12 pounds, 12 tons, etc. In one respect this complicates our operation, because we must worry

about using the appropriate units. But using units also provides a check to make sure we have done the problem correctly. As a simple example, let us calculate the number of ounces in 5 quarts. We know that there are 32 ounces in each quart. We can indicate this relationship in several ways:

$$32 \text{ oz} = 1 \text{ qt}; \quad 32 \text{ oz}/1 \text{ qt}; \quad 32 \text{ oz}/\text{qt}.$$

This tells us that every time we have an answer in the units oz/qt, the numbers must be in the ratio of 32/1. If, then, we want to know the number of ounces in 5 quarts, that is

$$? \text{ oz} = 5 \text{ qt} \quad \text{or} \quad ? \text{ oz}/5 \text{ qt},$$

we know that the oz/qt ratio must equal 32/1 or

$$\frac{? \text{ oz}}{5 \text{ qt}} = \frac{32 \text{ oz}}{1 \text{ qt}}.$$

To solve the problem, we say

$$? \text{ oz} = \frac{32 \text{ oz}}{1 \text{ qt}} \times 5 \text{ qt}.$$

We cancel the units and multiply, and we get $5 \times 32 \text{ oz} = 160 \text{ oz}$ as our answer.

Let us say that we knew how to set up the proportion

$$\frac{? \text{ oz}}{5 \text{ qt}} = \frac{32 \text{ oz}}{1 \text{ qt}},$$

but we solved the proportion incorrectly as

$$? \text{ oz} = \frac{32 \text{ oz}}{1 \text{ qt}} \times \frac{1}{5 \text{ qt}}.$$

Without bothering to carry out the mathematics, we can see that the solution is incorrect merely by solving the problem using the units. Thus

$$\text{oz} = \frac{\text{oz}}{\text{qt}} \times \frac{1}{\text{qt}} = \frac{\text{oz}}{\text{qt}^2}$$

is clearly incorrect and points out our error.

Problems using the ideal-gas equation are greatly simplified by the use of this technique of solving the problem in terms of the units.

Example A–1. In the equation $PV = nRT$, P = pressure in atm, V = volume in ℓ, n = moles, R = gas constant in ℓ-atm/mole-°K, T = temperature in °K.

Let us say that we wish to solve for the pressure. We can write

$$P = \frac{nRT}{V}.$$

Substituting the units and canceling gives us

$$P = \frac{\text{moles} \dfrac{\ell\text{-atm}}{\text{mole-}°K} °K}{\ell} = \text{atm}.$$

This quick check shows us that the problem is set up correctly, since atm is the desired unit for the pressure P.

Example A–2. Calculate the density of CO_2 at STP $(0\,°K, 1\,atm)$.

It is not immediately clear that the ideal-gas equation relates to density, but we can look a bit closer. The units of density of a gas are g/ℓ, and the gas equation has the volume in liters but no gram units. If we recall, though, that a mole is equal to the number of grams per formula weight or molecular weight, we can write

$$PV = nRT,$$

$$n = \text{moles} = g/\text{MW},$$

$$PV = \frac{g}{\text{MW}}RT.$$

We can now solve the equation for density (g/V):

$$gRT = \text{MW}\ PV,$$

$$g = \frac{\text{MW}\ PV}{RT},$$

$$g/V = \frac{\text{MW}\ P}{RT}.$$

We can check our solution by inserting the appropriate units and canceling:

$$g/\ell = \frac{(\text{MW})\ (\cancel{\text{atm}})}{\dfrac{\ell\text{-}\cancel{\text{atm}}}{°K\text{-}(g/\text{MW})}\times °K} = \frac{\text{MW}/\ell}{(g/\text{MW})} = \frac{\cancel{\text{MW}}}{\ell}\ \frac{g}{\cancel{\text{MW}}}, \qquad g/\ell = g/\ell.$$

Since our solution of the gas equation is correct, we can substitute the numbers and do the arithmetic:

$$\text{Density of } CO_2 = \frac{(44\ \text{MW})(1\ \text{atm})}{\left(0.082\ \dfrac{\ell\text{-atm}}{°K\text{-}(g/\text{MW})}\right)(273°K)}$$

$$= \frac{44}{0.082 \times 273} = 2.0\ g/\ell.$$

LOGARITHMS

In order to do some problems in Chapter 13 involving pH, you need to be able to consult a table of logarithms (see Appendix D). Therefore we present a brief review of logarithms.

The *common logarithm* of a number is equal to the exponential to which 10 must be raised to get the number. Thus

$$\log 100 = 2,\ \text{since } 10^2 = 100,$$
$$\log 1000 = 3,\ \text{since } 10^3 = 1000,$$
$$\log 1 = 0,\ \text{since } 10^0 = 1,$$
$$\log 0.0001 = -4,\ \text{since } 10^{-4} = 0.0001.$$

Or, generally:

$$\log x = y \quad \text{means} \quad 10^y = x.$$

If we have an aqueous solution in which $[H^+] = 1 \times 10^{-5}$, then

$$pH = -\log [H^+], \quad \log [H^+] = \log 10^{-5}.$$

Since 10 must be raised to the -5 power to get 10^{-5},

$$\log [H^+] = -5, \quad pH = -(-5) = 5.$$

But what if $[H^+]$ does not equal a simple power of 10, but a more complicated number? The following examples should illustrate the approach to be used.

Example A–3. What is the pH of a solution with a hydrogen ion concentration of $[H^+] = 1.84 \times 10^{-5}$?

$$pH = -\log [H^+] = -\log 1.84 \times 10^{-5}.$$

Recalling that the logarithm of a product of two numbers is equal to the sum of the logarithms of the two numbers, we write

$$\log 1.84 \times 10^{-5} = \log 1.84 + \log 10^{-5}.$$

We know that $\log 10^{-5} = -5$ but we must get $\log 1.84$ from a table of logarithms. Since the number is between 1 and 10, we can read the logarithm directly from the table; $\log 1.84 = 0.2648$. Therefore

$$\log 1.84 \times 10^{-5} = 0.265 - 5 = -4.735$$

and

$$pH = -(4.74) = 4.74.$$

Example A–4. Calculate the pH of a $0.0045 M$ HCl solution.

Since $[H^+] = 0.0045$ in the solution of completely dissociated HCl,

$$pH = -\log [H^+] = \log 0.0045.$$

If we are to be able to read logs directly from the table, the number should be between 1 and 10. Thus

$$\log 0.0045 = \log 4.5 \times 10^{-3}$$
$$= \log 4.5 + \log 10^{-3} = 0.653 - 3 = -2.347.$$

Therefore

$$pH = -(-2.35) = 2.35.$$

Example A–5. What is the hydrogen ion concentration of a solution with a measured pH $= 8.67$?

$$pH = -\log [H^+] = 8.67; \quad \log [H^+] = -8.67.$$

It is easiest for us to determine the antilog of -8.67 by rewriting it

$$\log [H^+] = -8.67 = 0.33 - 9.$$

From the tables, we find the antilog of 0.33, and

$$[H^+] = 2.14 \times 10^{-9}.$$

9 00
8.67
‾‾‾‾
33

, 3̶3 - 9

-18.8
$= 1.88 \times 10^{1}$

19.00
18.80
‾‾‾‾
.20

.20 - 19
1.59×10^{-18}

Appendix B: THE METRIC SYSTEM

The system of measurement which is used exclusively in science, and by laymen in most of the world, is the *metric system*. In the metric system, the unit of length is the meter, the unit of volume is the liter, and the unit of mass is the gram. Larger or smaller measurements are designated by prefixes. For example, a milligram (mg) is one thousandth of a gram, a microliter ($\mu\ell$) is one millionth of a liter, and a kilocalorie (kcal) is one thousand calories. Table B–1 lists the prefixes commonly used in the metric system and Table B–2 lists some common conversions between the metric and English systems.

Table B–1. Prefixes used in the metric system of measurement

micro (μ)	1/1,000,000
milli (m)	1/1000
centi (c)	1/100
deci (d)	1/10
kilo (k)	1000

Table B–2. Selected conversions between the English and metric systems of measurement

1 centimeter (cm)	=	0.394 in.
1 meter (m)	=	3.281 ft
1 foot (ft)	=	30.48 cm
1 inch (in.)	=	2.54 cm
1 gram (g)	=	0.03527 oz
1 kilogram (kg)	=	2.205 lb
1 ounce (avoirdupois) (oz)	=	28.35 g
1 pound (avoirdupois) (lb)	=	453.6 g
1 cubic centimeter (cu cm, cc, or cm^3)	=	0.06102 cu. in. (in^3)
1 liter (ℓ)	=	61.03 cu. in.
1 liter (ℓ)	=	1.0567 qt
1 quart (qt)	=	0.9464 ℓ

Appendix C: NOMENCLATURE

NOMENCLATURE OF INORGANIC COMPOUNDS

The naming of compounds can get confusing. Often a compound may be referred to by more than one name. For example, the correct name for $NaHCO_3$ is sodium hydrogen carbonate, and yet you find it in the chemical catalogs under the name sodium bicarbonate. In this section we shall consider some rules which will enable us to systematically name most of the compounds of interest to a beginning student.

NAMING IONIC COMPOUNDS

Ionic compounds are named by indicating, first, the names of the positive ion, and then the negative ion. Positive ions containing only one atom (monatomic) are named the same as the parent element:

$$Na^+ \text{ sodium}, \quad Ba^{2+} \text{ barium}, \quad Ga^{3+} \text{ gallium}.$$

When a metal forms more than one ion, the ions are distinguished by using Roman numerals to indicate the oxidation state:

$$Fe^{2+} \text{ iron(II)}, \quad Pb^{2+} \text{ lead(II)}$$
$$Fe^{3+} \text{ iron(III)}, \quad Pb^{4+} \text{ lead(IV)}$$

Monatomic negative ions are named by replacing the end of the name of the element with the suffix *ide*:

$$N^{3-} \text{ nitride}, \quad O^{2-} \text{ oxide}, \quad F^- \text{ fluoride}, \quad H^- \text{ hydride}.$$

Polyatomic anions are more complex and the names are less systematic:

OH^-	hydroxide	CO_3^{2-}	carbonate
NO_3^-	nitrate	ClO_4^-	perchlorate
NO_2^-	nitrite	ClO_3^-	chlorate
SO_4^{2-}	sulfate	ClO_2^-	chlorite
SO_3^{2-}	sulfite	ClO^-	hypochlorite
PO_4^{3-}	phosphate	MnO_4^-	permanganate
HCO_3^-	hydrogen carbonate	HSO_3^-	hydrogen sulfite

If these polyatomic ions have an atom which can exhibit two oxidation states, the higher oxidation state is given the suffix *-ate* (nitr*ate*) and the lower oxidation state the suffix *-ite* (nitr*ite*). Polyatomic anions which have a nonmetal associated with oxygens (*oxyanions*), in which the nonmetal exhibits more than two oxidation states, take the prefix *per* to indicate the highest oxidation state (*per*chlor*ate*, *per*mangan*ate*) and the prefix *hypo* (*hypo*chlor*ite*) to indicate the lowest oxidation state. If hydrogen is present in the oxyanion, it is stated in the

name: HSO_3^-, hydrogen sulfite. To name ionic compounds (or other metal–nonmetal compounds), we name the positive ion first and the negative ion second. The following examples illustrate the method:

NaBr	sodium bromide	$PbCl_2$	lead(II) chloride
$MgCl_2$	magnesium chloride	$Fe_2(SO_4)_3$	iron(III) sulfate
$KHSO_4$	potassium hydrogen sulfate	SrI_2	strontium iodide
$Ca(OBr)_2$	calcium hypobromite	$InCl_3$	indium(III) chloride
$LiClO_4$	lithium perchlorate	NaH	sodium hydride
Cr_2O_3	chromium(III) oxide	ZrO_2	zirconium oxide

NAMING BINARY COMPOUNDS OF NONMETALS

Compounds formed from two nonmetal atoms are named by a method analogous to that used for ionic compounds. The more electropositive atom is named as the element and placed first. The electronegative atom is given the ending -ide and placed second:

HBr, hydrogen bromide; H_2S, hydrogen sulfide.

If more than one atom ratio can be formed between the atoms present, the prefixes di- (two), tri- (three), tetra- (four), penta- (five), etc., are used. (The a may be dropped preceding a vowel.)

N_2O_5	dinitrogen pentoxide	OF_2	oxygen difluoride
CO_2	carbon dioxide	N_2O_4	dinitrogen tetroxide
SO_3	sulfur trioxide	Na_2HPO_4	disodium hydrogen phosphate
$BrCl_3$	bromine trichloride	NaH_2PO_4	sodium dihydrogen phosphate

NAMING OXYGEN ACIDS

Acids containing oxyanions can be named by replacing the -ate ending by -ic acid or the -ite ending by -ous acid, as follows:

H_2CO_3	carbonic acid	$HClO_2$	chlorous acid
H_2SO_4	sulfuric acid*	$HClO_3$	chloric acid
H_2SO_3	sulfurous acid*	$HClO_4$	perchloric acid
HClO	hypochlorous acid	HNO_3	nitric acid

There are many more rules to cover other types of inorganic compounds, but the above will suffice for our purposes. Remember that there are many compounds that are not named systematically, but are known by common names. Some examples are:

H_2O,	water	NH_3,	ammonia
BH_3,	borane	PH_3,	phosphine
N_2H_4,	hydrazine	SbH_3,	stibine

---

*   Sulfuric and sulfurous acids represent slight exceptions to the rule.

NOMENCLATURE OF ORGANIC COMPOUNDS

The problem of naming the tremendous number of organic compounds is even more complex than naming inorganic compounds. We need not go through the methods of systematically naming the more complex organic molecules here. We shall give the details of the naming procedures only for the saturated hydrocarbons, the alkanes.

Table C–1. Names for some alkanes

CH_4	methane	C_6H_{14}	hexane
C_2H_6	ethane	C_7H_{16}	heptane
C_3H_8	propane	C_8H_{18}	octane
C_4H_{10}	butane	C_9H_{20}	nonane
C_5H_{12}	pentane	$C_{10}H_{22}$	decane

Table C–1 lists the names of the alkanes containing one through ten carbon atoms. These names are the basis for naming the alkanes. We must also define an *alkyl group* as an alkane from which we have removed one hydrogen in order to make possible bonding to another atom. The name of the alkyl group is obtained by replacing the *-ane* ending of the parent alkane by *-yl*. Thus from methane comes methyl, from ethane comes ethyl, etc. Table C–2 shows four simple alkyl groups.

Table C–2. Names for some simple alkyl groups

$CH_3—$	methyl	$CH_3CH_2CH_2—$	propyl
$CH_3CH_2—$	ethyl	$CH_3CH_2CH_2CH_2—$	butyl

The rules for naming alkanes are:

1) Determine the longest continuous carbon chain in the molecule and name it as an alkane according to the number of carbon atoms.

2) Write in alphabetical order the names of alkyl groups attached to the carbon chain. In the case of more than one group of the same type, indicate the number with the prefixes *di-*, *tri-*, *tetra-*, etc.

3) Prefix each alkyl group with a number indicating the point of attachment to the main carbon chain. Choose the numbers by numbering the chain from either end so that the attached groups have the lowest numbers possible.

The following examples illustrate the step-by-step application of the rules.

Example C–1. Name the following compound:

$$\underset{\underset{\displaystyle CH_3}{|}}{\overset{\overset{\displaystyle CH_3}{|}}{CH_3CH_2CHCHCH_2CH_3}}$$

1) There are several continuous chains of carbon atoms in the molecule, as indicated by the colored atoms.

$$\underset{\underset{\displaystyle CH_3}{|}}{\overset{\overset{\displaystyle CH_3}{|}}{CH_3CH_2CHCHCH_2CH_3}} \qquad\qquad \underset{\underset{\displaystyle CH_3}{|}}{\overset{\overset{\displaystyle CH_3}{|}}{CH_3CH_2CHCHCH_2CH_3}}$$

$$\underset{\underset{\displaystyle CH_3}{|}}{\overset{\overset{\displaystyle CH_3}{|}}{CH_3CH_2CHCHCH_2CH_3}} \qquad\qquad \underset{\underset{\displaystyle CH_3}{|}}{\overset{\overset{\displaystyle CH_3}{|}}{CH_3CH_2CHCHCH_2CH_3}}$$

$$\underset{\underset{\displaystyle CH_3}{|}}{\overset{\overset{\displaystyle CH_3}{|}}{CH_3CH_2CHCHCH_2CH_3}} \qquad\qquad \underset{\underset{\displaystyle CH_3}{|}}{\overset{\overset{\displaystyle CH_3}{|}}{CH_3CH_2CHCHCH_2CH_3}}$$

We see that the longest continuous chain contains six carbon atoms, so we apply the name *hexane*.

2) We can see that the hexane chain has two methyl groups attached, so we must add *dimethyl* to the name.

$$\underset{\underset{\displaystyle CH_3}{|}}{\overset{\overset{\displaystyle CH_3}{|}}{CH_3CH_2CHCHCH_2CH_3}} \qquad \text{dimethylhexane}$$

3) In this instance it does not matter whether the hexane chain is numbered from the right or from the left. The methyl groups are attached at the 3 and 4 positions in each case.

$$\underset{\underset{\displaystyle CH_3}{|}}{\overset{\overset{\displaystyle CH_3}{|}}{\overset{1}{C}H_3\overset{2}{C}H_2\overset{3}{C}H\overset{4}{C}H\overset{5}{C}H_2\overset{6}{C}H_3}} \qquad \text{3,4-dimethylhexane}$$

Example C–2. Name the following compound:

a)

$$\begin{array}{c}
CH_3 \\
| \\
CH_2 \\
| \\
CH_3 \quad CH-CH_2-CH_2-CH_3 \\
| \qquad | \\
CH_3-CH_2-C\!\!-\!\!-\!\!-C-CH_2-CH_3 \\
| \qquad | \\
CH_3 \quad CH_3
\end{array}$$